Laboratory Textbook of Anatomy & Physiology

Second Edition

Michael G. Wood, M.S.

Del Mar College, Corpus Christi, Texas

Prentice
Hall

Prentice Hall, Upper Saddle River, New Jersey 07458

Library of Congress Cataloging-in-Publication Data

Wood, Michael G.
   Laboratory textbook of anatomy and physiology / Michael G. Wood.
      p.   cm.
   Previously published: 1998
   ISBN 0-13-019694-0
   1. Human physiology--Laboratory manuals.   2. Human anatomy--Laboratory manuals.
   I. Title: Laboratory textbook of anatomy and physiology.   II. Title

   QP44.W66   2001
   612'.0078--dc21

                                                          00-050933

Safety Notification

Some of the exercises in this laboratory text involve the use of chemicals, heat, glass, sharp objects, and body tissues and fluids. If handled or used improperly, these materials may be hazardous. Safety precautions are outlined in Exercise One of the text and should be followed by students at all times. Your school may have additional rules and regulations pertaining to laboratory safety. Always ask your laboratory instructor if you are unsure of a procedure.

Editor-in-Chief for Biology: *Sheri Snavely*
Senior Acquisitions Editor: *Halee Dinsey*
Project Manager: *Don O'Neal*
Special Projects Manager: *Barbara A. Murray*
Manufacturing Buyer: *Trudy Pisciotti*
Composition and Production Editorial: *Carlisle Publishers Services*
Illustrators: *William C. Ober, M.D.; Claire Garrison, R.N.*
Cover photograph: *Fotopic/Omni-Photo Communications, Inc.*
Special thanks to Ward's Natural Science

Notice: Our knowledge in clinical sciences is constantly changing. The author and the publisher of this volume have taken care that the information contained herein is accurate and compatible with the standards generally accepted at the time of publication. Nevertheless, it is difficult to ensure that all information given is entirely accurate for all circumstances. The author and the publisher disclaim any liability, loss, or damage incurred as a consequence, directly or indirectly, of the use and application of any of the contents of this volume.

 © 2001, 1998 by Prentice-Hall, Inc.
Upper Saddle River, New Jersey 07458

Printed in the United States of America

10   9   8   7

ISBN 0-13-019694-0

Prentice-Hall International (UK) Limited, *London*
Prentice-Hall of Australia Pty. Limited, *Sydney*
Prentice-Hall of Canada, Inc., *Toronto*
Prentice-Hall Hispanoamericana, S. A., *Mexico*
Prentice-Hall of India Private Limited, *New Delhi*
Prentice-Hall of Japan, Inc., *Tokyo*
Pearson Education Asia Pte. Ltd.
Editora Prentice-Hall do Brasil, Ltda., *Rio de Janeiro*

Brief Contents

Detailed Table
of Contents

Preface

This laboratory text and study guide is designed for students with a limited background in science who are enrolled in a two semester sequence anatomy and physiology course. The manual is written to correspond to all popular anatomy and physiology texts, especially Frederic Martini's *Fundamentals of Anatomy and Physiology,* fifth edition. The manual accurately presents background material and laboratory procedures for each exercise. Self-references provide students relevancy and perspective of the material. The laboratory text is designed around the following elements:

ORGANIZATION

The laboratory text is arranged into 51 exercises which are split into anatomical and physiological studies. Each exercise consists of student objectives, a terminology guide, materials list, laboratory activities, a laboratory report, and review list called the QuickCheck. An outline approach enables faculty to easily assign specific sections of an exercise. Large systems, such as the skeletal, muscular, and nervous systems, are subdivided into several exercises, the first being an overview exercise that introduces the major anatomical organization. Programs with limited laboratory time may use the overview exercises for a hands-on summary of these organ systems that can be completed during a short laboratory period.

STUDENT ACTIVITIES

The text encourages students to become more involved with their learning through self-references, labeling figures, recording observations, and completing laboratory reports. Involvement in these activities will assist students' understanding of human form and function.

STUDENT OBJECTIVES

Each exercise includes objectives to identify major themes students should master during the exercise. Laboratory reports at the conclusion of each exercise test for mastery of each objective with a variety of labeling and written assignments.

WORD POWER AND PRONUNCIATION GUIDES

To promote student awareness in the language of science, each exercise opens with a "Word Power" list of relevant prefixes, suffixes and root words. In the body of each exercise, a pronunciation guide is included in parentheses after many terms. Each word is divided into syllables by hyphens, and the main accented syllable of multisyllabic words are in capital letters. An overbar above letters indicates long vowel sounds:

"a" as in "tray" (trā)
"e" as in "ear" (ēr)
"i" as in "fly" (flī)
"o" as in "bone" (bōn)
"u" as in "you" (ū)

MATERIALS

A list of required laboratory equipment, models, prepared microscope slides, and other materials is included at the beginning of each exercise and activity. Faculty will find these lists useful in preparing the laboratory. Students can use the lists to determine the materials necessary to accomplish the laboratory activities.

LABORATORY ACTIVITIES

Each section of an exercise concludes with a series of laboratory observations and labeling activities that provide concise instructions and procedures to guide students through the material. References to laboratory models, microscopic specimens, and text figures clearly delineate the assignments required to master the student objectives.

BIOPAC ALTERNATIVE EXERCISES

New to the second edition are three alternate exercises using the Biopac Student Laboratory System, an integrated suite of hardware and software that provides students with powerful tools for physiological studies. The probes and transducers are designed for quick set-ups that easily configure for various investigations such as muscle contraction (EMG), the electrocardiogram ECG, and pulse. The manual includes step by step instructions to guide students in the set up and use of the hardware and software. The Biopac Student Laboratory System works in both Macintosh™ and Windows™ operating environments.

ART PROGRAM

A major strength of this laboratory manual is the outstanding full-color art program with illustrations by William Ober and Claire Garrison and photographs by Ralph Hutchings. The art is designed to guide students from familiar to detailed structures. The text cites the art at appropriate times to entwine reading and visual learning modalities. Figures include detailed legends, reference icons, pin-point labeling, and multidimensional view of the human body. Numerous figures incorporate written labeling assignments as part of the laboratory activities.

CLINICAL APPLICATIONS

Many exercises include relevant discussions of disease and injury in boxed clinical application sections. Thought-provoking critical thinking questions are featured in many of these boxed sections.

DISSECTIONS

Dissection gives students perspectives on the texture, scale, and relationships of anatomy. The dissection specimens presented in this manual are the cat and sheep organs. Exercise 23 details preparation and dissection of the cat muscular system. Dissection of other cat organ systems are included in the corresponding anatomy exercises, as are sheep brain, eye, heart, and kidney dissections. Safety guidelines and disposal methods are incorporated into each dissection exercise.

LABORATORY SAFETY

Correct usage of instrumentation and safe handling and disposal of chemical and biological wastes are emphasized in appropriate exercises throughout the manual. Exercise 1 covers laboratory safety and should be completed by all students at the beginning of a semester.

QUICKCHECKS

QuickChecks are review lists of the important terminology and structures presented in each exercise. Students may use the lists to review laboratory models and as terminology lists to practice spelling. Faculty will find the lists useful to post assigned laboratory sections or structures by highlighting words on the QuickChecks.

LABORATORY REPORTS

Comprehensive laboratory reports are located at the conclusion of each exercise. The reports include a variety of tasks such as matching, short-answer questions, fill-in-the-blanks, and drawings to reinforce the laboratory objectives. Instructors may utilize the reports for quizzes and out-of-class assignments. The page layout of the manual permits the removal of the laboratory reports without loss of other material.

ACKNOWLEDGMENTS

I thank the talented and creative individuals of the Prentice Hall team. I especially credit Don O'Neal, Project Manager, for his expertise in organizing and coordinating the enormous details of the project during a tight production schedule. Halee Dinsey, Senior Acquisitions Editor, ensured that all resources were available to me, Don, and the production staff. Barbara Murray, Special Projects Manager, provided the production house with art and text files. I thank Terry Routley and her fine team at Carlisle Publishers Services for their creative layout and attention to detail. The Carlisle team and I used soft-copy or digital editing for the second edition production and I am appreciative of their willingness to incorporate this technology into their process.

Foremost, I am grateful to Frederic H. Martini, author of *Fundamentals of Anatomy and Physiology,* fifth edition, and *Human Anatomy,* third edition, with Michael J. Timmons and Michael McKinley, for their insight and creativity in visualizing anatomical and physiological concepts. Together, with illustrations by William Ober and Claire Garrison and photographs by Ralph Hutchings, they have created the intelligible art I have fortunately had access to.

Many individuals made significant contributions to the first edition text which are now incorporated into the new version. Thank you again to Edward J. Greding Jr., Anil Rao, Joe Wheeler, Michael Timmons, and Michael McKinley for their generosity. I am grateful to Esmeralda Salazar and Gerardo Cobarruvias from Del Mar College for their technical talent in the photography studio. I also thank my colleague Albert Drumright, III, for our ongoing discussions of biology and teaching. I thank Biopac Systems, Inc. for their partnership with Prentice Hall and assistance in incorporating their state-of-the-art instrumentation into the second edition. I also thank Mike Deniz at Ward's Natural Science for developing the materials cross-reference guide.

My gratitude is extended to the many students who have provided suggestions and comments to me over the years. Reviewers and users of the first edition manual provided insightful and experienced-based suggestions, many of which are incorporated into the second edition. I thank the reviewers for their time and devotion to the project:

Terri DiFiori, Pasadena City College
Michael A. Dorset, Cleveland State Community College
Linda Griffin Gingerich, St. Petersburg Junior College
Theresa Hornstein, Lake Superior College
Stephen Lebsack, Linn-Benton Community College
Wendy Rounds, Harford Community College
Kelly J. Sexton, Northlake College

Most importantly, I thank my wife Laurie, and children, Abi and Beth, for their support and encouragement during this project.

Any errors or omissions in this second edition publication are my responsibility and are not a reflection of the editorial and review team. Comments from faculty and students are welcomed and may be directed to me at the addresses below. I will consider each suggestion in the preparation of the third edition.

Michael G. Wood
Del Mar College
101 Baldwin Blvd.
Corpus Christi
Texas, 78404
mwood@delmar.edu

Dedication

This second edition text is dedicated to my parents,

Ernest M. Wood and Janis G. Wood;

life-long learners and lovers of the natural world.

Laboratory Safety

OBJECTIVES

On completion of this exercise, you should be able to:

- Locate all safety equipment in the laboratory.
- Show how to safely handle glassware, including insertion and removal of glass rods used with stoppers.
- Demonstrate how to safely clean up and dispose of broken glass.
- Demonstrate how to safely plug in and unplug electrical devices.
- Explain how to protect yourself from and dispose of body fluids.
- Demonstrate how to mix solutions and measure chemicals safely.
- Describe the potential dangers of each lab instrument.
- Discuss disposal techniques for glass, chemical, body fluids, and other hazardous materials.

INTRODUCTION

Experiments and exercises in the anatomy and physiology laboratory are, by design, safe. Some of the hazards are identical to risks found in your home such as broken glass and the chance of electrical shock. The major hazards can be grouped into the following categories: electrical, chemical, body fluids, preservatives, and instrumentation. The following is a discussion of the hazards each category poses and a listing of safety guidelines students should follow to prevent injury to themselves and others while in the lab. Proper disposal of biological and chemical wastes ensures that these contaminants will not be released into your local environment.

A. Laboratory Safety Rules

The following guidelines are necessary to ensure that the laboratory will be a safe environment for students and faculty alike.

1. No unauthorized persons are allowed in the laboratory. Only students enrolled in the course and faculty are to enter the laboratory.
2. No unauthorized experiments or variations on experiments in this manual or class handouts are allowed without the faculty's consent.
3. No smoking, eating, gum chewing, or drinking is allowed in the laboratory.
4. Always wash your hands before and after each laboratory exercise involving chemicals, preserved materials, or body fluids, and after cleaning up spills.
5. Shoes must be worn at all times while in the laboratory.
6. Be alert to unsafe conditions and actions of other individuals in the laboratory. Call attention to those activities. Someone else's accident can be as dangerous to you as one that you might cause.
7. Glass tubes called pipettes are commonly used to measure and transfer solutions. Never pipette a solution by mouth. Always use a pipette bulb.

8. Immediately report all spills and injuries to the laboratory faculty.
9. You must inform the laboratory faculty of any medical condition that may limit your activities in laboratory.

B. Location of Safety Equipment

Write the location of each piece of safety equipment as your instructor explains how and when to use each item.

nearest telephone _____

first aid kit _____

fire exits _____

fire extinguisher _____

eye wash station _____

chemical spill kit _____

fan switches _____

biohazard container _____

C. Glassware

Glassware is perhaps the most dangerous item in the laboratory. Broken glass has the sharpest edges known, and it must be cleaned up and disposed of safely. Other glassware-related accidents can occur when a glass rod or tube is broken while attempting to insert it into a cork or stopper.

Broken glass

- Broken glass should be swept up immediately. Never use your hands to pick up broken glass. A whisk broom and a dustpan should be used to sweep the area of all glass shards.
- Broken glass and other sharp objects should be packaged in a box that will keep the glass from falling out. Tape the box closed and write "**BROKEN GLASS INSIDE**" in large letters across it. This will alert custodians and other waste collectors to the hazard inside. Your laboratory instructor will arrange for disposal of the sealed box.

Inserting glass into a stopper

- Never force a glass rod, tube, or pipette into a cork or rubber stopper. Use a lubricant such as glycerine or soapy water to ease the glass through the stopper.
- Always push on a glass tube or rod near the stopper you are trying to insert the glass through. This reduces the length of glass between the stopper and your hand and greatly decreases the chance of breaking the rod and jamming glass into your hand.

D. Electrical Equipment

The electrical hazards in the laboratory are similar to the hazards in your home. A few commonsense guidelines will almost eliminate the risk of electrical shock.

- Unravel an electrical cord completely before plugging it into an electrical outlet. Electrical cords are often wrapped tightly around the base of a microscope, and users often unravel just enough cord to plug in the microscope. Moving the focusing mechanism of the microscope may pinch the wrapped electrical cord and shock the user. Inspect the cord for frays, and the plug for secure connections.
- Do not force an electrical plug into an outlet. If the plug does not easily fit into the outlet, inform your laboratory instructor.

- Unplug all electrical cords by pulling on the plug, not by tugging on the cord. Many electrical accidents are caused by users pulling a cord out of a plug, leaving the plug in the outlet.
- Never plug in or unplug an electrical device in a wet area.

E. Body Fluids

Three body fluids are most frequently used in the laboratory; saliva, urine, and blood. Because body fluids can harbor infectious organisms, safe handling and disposal procedures must be followed to prevent infecting yourself and others.

- Work only with your own body fluids. It is beyond the scope of this course to collect and experiment on body fluids from another individual.
- Never allow a body fluid to touch your unprotected skin. Always wear gloves and safety glasses when working with body fluids—even when using your own fluids.
- Always assume that a body fluid can infect you with a disease. Putting this safeguard into practice will prepare you for working in a clinical setting where you may be responsible for handling body fluids from the general population.
- Cleanup all body fluid spills with a 10% bleach solution or a commercially prepared disinfectant labeled for this purpose. Always wear gloves during the cleanup and dispose of contaminated wipes in a biohazard container.

F. Chemical Hazards

Most chemicals used in laboratories are safe. Following a few simple guidelines will protect you from chemical hazards.

- Read the label describing the chemical you are going to work with. Be aware of chemicals that may irritate your skin or stain your clothing. Chemical containers are usually labeled to show contents and potential hazards related to the contents. Most laboratories or chemical stockroom managers keep copies of technical chemical specifications called Material Safety Data Sheets (M.S.D.S.). These publications from the chemical manufacturer detail the proper usage of the chemical and known adverse effects the chemical may cause. All individuals have a federal right to inspect these documents. Ask your laboratory instructor for more information about the M.S.D.S.
- Never touch a chemical with unprotected hands. Wear gloves and safety glasses when weighing and measuring chemicals and during all experimental procedures involving chemicals.
- Always use an appropriate spoon or spatula for retrieving dry chemical samples from a large storage container. Do not shake chemicals out of the jar. This may result in your dumping the entire container of chemical onto yourself and your workstation.
- If pipetting is required, always use a pipette bulb and never pipette by mouth. Your instructor will demonstrate how to use the particular type of pipette bulb available in the laboratory.
- When pouring an acid or other solution kept in a large container, always pour the approximate amount required into a smaller beaker first. Use this smaller container to fill your glassware with the solution. Attempting to pour from a large storage container may result in a spill that contacts the skin and clothing.
- Do not return the unused portion of a chemical to its original container. This will prevent possible contamination of the material in the original container. Dispose of the excess chemical as directed by your laboratory instructor. Do not pour unused or used chemicals down the sink unless specifically directed by your laboratory instructor.

- When mixing solutions, always add a chemical to water, never add water to the chemical. By diluting chemicals in water, you reduce the chance of a strong chemical reaction occurring.

G. Instrumentation

You will use a variety of scientific instruments in the anatomy laboratory. Safety guidelines for specific instruments are included in the appropriate exercises. This discussion concerns the instruments most frequently used in laboratory exercises.

- **Microscopes** The microscope is the main instrument you will use in the study of anatomy. Exercise 4 of this manual is devoted to the use and care of the microscope. A few simple safety rules will prevent injury to yourself or damage to the microscope.
  1. Always carry a microscope with two hands and do not swing the instrument as you carry it.
  2. Use only the special lens paper and cleaning solution provided by your laboratory instructor to clean the microscope lenses. Other papers and cloths may scratch the optical coatings on the lenses. Unapproved cleaning agents may act as a solvent and dissolve the adhesives used in the lenses.
  3. Always unwrap the electrical cord completely before plugging in the microscope.
  4. Unplug the microscope by pulling on the plug, not by tugging on the electrical cord.

- **Dissection Tools** Working with sharp blades and points always presents the possibility of a cutting injury. Remember to always cut away from yourself and not to force a blade through a tissue. Use small knife strokes for increased blade control rather than large cutting motions. Always use a sharp blade and dispose of used blades in a specially designated "sharps" container. Carefully wash and dry all instruments upon completion of each dissection.

  Special care is necessary while changing disposable scalpel blades. Your laboratory instructor may demonstrate the proper technique for blade replacement. First, wash the used blade before your attempt to remove it from the handle. Examine the handle and blade and determine how the blade fits onto the handle. Do not force the blade off the handle. If you have difficulty changing blades, do not hesitate to ask your instructor for assistance. Discard the used blade in the appropriate container for sharp objects.

- **Water baths** Water baths are used to incubate laboratory samples at specific temperatures. Potential hazards involving water baths include electrical shock due to contact with water, and burn-related injuries caused by touching hot surfaces or spilling hot solutions. **Electrical hazards** are minimized by following the safety rules concerning plugging and unplugging electrical devices. **Burns** are avoided by using tongs to immerse or remove samples from a water bath. Point the open end of all glassware that contains samples away from yourself and others. If the sample boils, it could splatter out and burn your skin. Use racks in the water bath to support all glassware and place hot samples removed from a water bath in a cooling rack. Monitor the temperature and water level of all water baths. Excessively high temperatures increase the chance of burns and usually ruin an experiment. When using boiling water baths, add water frequently and do not allow all the water to evaporate.

- **Microcentrifuge** A microcentrifuge is used in blood and urine analyses. The instrument spins at speeds exceeding thousands of revolutions per minute. Although the moving parts are housed in a protective casing, it is important to keep all loose hair, clothing, and jewelry away from the instrument. Never open the safety lid while the centrifuge is on or spinning. *Do not* attempt to stop a spinning centrifuge with your hand. The instrument has an internal braking mechanism that stops the centrifuge safely.

H. Preservatives and Fixatives

Most animal and tissue specimens used in the laboratory have been treated with chemicals to prevent decay. These preservatives are irritants and should not contact your skin or your mucous membranes (linings of the eyes, nose, and mouth, and urinary, digestive, and reproductive openings). The following guidelines will protect you from these hazards:

- If you are pregnant, limit your exposure to all preservatives. Discuss the laboratory exercise with your laboratory instructor. Perhaps you can observe rather than perform the dissection.
- Always wear gloves and safety glasses when working with preserved material.
- Your laboratory may be equipped with exhaust fans to ventilate preservative fumes during dissections. Do not hesitate to ask your lab instructor to turn on the fans if the odor of the preservatives becomes bothersome.
- Many preservatives are toxic and require special handling techniques. Drain the preservative from the specimen before the dissection or observation. Pour the fluid into the specimen's storage container or into a container provided by your lab instructor. Do not pour the preservative down the drain.
- Promptly wipe up all spills and clean your work area when you have completed the dissection. Keep your gloves on during the cleanup and dispose of gloves and paper towels in the proper biohazard container.
- Do not dispose of biological tissues or specimens in the general trash. All preserved materials must be placed in a biohazard container for proper disposal. Specimens should be wrapped in large plastic bags that are filled with an absorbent material such as cat litter. All sealed bags should be placed in a large cardboard box for pickup by a hazardous waste company for incineration.
- To dispose of preservatives, such as Formalin, a central storage container is maintained until it is collected by a waste management company. Formalin solutions contain formaldehyde, a cancer-causing agent, and must be disposed of by trained individuals. Under no circumstances should preservatives be discarded down the drain.

I. Disposal of Chemical and Biological Wastes

To safeguard the environment and individuals employed in waste collection, it is important to dispose of all potentially hazardous wastes in specially designed containers. State and federal guidelines detail the storage and handling procedures for chemical and biological wastes. Your laboratory instructor will manage the wastes produced in this course.

- **Body Fluids** Objects contaminated with body fluids are considered a high-risk biohazard and must be disposed of properly. Special biohazard containers will be available during exercises that involve body fluids. For example, during the experiment on blood typing, all used blood lancets, slides, alcohol wipes, gloves, and paper towels will be disposed of in a clearly marked biohazard container. A special biohazard sharps container may be provided to dispose of all sharp objects such as glass, needles, and lancets.

  When full, biohazard containers and their contents are sterilized by autoclaving at a high temperature and pressure to kill all pathogens associated with the body fluids. The biohazard container is then disposed of according to state and federal regulations (usually incinerated by a waste disposal company).

- **Chemical Waste** Most chemicals used in undergraduate laboratories are relatively harmless and may be diluted in water and disposed of by pouring down the drain. Your instructor will indicate during each lab session which chemicals can be discarded in this manner. Other chemicals should be stored in a waste container until a waste disposal company is notified to make a collection.
- **Preservatives and Preserved Specimens** The tissues of most biological specimens are fixed in chemicals to prevent decay. These chemicals are usually irritating to the skin and mucous membranes of your body. To dispose of preservatives, a central storage container is maintained until a waste management company is called to collect the wastes. Under no circumstances should preservative waste be discarded by pouring it down the drain.

  Like preservatives, all tissues and specimens fixed in preservatives require special storage and disposal. Specimens for disposal should be wrapped in large plastic bags filled with an absorbent material such as cat litter. All sealed bags are placed in a large cardboard box for pickup by a hazardous waste company which will incinerate the specimens.

LABORATORY SAFETY
Laboratory Report Exercise 1

Short-Answer Questions

1. Discuss how to protect yourself from body fluids such as saliva and blood.

wear gloves, safety glasses

2. Why should you consider a body fluid capable of infecting you with a disease?

They can harbor infectious organisms

3. Describe how to dispose of materials contaminated with body fluids.

always wear rubber gloves + dispose of contaminated wipes in biohazard container. Clean spills w/ 10% bleach solution).

4. Explain how to safely plug and unplug an electrical device.

1. unravel cord completely
2. Do Not force plug in outlet
3 Unplug by pulling plug Not cord
4. Never plug or unplug electrical device in wet area.

5. Discuss how to protect yourself from preservatives used on biological specimens.

1.) Always wear gloves + safty glasses.
2.) Use exhaust fans for fumes.
3.) Drain preservatives before dissection (into specimen container)
4.) wipe spills promptly - clean area after dissection, dispose of gloves + paper towels in Biohazard container
5) Wrap specimens in plastic bag w/ absorbent material (cat litter)

6. Why are special biohazard containers used for biological wastes?

to protect our environment + the employees of waste management comp. There are state + Fed. guidelines for handling Chemical + Biological Waste.

7. Explain how to clean up broken glass.

Immediately sweep area (Never use hands) with broom + dustpan put glass in box, tape box shut, write "BROKEN Glass Inside" across it.

8. List the location of the following safety items in the laboratory:

first aid kit _____

nearest telephone _____

eye wash station _____

fire exit _____

fire extinguisher _____

chemical spill kit _____

fan switches _____

biohazard container _____

9. Your instructor informs you that a chemical is not dangerous. How should you dispose of the chemical?

10. What precautions should you take while using a centrifuge?

Keep hair, clothing + jewelry away, Never open when on or spinning
Do not try to stop spinning it has internal braking mechanism.

11. How are preservatives, such as Formalin, correctly discarded?

Formalin contains formaldehyde, a cancer causing agent
It must be disposed of by trained people. A central
storage container is used untill collected by waste management
company.

12. Discuss how to safely measure and mix chemicals.

1. Read labels
2. Aways wear gloves + safty glasses.
3. Use appropriate spoon or spatula for dry chemicals
   Do not shake chemicals
4. always use pipette bulb.
5. Pour from large container to smaller beaker first
6. Do not return unused portion of chemical to
   original container - follow instuctors directions
   (Do not pour down drain — unless told by
   instructor.)

EXERCISE 2

Introduction to the Human Body

OBJECTIVES

On completion of this exercise, you should be able to:

- Define anatomy and physiology and discuss the specializations of each.
- Describe each level of organization of the body.
- Describe anatomical position and its importance in anatomical studies.
- Use directional terminology to describe the relations of your surface anatomy.
- Describe and identify the major planes and sections of the body.
- Locate all abdominopelvic quadrants and regions on laboratory models.
- Locate the organs of each organ system.
- Describe the main function of each organ system.

WORD POWER

superior (super—over, above)
inferior (infer—low, underneath)
anterior (anter—in front of)
posterior (post—behind)
dorsal (dors—back)
ventral (vent—underside, belly)
cephalic (cepha—head)

caudal (caud—tail)
medial (medi—middle)
lateral (later—side)
proximal (proxim—near)
distal (dist—distance)
transverse (trans—across, through)

MATERIALS

torso model
sectioned objects
objects for level of organization

INTRODUCTION

Knowledge of what lies beneath the skin and how it works has been slowly amassed over a span of nearly 3,000 years. Obviously any logical practice of medicine depends on an accurate knowledge of human anatomy. Yet people have not always realized this. Through most of human history, corpses have been viewed with superstitious awe and dread. Observations of anatomy by dissection were illegal, and medicine therefore remained an elusive practice that often harmed rather than cured the unfortunate patient. Yet, despite superstitions and prohibitions, there have been, throughout the ages, scientists who wanted to know the human body as it really is rather than how it was imagined to be.

The founder of anatomy was the Flemish anatomist and physician Andreas Vesalius (1514–1564). Vesalius set about to accurately describe human structure. In 1543 he published his monumental work *De Humani Corporis Faberica* (*On the Structure of the Human Body*), the first meaningful text on human anatomy. In this work he corrected over 200 errors of earlier anatomists and produced drawings that are still useful today. Vesalius' work laid the foundation for all future knowledge of the human body. Imagining the body's internal structure at last became unacceptable in medical literature.

Many brilliant anatomists and physiologists since Vesalius' time have contributed significantly to the understanding of human form and function. Your campus and community libraries have numerous books on the history of medicine and science.

A. Organization of the Body

Anatomy is the study of body structures. Early anatomists described the body's **gross anatomy**, the large parts such as muscles and bones. As knowledge of the body advanced and scientific tools permitted more detailed observations, the field of anatomy began to diversify into areas such as **microanatomy**, the study of microscopic structures, and **cytology**, the study of cells. **Histology** is the science of tissues, groups of cells coordinating their effort toward a common function.

Physiology is the study of organ function. As in anatomy, a physiologist may investigate organ functions from a molecular, cellular, histological, or more complex level. Physiology may be considered the work that cells must do to keep the body stable and operating efficiently. **Homeostasis** (hō-mē-ō-STĀ-sis) is the maintenance of a relatively steady internal environment through physiological work. Stress, inadequate diet, and disease disrupt the normal physiological processes and may lead to serious health problems or death of the individual.

Anatomists and physiologists study the body from many different levels of structural and functional detail. These **levels of organization** are reflected in the fields of specialization in anatomy and physiology. Each higher level increases in structural and functional complexity, progressing from chemicals to cells, tissues, organs, and organ systems which function to maintain the organism. Figure 2.1 illustrates the levels of organization with the cardiovascular system as an example. The smallest level of organization is the **chemical and molecular level**. Atoms, such as carbon and hydrogen, bond together and form molecules. The heart contains protein molecules that are involved in contraction of the cardiac muscle. Molecules are organized into cellular structures, called organelles, which have distinct shapes and functions. The organelles collectively comprise the next level of organization, the **cellular level**. Cells are the fundamental level of biological organization because it is cells, not molecules, that are alive. Living organisms can reproduce, grow, and regulate internal chemical reactions to maintain homeostasis. Different types of cells working together constitute the **tissue level**. Although tissues lack a distinct shape, they are distinguishable by cell type, such as the cardiac muscle cells of the heart. Tissues function together at the **organ level**. Each organ has a distinct three-dimensional shape and a broader range of functions than individual cells or tissues. The **organ system level** includes all the organs interacting to accomplish a common goal. The heart and blood vessels comprise the cardiovascular system and physiologically work to move blood through the body. All organ systems make up the individual or the **organism level**.

LABORATORY ACTIVITY   LEVELS OF ORGANIZATION

MATERIALS

A variety of objects each representing a level of organization

PROCEDURE

Classify each object set on display by your laboratory instructor as to the level of organization it represents. Write your answers in the spaces provided.

- Chemical and molecular level _____
- Cellular level _____
- Tissue level _____
- Organ level _____
- Organ system level _____
- Organism level _____

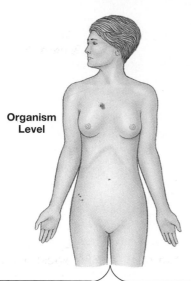

Organism Level

Figure 2.1
Levels of Organization.
Interacting atoms form molecules that combine in the protein fibers of heart muscle cells. These cells interlock, creating heart muscle tissue that constitutes most of the walls of a three-dimensional organ, the heart. The heart is one component of the cardiovascular system, which also includes the blood and blood vessels. The combined organ systems form an organism, a living human being.

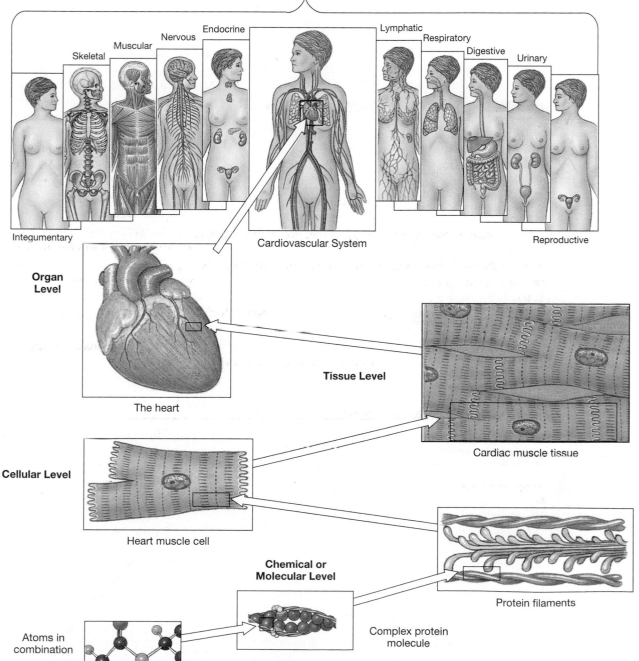

Organ System Level

Integumentary

Skeletal

Muscular

Nervous

Endocrine

Cardiovascular System

Lymphatic

Respiratory

Digestive

Urinary

Reproductive

Organ Level

The heart

Tissue Level

Cardiac muscle tissue

Cellular Level

Heart muscle cell

Chemical or Molecular Level

Protein filaments

Complex protein molecule

Atoms in combination

11

STUDY TIP

You will benefit more from your laboratory studies if you prepare for each laboratory meeting. Before class, read the appropriate exercise(s) in this manual and complete the labeling of as many of the figures as possible. If possible, relate the laboratory exercises to the theory concepts in the lecture textbook. Approaching the laboratory in this manner will maximize your hands-on time with laboratory materials and improve your understanding of the subject you are studying.

B. Anatomical Position

The human body can bend and stretch in a wide variety of directions. Although this flexibility allows us to move and manipulate objects in our environment, it can cause difficulty when describing and comparing structures. For example, what is the correct relation between the wrist and the elbow? If your arm is raised above your head, you might reply that the wrist is above the elbow. With your arms at your sides, you would respond that the wrist is below the elbow. Each response appears correct, but which is the proper anatomical relation?

To avoid confusion, the body is always referred to in a universal position called **anatomical position**. In anatomical position, the individual is standing erect with the feet pointed forward, the eyes straight ahead, and the palms of the hands facing forward with the arms at the sides (see Figure 2.2). Stand up and assume the anatomical position. The entire front of your body should be oriented forward. Notice that the position is not the natural posture of the body. The arms and hands have been rotated forward to bring the forearms and palms to the front. When lying on the back in anatomical position, an individual is said to be **supine** (soo-PĪN); when lying face down, **prone**.

When observing a person in anatomical position, his or her right is on your left and his or her left is on your right. For example, to shake someone's hand, your right hand crosses over to meet his or her right hand. The positioning of right and left is important to remember when viewing structures that differ structurally between right and left sides.

C. Directional Terminology

Imagine attempting to give someone directions if you could not use terms like "north" and "south" or "left" and "right." These words have a unique meaning and guide the traveler toward his or her destination. Describing the body also requires specific terminology. Expressions such as "near," "close to," or "on top of" are too vague for anatomical descriptions. To prevent misunderstandings, concise terms are used to describe the location and associations of anatomy. These terms have their roots in the Greek and Latin languages. Figure 2.2 and Table 2.1 display the most frequently used directional terms. Notice that most of the directional terms can be grouped into opposing pairs or antonyms.

Superior and **inferior** are used to describe vertical positions of structures. "Superior" means above and "inferior" means below another. For example, the head is superior to the shoulders, and the knee is inferior to the hip. Remember that the body is always assumed to be in anatomical position. **Cranial** refers to the head or toward the head region. The cranial cavity is the internal space of the skull.

Anterior and **posterior** refer to the front and the back positions. "Anterior" means in front of or forward. The anterior surface of the body includes all front surfaces including the palms of the hand. Structures on the back surface of the body are said to be posterior. The heart is posterior to the breastbone and anterior to the spine.

In four-legged animals, anatomical position is with all four limbs on the ground, and therefore the meanings of some directional terms change. "Superior" now refers to the back or dorsal surface, and "inferior" relates to the belly or **ventral** surface. **Cephalic** means toward the front or anterior, and **caudal** refers to posterior structures.

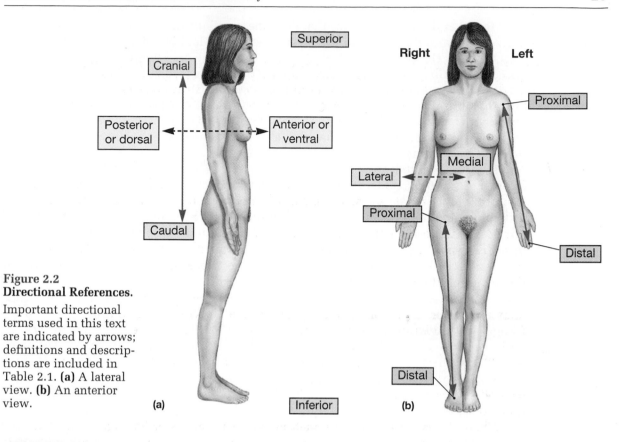

Figure 2.2
Directional References.

Important directional terms used in this text are indicated by arrows; definitions and descriptions are included in Table 2.1. **(a)** A lateral view. **(b)** An anterior view.

TABLE 2.1	Directional Terms (see Figure 2.2)	
Term	*Region or Reference*	*Example*
Anterior	The front; before	The navel is on the *anterior* surface of the trunk.
Ventral	The belly side (equivalent to anterior when referring to human body)	In humans, the navel is on the *ventral* surface.
Posterior	The back; behind	The shoulder blade is located *posterior* to the rib cage.
Dorsal	The back (equivalent to posterior when referring to human body)	The *dorsal* body cavity encloses the brain and spinal cord.
Cranial or cephalic	The head	The *cranial*, or *cephalic*, border of the pelvis is on the side toward the head rather than toward the thigh.
Superior	Above; at a higher level (in human body, toward the head)	In humans, the cranial border of the pelvis is *superior* to the thigh.
Caudal	The tail (coccyx in humans)	The hips are *caudal* to the waist.
Inferior	Below; at a lower level	The knees are *inferior* to the hips.
Medial	Toward the body's longitudinal axis; toward the midsagittal plane	The *medial* surfaces of the thighs may be in contact; moving medially from the arm across the chest surface brings you to the sternum.
Lateral	Away from the body's longitudinal axis; away from the midsagittal plane	The thigh articulates with the *lateral* surface of the pelvis; moving laterally from the nose brings you to the eyes.
Proximal	Toward an attached base	The thigh is *proximal* to the foot; moving proximally from the wrist brings you to the elbow.
Distal	Away from an attached base	The fingers are *distal* to the wrist; moving distally from the elbow brings you to the wrist.
Superficial	At, near, or relatively close to the body surface	The skin is *superficial* to underlying structures.
Deep	Farther from the body surface	The bone of the thigh is *deep* to the surrounding skeletal muscles.

Medial and **lateral** are terms that describe positions relative to the body's midline, the vertical middle of the body or of a structure. "Medial" refers to the middle, and structures on the periphery or away from the midline are lateral. In anatomical position, the thumb is lateral to the ring finger. The nose is medial to the eyes.

Proximal refers to parts near or close to another structure. **Distal** describes structures that are away or distant from other structures. These terms are frequently used to describe the proximity of a structure to its point of attachment on the body. The upper leg bone, the femur, has a proximal region where it attaches to the hip, and a distal portion toward the knee.

Superficial and **deep** describe layered structures. "Superficial" refers to parts on or close to the surface. Underneath an upper layer are deep or bottom structures. The skin is superficial to the muscular system, and bones are usually deep to the muscles.

Some directional terms seem to be interchangeable, but there is usually a precise term for each description. For example, "superior" and "proximal" both describe the upper region of arm and leg bones. When discussing the point of attachment of a bone, "proximal" is the most descriptive term. When comparing the location of a bone relative to an inferior bone, the term "superior" is used.

LABORATORY ACTIVITY   DIRECTIONAL TERMINOLOGY

MATERIALS

torso models, charts, anatomical models

PROCEDURES

1. Review each directional term presented in Figure 2.1 and Table 2.1.
2. Utilize the available laboratory models and your body to practice using directional terms while comparing anatomy. The Laboratory Report at the end of this exercise may be used as a guide for comparisons.

D. Planes and Sections

To observe the body's internal organization, it is cut or **sectioned**. Most structures, such as the trunk, knee, arm, and eyeball can be sectioned. The imaginary line of a section is called a **plane**. The orientation of the plane of section determines the shape and appearance of the exposed internal design. Imagine cutting a soda straw crosswise (transversely), and another straw lengthwise (sagittally). The transverse section produces a circle, and the sagittal section produces a long U shape. Figure 2.3 presents the appearance of two common objects in various planes of section.

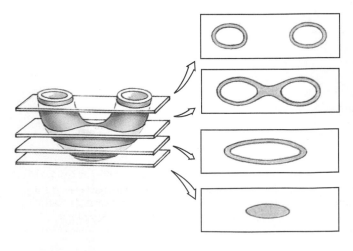

Figure 2.3
Sectional Planes and Visualization.

Taking a series of sections through an object provides detailed information about its three-dimensional structure. This process is called serial reconstruction. Notice how the sectional views change; although it is a simple tube, a piece of elbow macaroni can look like a pair of tubes, a dumbbell, an oval, or a solid, depending on where the section is taken. The effects of sectional plane should be kept in mind when looking at slides under the microscope.

Three major types of sections are employed in the study of anatomy, two vertical sections and one transverse section (see Figure 2.4 and Table 2.2). **Transverse** sections are perpendicular to the vertical orientation of the body. These sections are often called cross sections because they go across the body axis. Superior and inferior structures are divided by transverse sections. **Vertical** sections are parallel to the axis of the body and

Figure 2.4
Planes of Section.

The three primary planes of section, defined and described in Table 2.2.

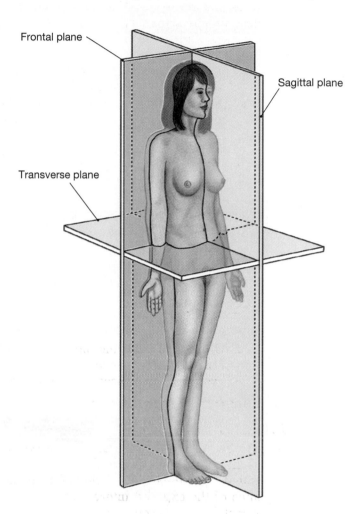

Frontal plane

Sagittal plane

Transverse plane

TABLE 2.2	Terms That Indicate Planes of Section (see Figure 2.4)		
Orientation of Plane	**Plane**	**Directional Reference**	**Description**
Perpendicular to long axis	Transverse or horizontal	Transversely or horizontally	A *transverse*, or *horizontal*, *section* separates superior and inferior portions of the body.
Parallel to long axis	Sagittal	Sagittally	A *sagittal section* separates right and left portions. You examine a sagittal section, but you section sagittally.
	Midsagittal		In a *midsagittal section* the plane passes through the midline, dividing the body in half and separating right and left sides.
	Parasagittal		A *parasagittal section* misses the midline, separating right and left portions of unequal size.
	Frontal or coronal	Frontally or coronally	A *frontal*, or *coronal*, *section* separates anterior and posterior portions of the body; *coronal* usually refers to sections passing through the skull.

include sagittal and frontal sections. **Sagittal** sections divide parts into right and left portions. A **midsagittal** section equally divides structures, and **parasagittal** sections produce nearly equal divisions. A **frontal** or **coronal** section separates anterior and posterior structures.

LABORATORY ACTIVITY PLANES AND SECTIONS

MATERIALS

sectioned anatomical models

PROCEDURES

1. Review each plane and section shown in Table 2.2 and Figures 2.3 and 2.4.
2. Identify the sections on models and other material presented by your lab instructor.
3. Cut several common objects, such as an apple and a hot dog, along their sagittal and transverse planes. Compare the exposed arrangement of the interior.

E. **Regional Terminology**

Approaching the body from a regional perspective simplifies the learning of anatomy. Many internal structures are named after the overlying surface structures. Surface features of the body are used as anatomical landmarks to assist in locating internal structures. For example, the knee is called the popliteal region, and the major artery in the knee is the popliteal artery. Figure 2.5 and Table 2.3 present the major regions of the body.

The head is referred to as the **cephalon** and contains the **cranium** or skull and the **face**. The neck is the **cervical** region. The main part of the body is the **trunk** that attaches the neck, arms, and legs. The upper trunk is the chest or **pectoral** region.

TABLE 2.3	Regions of the Human Body (see Figure 2.5)
Structure	*Region*
Cephalon (head)	Cephalic region
Cervicis (neck)	Cervical region
Thoracis (thorax, or chest)	Thoracic region
Brachium (arm)	Brachial region
Antebrachium (forearm)	Antebrachial region
Carpus (wrist)	Carpal region
Manus (hand)	Manual region
Abdomen	Abdominal region
Lumbus (loin)	Lumbar region
Gluteus (buttock)	Gluteal region
Pelvis	Pelvic region
Pubis (anterior pelvis)	Pubic region
Inguen (groin)	Inguinal region
Femur (thigh)	Femoral region
Crus (anterior leg)	Crural region
Sura (calf)	Sural region
Tarsus (ankle)	Tarsal region
Pes (foot)	Pedal region
Planta (sole)	Plantar region

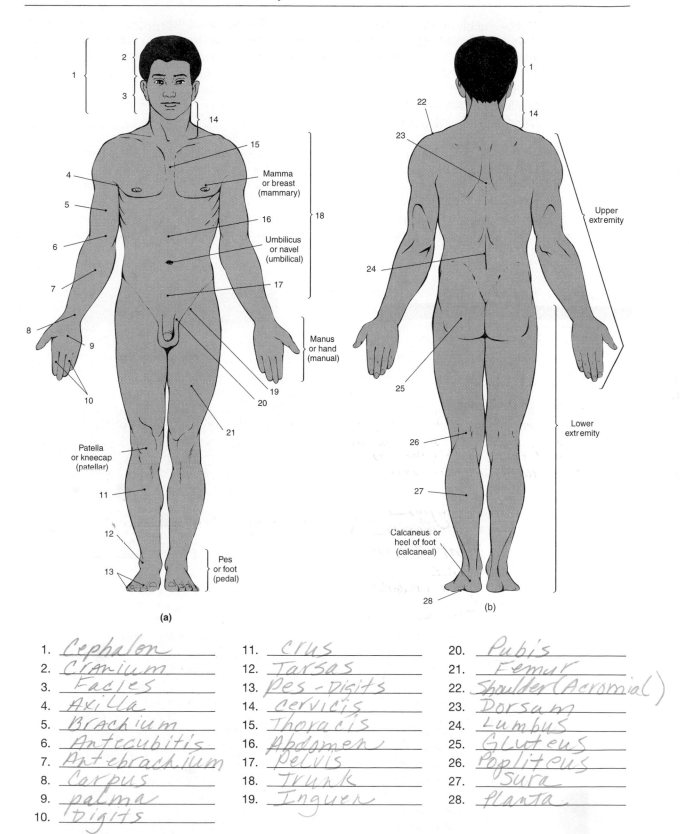

Figure 2.5
Regional Terminology.

1. Cephalon
2. Cranium
3. Facies
4. Axilla
5. Brachium
6. Antecubitis
7. Antebrachium
8. Carpus
9. Palma
10. Digits

11. Crus
12. Tarsas
13. Pes - Digits
14. Cervicis
15. Thoracis
16. Abdomen
17. Pelvis
18. Trunk
19. Inguen

20. Pubis
21. Femur
22. Shoulder (Acromial)
23. Dorsum
24. Lumbus
25. Gluteus
26. Popliteus
27. Sura
28. Planta

Below the chest is the **abdominal** region which narrows inferiorly at the **pelvis**. The posterior surface of the trunk, the **dorsum**, includes the **loin** or lower back and the **gluteal** region of the buttocks. The side of the trunk below the ribs is the **flank**.

The shoulder or **scapular** region attaches the **brachium** to the trunk and forms the **axilla**, the armpit. The **antebrachium** is the forearm. Between the brachium and antebrachium is the **antecubitis** region, the elbow. The wrist is called the **carpus**, and the anterior surface of the hand is the **palm**.

The **pelvis** attaches the leg to the trunk at the **groin**. The upper leg is the **thigh**, the knee is the **popliteal** (pop-LIT-ē-al) region, and the lower leg is the **calf**. **Tarsus** refers to the ankle, and the sole of the foot is the **plantar** surface.

The position of internal abdominal organs is simplified by partitioning the trunk into four equal **quadrants**. Observe in Figure 2.6a the midsagittal and transverse planes used to delineate the quadrants. Quadrants are used to describe the positions of organs. The stomach, for example, is mostly located in the left upper quadrant.

Figure 2.6
Abdominopelvic Quadrants and Regions.

(a) Abdominopelvic quadrants divide the area into four sections. These terms, or their abbreviations, are most often used in clinical discussions. (b) Abdominopelvic regions divide the same area into nine sections, providing more-precise regional descriptions. (c) Overlapping quadrants and regions and the relationship between superficial anatomical landmarks and underlying organs.

1. *Right hypochondriac region*
2. *Epigastric Region*
3. *Left hypochondriac region*
4. *Rt. Lumbar region*
5. *Umbilical Region*
6. *Left Lumbar region*
7. *Rt. Inguinal*
8. *hypogastric (Pubic)*
9. *Left Inguinal*

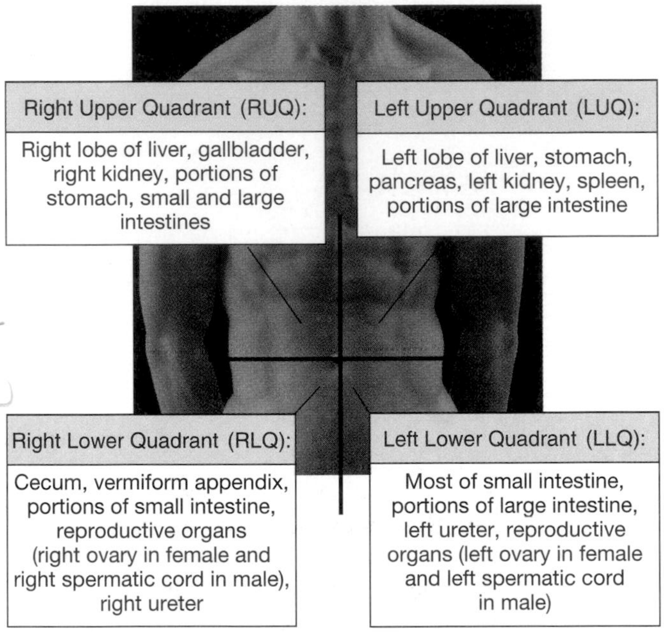

Right Upper Quadrant (RUQ):

Right lobe of liver, gallbladder, right kidney, portions of stomach, small and large intestines

Left Upper Quadrant (LUQ):

Left lobe of liver, stomach, pancreas, left kidney, spleen, portions of large intestine

Right Lower Quadrant (RLQ):

Cecum, vermiform appendix, portions of small intestine, reproductive organs (right ovary in female and right spermatic cord in male), right ureter

Left Lower Quadrant (LLQ):

Most of small intestine, portions of large intestine, left ureter, reproductive organs (left ovary in female and left spermatic cord in male)

(a)

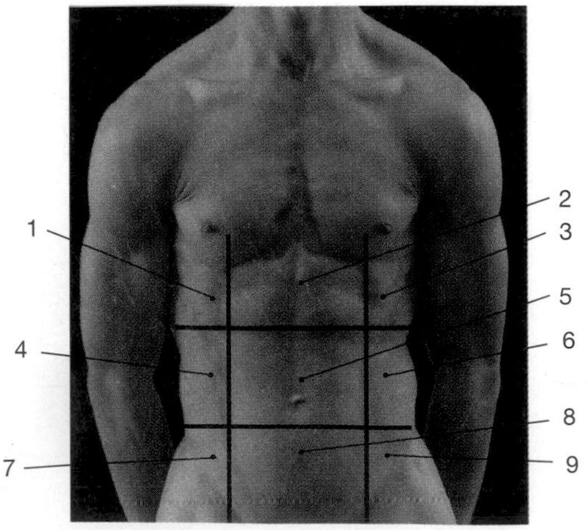

(b)

Liver
Gallbladder
Large intestine
Small intestine
Vermiform appendix
Stomach
Spleen
Urinary bladder

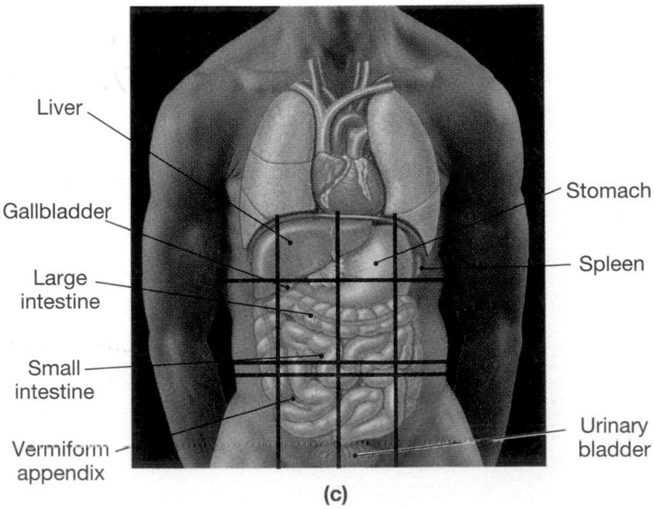

(c)

For more detailed descriptions, the abdominal surface is divided into nine **abdominopelvic regions**, shown in Figure 2.6b. Four planes are used to define the regions, two vertical and two transverse planes arranged in the familiar tic-tac-toe pattern. The vertical planes, called the right and left **lateral** planes, are slightly medial to the nipples. They divide the trunk into three nearly equal vertical regions. A pair of transverse planes cross the vertical planes to isolate the nine regions. The **transpyloric** plane is superior to the umbilicus (navel) at the level of the pylorus, the lower region of the stomach. The **transtubercular** plane is inferior to the umbilicus and crosses the abdomen at the level of the superior hips.

The nine abdominopelvic regions are as follows. The **umbilical** region surrounds the umbilicus or navel. Lateral to this region are the right and left **lumbar** regions. Above the umbilicus is the **epigastric** region containing the stomach and much of the liver. The right and left **hypochondriac** regions are lateral to the epigastric region. Inferior to the umbilical region is the **hypogastric** or **pubis** region. The right and left **iliac,** or inguinal, regions border the hypogastric region laterally.

LABORATORY ACTIVITY   REGIONAL REFERENCES

MATERIALS

> torso models
> charts

PROCEDURES

1. Review and label the regional terminology in Figure 2.5 and Table 2.3.
2. Use a lab model or yourself and identify the regional anatomy as presented in Figure 2.5.
3. Identify each abdominopelvic quadrant and region in Figure 2.6b and on laboratory models.

F. Introduction to Organ Systems

The human body comprises 11 organ systems, each of which is responsible for a specific function. Most anatomy and physiology courses are designed to progress through the lower levels of organization first and then examine each organ system. Because organ systems work together to maintain the organism, it is important that you have a basic understanding of the function of each one. Examine Figure 2.7, an introduction to organ systems, and learn the major organs and the basic function of each organ system.

LABORATORY ACTIVITY   IDENTIFICATION OF ORGAN SYSTEMS

MATERIALS

> torso model with internal organs
> anatomical models
> charts

PROCEDURES

1. Locate the principal organs of each organ system on lab models.
2. On your body, identify the general location of as many organs as possible.

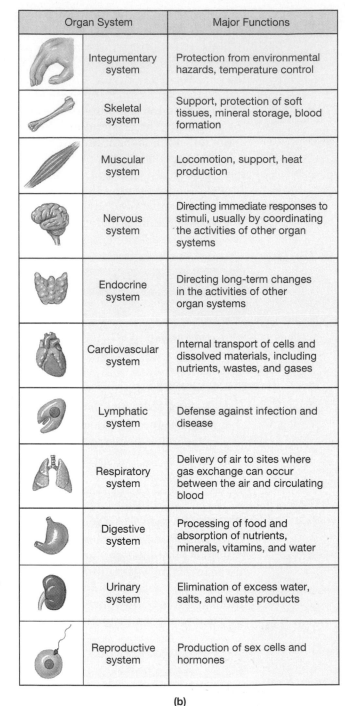

Organ System		Major Functions
	Integumentary system	Protection from environmental hazards, temperature control
	Skeletal system	Support, protection of soft tissues, mineral storage, blood formation
	Muscular system	Locomotion, support, heat production
	Nervous system	Directing immediate responses to stimuli, usually by coordinating the activities of other organ systems
	Endocrine system	Directing long-term changes in the activities of other organ systems
	Cardiovascular system	Internal transport of cells and dissolved materials, including nutrients, wastes, and gases
	Lymphatic system	Defense against infection and disease
	Respiratory system	Delivery of air to sites where gas exchange can occur between the air and circulating blood
	Digestive system	Processing of food and absorption of nutrients, minerals, vitamins, and water
	Urinary system	Elimination of excess water, salts, and waste products
	Reproductive system	Production of sex cells and hormones

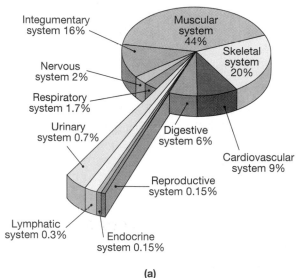

(a)

(b)

Figure 2.7
An Introduction to Organ Systems.

(a) The relative percentage each system contributes to total body weight. **(b)** The major functions of each organ system.

INTRODUCTION TO THE BODY CHECKLIST

This is a list of bold terms presented in Exercise 2. Use this list for review purposes after you have completed the laboratory activities.

SCIENCES OF ANATOMY AND PHYSIOLOGY

anatomy, gross anatomy
microanatomy, cytology, histology
physiology, homeostasis

LEVELS OF ORGANIZATION

• Figure 2.1
chemical and molecular level
cellular level, tissue level
organ level, organ system level
organism level

ANATOMICAL POSITION AND DIRECTIONAL TERMINOLOGY

• Figure 2.2 and Table 2.1
anatomical position
supine, prone
superior, inferior
anterior, posterior
cranial, cephalic
dorsal
ventral
caudal
medial, lateral
proximal, distal
superficial, deep

PLANES AND SECTIONS

• Figures 2.3 and 2.4, Table 2.2
section, plane
transverse and vertical sections
sagittal, midsagittal, and parasagittal sections
frontal and coronal sections

REGIONAL TERMINOLOGY

• Figure 2.5 and Table 2.3
cephalon, cranium, face
cervical, trunk, pectoral
abdominal, pelvis

dorsum
loin, gluteal, groin, flank
scapular, axilla
brachium, antebrachium, antecubitus
carpus, palm
pelvis, thigh
popliteal, calf
tarsus, plantar surface

ABDOMINOPELVIC QUADRANTS AND REGIONS

• Figure 2.6
abdominopelvic quadrants
midsagittal plane
transverse plane
upper right and upper left quadrants
lower right and lower left quadrants
abdominopelvic regions
lateral planes, transpyloric plane
transtubercular plane
umbilical region, epigastric region
right and left hypochondriac regions
right and left lumbar regions
hypogastric (pubis) region
right and left iliac regions

INTRODUCTION TO ORGAN SYSTEMS

• Figure 2.7
integumentary system
skeletal system
muscular system
nervous system
endocrine system
cardiovascular system
lymphatic system
respiratory system
digestive system
urinary system
reproductive system

INTRODUCTION TO THE BODY
Laboratory Report Exercise 2

A. Matching

Match each directional term listed on the left with the correct description on the right.

1. _D_ anterior
2. _A_ lateral
3. _G_ proximal
4. _H_ inferior
5. _I_ posterior
6. _J_ medial
7. _B_ distal
8. _C_ superficial
9. _F_ superior
10. _E_ deep

A. to the side
B. away from a point of attachment
C. close to the body surface
D. front
E. away from the body surface
F. above, on top of
G. toward a point of attachment
H. below, a lower level
I. back
J. to the middle

B. Short-Answer Questions

1. Describe the levels of organization in the body.

2. Why is anatomical position important when describing structures?

3. List the nine abdominopelvic regions and the location of each.

4. Compare the study of anatomy with that of physiology.

C. Directional and Regional Terminology

Use the correct directional term to show the relationship between the following structures:

1. The chin is _Inferior_ to the nose.
2. The brachium is _Superior_ to the antecubitis.
3. The thumb is _Lateral_ to the ring finger.
4. The skin is _Superficial_ to the muscles.
5. The trunk is _Superior_ to the pubis.
6. The ring finger is _Inferior_ to the little finger. _Lateral_
7. The upper humerus is _Inferior_ to the elbow.
8. The thigh near the knee is _Inferior_ to the upper thigh.
9. The ears are _Caudial_ to the eyes.
10. The buttock is _Posterior_ to the pubis.

D. Drawing

1. Draw a picture of a doughnut sectioned by two perpendicular planes.

2. Sketch the body trunk and the planes that designate the abdominopelvic regions.

Body Cavities and Membranes

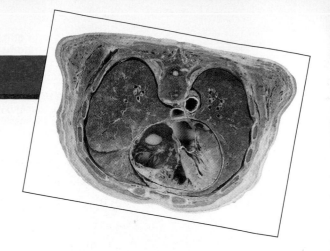

OBJECTIVES

On completion of this exercise, you should be able to:

- Describe and identify the location of the dorsal cavity and its divisions.
- Describe and identify the location of the ventral cavity and its divisions.
- Describe and identify the serous membranes of the body.

WORD POWER

mediastinum (media—middle)
serous (seri—serum)
visceral (viscera—internal organ)
parietal (pariet—wall)

pleural (pleuro—rib)
retroperitoneal (retro—behind, in back)

MATERIALS

torso model
anatomical models
charts
articulated skeleton

INTRODUCTION

Body cavities are internal spaces that house internal organs such as the brain in the cranium and the digestive organs in the abdomen. The walls of a body cavity support and protect the soft organs. Large cavities have smaller cavities that consist of membrane partitions to contain each organ. Most of the room in a cavity is occupied by the enclosed organ. Some of the body cavities are thin sacs, such as around the heart, lungs, and intestines. There are two major body cavities, the dorsal body cavity and the ventral body cavity. The dorsal body cavity contains the central nervous system, and the ventral body cavity houses the thoracic and abdominopelvic organs.

Organs in the ventral body cavity are covered with a double-layered **serous (SER-us) membrane.** These membranes isolate organs and reduce friction and abrasion on the organ surface. The heart, lungs, and the stomach and intestines are protected with serous membranes. In the dorsal body cavity, the brain and spinal cord are contained within the meninges, a protective three-layered membrane. The meninges will be studied later with the nervous system.

A. Dorsal Body Cavity

The **dorsal body cavity** contains the brain and spinal cord. It is subdivided into the **cranial** and the **spinal cavities**, shown in Figure 3.1. The cranial cavity is the space inside the skull where your brain is found. This cavity is formed by bones of the skull and functions to protect the delicate brain. The spinal cavity is a canal that passes through the spine or vertebral column. The spinal cord is located in the spinal cavity. The vertebrae have holes called vertebral foramina that collectively form the walls of the spinal cavity. The cranial and spinal cavities are interconnected at the base of the skull where the spinal cord joins the brain.

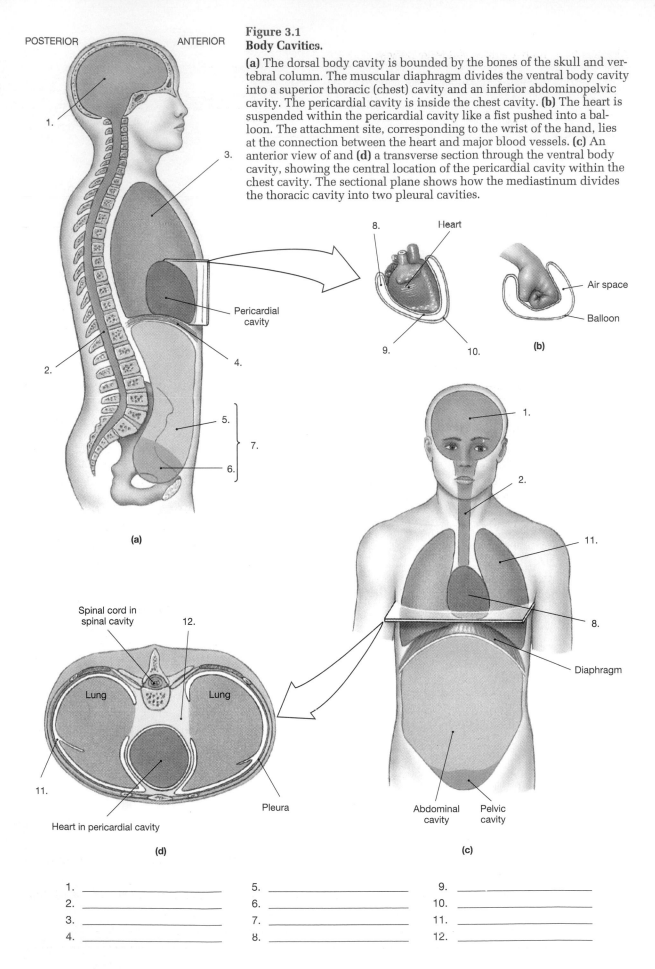

Figure 3.1
Body Cavities.

(a) The dorsal body cavity is bounded by the bones of the skull and vertebral column. The muscular diaphragm divides the ventral body cavity into a superior thoracic (chest) cavity and an inferior abdominopelvic cavity. The pericardial cavity is inside the chest cavity. **(b)** The heart is suspended within the pericardial cavity like a fist pushed into a balloon. The attachment site, corresponding to the wrist of the hand, lies at the connection between the heart and major blood vessels. **(c)** An anterior view of and **(d)** a transverse section through the ventral body cavity, showing the central location of the pericardial cavity within the chest cavity. The sectional plane shows how the mediastinum divides the thoracic cavity into two pleural cavities.

8.
Heart

9.
10.

(b)
Air space
Balloon

Pericardial
cavity

4.

5.
7.
6.

(a)

1.
2.
11.
8.
Diaphragm

Spinal cord in
spinal cavity
12.

Lung
Lung

11.
Pleura
Heart in pericardial cavity

Abdominal
cavity
Pelvic
cavity

(d)
(c)

1. _____
2. _____
3. _____
4. _____

5. _____
6. _____
7. _____
8. _____

9. _____
10. _____
11. _____
12. _____

LABORATORY ACTIVITY   THE DORSAL BODY CAVITIES

MATERIALS

torso models
articulated skeleton
charts

PROCEDURES

1. Label and review each component of the dorsal body cavity in Figure 3.1.
2. Locate the dorsal body cavity and each subdivision on your body.
3. Identify each cavity on the available laboratory models and an articulated skeleton.
4. Identify the organ(s) of each body cavity on the lab models.

B. Ventral Body Cavity

The ventral body cavity is the entire space of the body trunk anterior to the vertebral column and posterior to the sternum and the abdominal muscle wall. Using Figure 3.1 as a guide, trace the outline of the ventral body cavity on the anterior surface of your body. This large cavity is divided into two major cavities, the **thoracic cavity** and the **abdominopelvic cavity**. These cavities, in turn, are further subdivided into the specific cavities that surround individual organs. The thoracic cavity, or chest cavity, contains the heart, lungs, trachea, larynx, esophagus, thymus gland, and many large blood vessels. The walls of the thoracic cavity are muscle and bone. The **pericardial cavity** surrounds the heart. Each lung is contained within a **pleural cavity**. Notice in Figure 3.1c that the middle of the thoracic cavity contains an area called the **mediastinum** (mē-dē-as-TĪ-num) which contains the heart, the large vessels of the heart, the thymus gland, the trachea, and the esophagus. The lungs are located outside the mediastinum.

The **abdominopelvic cavity** is separated from the thoracic cavity by a dome-shaped muscle, the diaphragm. Locate the approximate position of the diaphragm on your anterior surface and in the various views of Figure 3.1. The abdominopelvic cavity is the space between the diaphragm and the floor of the pelvis. This cavity is subdivided into the abdominal cavity and the pelvic cavity. The **abdominal cavity** contains most of the digestive system organs such as the liver, gallbladder, stomach, pancreas, kidneys, and small and large intestines. The **pelvic cavity** is the small cavity enclosed by the pelvic girdle of the hips. This cavity contains internal reproductive organs, parts of the large intestine, the rectum, and the urinary bladder.

LABORATORY ACTIVITY   THE VENTRAL BODY CAVITY

MATERIALS

torso models
articulated skeleton
charts

PROCEDURES

1. With your finger, trace the location of the various divisions of the ventral body cavity on the anterior of your trunk.
2. Locate each body cavity on laboratory models, mannequins, charts, and skeleton.
3. Identify the organ(s) of each body cavity on the lab models.
4. Label the components of the abdominopelvic cavities in Figure 3.1.

C. Serous Membranes

The heart, lungs, and intestines are encased in specialized **serous membranes**, double-layered membranes with a lubricating fluid in the minuscule cavity between the membranes. Directly attached to the exposed surface of an internal organ is the **visceral (VIS-er-al) layer** of the serous membrane. The **parietal (pah-RĪ-e-tal) layer** is superficial to the visceral layer and lines the wall of the body cavity. The serous fluid between these layers is a lubricant that reduces friction and abrasion as the enclosed organ moves.

Figure 3.1a highlights the anatomy of the serous membrane of the heart, the **pericardium**. It is composed of an outer fibrous sac, the **parietal pericardium**, which anchors the heart and prevents its overexpansion. The **visceral pericardium** is attached to the surface of the heart muscle. Between the serous layers is the **pericardial cavity**. Imagine pushing your fist into a water balloon; your fist is the heart and the balloon is the serous membrane. The balloon immediately surrounding your hand represents the visceral pericardium, and the outer layer of the balloon is the parietal pericardium. The water-filled space is the pericardial cavity with water representing serous fluid.

Each lung is isolated in a separate pleural cavity. The **parietal pleura (PLOO-rah)** lines the thoracic wall, and the **visceral pleura** is attached to the surface of the lung. Because each lung is contained inside a separate cavity, a puncture wound on one side of the chest usually only collapses the corresponding lung.

Most of the digestive organs are encased in the **peritoneum (per-i-tō-NĒ-um)**, the serous membrane of the abdomen. The **parietal peritoneum** has numerous folds that wrap around and attach the abdominal organs to the posterior abdominal wall. The **visceral peritoneum** lines the organ surface. The **peritoneal cavity** is between the parietal and visceral peritoneal layers. The peritoneum has many blood vessels, lymphatic vessels, and nerves which support the digestive organs. The kidneys are **retroperitoneal** and are located outside the peritoneum.

Serous membranes may become inflamed and infected as a result of bacterial invasion or damage to the underlying organ. Fluids often build up in the cavity of the serous membrane causing additional complications. **Peritonitis** is an infection of the peritoneum. Ulceration, rupture, or a puncture wound to the digestive tract permits intestinal bacteria to contaminate the peritoneum. **Pleuritis** or **pleurisy** is an infection of the pleural membrane. Breathing becomes painful as the dry serous membrane layers rub against one another. **Pericarditis** is an infection of the pericardium.

What would happen if the appendix burst and intestinal contents leaked into the peritoneal cavity? Would there be an immediate effect on the kidneys?

LABORATORY ACTIVITY   SEROUS MEMBRANES

MATERIALS

torso model

charts

PROCEDURES

1. Review and label each serous membrane in Figure 3.1.
2. Identify the pericardium, pleura, and peritoneum on the models and charts.
3. Identify the organ(s) of each body cavity on the lab models.
4. Label the layers of each serous membrane in Figure 3.1.

BODY CAVITIES AND MEMBRANES CHECKLIST

This is a list of bold terms presented in Exercise 3. Use this list for review purposes after you have completed the laboratory activities.

BODY CAVITIES

- Figures 3.1 and 3.2
dorsal body cavity
cranial cavity
spinal cavity
ventral body cavity
thoracic cavity
pleural cavity
pericardial cavity
mediastinum
abdominopelvic cavity
abdominal cavity
peritoneal cavity
pelvic cavity

SEROUS MEMBRANES

- Figures 3.1 and 3.2
parietal layer of serous membranes
    serous fluid
visceral layer of serous membranes
pericardium
    parietal pericardium
    visceral pericardium
pericarditis
pleura
    parietal pleura
    visceral pleura
pleuritis (pleurisy)
peritoneum
    parietal peritoneum
    visceral peritoneum
peritonitis

BODY CAVITIES AND MEMBRANES
Laboratory Report Exercise 3

A. Fill in the Blanks

1. Your heart is located in a small cavity called the _____ which is located in a larger cavity, the _____.

2. Your intestines are surrounded by a double membrane called a _____ membrane.

3. Your kidneys are _____ because they are located outside the _____.

4. A plane at the top of the hips separates the abdominal cavity from the _____.

5. The inner membrane layer surrounding the heart is the _____.

6. The brain and spinal cord are contained in the _____.

7. A lubricating substance in body cavities is called _____.

8. The large medial area of your chest is called the _____.

9. The muscle that divides the ventral body cavity is the _____.

10. The outer layer of a serous membrane is the _____ layer.

B. Matching

Match each term on the left with the correct description on the right.

1. ___D___ serous membrane   A. in reference to an internal organ
2. ___A___ parietal   B. serous membrane of abdomen
3. ___E___ pleura   C. in reference to a body wall
4. ___C___ visceral   D. double-layered protective membrane
5. ___F___ pericardium   E. serous membrane of lungs
6. ___B___ peritoneum   F. serous membrane of heart

C. Labeling

Label the cavities in Figure 3.2.

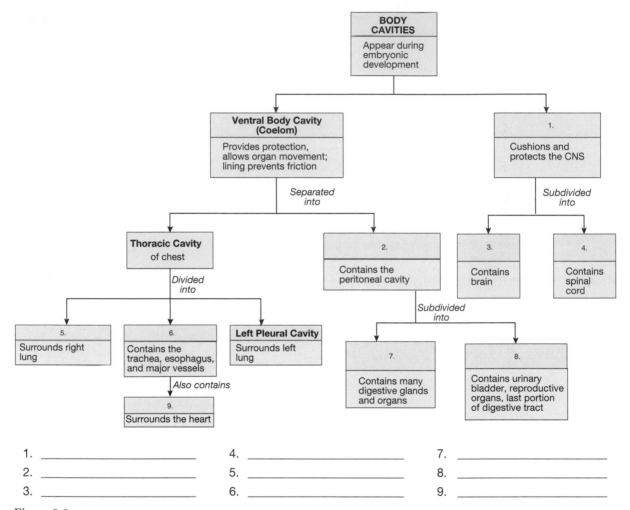

1. _____   4. _____   7. _____
2. _____   5. _____   8. _____
3. _____   6. _____   9. _____

Figure 3.2
Relationships of the Various Body Cavities.

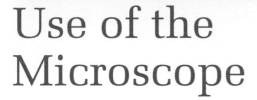

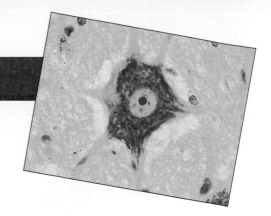

Use of the Microscope

OBJECTIVES

On completion of this exercise, you should be able to:

- Describe how to properly carry, clean, and store a microscope.
- Identify the parts of the microscope.
- Focus the microscope on a specimen and adjust the magnification.
- Adjust the light source of the microscope.
- Calculate the total magnification for each objective lens.
- Measure the field diameter at each magnification.
- Make a wet mount slide.

WORD POWER

oculus (the eye, ocular lens)
aper- (open, aperture)
magni- (large, magnification)

MATERIALS

compound microscope
clean microscope slides
microscope coverslips
dropper bottle of water

scissors and newspaper
millimeter graph slides
practice slides

INTRODUCTION

As a student of anatomy and physiology, you will explore the organization and structure of cells, tissues, and organs. The basic research tool for your observations will be the microscope. The instrument is easy to use once you learn its parts and how to adjust them to produce a clear image of a specimen. It is important that you complete each activity in this exercise and that you are able to use the microscope effectively by the end of the laboratory period.

The compound microscope uses several lenses to direct a narrow beam of light through a thin specimen mounted on a glass slide. Focusing knobs move the lenses to bring the specimen into **focus** within the round viewing area of the lenses, the **field of view**. Lenses **magnify** objects so they appear larger than they are in life. As magnification increases, the viewer sees more details of objects that are closer together. It is this increase in **resolution** that makes the microscope a powerful observational tool.

A. Care and Handling of the Microscope

The microscope is a precision scientific instrument with delicate optical components. Observe the following guidelines as you use the microscope:

1. Carry the microscope with two hands, one hand on the arm and the other hand supporting the base. (See Figure 4.1 for parts of the microscope.) Do not swing the microscope as you carry it to your laboratory bench. You may bump the instrument or cause a lens to fall out.

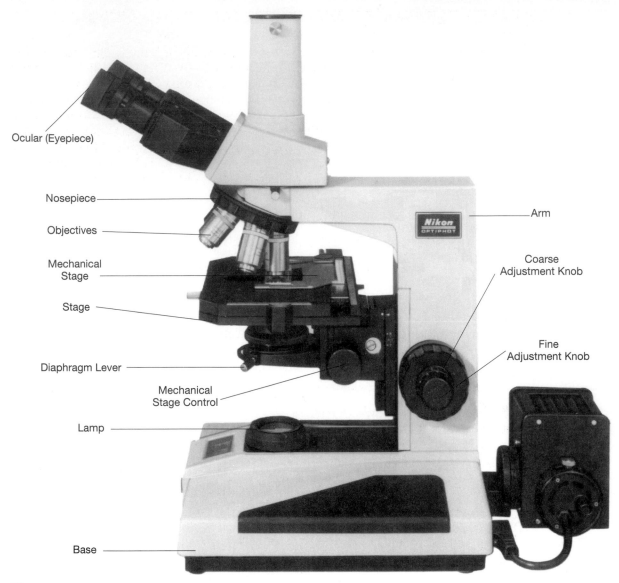

Figure 4.1
Parts of the Compound Microscope.

2. Completely unravel the electric cord on microscopes with a built-in light source.

3. To clean the lenses, use only the special lens cleaning fluid and lens paper provided by your laboratory instructor. Facial tissue paper is made of small wood chips and fibers that will damage the special optical coating on the lenses.

4. Store the microscope with the cord wrapped neatly, the low-power objective lens in position, and the stage in the uppermost position. Return the microscope to the storage cabinet or cover it with a dust cover.

B. Parts of the Compound Microscope

Get a microscope from the storage cabinet in the laboratory and find each part as it is described. Your laboratory may be equipped with a different microscope than the one shown in Figure 4.1. Your laboratory instructor will discuss the type of microscope you will use.

Ocular lens

The ocular lens is the eyepiece where the user places his or her eye(s) to observe the specimen. The magnification of most ocular lenses is 10×. This results in an image 10 times larger than the actual size of the specimen. **Monocular** microscopes have a single ocular lens; **binocular** microscopes have two ocular lenses, one for each eye. It is necessary to adjust binocular microscopes so that the two eyepieces form a single image. This is easily accomplished by adjusting the distance between the ocular lenses so that it is the same as the distance between your pupils. Move the body tubes apart and look into the microscope. If two images are visible, slowly move the body tubes closer together until a single circle, the **field of view**, is seen with both eyes.

Nosepiece

The nosepiece is a rotating mechanism at the base of the cylindrical body tube with several objective lenses of different lengths. Turning the nosepiece moves an objective into place over the specimen.

Objective lenses

The objective lenses are mounted on the nosepiece. Magnification is determined by the user's choice of objective lens. The longer the objective lens, the greater its magnifying power.

Stage

The stage is a flat, horizontal shelf under the objectives that supports a glass specimen slide. The center of the stage has an **aperture** or hole through which light passes to illuminate the specimen on the slide. A pair of **stage clips** holds the slide steady while you move the slide by hand. Some microscopes have a **mechanical stage** that moves the slide with more precision than is possible manually. The mechanical stage has two **adjustment knobs** on the side that move a slide over the stage's platform. One knob moves the slide horizontally, and the other moves it vertically.

Coarse (gross) focus adjustment

The coarse focus adjustment is the large dial on the side of the microscope that moves the objective lenses to produce a sharp, clear image. Use the coarse focus knob *only at low magnification* to initially focus a specimen.

Fine focus adjustment

The small dial on the side of the microscope is the fine focus adjustment. This knob moves the objective lens for precision focusing after gross focus is achieved. The fine focus knob is used at all magnifications and is the *only focusing knob* used at magnifications greater than low power.

Condenser

The condenser is the small lens under the stage that condenses or narrows the beam of light and directs the light through the slide specimen. A **condenser adjustment knob** moves the condenser vertically. For your studies, the condenser should always be in the upper position next to the aperture of the stage.

Iris diaphragm

The iris diaphragm is a series of flat metal plates found at the base of the condenser. The plates slide together and create an aperture that regulates the amount of light that can pass into the condenser. Most microscopes have a small **diaphragm lever** extending from the iris to open or close the diaphragm to adjust the light for optimal contrast and minimal glare.

Light source

Most microscopes have a built-in **substage** light source with a rheostat dial to control the intensity of the light. Microscopes without an electric light source use a **mirror** to reflect surrounding light into the condenser.

Body tube

The body tube is the cylindrical tube that supports the ocular and extends down to the nosepiece.

Arm

The supportive frame of the microscope is called the arm. It joins the body tube to the base. The microscope is correctly carried with one hand on the arm and the other on the base.

Base

The base is the broad, flat lower support of the microscope.

C. Microscope Basics

Three basic adjustments allow the user to control the appearance and quality of an image under the microscope: (1) illumination controls, (2) objective lens selection, and (3) focusing knobs.

Illumination control

Imagine shooting a roll of film and overexposing the pictures. Contrast between light and dark is lost, and details are washed out. Too much light in the microscope also compromises an image. Light intensity is controlled to produce high contrast and low glare. Light is also condensed into a narrow beam to pass through the specimen. Correct illumination in the microscope not only results in a clearer image but also reduces eyestrain.

In microscopes with an electric light source, the intensity or brightness of light is regulated by a rheostat that controls the output of an electric bulb. Some microscopes use a mirror instead of an electric light source to reflect light toward the specimen. Changing the angle of the mirror varies the light intensity. Whatever the light source, the light rays pass through the condenser lens and narrow into a beam to illuminate the specimen. An iris regulates the amount of light passing into the condenser, much as the iris in the eye controls the light striking the lens. As magnification increases, more light is necessary to fully illuminate the specimen.

Lens selection

Magnification is how large an image appears compared to the specimen's natural size. An object viewed at 10× appears 10 times larger than life size (1×). To change magnification, turn the nosepiece and rotate a different objective into place over the specimen. Figure 4.2 illustrates the optical components of a compound microscope.

A simple lens system, like a hand-held magnifying glass, uses a single lens to magnify an object. Microscopes use a **compound lens** system with each lens consisting of

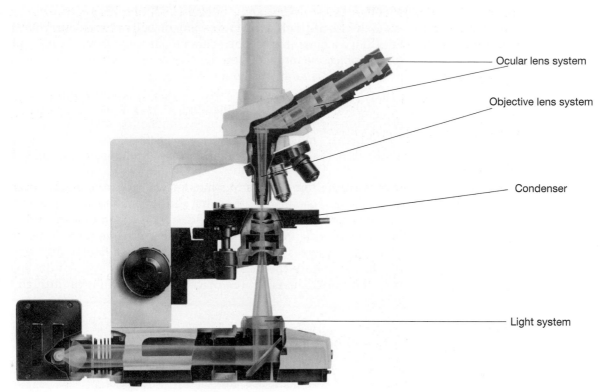

Figure 4.2
The Optical Components of a Compound Microscope.

many pieces of optical glass. The ocular lens, or eyepiece, is usually a 10× lens. Student microscopes typically have three objective lenses: 4×, 10×, and 40×. The magnification is stamped on the barrel of each objective lens. To calculate the total magnification of the microscope at a particular lens setting, the ocular lens is multiplied by the objective lens. For example, a 10× ocular used with a 10× objective lens produces a total magnification of 100×.

Notice that the higher-power objectives are longer than the low-power objective. As higher-power objectives are used, the distance between the specimen and the lens decreases. Therefore, always rotate the low-power objective into place before inserting or removing a slide from the stage. This provides ample working distance between the lens and the stage to adjust the slide. Always view a slide starting at low power. You will see more of the specimen and can quickly select areas on the slide for detailed studies at a higher magnification.

Determine the magnification of the ocular and each objectives lens on the microscope. Complete Table 4.1 by calculating the total magnification for each combination of ocular and objective lenses of your microscope in lab.

TABLE 4.1	**Total Magnification**		
Ocular Lens	*Objective Lens*	*Total Magnification*	*Working Distance (estimate in mm)*

The focusing mechanism

The coarse and fine focus knobs are both used to move an objective lens closer to or further from the slide to achieve a clear image. The coarse focus knob is used only to initially move the low-power objective into focus on the specimen. While looking into the ocular lens, move the coarse focus knob until a focused image of the specimen is visible in the field of view. Minimal focusing is needed at higher magnifications because the space between the objective lens and the specimen is very small. Using a coarse focus at higher magnifications would move the objective lens too much and possibly force it into the slide. Further, most microscopes are **parfocal**, and once focus is achieved at low magnification, the image remains in focus as high magnification objectives are used.

Another reason for using the fine focus adjustment at higher powers is to observe individual layers of a thick slide specimen. Although a slide specimen may appear very thin, it usually has many layers of cells. **Depth of field** is the portion of a specimen in focus, the layer of the slide currently in focus. This focal depth is greatest at low power and decreases as magnification increases. The reduction in depth of field at high power requires the use of fine focus to move the focal depth and scan the layers of the specimen. By slowly turning the fine focus knob, the objective lens will move through the layers with a plane of focus.

USE OF THE MICROSCOPE

1. If the microscope has an internal light source, plug in the microscope and turn the lamp on. If the microscope does not have a built-in light source, adjust the mirror to reflect light into the condenser.
2. To prevent the possibility of breaking a slide, always start with the low-power objective. Rotate the nosepiece to swing the low-magnification objective lens into position over the aperture. This provides maximum clearance between the objective and the stage for the placement of a slide.
3. Place the slide on the stage of the microscope and use the stage clips or mechanical slide mechanism to secure the slide. Move the slide so the specimen is over the stage aperture.
4. To focus on the specimen, first move the low-power objective to its lowest position. Next, look into the ocular and raise the low-power objective lens by slowly turning the coarse focus knob. The image should come into focus. Note that some microscopes are focused by moving the stage rather than the objectives. To focus a specimen on this type of microscope, move the stage closer to the objective lens by turning the coarse focus knob.
5. Once the image is clear, use the fine focus knob to examine the detailed structure of the specimen.
6. To examine part of the specimen at a higher magnification, center that part of the image over the aperture before changing to a higher-magnification objective lens. This keeps the specimen in the smaller field of view at the high magnification.
7. Do not move the focusing knobs before increasing magnification. Most microscopes are parfocal and are designed to stay in focus when a different objective lens is selected. After changing magnification, use the fine focus knob only to adjust the lens.
8. On completion of your observations, reset the microscope to low magnification, remove the slide from the stage, and store the microscope.

LABORATORY ACTIVITY   USING THE MICROSCOPE

I. Newspaper Print Observation

MATERIALS

compound microscope, slide, and coverslip
newspaper cut into small pieces
dropper bottle containing water

Figure 4.3
Coverslip Placement.

Touch the water or stain with the edge of the coverslip, and then lower the coverslip flat onto the slide. Use a paper towel to absorb excess water that has leaked out from under the coverslip.

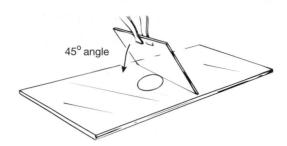

PROCEDURES

1. Make a **wet mount slide** of a small piece of the newsprint as follows:
   a. Obtain a slide, a coverslip, and a small piece of newspaper print.
   b. Place the newsprint on the slide and add a small drop of water to it.
   c. Put the coverslip over the newsprint as shown in Figure 4.3. The coverslip will keep the lenses dry.

2. Select the low-magnification objective lens and place the slide on the stage. What is the total magnification of your microscope at low magnification? _____.

3. Use the coarse focus knob to move the objective as close to the specimen as possible without touching the slide.

4. Move the slide until the newspaper print is directly over the aperture of the stage. Look into the ocular lens and slowly turn the coarse focus knob until you see the fibers of the newspaper. Once they are in focus, adjust the light source for optimal contrast and resolution.

5. Use the fine focus knob to bring the image into crisp focus. Remember, the microscope you are using is a precise instrument and produces a clear image when adjusted correctly. If your specimen is still not in focus, return to the box entitled "Use of the Microscope."

6. Follow each procedure below and record your observations in the space provided.
   a. Once the image is correctly focused, locate the letter "a" or the letter "e." Describe the ink and the paper fibers. _____.
   b. Move the slide toward you. In which direction does the image of the letter move? _____.
   c. Move the slide to the left. In which direction does the image of the letter move? _____.
   d. Is the image of the letter oriented in the same direction in which it is placed on the slide? _____.

II. Depth of Field Observation

To show depth of field, you will examine a slide of overlapping colored threads. Notice how the threads are layered and how much of the slide is in focus at each magnification.

MATERIALS

compound microscope

slide of colored threads

(If colored thread slides are unavailable, make a slide using hairs from different students.)

PROCEDURES

1. Move the low-power objective into position and place the slide on the stage with the threads over the aperture.

2. Use the coarse focus knob to bring the threads into focus. Move the slide to view the area where the threads overlap. Turn the nosepiece to select the medium-power objective lens.

3. Use the fine focus adjustment and focus through the overlapping threads. Determine which thread is on top, in the middle, and on the bottom. Write your observations below.

   Top thread   _____.
   Middle thread   _____.
   Bottom thread   _____.

III. Additional Practice with the Microscope

MATERIALS

compound microscope
practice slides (preferably slides with a tissue that is visible to the unaided eye)

PROCEDURES

1. Start with low magnification and locate the stained cells on the practice slide.
2. Increase magnification and examine individual cells.
3. Draw a sketch of the tissue at low and medium powers in the spaces below.

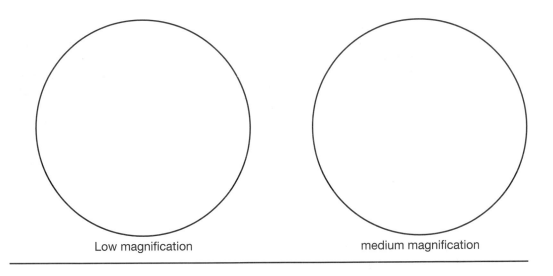

Low magnification   medium magnification

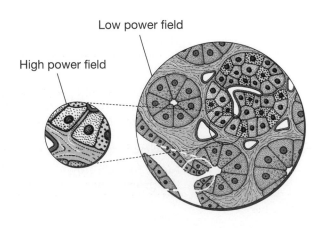

Figure 4.4
**The Relation between Magnification
and Field Diameter.**

Figure 4.5
Calculation of Field Diameter Using
Millimeter Graph Paper. This
Sample Field is Approximately
3.5 mm in Diameter.

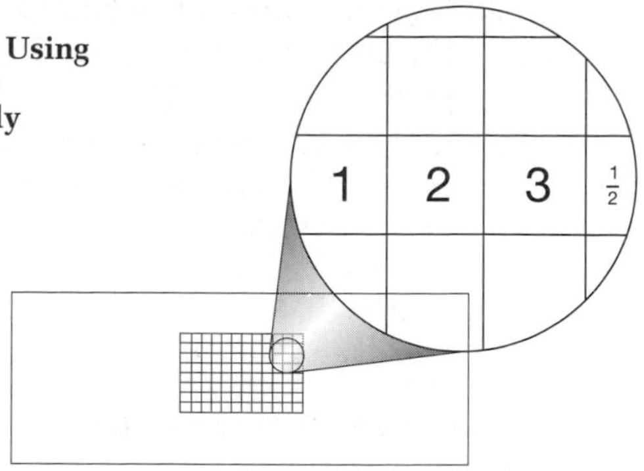

D. Relation Between Magnification and Field Diameter

The area or circle you see when looking through the microscope is called the field of view. At low magnification, the diameter of the field is large and most of the slide specimen is visible. As magnification increases, the **field diameter** decreases. At a higher power, the objective lens is closer to the slide and magnifies a smaller area of the slide. This effect can be shown by making a "ring" with your thumb and index finger and placing it over an eye, much like an ocular lens. Now look at your lab manual from 2 feet away. You can see the entire manual at this "low magnification." Now move just a few inches away from the manual. In this "high-magnification" view the field is much smaller and you can see only part of a single page. Figure 4.4 reviews the relationship between magnification and field diameter.

Field diameter at low and medium magnifications can be measured using millimeter graph paper glued to a microscope slide. By aligning a millimeter on the graph slide with the edge of the field and then counting the number of millimeter squares across the field, the diameter is found; see Figure 4.5. Knowing the diameter of a field enables a user to estimate the actual size of an object. For example, if the field diameter is 4 mm and an object occupies one-half the field, the object is approximately 2 mm in size.

LABORATORY ACTIVITY   ESTIMATING THE FIELD DIAMETER

MATERIALS

compound microscope
graph paper slides

PROCEDURES

1. Obtain a microscope slide with graph paper glued on it.
2. Focus on the graph paper at low magnification. Move the slide to where the edge of a millimeter lines up with the edge of the field.
3. Count the number of millimeters across the field to measure the field diameter.
4. Repeat the measurement at medium power. Write your data in Table 4.2.

TABLE 4.2	Relation between Magnification and Field Diameter	
Objective	*Total Magnification**	*Field Diameter*
Low power	_____	_____
Medium power	_____	_____
High power	_____	(estimate) _____

*See Table 4.1.

USE OF THE MICROSCOPE CHECKLIST

This is a list of the bold terms presented in Exercise 4. Use this list for review purposes after you have completed the laboratory activities.

PARTS OF THE MICROSCOPE

- Figures 4.1 and 4.2

compound lens
ocular lens
monocular microscope
binocular microscope
body tube
arm
base
nosepiece
objective lenses
stage
aperture
mechanical stage
(mechanical stage) adjustment knobs
stage clips
coarse focus knob
fine focus knob

condenser
condenser adjustment knob
iris diaphragm
diaphragm lever
light source
substage light
mirror

MICROSCOPE TERMINOLOGY AND SKILLS

focus
parfocal
resolution
field of view
total magnification
field diameter
depth of field
wet mount slide

Name _____   Date _____

Section _____   Number _____

USE OF THE MICROSCOPE
Laboratory Report Exercise 4

A. **Short-Answer Questions**

   1. Explain what parts of the microscope are used to regulate the intensity and contrast of light. What does each of these parts do?

   2. How is magnification controlled in a microscope?

   3. How is total magnification changed if field diameter is decreased?

   4. Why should you always view a slide at low power first?

   5. Briefly explain how to care for a microscope.

   6. Describe when to use coarse focus and when to use fine focus.

B. Matching
    Match the part of the microscope on the left with the correct description on the right.

1. _____ ocular   A. used for precise focusing
2. _____ aperture   B. lower support of microscope
3. _____ body tube   C. narrows the beam of light
4. _____ mechanical stage   D. hole in stage
5. _____ fine focus   E. used only at low power
6. _____ base   F. has knobs to move slide
7. _____ objective lens   G. special paper for cleaning
8. _____ coarse focus   H. eyepiece
9. _____ condenser   I. holds the ocular lens
10. _____ lens paper   J. lens on nosepiece

C. Labeling
    Label the parts of the microscope in Figure 4.6.

1. _____
2. _____
3. _____
4. _____
5. _____
6. _____
7. _____
8. _____
9. _____
10. _____

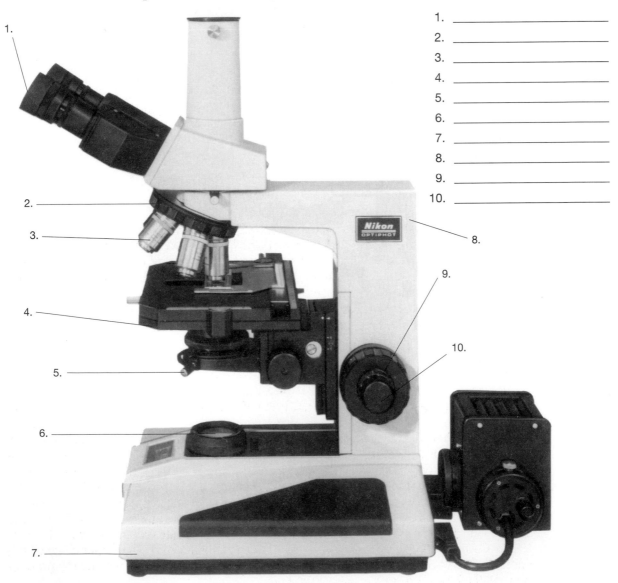

Figure 4.6
Parts of the Compound Microscope.

Anatomy of the Cell and Cell Division

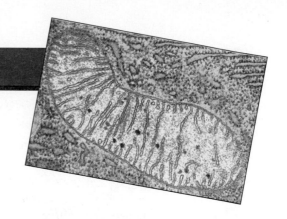

OBJECTIVES

On completion of this exercise, you should be able to:

- Identify organelles of the cell on charts, models, and other lab material.
- State a function of each organelle.
- Discuss a cell's life cycle, including the stages of interphase and mitosis.
- Identify the stages of mitosis using the roundworm (*Ascaris*) slide.

WORD POWER

cell (cell—small room)
cytoplasm (plasm—something molded)
nucleus (nucle—small nut)
cytokinesis (kinesis—movement)

prophase (pro—before)
metaphase (meta—between)
anaphase (an—not)
telophase (telo—end)

MATERIALS

cell models
mitosis models
microscope
prepared slides:
   whitefish blastula

INTRODUCTION

Cells were first described in 1665 by a British scientist named Robert Hooke. Hooke examined a thin slice of tree cork with a microscope and observed open spaces in the cork that he called **cells**. Over the next two centuries, scientists examined cells from plants and animals and formulated the cell theory which states that (1) all plants and animals are composed of cells, (2) all cells come from preexisting cells, (3) cells are the smallest living units that perform physiological functions, (4) each cell works to maintain itself at the cellular level, and (5) homeostasis is the result of the coordinated activities of all the cells in an organism.

Your cells are descendants of your parents' sperm and egg cells that combined to create your first cell, the zygote. You are now composed of an estimated 75 trillion cells, more than you could count in your lifetime. These cells must coordinate their activities to maintain homeostasis for your entire body. If a population of cells becomes dysfunctional, disease may result. Some organisms, like amebas, are composed of a single cell which performs all functions necessary to keep them alive. In multicellular organisms like us, cells are diversified to perform specific functions. This specialization leads to dependency among cells. For example, muscle cells are responsible for movement of the body. Movement requires a large amount of energy, and the muscle cells rely on the cells of the cardiovascular system to supply the muscle with oxygen and nutrients.

Although the body has a variety of cell types, a generalized or composite cell will be used in describing cell structure. All cells have an outer boundary, the cell membrane. Cells also have a nucleus and other internal structures called organelles. In this exercise you will examine the structure of the cell and how cells reproduce to produce more cells for growth and repair of the body.

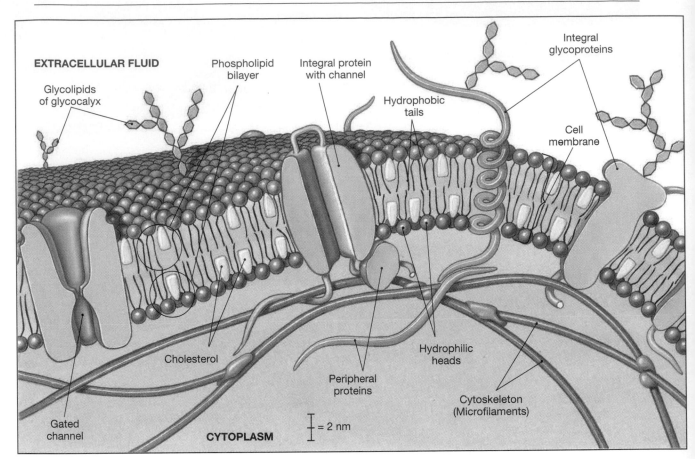

EXTRACELLULAR FLUID

Glycolipids
of glycocalyx

Phospholipid
bilayer

Integral protein
with channel

Hydrophobic
tails

Integral
glycoproteins

Cell
membrane

Cholesterol

Peripheral
proteins

Hydrophilic
heads

Cytoskeleton
(Microfilaments)

Gated
channel

CYTOPLASM

= 2 nm

Figure 5.1
The Cell Membrane.

A. The Cell Membrane

The **cell membrane**, also called the plasma membrane, is the physical boundary of the cell, separating the **extracellular fluid** surrounding the cell from the **cytosol**, the intracellular liquid. The cell membrane regulates the movement of ions, molecules, and other substances into and out of the cell. In Exercise 6 you will study transport of materials across cell membranes.

Cells interact with other cells and with the surrounding environment across the cell membrane. Hormones, enzymes, and other regulatory molecules bind to specific receptor proteins in the cell membrane. Muscle and nerve cell membranes are unique excitable membranes and produce electrical currents called action potentials. Some cells have **microvilli**, or folds in the cell membrane, which increase the surface area for absorption of materials into the cell.

The cell membrane is composed of a **phospholipid bilayer**, a double layer of phospholipid molecules (see Figure 5.1). Each phospholipid molecule consists of a **hydrophilic** (water-loving) **head** and two **hydrophobic** (water-fearing) **tails**. In a cell membrane, the phospholipids are arranged in a double sheet of molecules with the hydrophilic heads facing the watery internal and external environments of the cell. The hydrophobic tails are sandwiched between the phospholipid heads. Other molecules in the cell membrane, such as cholesterol and glycolipids, are structural components of the cell membrane.

Proteins are the second major component of the cell membrane. Floating like icebergs in the phospholipid bilayer are a variety of **integral proteins**. Some of these proteins have **channels** that regulate the passage of specific ions through the membrane.

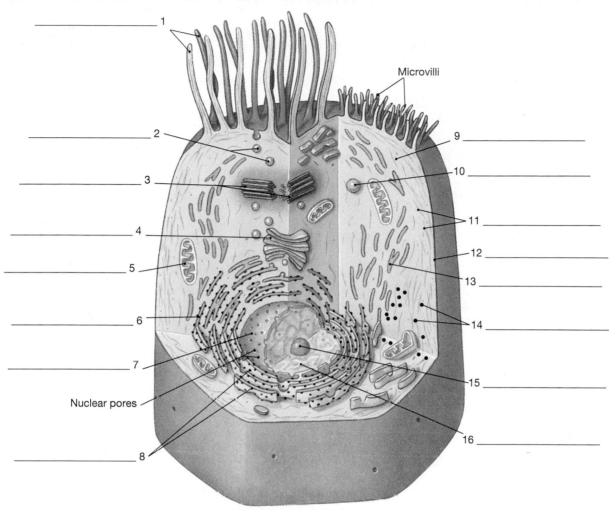

Figure 5.2
The Anatomy of a Composite Cell.

Loosely attached to the external and internal surfaces of the membrane are **peripheral proteins**. Membrane proteins occur in a wide variety and have diverse functions. Receptor proteins are sensitive to specific chemicals in the extracellular fluid. Other membrane proteins and glycoproteins function in cell-to-cell recognition and as enzymes for intracellular and extracellular reactions.

B. Organelles

Within the cell are **organelles (or-gan-ELZ)**, the functional pieces of biological machinery. As represented in Figure 5.2 and Table 5.1, each organelle has a distinct anatomical organization and is specialized for a specific function. Organelles are suspended in the cytosol. The organelles and cytosol together compose the **cytoplasm** of the cell.

Organelles are grouped into two broad classes: membrane-bound organelles and non-membrane-bound organelles. **Membrane-bound** organelles are enclosed in a lipid membrane that isolates the organelle from the cytosol. The nucleus, endoplasmic reticulum, Golgi apparatus, vesicles, and mitochondria are examples of membrane-bound organelles. **Non-membrane-bound** organelles lack an outer membrane and are directly exposed to the cytosol. Ribosomes, centrioles, the cytoskeleton, cilia, and flagella are non-membrane-bound organelles.

TABLE 5.1	Summary of Cell Organelles and Their Function		
Appearance	*Structure*	*Composition*	*Function(s)*
	CELL MEMBRANE	Lipid bilayer, containing phospholipids, steroids, and proteins	Isolation; protection; sensitivity; support; control of entrance /exit of materials
	CYTOSOL	Fluid component of cytoplasm	Distributes materials by diffusion
	NONMEMBRANOUS ORGANELLES		
	Cytoskeleton: Microtubule Microfilament	Proteins organized in fine filaments or slender tubes	Strength and support; movement of cellular structures and materials
	Microvilli	Membrane extensions containing microfilaments	Increase surface area to facilitate absorption of extracellular materials
	Centrosome Centriole	Cytoplasm containing two centrioles, at right angles: Each centriole is composed of 9 microtubule triplets	Essential for movement of chromosomes during cell division; organization of microtubules in cytoskeleton
	Cilia	Membrane extensions containing microtubule doublets in a 9 + 2 array	Movement of materials over cell surface
	Ribosomes	RNA + proteins; fixed ribosomes bound to rough endoplasmic reticulum, free ribosomes scattered in cytoplasm	Protein synthesis
	MEMBRANOUS ORGANELLES		
	Endoplasmic reticulum (ER)	Network of membranous channels extending throughout the cytoplasm	Synthesis of secretory products; intracellular storage and transport
	Rough ER (RER)	Has ribosomes bound to membranes	Modification and packaging of newly synthesized proteins
	Smooth ER (SER)	Lacks attached ribosomes	Lipid and carbohydrate synthesis
	Golgi apparatus	Stacks of flattened membranes (cisternae) containing chambers	Storage, alteration, and packaging of secretory products and lysosomal enzymes
	Lysosomes	Vesicles containing digestive enzymes	Intracellular removal of damaged organelles or of pathogens
	Peroxisomes	Vesicles containing degradative enzymes	Catabolism of fats and other organic compounds; neutralization of toxic compounds generated in the process
	Mitochondria	Double membrane, with inner membrane folds (cristae) enclosing important metabolic enzymes	Produce 95% of the ATP required by the cell
Nuclear pore	**NUCLEUS**	Nucleoplasm containing nucleotides, enzymes, nucleoproteins, and chromatin; surrounded by double membrane (nuclear envelope)	Control of metabolism; storage and processing of genetic information; control of protein synthesis
Nuclear envelope	**Nucleolus**	Dense region in nucleoplasm containing DNA and RNA	Site of rRNA synthesis and assembly of ribosomal subunits

STUDY TIP

Some students have difficulty integrating concepts of anatomy and physiology. Anatomy becomes a memorization process, and physiology a confusing hurdle. Because form (anatomy) and function (physiology) are intertwined, it is helpful to link a function to each anatomical structure. As you identify organelles on cell models, consider the function of each one.

The **nucleus** controls all the activities of the cell. Protein synthesis, gene action, cell division, and metabolic rate are all regulated by the nucleus. The nucleus usually appears as a dark-stained structure in the cell. The stained material is **chromatin**, uncoiled chromosomes consisting of DNA and protein molecules. A **nuclear envelope** surrounds the nuclear material and contains pores through which instruction molecules from the nucleus pass into the cytosol. A darker-stained **nucleolus** produces ribosomal RNA molecules for the creation of ribosomes.

Ribosomes orchestrate protein synthesis. Instructions for making each protein are copied from DNA onto RNA molecules that carry the instructions out of the nucleus to the ribosomes. Each ribosome consists of two ribosomal RNA subunits that clamp around the messenger RNA molecule and coordinate the synthesis of proteins. Ribosomes occur **free** in the cytoplasm or **attached** to a membranous system, the endoplasmic reticulum (ER). Each ribosome is composed of a small and a large subunit molecule.

Surrounding the nucleus is the **endoplasmic reticulum (en-dō-PLAZ-mik re-TIK-ū-lum)**. Two types of ER occur, **rough ER** with attached ribosomes on the surface of the membrane, and **smooth ER** which lacks ribosomes. Generally, the endoplasmic reticulum functions in the synthesis of organic molecules, transport of materials within the cell, and storage of molecules. Materials in the ER may pass into the Golgi apparatus for eventual transport out of the cell. Proteins produced by ribosomes on the rough ER surface enter the ER and assume their complex folded shape. Smooth ER is involved in the synthesis of many organic molecules such as cholesterol and phospholipids. In reproductive cells, smooth ER produces sex hormones. The smooth ER in muscle and liver cells synthesizes and stores glycogen. Intracellular calcium ions are stored in the smooth ER in muscle, nerve, and other types of cells.

The **Golgi (GŌL-jē) apparatus** is a series of flattened saccules adjoining the endoplasmic reticulum. The ER can pass protein molecules to the Golgi apparatus for modification and secretion. Cell products such as mucus are synthesized, packaged, and secreted by the Golgi apparatus. In a process called **exocytosis**, small **secretory vesicles** pinch off the saccules, fuse with the cell membrane, and then rupture to release their contents. The phospholipid membranes of the empty vesicles contribute to the renewal of the cell membrane.

One type of vesicle produced by the Golgi apparatus is the **lysosome (LĪ-sō-sōm)**. These vesicles are filled with powerful enzymes that digest worn out cell components and destroy microbes. As certain organelles such as mitochondria become worn out, lysosomes dissolve the organelle and some of the materials are used to rebuild the organelles. White blood cells trap bacteria with cell membrane extensions and pinch the membrane inward to release a vesicle inside the cell. Lysosomes fuse with the vesicle and release enzymes to digest the bacteria. Injury to a cell may result in the rupture of lysosomes, followed by destruction or autolysis of the cell. Autolysis is implicated in the aging of cells owing to the accumulation of lysosomal enzymes in the cytosol.

Mitochondria (mī-tō-KON-drē-uh) convert energy into useful forms for the cell. Each mitochondrion is wrapped in a double-layered membrane. The inner membrane is folded into fingerlike projections called **cristae**. The middle region of the inner membrane between the cristae is the **matrix**. To provide cellular energy, molecules from nutrients are passed along a series of metabolic enzymes in the cristae to produce ATP, the energy currency of cells. The abundance of mitochondria varies greatly among cell types. Mature red blood cells lack mitochondria and specialize in the transport of dissolved blood gases. Muscle and nerve cells have dense numbers of mitochondria which supply energy for contraction and generate nerve impulses, respectively.

CLINICAL APPLICATION

Mitochondria are unique in that they contain **mitochondrial DNA** and evidently do not rely on the DNA in the cell's nucleus for replication instructions to produce more mitochondria. Your mitochondria were inherited from your mother's egg; the sperm contributes no mitochondria and few other organelles to the offspring. Suppose a woman inherited a mitochondrial disorder from her mother. Could she pass this trait on to a son? In general, how would the health of a cell be affected if the mitochondria were not fully functional?

Microtubules are hollow tubes constructed from the protein tubulin. These protein pipes are found in all cells and function to anchor organelles and give rigidity to the cell membrane. The cytoskeleton, centrioles, cilia, and flagella are microtubules.

Centrioles are paired organelles composed of microtubules; the **centrosome** is the area surrounding the centrioles. Centrioles are involved in cell division. When a cell divides, the centrioles separate to opposite poles and a series of spindle fibers radiate from them. The spindle fibers pull the chromosomes apart to give each forming cell a full complement of genetic instructions. Other microtubules occur in cells; all cells have a **cytoskeleton** for structural support. Many cells of the respiratory and reproductive system have **cilia**, short, hairlike projections extending from the plasma membrane. One type of human cell, the spermatozoon, has a single, long **flagellum (fla-JEL-um)** for locomotion.

LABORATORY ACTIVITY   IDENTIFICATION OF CELL ORGANELLES

MATERIALS

cell models and charts

PROCEDURES

1. Label the organelles in Figure 5.2 and review organelles in Table 5.1.
2. Identify the organelles on a laboratory cell model.
3. Complete Table 5.2 by filling in the blanks.

TABLE 5.2   Summary of cell organelles and their functions

Organelle	Organelle Structure	Organelle Function
1. Mitochondria	_____	_____
2. _____	_____	Contains digestive enzymes
3. _____	Phospholipid bilayer with embedded proteins	_____
4. Endoplasmic reticulum	_____	_____
5. _____	_____	Exocytosis of cellular products
6. _____	Composed of two subunit molecules	_____
7. _____	_____	Regulates cell activities

C. Cell Division

Cells must reproduce for growth and tissue repair. During cell reproduction, a cell divides its genes equally and then splits into two identical cells. The divisional process involves two major "events": mitosis and cytokinesis. During **mitosis (mī-tō-sis)**, the genetic material of the nucleus condenses into chromosomes and is equally divided between the forming cells. Toward the end of mitosis, **cytokinesis (sī-tō-ki-NĔ-sis)** separates the cytoplasm to produce the two daughter cells. The daughter cells have the same number of chromosomes as the parental cell.

Examine the cell life cycle in Figure 5.3. Most of the time a cell is not dividing and is in **interphase**. Interphase is not a resting period for the cell; it is involved in various functions and prepares for the next cell division. Distinct phases occur during interphase, each related to the cell's activity. During the **G_0 phase** the cell performs basic functions. The **G_1 phase** is a time for protein synthesis, growth, and replication of organelles. Replication of genetic material, DNA, occurs during the **S phase**. After replication each chromosome has two strands of DNA, the original strand and an identical copy. Each strand is called a **chromatid**, and matching chromatids are held together by a **centromere**. The **G_2 phase** is another time for protein synthesis. The **M phase** is the time of mitosis, during which the nuclear material is divided. The double-stranded chromosomes are organized in the middle of the cell, and a thin microtubule called a **spindle fiber** attaches to each chromatid. When the centromeres split, each chromosome is equally divided by the spindle fibers dragging the corresponding chromatids to opposite sides of the cell. To complete the divisional process, **cytokinesis** pinches the dividing cell into two separate **daughter cells**.

Figure 5.3
The Cell Life Cycle.

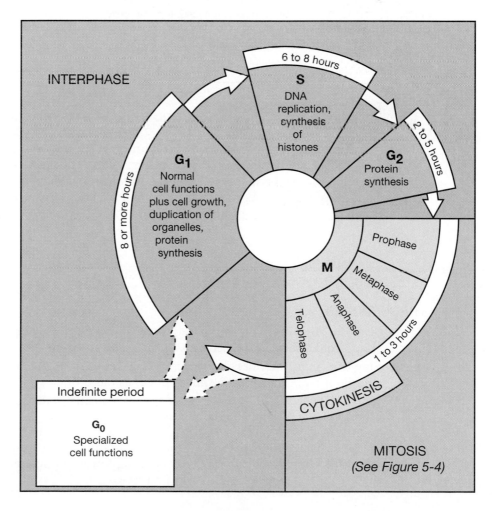

Mitosis starts with **prophase** as chromosomes become visible in the nucleus, see Figure 5.4. In early prophase the chromosomes are long and disorganized, but as prophase continues, they shorten and begin to move toward the middle of the cell. In the cytoplasm, the **centrioles** begin to separate to opposite sides of the cell. Between the centrioles, a microtubule apparatus called the **spindle fibers** spans across the cell.

Metaphase occurs as chromosomes line up in the middle of the cell at the **metaphase plate**. Spindle fibers extend across the cell from one pole to the other and attach to the centromeres of the chromosomes. The cell is now prepared to partition the genetic material and give rise to two new cells.

Anaphase is the separation of the chromosomes. The chromatids of each chromosome separate as the spindle fibers pull them apart and drag them toward opposite poles of the cell. Once apart, individual chromatids are considered chromosomes. Toward the end of anaphase, cytokinesis begins and the cell membrane pinches inward and distributes the cytoplasm and chromosomes to make two cells. Cytokinesis continues into the next stage of mitosis, telophase.

Mitosis ends with **telophase** as each batch of chromosomes unwind and are enclosed in a nuclear membrane. Each daughter cell has a nucleus with a complete set of genes, and a set of each of the organelles. Telophase ends as cytokinesis completes the separation of identical daughter cells. These cells are in interphase and, depending on their cell type, may divide again.

CLINICAL APPLICATION

A tumor is a mass of cells caused by uncontrolled cell division. The mass replaces normal cells, and cellular and tissue functions are compromised. If **metastasis** (me-TAS-ta-sis) or spreading of the abnormal cells occurs, secondary tumors may develop. Cells that metastasize are often malignant and can cause cancer.

LABORATORY ACTIVITY   IDENTIFICATION OF MITOTIC STAGES

The whitefish blastula is ideal for mitosis studies. A blastula is a stage in early development when the embryo is a rapidly dividing mass of cells. The slide is made by sectioning a whitefish embryo and staining the thin specimens. Each section has cells in various stages of mitosis. As cells complete their division, the newly formed daughter cells divide and the embryo grows in size and, eventually, complexity.

MATERIALS

compound microscope
whitefish blastula slide

PROCEDURES

1. Scan the slide at low power and observe the numerous cells of the blastula. The slide usually has several sections.
2. Slowly scan a group of cells at medium power to locate a nucleus, centrioles, and spindle fibers. The chromosomes appear as dark, thick structures in the cell.
3. Using Figures 5.4 and 5.5 as references, locate cells in the following phases:
   a. Interphase with a distinct nucleus
   b. Prophase with disorganized chromosomes
   c. Metaphase with equatorial chromosomes attached to spindle fibers
   d. Anaphase with chromosomes separating toward opposite poles

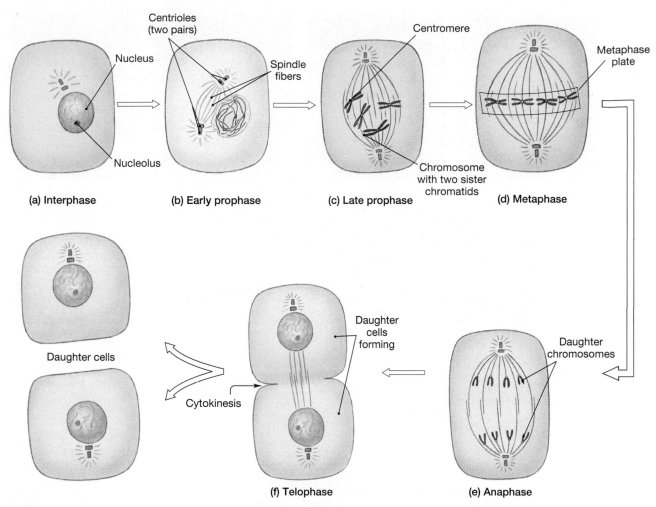

(a) Interphase   (b) Early prophase   (c) Late prophase   (d) Metaphase

Daughter cells   (f) Telophase   (e) Anaphase

Cytokinesis

Figure 5.4
Mitosis.

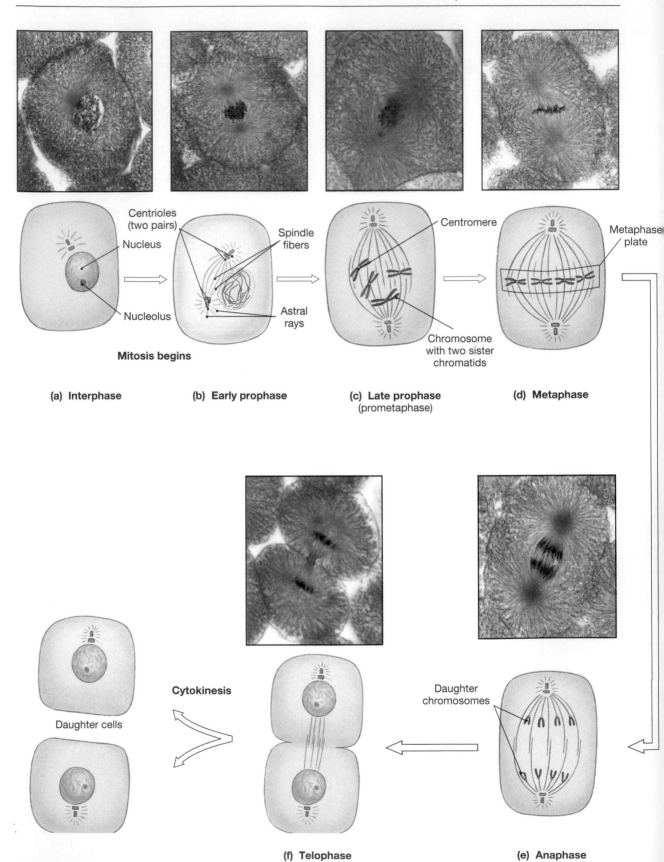

Figure 5.5
Mitosis in the whitefish blastula.

    e. Telophase with nuclear membranes forming around each set of genetic material

    f. Observe cytokinesis in late anaphase and telophase.

4. Draw and label cells in each stage of mitosis in the space provided.

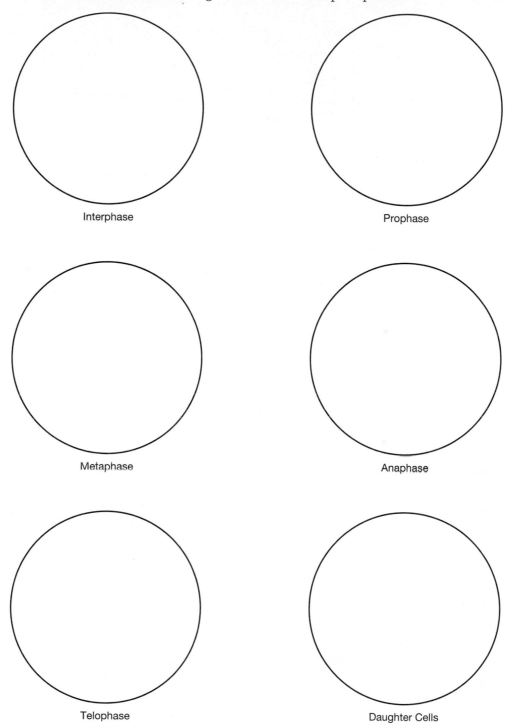

Interphase   Prophase

Metaphase   Anaphase

Telophase   Daughter Cells

ANATOMY OF THE CELL
AND CELL DIVISION CHECKLIST

This is a list of bold terms presented in Exercise 5. Use this list for review purposes after you have completed the laboratory activities.

CELL MEMBRANE

• Figure 5.1 and Table 5.1
plasma (cell) membrane
phospholipid bilayer
hydrophilic heads
hydrophobic tails
integral proteins
channels
peripheral proteins
microvilli

ORGANELLES

• Figure 5.2 and Table 5.1
cell
organelle
extracellular fluid
cytosol
cytoplasm
membrane-bound organelles
non-membrane-bound organelles
nucleus
chromatin
nuclear envelope
nucleolus
free ribosomes
attached ribosomes
smooth endoplasmic reticulum
rough endoplasmic reticulum
Golgi apparatus
secretory vesicles

exocytosis
lysosomes
mitochondria
cristae
matrix
microtubules
centrioles
centrosome
cytoskeleton
cilia
flagellum

CELL DIVISION

• Figures 5.3, 5.4, and 5.5
mitosis
interphase
G_0 phase
G_1 phase
S phase
DNA replication
G_2 phase
M phase
prophase
chromatids
centromere
centrioles
spindle fibers
metaphase
metaphase plate
anaphase
cytokinesis
telophase
daughter cells

ANATOMY OF THE CELL AND CELL DIVISION
Laboratory Report Exercise 5

A. Matching
Match each cellular structure listed on the left with the correct description on the right.

1. _____ cell membrane
2. _____ centrioles
3. _____ ribosome
4. _____ smooth ER
5. _____ chromatid
6. _____ lysosomes
7. _____ integral protein
8. _____ cytoplasm
9. _____ cristae
10. _____ hydrophilic heads
11. _____ cytosol
12. _____ cilia

A. copy of a chromosome
B. component of cell membranes
C. short, hairlike cellular extensions
D. part of phospholipid molecule
E. intracellular fluid
F. involved in mitosis
G. folds of the inner mitochondrial membrane
H. composed of a phospholipid bilayer
I. stores calcium ions in muscle cells
J. site for protein synthesis
K. vesicle with powerful digestive enzymes
L. intracellular fluid and the organelles

B. Fill in the Blanks
1. Replication of genetic material results in chromosomes consisting of two _____.
2. A cell in metaphase has chromosomes located in the _____ of the cell.
3. Dividing the cytoplasm to produce two daughter cells is called _____.
4. Double-stranded chromosomes separate during _____ of mitosis.
5. During interphase, DNA replication occurs in the _____ phase.
6. Microtubules called _____ attach to chromatids and pull them apart.
7. Chromosomes become visible during _____ of mitosis.
8. The last stage of mitosis is _____.
9. Division of the nucleus is _____.
10. Matching chromatids are held together by a _____.

C. Short-Answer Questions
1. What is the purpose of cell division? _____.
2. Which organelles are involved in protein synthesis? _____.
3. What is the function of the spindle fibers during mitosis? _____.
4. What structures in the cell membrane regulate ion passage? _____
   _____.

D. **Drawing**

  1. Draw and label the cell membrane. Include details of the phospholipid bilayer and integral proteins.

  2. Draw and label the internal organization of a mitochondrion.

  3. Draw and label a cell with four chromosomes during metaphase of mitosis. How will the chromosomes appear during anaphase and telophase?

E. **Form or Function?**

  In this section indicate whether each question is relative to the study of anatomy (a) or physiology (b).

  1. _____ A ribosome is composed of two subunit molecules.

  2. _____ The smooth endoplasmic reticulum synthesizes cholesterol.

  3. _____ Chromosomes are dark-stained structures in dividing cells.

  4. _____ Ions may pass through integral proteins in the cell membrane.

6

Cell Transport Mechanisms

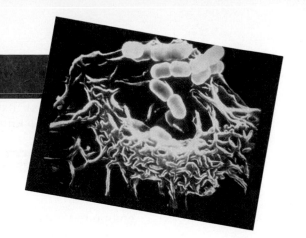

OBJECTIVES

On completion of this exercise, you should be able to:

- Describe the two main processes by which substances move into and out of cells.
- Explain Brownian movement and how it can be shown.
- Discuss diffusion, concentration gradients, and equilibrium.
- Describe the effect of isotonic, hypertonic, and hypotonic solutions on cells.
- Discuss the effects of temperature and solute concentration on the rate of diffusion and osmosis.

WORD POWER

diffusion (diffu—apart)
equilibrium (equil—equal)
gradient (grad—a hill)
osmosis (osmo—pushing)
solution (solu—to dissolve)
solute (solu—dissolved)
dialysis (dialy—to separate)
isotonic (iso—same)
hypertonic (hyper—excessive)
hypotonic (hypo—below)
hemolysis (lys—loose)

MATERIALS

compound microscope
microscope slides
coverslips
dropper bottle
waterproof ink
powdered cleanser
beaker, 250 ml
food color dye
petri dish and plain agar
cork hole bore
potassium permangenate
iodine crystals

ruler
dialysis tubing and clips
beaker, 500 ml
saturated starch solution
Lugol's solution
1%, 5%, and 10% sugar solution
blood (nonhuman)
wax pencil
isotonic solution (0.9% solute)
hypertonic solution (2% solute)
hypotonic solution (0.5% solute)
gram scale

INTRODUCTION

Cells are the functional living unit of your body. For cells to survive, materials must be transported across the cell membrane. Cells import nutrients, oxygen, hormones, and other regulatory molecules. Wastes and cellular products are exported to the extracellular fluid. Cell membranes are selectively permeable and regulate passage of certain materials. Proteins and other macromolecules are too big to pass through channels in the cell membrane. Movement of these molecules requires the use of carrier molecules in a process called **active transport** which consumes cell energy. Smaller molecules, such as water and many ions, cross the membrane without assistance from the cell. This movement is called **passive transport** and requires no energy expenditure by the cell. Diffusion and osmosis are the primary passive processes in the body and will be studied in this laboratory exercise.

A. Brownian Movement

Molecules are in a constant state of **Brownian movement**, motion that causes molecules to vibrate and bump into adjacent molecules. Collisions occur more frequently when molecules are packed closer together. Because of collisions, molecules spread out and move toward an equal distribution rather than remain packed together. Brownian movement supplies the **kinetic energy** of motion for passive transport mechanisms.

LABORATORY ACTIVITY   OBSERVATION OF BROWNIAN MOVEMENT

MATERIALS

> compound microscope
> microscope slide and coverslip
> small dropper bottle with eyedropper
> tap water
> waterproof ink
> powdered cleanser (Ajax™, Comet™, etc.)

PROCEDURE

1. Fill the dropper bottle three-fourths full of tap water and add a small amount of cleanser powder and waterproof ink. Add 10–15 ml of additional tap water.
2. Shake the bottle gently to mix the contents. Place a drop of the mixture on a microscope slide and place a coverslip over the drop.
3. Focus on the slide and locate the small granules of cleanser. Observe how the particles vibrate and occasionally collide with one another.

RESULTS

1. Why does Brownian movement occur?

   _____

   _____

2. How would temperature affect Brownian movement?

   _____

   _____

   _____

B. Diffusion

Diffusion is the net movement of substances from a region of greater concentration to a region of lesser concentration. Simply put, diffusion is the spreading out of substances owing to collisions between vibrating molecules. Diffusion occurs throughout the body, in extracellular fluid, across cell membranes, and in the cytosol of cells. Examples of diffusion include oxygen moving from the lungs into pulmonary capillaries, odor molecules moving through the nasal lining to reach the olfactory cells, and movement of ions in and out of nerve cells to produce electrical impulses. Molecules diffuse through cells by two basic mechanisms: lipid-soluble molecules diffuse through the phospholipid bilayer of the membrane, and small water ions and molecules pass through the channels of integral proteins. Molecules like proteins are too large to enter the membrane channels and do not diffuse across the membrane.

Cells cannot directly control diffusion; it is a passive transport process much like a ball rolling downhill. If a substance is unequally distributed, a **concentration gradient** exists and one region will have a greater concentration of the substance than other regions. The substance will diffuse until an equal distribution occurs, at a point called **equilibrium**. Figure 6.1 presents the process of diffusion of an ink drop in water. Notice in step 1 that the molecules in the ink drop are concentrated together before they are placed in the water. Once in the water, these molecules disperse as they bump into other ink molecules and water molecules (see step 2). Eventually, the molecules will become evenly distributed in equilibrium, as illustrated in step 3.

Figure 6.1
Diffusion.

Placing a colored sugar cube in a glass of water establishes a steep concentration gradient. As the cube dissolves, many sugar molecules are in one location and none are elsewhere. As diffusion occurs, the molecules spread through the solution. Eventually, diffusion eliminates the concentration gradient. The sugar cube has dissolved completely, and the molecules are distributed evenly. Molecular motion continues, but there is no directional movement.

Figure 6.2 summarizes diffusion across cell membranes. Notice that molecules diffuse by two basic mechanisms: lipid-soluble molecules diffuse through the phospholipid bilayer of the membrane, and water ions and small molecules diffuse through the channels of integral proteins. If necessary, review the structure of the cell membrane in Exercise 5.

LABORATORY ACTIVITY   DIFFUSION EXPERIMENTS

Experiment I—Diffusion of a Liquid in a Liquid
MATERIALS

       beaker, 250 ml or larger
       food coloring dye

PROCEDURE

1. Fill the beaker three-fourths full with tap water. Leave it undisturbed for several minutes to let the water settle.
2. Carefully add several drops of food coloring to the water.
3. Observe diffusion of the dye for several minutes.

RESULTS

1. Why does the dye diffuse in water?

_____

_____

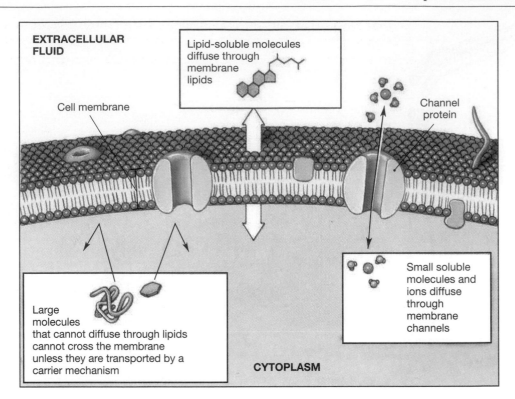

Figure 6.2
Diffusion across the Cell Membrane

2. How would water temperature affect the diffusion rate?

_____

_____

3. What is the energy source for diffusion?

_____

_____

Experiment II—Diffusion of a Solid in a Gel
MATERIALS

    petri dish with plain agar (gel)
    cork bore (or soda straw)
    potassium permangenate crystals
    iodine crystals
    ruler

PROCEDURE

1. Punch two small holes in the agar, approximately equal distance from the center of the petri dish.

2. Place equal amounts of each crystal in separate holes. Do not spill crystals on any other part of the petri dish.

   3. After 20–30 min, measure the distance each crystal has diffused. Measure from the edge of the hole to the outer boundary of the diffused crystals.

   4. Dispose of the petri dish as instructed by your laboratory instructor.

RESULTS

   1. Which crystal diffuses further?

   _____

   _____

   2. Which crystal has a heavier molecular weight? How does molecular weight affect the diffusion rate?

   _____

   _____

   _____

C. Osmosis

Osmosis (oz-MŌ-sis) is the net movement of water through a selectively permeable membrane, from a region of greater concentration to a region of lesser concentration. With an understanding of diffusion, one can define osmosis as the diffusion of water through a selectively permeable membrane. Osmosis occurs when solutions of differing solute concentration are separated by a selectively permeable membrane. A **solution** is the result of dissolving a **solute** in a **solvent**. In a 1% salt solution, salt is the solute that occupies 1% of the volume and the solvent, water, fills the remaining 99% of the container. For osmosis to occur, there must be a difference in water concentrations across the cell membrane. As solute concentration increases, the available space for water decreases. This establishes the concentration gradient for osmosis.

STUDY TIP

Only the water moves across the membrane during osmosis. If the solute could pass through the membrane, it would move to equilibrium and eliminate the concentration gradient of water. If the solutes are in equilibrium, then so is the water and osmosis will not occur.

Examine Figure 6.3 presenting osmosis. The figure illustrates a beaker divided by a selectively permeable membrane. Each side of the beaker has identical molecules but in different concentrations. The small dots represent water, and the larger circles are solutes that cannot cross the membrane. Side A of the beaker has more water than side B, the side with more solute and less water. In step 1, the water level between the sides of the membrane is equal. As water moves from side A to side B, the volume of side B increases until equilibrium is reached in step 2a. Water and solute concentrations are now equal between the sides of the membrane.

Solutions have an **osmotic pressure** due to the presence of solutes. The greater the solute concentration, the greater the osmotic pressure of the solution. During osmosis, the solution with the greatest osmotic pressure causes water to move toward it. In effect, "water follows solute" and osmotic pressure is a "pulling pressure" that draws water toward the higher solute concentration. Notice in step 2b that a force applied to side B will stop osmosis if that pressure is equal to the osmotic pressure causing the osmosis. Drinking water is often purified by reverse osmosis, applying pressure to water on one side of a membrane to force it through a filter while leaving solutes behind. The resulting water has less solute and is purified for drinking.

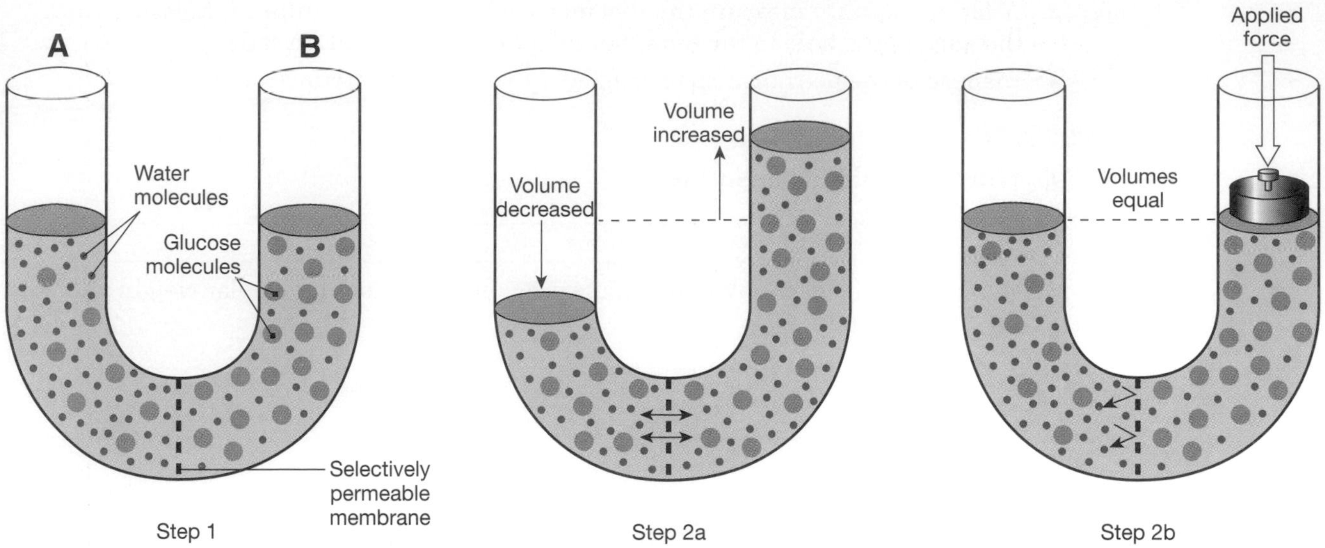

Figure 6.3
Osmosis.

Step 1: Two solutions containing different solute concentrations are separated by a selectively permeable membrane. Water molecules (small blue dots) begin to cross the membrane toward solution B, the solution with the higher concentration of solutes (larger pink circles). *Step 2a:* At equilibrium, the solute concentrations on the two sides of the membrane are equal. The volume of solution B has increased at the expense of that of solution A. *Step 2b:* Osmosis can be prevented by resisting the volume change. The osmotic pressure of solution B is equal to the amount of hydrostatic pressure required to stop the osmotic flow.

CLINICAL APPLICATION—DIALYSIS

Dialysis is a passive process similar to osmosis except that, besides water, solute particles can pass through a selectively permeable membrane. The large particles are unable to cross the membrane, and thus particles may be separated by size during dialysis. Dialysis does not occur in the body, yet it is used in **kidney dialysis** to remove wastes from the blood by employing the passive processes of diffusion and osmosis. Blood from an artery passes into thousands of minute selectively permeable tubules in a dialysis cartridge. A dialyzing solution with the same concentration of materials to remain in the blood (nutrients and certain electrolytes) is pumped into the cartridge to flow over the tubules. As blood flows through the tubules, wastes diffuse from the blood, through the selectively permeable tubules, and into the dialyzing solution. Once waste levels in the blood have been reduced to a safe level, the patient is disconnected from the dialysis apparatus.

LABORATORY ACTIVITY   OSMOSIS EXPERIMENTS

Experiment I—Dialysis

This laboratory experiment will demonstrate the movement of materials through dialysis tubing. The dialysis tubing, like a cell membrane, is selectively permeable. Small pores in the tubing allow the passage of small particles yet restrict the passage of large molecules. To show dialysis, the tubing is folded into a bag and filled with a concentrated starch solution. The bag is then placed in a beaker of iodine solution (see Figure 6.4). To determine if water has osmosed, the dialysis bag is weighed before and after the experiment. Any change in weight can be attributed to a change in water volume in the bag. When iodine and starch combine, a dark-blue color is produced. Diffusion of iodine and starch will be detected by this chemical test.

Figure 6.4
Osmosis Setup Using Dialysis
Membrane.

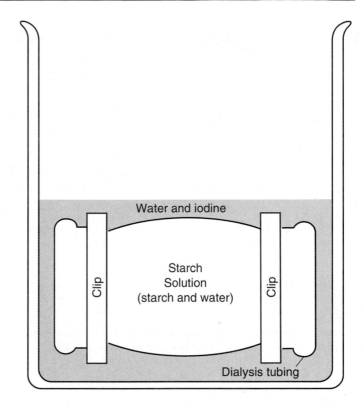

MATERIALS

one strip of dialysis tubing 15 cm (6″) long and two clips

gram scale

500-ml beaker

distilled water (DW)

5% starch solution

Lugol's (iodine) solution

PROCEDURE

1. Add approximately 100 ml of distilled water to the beaker. Soak the dialysis tubing in the water for 3–4 minutes and then remove it from the beaker.

2. Fold one end of the tubing over and seal it securely with a tubing clip. Rub the unclipped side of the membrane between your fingers to open the tubing.

3. Fill the bag approximately three-quarters full with starch solution. Fold and clip the bag securely without trapping too much air inside.

4. Rinse the bag to remove traces of starch solution. Dry and weigh the bag and then submerge it completely in the beaker of water. Record your measurement in Table 6.1.

5. Add enough Lugol's solution (iodine) to discolor the beaker water. Complete and record the initial observations listed in Table 6.1.

6. After 60 minutes:
   a. Without disturbing the setup, examine the beaker and dialysis bag and decide if starch or iodine diffused. Record your observations in Tables 6.1 and 6.2.
   b. Remove the bag from the beaker and dry and weigh it. Record your measurement and observations in Tables 6.1 and 6.2.

RESULTS

Complete Tables 6.1 and 6.2 and answer the following questions.

1. Which solution had the greater osmotic pressure, the starch solution in the bag or the iodine solution in the beaker?

   _____

   _____

2. Use osmotic pressure to explain the osmosis of water in this experiment.

   _____

   _____

3. How did you detect whether starch and iodine moved?

   _____

   _____

   _____

TABLE 6.1   Dialysis Experiment Observations

Dialysis Bag	Initial Observations	Final Observations
Weight of bag	_____	_____
Shape of bag	_____	_____
Color of starch solution	_____	_____
Color of beaker water	_____	_____

TABLE 6.2   Dialysis Experimental Results

Substance	Movement (In, out, none)	Process (Diffusion, osmosis)
Water	_____	_____
Starch	_____	_____
Iodine	_____	_____

Experiment II—Concentration Gradients and Osmotic Rate

This experiment demonstrates the relationship between concentration gradient and rate of osmosis. A series of dialysis bags will be filled with different sugar solutions and weighed. After soaking in distilled water for 60 min, the bags will be reweighed and the measurements graphed.

MATERIALS

   three strips of dialysis tubing 15 cm (6″) in length
   six dialysis tubing clips (or string)
   three beakers, 500 ml

tap water

1%, 5%, and 10% sugar solutions

PROCEDURE

1. Add approximately 100 ml of distilled water to the beaker. Soak three strips of dialysis tubing in the water for 3–4 min to loosen the tubing. Remove dialysis tubing from beaker.

2. Prepare each bag as follows:
   a. Fold 20 mm of membrane over on one side and clip securely with a tubing clip.
   b. Rub the unclipped side of membrane between your fingers to open the tubing.
   c. Fill one dialysis bag with the 1% sugar solution and clip the bag closed securely without trapping too much air inside.
   d. Repeat step c for the two remaining bags. Fill one with the 5% solution and the other with the 10% solution. You should now have prepared three bags with 1%, 5%, and 10% solutions.

3. Rinse each bag to remove any sugar solution, and dry and weigh each bag individually. Completely submerge each bag in a separate beaker of distilled water. Record your initial weight measurements in Table 6.3.

4. After 60 min, remove, dry, and weigh the bags. Record the final weights in Table 6.3.

TABLE 6.3	Osmosis Experimental Data	
Dialysis Bag	*Initial Weight*	*Final Weight*
1% sugar	_____	_____
5% sugar	_____	_____
10% sugar	_____	_____

RESULTS

1. Which solution had the greatest osmotic pressure? Compared to the other solutions, what was the result of the higher pressure?

_____

_____

2. Describe the relationship between concentration and rate of osmosis.

_____

_____

3. Construct a graph that illustrates how solute concentration affected osmotic rate. Graph the increase in weight for each bag. Plot the increase in weight for each bag on the vertical axis (Y axis) for each sugar solution listed on the horizontal axis (X axis).

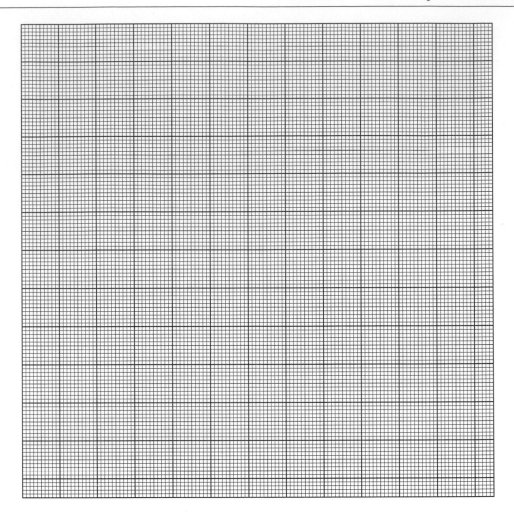

Experiment III—Observation of Osmosis in Cells

To observe osmosis in living cells, solutions of various concentrations will be added to drops of blood (see Figure 6.5). A solution with the same solute concentrations as a cell is an **isotonic solution**. If the solute concentrations are the same, the solvent concentrations will be equal, too. Blood has a solute concentration of 0.9% and a water volume of 99.1%. A solution with more solute (and therefore less solvent) than a cell is a **hypertonic solution**, and a solution having less solute than a cell is a **hypotonic solution**. Note that the cell is the reference point; solutions are compared to the cell. If a cell loses water as a result of osmotic movement, it will shrink or **crenate**. In a hypotonic solution, a cell will gain water and perhaps burst or **lyse**. Blood cells undergo **hemolysis** in hypotonic solutions.

Your laboratory instructor may choose to use plant cells rather than blood cells to study **tonicity**, the effect of solutions on cells. Leaves from the aquatic plant *Elodea* are placed in different solutions to promote osmosis. Plant cells have a thick outer cell wall which provides structural support for the plant. Pushed against the inner surface of the cell wall is the cell membrane. To study osmosis in plant cells, observe the distribution of the cell's organelles and attempt to locate the cell membrane. In hypertonic solutions, for example, the plant cell loses water and the cell membrane shrinks away from the cell wall.

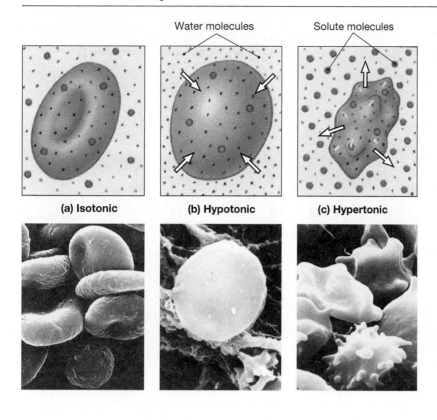

Figure 6.5
Osmotic Flow Across Cell Membranes.

(a) Because these red blood cells are immersed in an isotonic saline solution, no osmotic flow occurs and the cells have their normal appearance. **(b)** Immersion in a hypotonic saline solution results in the osmotic flow of water into the cells. The swelling may continue until the cell membrane ruptures. **(c)** Exposure to a hypertonic solution results in the movement of water out of the cells. The red blood cells shrivel and become crenated.

MATERIALS

blood (supplied by instructor) or live aquatic plant (*Elodea*)

microscope slides and coverslips, microscope, eye droppers, wax pencil

0.90% saline solution (isotonic)

2.0% saline solution (hypertonic)

distilled water (hypotonic)

gloves and safety glasses

PROCEDURE

1. With the wax pencil, label three microscope slides on the ends, "Iso," "Hypo," and "Hyper."

2. Put on safety glasses and disposable gloves before handling any blood.

3. Place a small drop of blood on each of the three microscope slides. **Do not touch the blood.** If you are using plant cells to observe osmosis, place a single *Elodea* leaf flat on the slide. Place a coverslip over each slide.

4. Add a drop of isotonic solution to the outer edge of the coverslip of the "Iso" slide. Repeat with the other slides and solutions.

5. Observe changes in cell shape as osmosis occurs. Compare your results with the cells in Figure 6.4.

6. Dispose of materials contaminated with blood in a biohazard waste box.

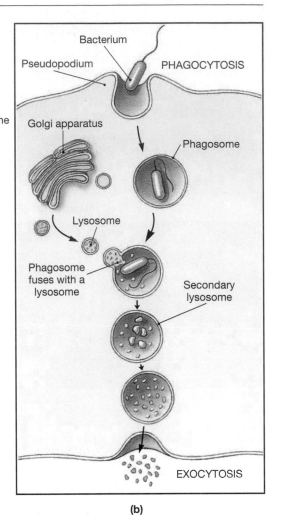

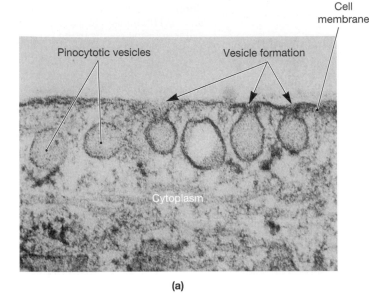

(a)

Figure 6.6
Pinocytosis and Phagocytosis.

(a) An electron micrograph showing pinocytosis at the bases of microvilli in a cell lining the intestinal tract. **(b)** Material brought into the cell by phagocytosis is enclosed in a phagosome and subsequently exposed to lysosomal enzymes. After nutrients are absorbed from the vesicle, the residue is discharged by exocytosis.

(b)

D. Active Transport Processes

Cells utilize carrier molecules to move nondiffusible materials through the cell membrane. Unlike passive processes, this movement may occur against the concentration gradient. Movement of this type is called **active transport** and requires the cell to use energy. While the passive processes of diffusion and osmosis may occur in both living and dead cells, only living cells can supply the necessary energy for active transport.

Endocytosis is the process of actively transporting materials into a cell. Figure 6.6 details **phagocytosis**, the movement of large particles into the cell. The figure shows a cell ingesting a microbe. The cell forms extensions of its cell membrane called **pseudopodia** to capture the microbe. When the pseudopodia touch, they fuse and trap the microbe in a membrane vesicle. Inside the cell, lysosomes surround and empty their powerful enzymes inside the vesicle and destroy the microbe. During **pinocytosis**, the cell invaginates a small area of the membrane and traps small particles and fluid. The forming vesicle continues to pinch inward.

Exocytosis is the active transport of materials out of the cell. An intracellular **vesicle** fills up with materials, fuses with the cell membrane, and releases its contents into the extracellular fluid. The Golgi apparatus secretes cell products by pinching off small secretory vesicles that fuse with the cell membrane for exocytosis. Cells also eliminate debris and excess fluids by exocytosis. Table 6.4 summarizes transport mechanisms across the cell membrane.

TABLE 6.4	Summary of Mechanisms Involved in Movement across Cell Membranes		
Mechanism	*Process*	*Factors Affecting Rate*	*Substances Involved (location)*
Diffusion	Molecular movement of solutes; direction determined by relative concentrations	Size of gradient; size of molecules; charge; lipid solubility, temperature	Small inorganic ions, lipid-soluble materials (all cells)
Osmosis	Movement of water molecules toward solution containing relatively higher solute concentration; requires selectively permeable membrane	Concentration gradient; opposing osmotic or hydrostatic pressure	Water only (all cells)
Filtration	Movement of water, usually with solute, by hydrostatic pressure; requires filtration membrane	Amount of pressure; size of pores in filter	Water and small ions (blood vessels)
Carrier-Mediated Transport			
Facilitated diffusion	Carrier proteins passively transport solutes across a membrane down a concentration gradient	Size of gradient, temperature and availability of carrier protein	Glucose and amino acids (all cells, but several different regulatory mechanisms exist)
Active transport	Carrier proteins actively transport solutes across a membrane regardless of any concentration gradients	Availability of carrier, substrate, and ATP	Na^+, K^+, Ca^{2+}, Mg^{2+} (all cells); other solutes by specialized cells
Secondary active transport	Carrier proteins passively transport two solutes, with one (normally Na^+) moving down its concentration gradient; the cell must later expend ATP to eject the Na^+	Availability of carrier, substrates, and ATP	Glucose and amino acids (specialized cells)
Vesicular Transport			
Endocytosis	Creation of membranous vesicles containing fluid or solid material	Stimulus and mechanics incompletely understood; requires ATP	Fluids, nutrients (all cells); debris, pathogens (specialized cells)
Exocytosis	Fusion of vesicles containing fluids and/or solids within the cell membrane	Stimulus and mechanics incompletely understood; requires ATP	Fluids, debris (all cells)

CELL TRANSPORT CHECKLIST

This is a list of bold terms presented in Exercise 6. Use this list for review purposes after you have completed the laboratory activities.

PASSIVE TRANSPORT

• Figures 6.1–6.5
Brownian movement
kinetic energy
diffusion
concentration gradient
equilibrium
osmosis
solution
solvent
solute
isotonic solution
hypertonic solution
hypotonic solution

osmotic pressure
crenation
hemolysis
dialysis

ACTIVE PROCESSES

• Figure 6.6 and Table 6.4
active transport
endocytosis
phagocytosis
pseudopodia
pinocytosis
exocytosis
vesicle

CELL TRANSPORT MECHANISMS
Laboratory Report Exercise 6

A. **Short-Answer Questions**

1. Why is a concentration gradient necessary for passive transport?

2. Describe the components of a 2% sugar solution.

3. A blood cell is placed in a 1.5% salt solution. Will osmosis, diffusion, or both occur across the blood cell's membrane? Why?

4. Discuss how temperature affects the rate of diffusion.

5. After a long soak in the tub, you notice that your skin has become wrinkled and your fingers and toes feel bloated. Describe why this occurs.

6. How does the molecular weight of a substance affect its diffusion rate?

B. Define

    1. osmotic pressure

    2. dialysis

    3. hypertonic solution

    4. hemolysis

    5. exocytosis

    6. Brownian movement

C. **Examples**

    Give an example of each of the following transport processes:

    1. Diffusion of a substance in the air

    2. Phagocytosis

    3. A hypertonic solution

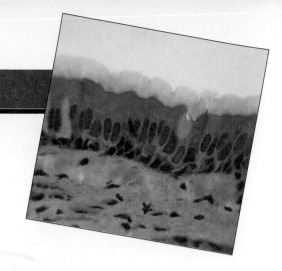

EXERCISE 7

Introduction to Tissues

OBJECTIVES

On completion of this exercise, you should be able to:

- Explain the major functions of each of the four groups of tissues.
- Describe the basic structure of an epithelium.
- Discuss the matrix of connective tissues.
- List the location of the three types of muscle tissue.
- Describe two cell types of the nervous system.

WORD POWER

epithelium (epi—above; thel—tender)
matrix (matria—where something is produced)
visceral (viscera—internal organ)
neuron (neuro—nerve)
glial (glia—glue)

MATERIALS

none
(microscopic examination of tissues is covered in the next four exercises)

INTRODUCTION

Histology is the study of tissues. A **tissue** is a group of similar cells working together to accomplish a specialized function. It may be difficult for us to relate how cells contribute to the life of the entire organism, but we readily see the effect of tissues in our body. Consider how much effort is focused on reducing fat tissue and exercising muscle tissue. An understanding of histology is vital for the study of organ function. The stomach, for example, plays major digestive roles in the secretion of digestive juices and the mixing and movement of food. Each of these functions is performed by a specialized tissue.

Figure 7.1 illustrates an overview of tissues of the body. Molecules and atoms combine to form cells that secrete materials into their surrounding extracellular fluid. The cells and their secretions compose the various tissues of the body. There are four major categories of tissues in the body: **epithelial**, **connective**, **muscle**, and **nerve**. Each category includes specialized tissues with specific locations and functions. Many tissues form an organ such as the stomach, a muscle, or a bone. Organs working together to accomplish major processes such as digestion, movement, and protection constitute an organ system.

During your microscopic observations of tissues in the following exercises, it is important to scan the entire slide to examine the tissue at low power. A slide may have several tissues, and you must survey the specimen to locate a particular tissue. Once the tissue is located, increase the magnification and observe the individual cells of the

Figure 7.1
An Orientation to the Tissues of the Body.

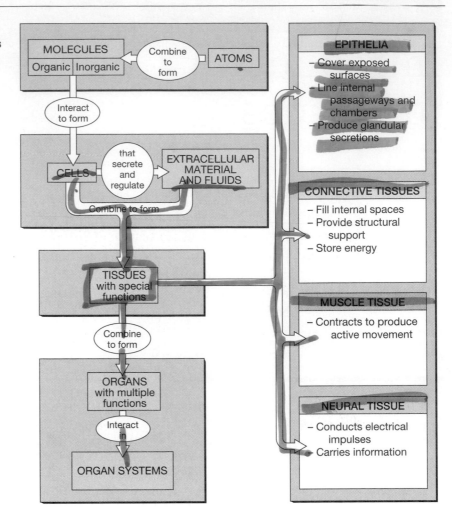

tissue. Take your time when studying a tissue; a quick glance through the microscope is not sufficient to learn about the tissue for identification on a laboratory examination.

A. Epithelium

Epithelium, or **epithelial tissue**, lines and covers surfaces. It is the only tissue you can directly see on your body; it forms the upper layers of the skin and lines passageways such as the nose, mouth, throat, anus, and lungs. Epithelium has a wide range of functions, each dependent on the type of cells in the tissue. On exposed surfaces, thick layers protect from abrasion. Thin, delicate epithelium provides surface areas for absorption, secretion, and diffusion of substances. In some epithelial tissues the cells are short-lived and stem cells must constantly produce new cells to replenish the tissue.

Epithelial cells are packed close together, leaving little room for other structures between the cells. The cells on the surface are exposed to a body cavity or the external environment. The bottom surface of epithelium is always attached to a deeper connective tissue layer by a double-layered **basement membrane**. The membrane layer next to the epithelium is the **basal lamina** which acts as a barrier to prevent materials from the underlying connective tissue from seeping into the epithelium. The membrane layer facing the connective tissue is called the **reticular lamina**. Connective tissue cells secrete protein fibers to give this layer strength. Figure 7.2 is a sectional view of the lining of the mouth. Glide your tongue over the inner surface of your cheeks and feel the lining epithelium of your oral cavity. Below this epithelium is a connective tissue layer.

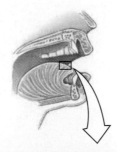

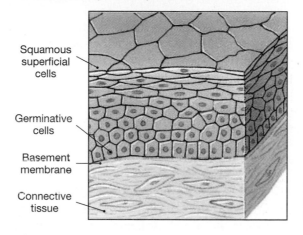

Stratified squamous epithelium ×310

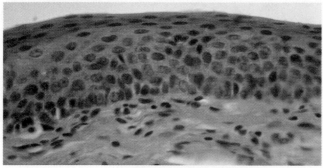

Squamous
superficial
cells

Germinative
cells

Basement
membrane

Connective
tissue

Figure 7.2
**Stratified Squamous Epithelium of
the Mouth.**

In Figure 7.2 how are the cells arranged in the epithelium compared to the connective tissue? Although not visible in the micrograph (photograph), where is the basement membrane located on the specimen in this figure?

_____

_____

Exercise 8 discusses epithelial tissues in detail and includes exercises for the microscopic examination of a variety of epithelia.

B. Connective Tissue

Connective tissue provides your body with protection, structural support, and attachment of other tissues. Connective tissues are the framework for soft organs like the liver, spleen, and intestines. Bone and cartilage are strong connective tissues that support the body. Unlike epithelium, connective tissue cells are loosely arranged and scattered in the tissue. Many types of cells can occur in a connective tissue. The tissue cells

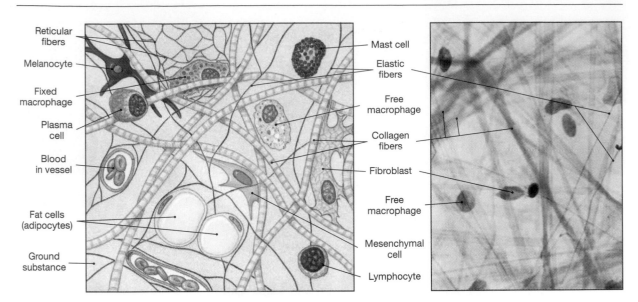

Figure 7.3
Loose Connective Tissue.

secrete protein **fibers** and a **ground substance**, collectively referred to as the **matrix,** which embed the cells. Fibers made by connective tissue cells include **collagen** fibers, which give the tissue strength, and **elastic** fibers for flexibility. **Reticular** fibers build a framework to support other tissues of internal soft organs.

Examine the connective tissue in Figure 7.3 and notice how the cells are widely spread among many protein fibers. The material between the cells is the matrix. Some connective tissues are composed of more than one type of cell. List the types of cells shown in Figure 7.3.

Connective tissues are grouped into three broad categories: connective tissue proper, fluid connective tissues, and supporting connective tissues. The ground substance determines the physical nature of each connective tissue group. **Connective tissue proper** is a group of connective tissues that have a matrix consistency similar to watery Jell-O. Fat, or adipose tissue, is classified as connective tissue proper. **Fluid connective tissues** have a liquid matrix and include blood and lymph tissues. The liquid matrix permits these tissues to flow freely in blood vessels or lymphatic vessels. Cartilage and bone tissues are **supporting connective tissues**. Cartilage has a gelatinous type of matrix and is very flexible and resistant to breaking. Bone has a solid matrix and provides a rigid framework for supporting other tissues of the body.

Exercise 9 discusses connective tissues in detail and includes exercises on the microscopic examination of a variety of connective tissues.

C. Muscle Tissue

Movement of the skeleton is an obvious function of muscle tissue, but have you thought about what causes those gurgling sounds after a large meal or how your eyes adjust to bright light after leaving a darkened room? These actions are a result of coordinated muscle contraction.

There are three types of muscle tissue, each named for its location in the body. **Skeletal** muscle is attached to bone and provides movement such as walking and moving your head. **Cardiac** muscle forms the walls of the heart. During the contraction

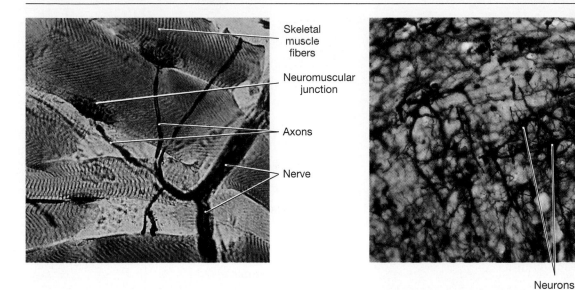

Figure 7.4
Skeletal Muscle Fiber and Nerve Branches.

Figure 7.5
Neurons of the Brain.

phase of the heartbeat, the heart pumps blood into blood vessels for transport of oxygen, nutrients, and other materials to the cells of the body. **Visceral** or **smooth** muscle tissue is found inside hollow organs such as the stomach, intestines, blood vessels, and uterus. Smooth muscle controls the movement of materials through the digestive system, the diameter of blood vessels, and the contraction of the uterus during labor. Figure 7.4 is a micrograph of several skeletal muscle cells, each called a **muscle fiber**. What is the function of the black nerve branches visible in the figure?

_____

_____

Exercise 10 discusses muscle tissue in detail and includes exercises on the microscopic examination of each type of muscle tissue.

D. **Nerve Tissue**

The nervous system is an elaborate communication system, relaying signals from cell to cell to regulate body functions. Nerve tissue consists of cells called **neurons** which produce and conduct electrical impulses in response to internal and external environmental changes. These impulses, called action potentials, regulate organ activity, muscle contraction, and memory. Neural tissue also includes supporting **glial** cells which protect, attach, and isolate neurons. The micrograph in Figure 7.5 shows a network of neurons from a part of the cerebrum of the brain where nerve commands to skeletal muscles are initiated. The neurons are the black cells with extensive branching and interconnections. Why are there so many interconnections between these neurons?

_____

_____

Exercise 11 discusses neural tissue in detail and includes exercises on the microscopic examination of nerve tissues.

INTRODUCTION TO TISSUES CHECKLIST

This is a list of bold terms presented in Exercise 7. Use this list for review purposes after you have completed the laboratory activities.

EPITHELIAL TISSUE

• Figures 7.1 and 7.2
epithelium
   basement membrane
     basal lamina
     reticular lamina

CONNECTIVE TISSUES

• Figures 7.1 and 7.3
ground substance
matrix
collagen fibers
elastic fibers
reticular fibers
connective tissue proper
fluid connective tissues
supporting connective tissues

MUSCLE TISSUE

• Figures 7.1 and 7.4
skeletal muscle
   muscle fibers
cardiac muscle
smooth muscle

NERVE TISSUE

• Figures 7.1 and 7.5
neurons
glial cells

INTRODUCTION TO TISSUES
Laboratory Report Exercise 7

A. Fill in the Blanks
1. The tissue that moves the skeleton is _____.
2. Supportive cells of the nervous system are _____ cells.
3. Strong fibers called _____ give connective tissue strength.
4. A _____ attaches epithelium to its underlying connective tissues.
5. The tissue group with widely scattered cells is _____.
6. Lining and covering tissues are classified as _____.
7. The muscle tissue of internal organs such as blood vessels is _____.

B. Completion
Complete the following table by listing the four groups of tissues and the distinguishing characteristics of each.

Tissue Group	Basic Description

C. Short-Answer Questions
1. What is the matrix of a tissue? Do all tissues have a matrix?

2. Give an example of each type of muscle tissue.

3. Describe how epithelium is attached to the body.

4. Give an example of each major group of connective tissues.

5. Describe the differences between cells and tissues.

EXERCISE 8

Epithelial Tissue

OBJECTIVES

On completion of this exercise, you should be able to:

- List the characteristics used to classify epithelia.
- Describe how epithelia are attached to the body.
- Describe the microscopic appearance of each type of epithelia.
- List the location and function of each epithelial tissue.
- Identify each tissue under the microscope.

WORD POWER

epithelium (epi—above; thel—tender)
lamina (lamin—thin sheet)
squamous (squam—a scale)
stratified (strat—layer)
pseudostratified (pseudo—false)
microvilli (micro—small; vill—hair)

MATERIALS

compound microscope
prepared microscope slides:
  simple squamous epithelium (lung or mesentary)
  simple cuboidal epithelium (kidney)
  simple columnar epithelium (digestive tract)
  stratified squamous epithelium (skin)
  transitional epithelium (urinary bladder)
  pseudostratified epithelium (trachea)

INTRODUCTION

Epithelial tissues are lining and covering tissues. The respiratory, digestive, reproductive, and urinary systems all have openings to the external environment and are lined with **epithelia (e-pi-THĒ-lē-a).** The entire body surface is covered with epithelia in the upper layer of the skin. Epithelia always have a free surface where cells are exposed to the external environment or to an internal passageway or cavity. Because epithelia are surface and lining tissues, they do not contain blood vessels; they are **avascular** tissues. Epithelial cells obtain nutrients and other necessary materials by diffusion from the underlying connective tissue. Glide your tongue over your inner cheeks and feel the epithelium.

Notice in Figure 8.1a how closely packed epithelial cells occur. The cells are attached to one another by strong intercellular connections. Some cell membranes of adjoining epithelial cells are glued together by protein molecules known as **cell adhesion molecules**. Special cell-to-cell unions, called **cell junctions**, interconnect cell membranes, and materials can move from one cell to the other through these connections. Four types of connections occur: gap junctions, tight junctions, intermediate junctions, and desmosomes. Use your lecture textbook to obtain more information about these cell junctions.

Epithelium is attached to the body by a **basement membrane** located between the epithelium and a connective tissue layer. Locate this membrane in Figure 8.1. Essentially, the basement membrane is two layers of cellular glue which anchor the

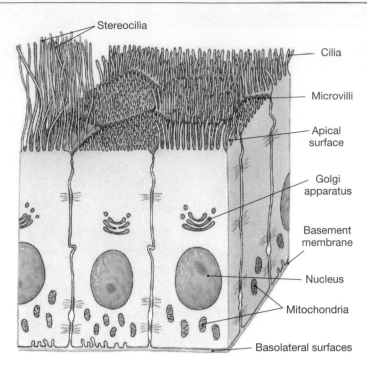

Figure 8.1
The Polarity of Epithelial Cells.
Many epithelial cells have an uneven distribution of organelles between the free surface (here, the top) and the basement membrane. In many cases, the free surface bears microvilli; in some cases, this surface has cilia or (very rarely) stereocilia. *(All three would not normally be on the same group of cells but are depicted here for purposes of illustration.)* In some epithelia, such as the lining of the kidney tubules, mitochondria are concentrated near the base of the cell, probably to provide energy for the cell's transport activities.

epithelium to an underlying connective tissue layer. The base of the epithelium is in contact with a layer called the **basement lamina** which contains glycoproteins and other protein molecules. These proteins prevent substances in the adjoining connective tissue from seeping into the epithelia. The basement membrane layer facing the connective tissue is the **reticular lamina**. Thick protein fibers in this layer entwine with the proteins in the basement lamina for secure attachment of the epithelium.

STUDY TIP

When observing epithelia microscopically it might be difficult to locate the basement membrane. Simple epithelia often have a conspicuous basement membrane. Find the free surface of the tissue and then look on the opposite edge of the cells. The basement membrane appears as a dark line between the epithelial cells and the connective tissue. You will not be able to distinguish between the basement lamina and reticular lamina with the microscopes available in your laboratory.

Epithelium has a wide range of functions. It protects the exposed surfaces of the body from excessive friction, prevents dehydration, and resists the invasion of microbes and chemicals into the body. All substances must pass through a layer of epithelium to enter the internal environment. Gases in the respiratory system, food being digested in

CELL LAYERS

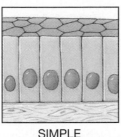

SIMPLE
EPITHELIUM

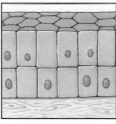

STRATIFIED
EPITHELIUM

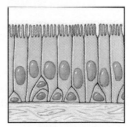

PSEUDOSTRATIFIED
EPITHELIUM

(a)

CELL SHAPES

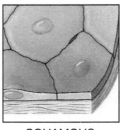

SQUAMOUS
EPITHELIUM

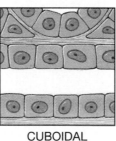

CUBOIDAL
EPITHELIUM

COLUMNAR
EPITHELIUM

(b)

Figure 8.2
Organization of Epithelium.
(a) All epithelium is attached to a basement membrane. Cells may occur in single or multiple layers. **(b)** Cell shape is also used to classify epithelium.

the intestines, and wastes filtered from the blood in the kidneys all pass through a lining of epithelium. Because epithelium is the external covering of the body, many sensory organs are found in these tissues. Epithelium also forms the secretory cells of glands.

Epithelial tissues are classified according to cell shape and cell organization (see Figure 8.2). These characteristics are used together to name a tissue.

Cell shape

Squamous (SKWĀ-mus) cells are irregularly shaped, flat, and scalelike. These cells, depending on how they are organized, function in protection or secretion and diffusion. **Cuboidal** cells are squarish and have a large central nucleus. They are found in the tubules of the kidneys and in many glands. The cube-shaped cells secrete material into the tubes or absorb material passing through the tubes. **Columnar** cells are taller than they are wide, like the columns of a building. The surface of the cell may be covered with microscopic plasma membrane extensions called **microvilli** which increase the cellular surface area for absorption and secretion. Other extensions are hairlike **cilia** which move mucus and remove debris. Many locations in the respiratory and digestive systems are lined with columnar cells. **Transitional** epithelium has a variety of cell shapes. This permits the tissue to stretch and shrink.

Cell organization

Simple epithelium is a single layer of cells. As in all epithelia, the cells are attached to an underlying connective tissue by the **basement membrane**. In simple tissues, all cells touch the basement membrane and are exposed to the upper surface of the tissue. Simple epithelium is found where a thin surface is required for diffusion, such as between the air sacs and the blood capillaries in the lungs. **Stratified** epithelium has layers of cells, one stacked upon another. The bottom cells are attached to the basement lamina, and only the upper layer of cells are exposed to the free surface of the tissue. Stratified epithelium is found in areas exposed to friction and abrasion. The upper layer of the skin, the epidermis, and the linings of the mouth, parts of the pharynx, and the anus are covered with stratified epithelium. **Pseudostratified** epithelium, as its name implies, is "falsely stratified." As in simple epithelium, the cells all touch the basement lamina, however, all cells do not reach the upper surface of the tissue. Cells grow to different heights, and taller cells grow over and cover shorter cells. Parts of the respiratory tract are lined with ciliated pseudostratified epithelium.

A. ### Simple Epithelia

Simple epithelia occur in a variety of cell shapes including **squamous**, **cuboidal**, and **columnar**. All cells in simple epithelia are attached to the basement lamina and reach the surface of the tissue. General functions of simple epithelia are diffusion, absorption, and secretion. Extensions of the plasma membrane, called **microvilli**, increase surface area for absorption. To protect the tissue at the exposed surface, **goblet cells** secrete mucus which coats the delicate epithelial cells. Other cells have hairlike **cilia** which sweep the mucus along the tissue surface to remove debris. Cilia are composed of microtubules which protrude from the cell. Figure 8.1b details the modifications of epithelial surfaces.

Simple squamous epithelium

Simple squamous epithelium (see Figure 8.3a) is found in serous membranes, in the lining of blood vessels and the heart, and in the air sacs (alveoli) of the lungs. When viewing a superficial preparation of this epithelium under the microscope, it appears as a sheet of cells, similar to ceramic tiles on a floor. Functions of simple squamous tissue include secretion and diffusion.

Simple cuboidal epithelium

Simple cuboidal epithelium preparations are often from the epithelia of kidney tubules or the follicles of the thyroid gland. Note in Figure 8.3b the cube-shaped cells arranged in a ring to make a tubule. On many slides the tubules have been transversely sectioned. Simple cuboidal epithelium functions in secretion and absorption.

Simple columnar epithelium

Simple columnar epithelium lines most of the digestive tract and the gallbladder. Figure 8.3c shows the lumen or passageway of the small intestine. The wall of the intestine is folded to increase the surface area for digestion and absorption of nutrients. Simple columnar epithelium covers the folded wall and is in direct contact with the contents of the intestine. Notice in the figure that the nuclei are uniformly located at the base of

Study Guide—Drawing
Chap. 4

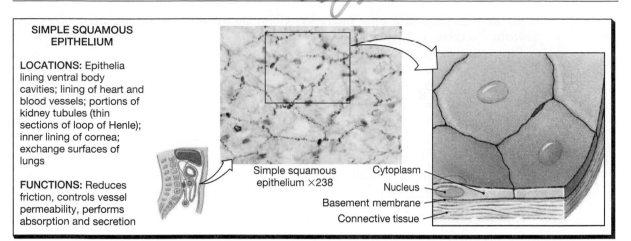

SIMPLE SQUAMOUS EPITHELIUM

LOCATIONS: Epithelia lining ventral body cavities; lining of heart and blood vessels; portions of kidney tubules (thin sections of loop of Henle); inner lining of cornea; exchange surfaces of lungs

FUNCTIONS: Reduces friction, controls vessel permeability, performs absorption and secretion

Simple squamous epithelium ×238

Cytoplasm
Nucleus
Basement membrane
Connective tissue

(a)

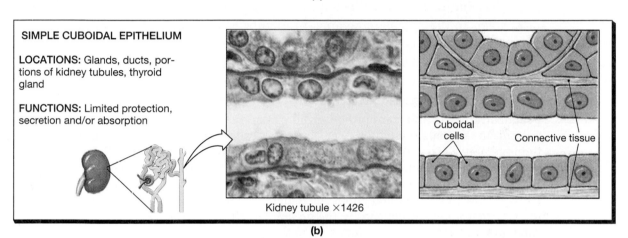

SIMPLE CUBOIDAL EPITHELIUM

LOCATIONS: Glands, ducts, portions of kidney tubules, thyroid gland

FUNCTIONS: Limited protection, secretion and/or absorption

Kidney tubule ×1426

Cuboidal cells
Connective tissue

(b)

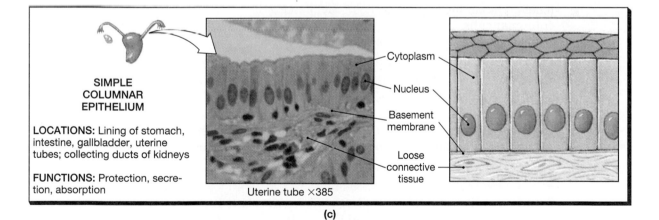

SIMPLE COLUMNAR EPITHELIUM

LOCATIONS: Lining of stomach, intestine, gallbladder, uterine tubes; collecting ducts of kidneys

FUNCTIONS: Protection, secretion, absorption

Uterine tube ×385

Cytoplasm
Nucleus
Basement membrane
Loose connective tissue

(c)

Figure 8.3
Simple Epithelia.

(a) A superficial view of the simple squamous epithelium that lines the periotoneal cavity. The three-dimensional drawing shows the epithelium in superficial and sectional view. (b) A section through the cuboidal epithelial cells of a kidney tubule. The diagrammatic view emphasizes structural details that permit classification of epithelium as cuboidal. (c) Micrograph showing the characteristics of simple columnar epithelium. In the diagrammatic sketch, note the relationships between the height and width of each cell: the relative size, shape, and location of nuclei; and the distance between adjacent nuclei. Compare with Figure 8.5.

the cells. The large oval cells interspersed among the columnar cells are goblet cells. The uterine tubes and sinuses of the skull are lined with ciliated columnar epithelium. In the uterine tubes the cilia transport released eggs to the uterus.

LABORATORY ACTIVITY   OBSERVATION OF SIMPLE EPITHELIA

MATERIALS

Prepared microscope slides of:

> simple squamous epithelium
> simple cuboidal epithelium
> simple columnar epithelium

PROCEDURE

1. Examine each simple epithelium under the microscope at low and medium powers.
2. Refer to the photomicrographs in Figures 8.3 and 8.4 and locate the featured structures.
3. Draw each tissue in the corresponding space in Figure 8.4.

B. Pseudostratified, Transitional, and Stratified Epithelia

Pseudostratified columnar epithelium lines the trachea and bronchi of the respiratory system. In the male reproductive system this tissue helps propel semen and is ciliated to move materials across the tissue surface. Although the tissue appears stratified, it is actually simple in cell organization. All cells touch the basement lamina, yet some do not reach the free tissue surface. Notice in Figure 8.5a that the nuclei are unevenly distributed, creating a stratified appearance. In between the columnar cells are goblet cells. In the lung, mucus produced by the goblet cells traps dust and other particles in the inhaled air. Cilia sweep the mucus up to the throat for disposal.

While at the microscope, use high magnification and the fine focus adjustment to locate cilia on the surface of the tissue. Large goblet cells are interspersed between the columnar cells. If possible, locate the basement membrane at the base of the tissue and the underlying connective tissue.

Transitional epithelium

Transitional epithelium lines organs, such as the urinary bladder, which must stretch and shrink. The cells have a variety of shapes and sizes, and not all cells contact the basement lamina (see Figure 8.5b). Most transitional tissue slides are prepared from relaxed transitional tissue, and thus the cells are stacked one upon another.

Stratified epithelium

Stratified epithelium is a multilayered tissue with only the bottom cellular layer in contact with the basement lamina and the upper cells exposed to the free surface. A variety of cell shapes occur in stratified epithelium, and the type of cells on the free surface layer is used to describe and classify the tissue. Exposed surfaces, such as the upper layer of the skin and the lining of the mouth, protect against abrasion by building up multiple cell layers of stratified squamous epithelium. Figure 8.5c highlights nonkeratinized stratified squamous of the mouth. This tissue must be kept moist to prevent cell dehydration. The stratified squamous epithelium of the upper skin is keratinized to reduce water loss.

Stratified squamous epithelium of the skin is usually stained red or purple on a microscope slide. Close observation within the stratified layer reveals that not all cells

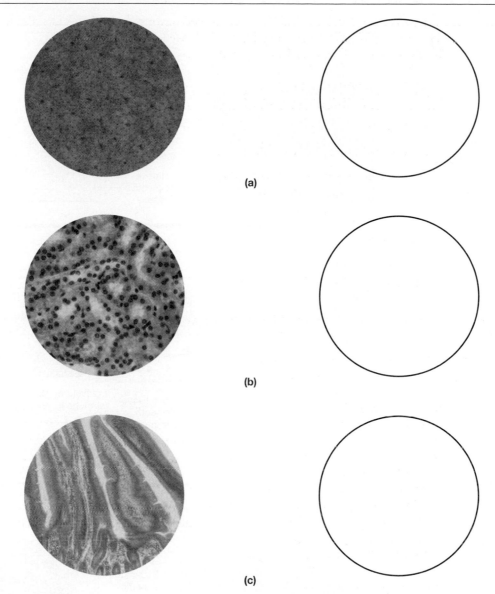

(a)

(b)

(c)

Figure 8.4
Simple Epithelia.

(a) Simple squamous epithelium (LM × 100) and sketch. **(b)** Simple cuboidal epithelium (LM × 400) and sketch. **(c)** Simple columnar epithelium (LM × 100) and sketch.

are squamous and that many cuboidal and columnar cells are distributed in middle layers. Cells near the surface of the tissue, however, are squamous. The bottom of the tissue is attached to connective tissue by the basement lamina.

CLINICAL APPLICATION—A BARRIER AGAINST INFECTION

Infectious organisms enter the body by penetrating the lining and covering epithelium. AIDS, caused by the HIV virus, is transmitted when the virus enters a wound or an abrasion in the epithelium. The concept of "safe sex" is to maintain barriers between your epithelia and other people's body fluids which may contain an infectious biological agent. The epithelial lining of the mouth, vagina, urethra, and rectum should all be protected to prevent exposure to a sexually transmitted disease.

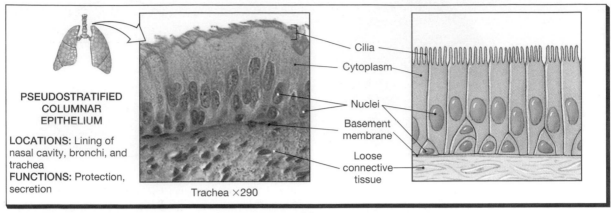

PSEUDOSTRATIFIED COLUMNAR EPITHELIUM

LOCATIONS: Lining of nasal cavity, bronchi, and trachea
FUNCTIONS: Protection, secretion

Cilia
Cytoplasm
Nuclei
Basement membrane
Loose connective tissue

Trachea ×290

(a)

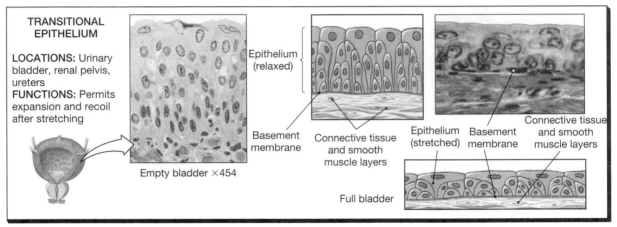

TRANSITIONAL EPITHELIUM

LOCATIONS: Urinary bladder, renal pelvis, ureters
FUNCTIONS: Permits expansion and recoil after stretching

Epithelium (relaxed)

Basement membrane
Connective tissue and smooth muscle layers

Epithelium (stretched)   Basement membrane   Connective tissue and smooth muscle layers

Empty bladder ×454

Full bladder

(b)

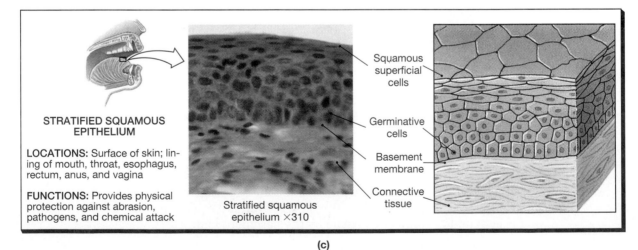

STRATIFIED SQUAMOUS EPITHELIUM

LOCATIONS: Surface of skin; lining of mouth, throat, esophagus, rectum, anus, and vagina

FUNCTIONS: Provides physical protection against abrasion, pathogens, and chemical attack

Squamous superficial cells
Germinative cells
Basement membrane
Connective tissue

Stratified squamous epithelium ×310

(c)

Figure 8.5
Stratified Epithelia.

(a) The pseudostratified ciliated columnar epithelium of the respiratory tract. Note the uneven layering of the nuclei. **(b)** At left, the lining of the empty urinary bladder, showing transitional epithelium in the relaxed state. At right, the lining of the full bladder, showing the effects of stretching on the arrangement of cells in the epithelium. **(c)** A sectional view of the stratified squamous epithelium that covers the tongue.

LABORATORY ACTIVITY MICROSCOPIC OBSERVATION OF PSEUDOSTRATIFIED, TRANSITIONAL, AND STRATIFIED EPITHELIA

MATERIALS

Prepared microscope slides:

> pseudostratified epithelium (trachea)
> transitional epithelium (urinary bladder)
> stratified squamous epithelium (skin)

PROCEDURE

1. Examine each epithelium under the microscope at low and medium powers.
2. Refer to the photomicrographs in Figures 8.5 and 8.6 and locate the featured structures.
3. Draw each tissue in the corresponding space in Figure 8.6.

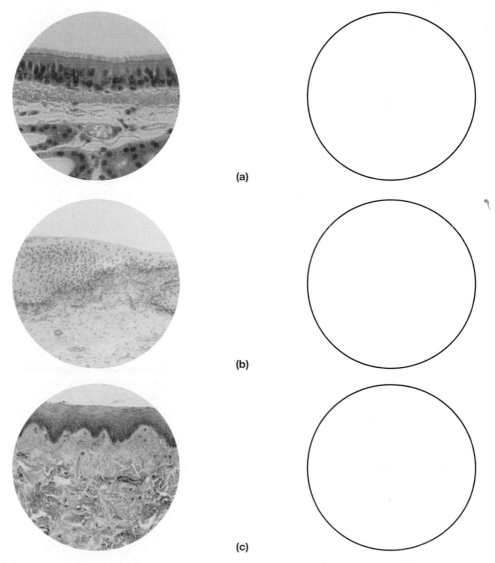

(a)

(b)

(c)

Figure 8.6
Pseudostratified, Transitional, and Stratified Epithelia.

(a) Pseudostratified epithelium (LM × 100) and sketch. **(b)** Transitional epithelium (LM × 400) and sketch. **(c)** Stratified squamous epithelium (LM × 100) and sketch.

EPITHELIAL TISSUE CHECKLIST

This is a list of bold terms presented in Exercise 8. Use this list for review purposes after you have completed the laboratory activities.

INTRODUCTION TO EPITHELIUM

avascular

cell adhesion molecules

cell junction

basement membrane

basement lamina

reticular lamina

cell shapes:

   squamous

   cuboidal

   columnar

cell organization:

   simple

   stratified

   pseudostratified

SIMPLE EPITHELIA

- **Simple squamous epithelium** (Figures 8.3 and 8.4)

  Description: Single layer of scalelike cells closely packed together.

  Location: Alveoli (air sacs) of lungs, serous membranes.

  Function: Diffusion and secretion.

- **Simple cuboidal epithelium** (Figures 8.3 and 8.4)

  Description: Cube-shaped cells organized into tubules. Large central nucleus prominent in each cell. Basement lamina may be visible.

  Location: Tubules of kidneys, follicles of thyroid gland.

  Function: Secretion and absorption.

- **Simple columnar epithelium (nonciliated)** (Figures 8.3 and 8.4)

  Description: Tall, narrow cells with nuclei in middle region of cells. Often contains numerous mucus-secreting goblet cells.

  Location: Lining of most of the digestive system.

  Function: Absorption of nutrients and secretion of digestive juices in digestive tract.

STRATIFIED AND OTHER EPITHELIA

- **Pseudostratified columnar epithelium (ciliated)** (Figures 8.5 and 8.6)

  Description: Single layer of cells, all touching basement lamina but not the upper surface; nuclei distributed throughout, causing stratified appearance. Tissue is ciliated in respiratory system.

  Location: Trachea of respiratory system, ducts of male reproductive system.

  Function: Lines passageways, cilia transport mucus from goblet cells in trachea, transport of semen in male.

- **Stratified squamous epithelium** (Figures 8.5 and 8.6)

  Description: Multiple layers of cells of various shapes, squamous cells on exposed surface.

  Location: Epidermis of skin, lining of mouth and anus.

  Function: Protection of underlying structures.

- **Transitional epithelium** (Figures 8.5 and 8.6)

  Description: Single layer of cells, all touching basement lamina but not the surface. Nuclei distributed to cause stratified appearance.

  Location: Lining of urinary bladder.

  Function: Stretches and retracts in bladder wall to fill and empty.

EPITHELIAL TISSUE

Laboratory Report Exercise 8

A. Matching

    1. _____ simple cuboidal   **A.** lines urinary bladder

    2. _____ stratified squamous   **B.** forms serous membranes

    3. _____ transitional   **C.** contains goblet cells

    4. _____ simple columnar   **D.** upper layer of skin

    5. _____ simple squamous   **E.** forms tubules in kidneys

B. Complete the Following Statements:

    1. Epithelium that occurs in a single layer is _____ epithelium.

    2. The tissue under epithelium is _____.

    3. Epithelium that stretches and relaxes is _____ epithelium.

    4. The noncellular layer of glycoproteins that attaches epithelium to connective tissue is the _____.

    5. Cells that secrete mucus are called _____.

C. Short-Answer Questions

    1. Describe the different kinds of cells and layering of epithelia.

    2. Compare the function of simple epithelia with that of stratified epithelia.

    3. What type of epithelia occurs where organs must stretch and expand?

    4. How is epithelium attached to the underlying tissues?

Connective Tissues

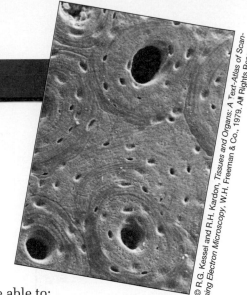

OBJECTIVES

On completion of this exercise, you should be able to:

- List the major groups and characteristics of connective tissues.
- Discuss the composition of the matrix of each type of connective tissue.
- List the location and function of each type of connective tissue.
- Identify each type of connective tissue and its cell and matrix structure under the microscope.

WORD POWER

matrix (matri—place of development)
reticular (reticul—network)
mesenchyme (enchyma—infusion)
fibroblast (blast—bud or sprout)
adipocyte (cyte—cell)
osteoclast (clast—to break)

macrophage (phage—to eat)
areolar (areol—small open space)
canaliculus (canali—canal)
chondroblast (chondro—cartilage)
perichondrium (peri—around)
osteon (osteo—bone)
lamella (lamella—small plate)

MATERIALS

compound microscope
prepared microscope slides:
  loose connective tissue
  adipose
  reticular tissue
  dense connective tissue (regular)
  blood smear
  hyaline cartilage
  elastic cartilage
  fibrocartilage
  compact bone

INTRODUCTION

Connective tissues provide the body with structural support and attachment of other tissues. Unlike those in epithelia, cells of connective tissue are widely scattered throughout the tissue. These cells produce and secrete protein **fibers** and a **ground substance** to form an extracellular **matrix**. The ground substance is composed mainly of glycoprotein and polysaccharide molecules which surround the cells as a thick liquid or a gelatinous material. Suspended in the ground matrix are **collagen** fibers, which give tissues strength, and **elastic** fibers, which provide flexibility. Leather is mostly collagen fibers from the dermis of animal skins that have been tanned and preserved. As we age, cells secrete fewer protein fibers into the matrix, resulting in brittle bones and wrinkled skin. **Reticular** fibers are found in reticular connective tissue and provide a framework for support of internal soft organs such as the liver and spleen.

The matrix of a connective tissue determines the physical nature of the tissue. Blood has a fluid matrix called blood plasma and can flow freely through vessels. Cartilage has a thick gel type of matrix and can slide easily over other structures. Bone has a solid matrix and provides the structural framework for the body.

All connective tissues develop from an unspecialized embryonic tissue called **mesenchyme**. Mesenchyme is capable of producing any type of adult connective tissue. If mesenchyme is located where bone is forming, it will differentiate (specialize) into bone cells.

blood vessels disorganize

95

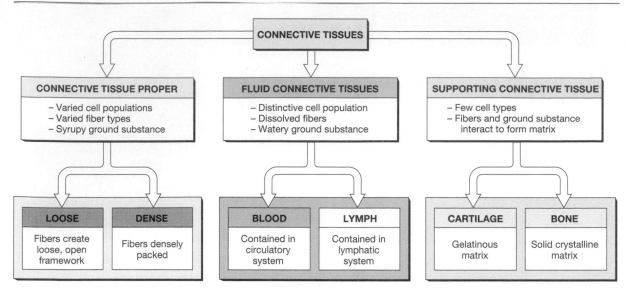

Figure 9.1
Major Types of Connective Tissue.

Connective tissues are classified into three broad groups distinguished primarily by their cellular composition and the characteristics of their extracellular matrix. **Connective tissue proper** has a thick fluid matrix and a variety of cell types. **Fluid connective tissues** are liquid tissues that flow through blood vessels or lymphatic vessels. **Supportive connective tissues** have a strong matrix and provide structural support for other tissues. Each tissue group is discussed below. Figure 9.1 and Table 9.1 present an overview of each connective tissue group.

TABLE 9.1   Summary of Connective Tissues

Tissue	Description
Connective Tissue Proper	**Syrup matrix, various cell types**
Loose connective tissue	Fibroblasts and mast cells, matrix with collagen and elastic fibers
Adipose tissue	Adipocytes with contents pushed to edge of cells
Reticular tissue	Network of reticular fibers supporting cells of liver, spleen, and lymph nodes
Dense connective tissue	Fibroblasts located between parallel bundles of yellow collagen fibers
Fluid Connective Tissues	**Fluid matrix, circulates in vessels**
Blood	Red blood cells transport respiratory gases, white cells provide immunity, platelets clot blood
Lymph	Fluid matrix with scattered lymphocytes
Supportive Connective Tissues	**Surrounded by outer cell-producing membrane, cells trapped in lacunae in matrix**
Hyaline cartilage	Perichondrium with chondroblasts, chondrocytes in lacunae
Elastic cartilage	Perichondrium with chondroblasts, chondrocytes in lacunae, elastic fibers visible in matrix
Fibrocartilage	Chondrocytes stacked up in columns within lacunae
Bone tissue	Concentric lamellae form osteons surrounding central canals, osteocytes trapped in lacunae, canaliculi interconnect osteocytes to central canal

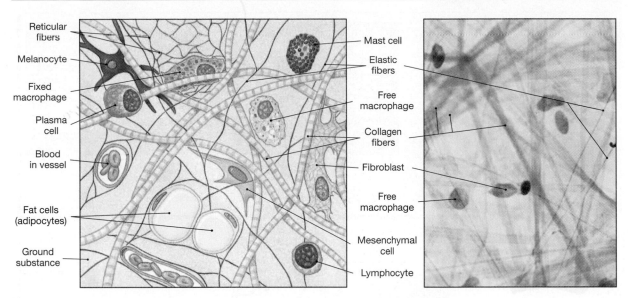

Reticular fibers
Melanocyte
Fixed macrophage
Plasma cell
Blood in vessel
Fat cells (adipocytes)
Ground substance

Mast cell
Elastic fibers
Free macrophage
Collagen fibers
Fibroblast
Free macrophage
Mesenchymal cell
Lymphocyte

Figure 9.2
The Cells and Fibers of Connective Tissue Proper.

A. Connective Tissue Proper

Connective tissue proper includes loose (areolar), adipose, reticular, and dense connective tissues. These tissues all have a matrix composed of a viscous (thick) ground substance with many extracellular fibers. As shown in Figure 9.2, connective tissues proper has a wide variety of cell types. **Fibroblasts (FĪ-bro-blasts)** are fixed (stationary) cells which secrete proteins that join other molecules in the matrix to form collagen and elastic fibers. Phagocytic **macrophage (MAK-rō-fā-jez) cells** patrol these tissues, ingesting microbes and dead cells. Macrophages are mobilized during an infection or injury, migrate to the site of disturbance, and phagocytize damaged tissue cells and microbes. **Mast cells** release histamines which cause an inflammatory response in damaged tissues. **Adipocytes** are fat cells with vacuoles for the storage of lipid. The description, location, and function of each connective tissue proper are presented below. Table 9.1 summarizes connective tissues.

Loose (areolar) connective tissue is distributed throughout the body. This tissue fills in spaces between structures for support and protection, much as packing material surrounds an object in a box. For example, muscles are separated from the surrounding anatomy by loose connective tissue. This tissue is very flexible and permits the muscle to move freely under the skin. Figure 9.3a details the organization of loose connective tissue. A variety of cells are scattered in the tissue, with fibroblasts and mast cells most numerous. Collagen and elastic fibers are clearly visible in the matrix.

Adipose or fat tissue is also distributed throughout the body and is abundant under the skin and in the buttocks, breasts, and abdomen. Fat cells are called **adipocytes (AD-i-pō-sīts)** and are packed close together as compared to other proper connective tissues (see Figure 9.3b). The distinguishing feature of adipose tissue is displacement of the cytoplasm due to the storage of lipids. When nutritional intake exceeds the energy requirements of the body, adipocytes fill their vacuoles with lipid droplets and swell in size. The vacuoles store so much lipid that the organelles and cytosol are pushed to the periphery of these cells, causing them to look like graduation rings.

LOOSE CONNECTIVE TISSUE

LOCATIONS: Beneath dermis of skin, digestive tract, respiratory and urinary tracts; between muscles; around blood vessels, nerves, joints

FUNCTIONS: Cushions organs; provides support but permits independent movement; phagocytic cells provide defense against pathogens

white Blood cells
inflamation occurs

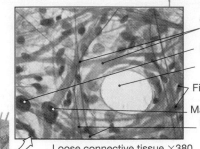
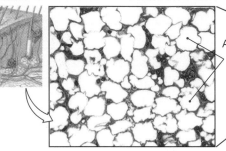
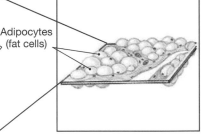

Collagen fibers
Mast cell
Fat cell
Fibroblasts
Macrophage
Elastic fibers

adipocyte

Loose connective tissue ×380

open area – Not alot of space

(a)

ADIPOSE TISSUE

LOCATIONS: Beneath skin, especially at sides, buttocks, breasts; behind eyeballs; around kidneys

FUNCTIONS: Provides padding and cushions shocks; insulates (reduces heat loss); stores energy reserves

Adipocytes (fat cells)

Adipose tissue ×133

more fat cells – empty – densely packed
with nuclei

(b)

DENSE CONNECTIVE TISSUES

LOCATIONS: Between skeletal muscles and skeleton (tendons); between bones (ligaments); covering skeletal muscles; capsules of visceral organs

FUNCTIONS: Provide firm attachment; conduct pull of muscles; reduce friction between muscles; stabilize relative positions of bones; help prevent overexpansion of organs such as the urinary bladder

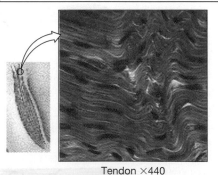
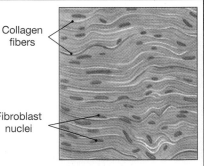

Collagen fibers

Fibroblast nuclei

Tendon ×440

fiber (network)

tendons

(c)

Figure 9.3
Loose and Dense Connective Tissue.

(a) All the cells of connective tissue proper are found in loose connective tissue.
(b) Adipose tissue is loose connective tissue dominated by adipocytes. In standard histological preparations the tissue looks empty because the lipids in the fat cells dissolve during the sectioning and staining procedures. **(c)** The dense regular connective tissue in a tendon. Notice the densely packed, parallel bundles of collagen fibers. The fibroblast nuclei can be seen flattened between the bundles.

Not strong
support Fibers

RETICULAR TISSUE

LOCATIONS: Liver, kidney, spleen, lymph nodes, and bone marrow.

FUNCTIONS: Provides supporting framework

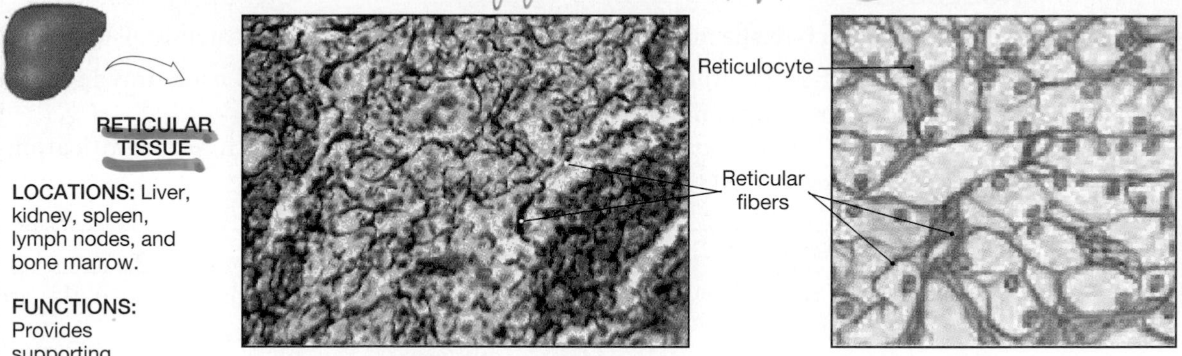

Reticular tissue ×330

Reticulocyte

Reticular fibers

**Figure 9.4
Reticular Tissue.**

CLINICAL APPLICATION—LIPOSUCTION

The surgical procedure called liposuction removes unwanted adipose tissue with a suction wand. The treatment is dangerous and may damage blood vessels or nerves near the site of fat removal. Overlying skin may appear pocketed and marbled after the procedure. Considering that connective tissues have mesenchyme cells, what would most likely occur if a liposuction patient consumed excessive calories after the surgery?

Dense connective tissue (regular) (Figure 9.3c) has thick, parallel bundles of collagen fibers with small fibroblasts between the collagen bundles. This strong tissue forms tendons between muscles and bones and ligaments between bones. Flat layers of dense connective tissue called fascia protect and isolate muscles from surrounding structures and allow muscle movement. On slides with limited stain, the profusion of collagen fibers makes this tissue appear yellow under the microscope.

Reticular tissue, shown in Figure 9.4, forms the internal framework for soft organs such as the spleen, liver, and lymphatic organs. The tissue is composed of an extensive network of reticular fibers with small, oval reticulocytes.

LABORATORY ACTIVITY   MICROSCOPIC OBSERVATIONS OF CONNECTIVE TISSUES PROPER

MATERIALS

compound microscope
prepared microscope slides:
    loose (areolar) connective tissue
    adipose tissue
    reticular tissue
    dense connective tissue

PROCEDURES

1. Examine each tissue under the microscope and locate each structure listed below.

2. Draw each tissue in the spaces provided in Figure 9.5. Label each drawing.

    a. Loose (areolar) connective tissue

    Scan a slide of loose connective tissue at low and medium magnifications. What type of fibers do you see?

    _____

    _____

    Identify the following cell types in this tissue: fibroblasts, mast cells, and macrophages. How can you distinguish each type?

    _____

    _____

    Draw a section of loose connective tissue with the microscope at medium magnification. Label both types of extracellular fibers and three types of cells.

    b. Adipose tissue

    Scan the slide of adipose tissue at low and medium magnifications. How are the cells arranged? Describe the distribution of cytoplasm in the cells.

    _____

    _____

    At high magnification observe individual adipocytes. Draw several cells in the space provided on the next page.

    c. Dense connective tissue (regular)

    Examine dense connective tissue at low and medium magnifications. Describe the composition of the matrix. How are the fibroblasts arranged in this tissue?

    _____

    _____

    Draw and label a section of dense connective tissue in the space below.

    d. Reticular tissue

    Observe reticular tissue at low, medium, and high magnifications. Draw a section of the tissue and label the reticular fibers and reticulocytes.

B. **Fluid Connective Tissues**

Fluid connective tissues include **blood** and **lymph** tissues. These tissues have a liquid matrix and circulate in blood vessels or lymphatic vessels. Blood is composed of cells collectively called the formed elements, which are supported in a fluid matrix called blood plasma. Protein fibers are usually absent from the fluid matrix of these circulating tissues. Blood, however, produces a fibrous net during the coagulation (clotting) process and has dissolved proteins to regulate viscosity. The formed elements are grouped into three general categories: erythrocytes, leukocytes, and platelets. Erythrocytes are red blood cells that function in the transport of blood gases. Leukocytes, or white blood cells, are the cells of the immune system and protect the body from infection. Upon injury to a blood vessel, platelets become sticky and form a plug to reduce bleeding. The most common cells of the lymphatic system are lymphocytes, white blood cells produced in lymphoid tissues. A detailed study of these tissues will be presented later in Exercises 36 and 40.

C. **Supportive Connective Tissues**

Supportive connective tissues have less diversity of cell types than connective tissue proper. **Cartilage** and **bone** supportive connective tissues contain a strong matrix of fibers capable of supporting body weight and stress. Cartilage is an avascular tissue lacking blood vessels. Materials diffuse through a gelatinous matrix to reach the carti-

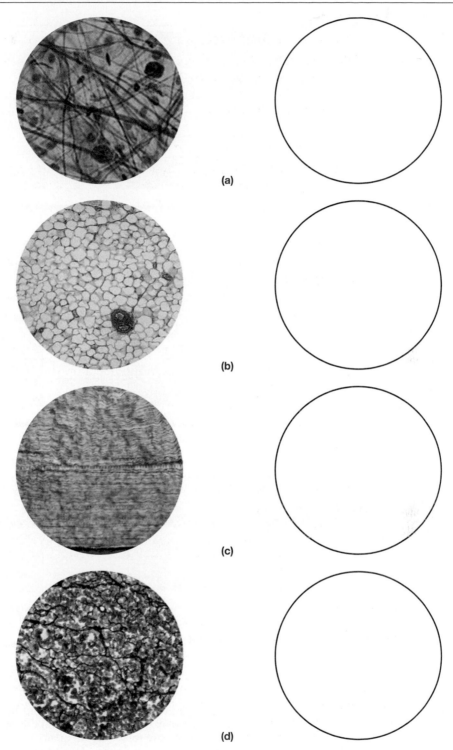

Figure 9.5
Connective Tissue Proper.

(a) Loose connective tissue (LM × 400) and sketch. **(b)** Adipose (LM × 100) and sketch. **(c)** Dense (regular) connective tissue (LM × 100) and sketch. **(d)** Reticular tissue (LM × 400) and sketch.

lage cells. Bone has a solid matrix composed of calcium phosphate and calcium carbonate salts. These salts crystallize on collagen fibers and form a hard material called hydroxyapatite. Small channels called **canaliculi** provide passageways through the solid matrix for exchange of materials between bone and blood tissues.

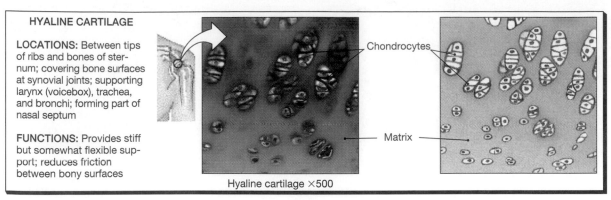

(a)

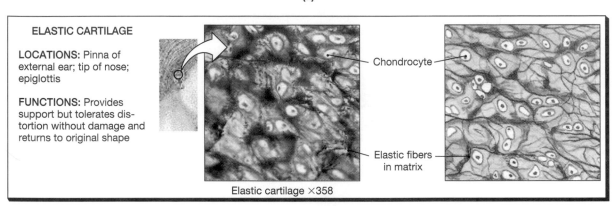

(b)

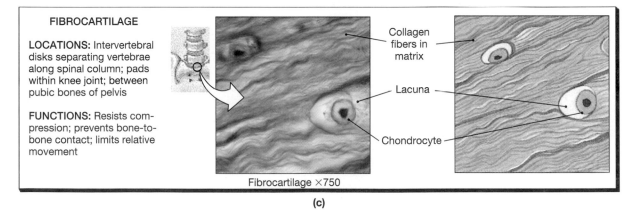

(c)

Figure 9.6
Types of Cartilage.

(a) Hyaline cartilage. Note the translucent matrix and the absence of prominent fibers. **(b)** Elastic cartilage. The closely packed elastic fibers are visible between the chondrocytes. **(c)** Fibrocartilage. The collagen fibers are extremely dense, and the chondrocytes are relatively far apart.

Supportive connective tissues are surrounded by a membrane that protects the tissue and supplies new tissue-producing cells. The **perichondrium (pe-rē-KON-drē-um)** surrounds cartilage and produces **chondroblasts (KON-drō-blasts)** which secrete the fibers and ground substance of the cartilage matrix. Eventually, chondroblasts become trapped in the matrix within small spaces called **lacunae (la-KOO-nē)** and lose the ability to produce additional matrix. These cells are then called **chondrocytes** and function in maintenance of the mature tissue. Examine Figure 9.6 and locate the features of cartilage.

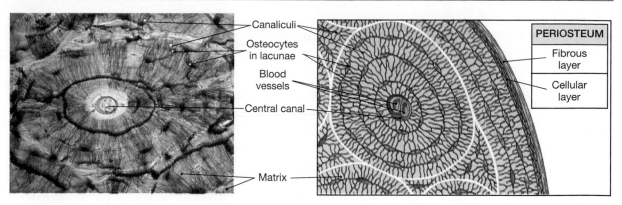

Figure 9.7
Bone.
The osteocytes in bone are usually organized in groups around a central space that contains blood vessels. For the photomicrograph, a sample of bone was ground thin enough to become transparent. Bone dust filled the lacunae and the central canal, making them appear dark.

As shown in Figure 9.7, bone tissue is surrounded by the **periosteum** which contains **osteoblasts (OS-tĕ-ō-blasts)** for bone growth and repair. Like chondroblasts, osteoblasts secrete the organic components of the matrix, become trapped in lacunae, and mature into **osteocytes**. Compact bone is characterized by the presence of columns of tissue called **osteons**. Each osteon surrounds a **central canal** containing blood vessels. Rings of matrix, called **lamellae**, surround the central canal. **Canaliculi (kan-a-LIK-ū-lē)** are small channels in the lamellae for diffusion of nutrients and wastes. Other bone cells, called **osteoclasts**, secrete small quantities of carbonic acid to dissolve portions of the bone matrix and release calcium ions into the blood for various chemical processes.

The supportive connective tissues are detailed below and are summarized in Table 9.1.

Hyaline (HĪ-uh-lin) cartilage is the most common cartilage in the body. It is located in most joints of the skeletal system, in the nasal septum, in the larynx, and in lower respiratory passageways. The gelatinous matrix provides flexible support and reduces friction between bones in a joint. Examine the micrograph of hyaline cartilage in Figure 9.6a. The tissue is distinguishable from other cartilages by the lack of stained fibers in the matrix. Hyaline cartilage contains extracellular elastic and collagen fibers, yet it does not pick up commonly used staining solutions. Along the outer perimeter of the cartilage is a perichondrium with chondroblasts. Chondrocytes occur deeper in the tissue and are surrounded by lacunae.

Elastic cartilage (Figure 9.6b), has many elastic fibers in the matrix and is therefore easily distinguished from hyaline cartilage. Chondrocytes are trapped in lacunae in the middle of the tissue, while the perichondrium and chondroblasts are found on the periphery. The elastic fibers permit considerable binding and twisting of the tissue. The pinna (flap) of the ear, the larynx, and the tip of the nose all contain elastic cartilage. Bend the pinna of one of your ears and notice how the elastic cartilage returns it to the original shape after distortion.

Fibrocartilage contains irregular collagen fibers which are visible in the matrix and chondrocytes which are stacked on one another much like coins (see Figure 9.6c). This cartilage is very strong and durable and functions to cushion joints and limit bone movement. The intervertebral discs of the spine, the pubic symphysis of the pelvis, and the pads in the knee joint are all exposed to stress forces of the skeleton. Fibrocartilage functions as a shock absorber in these areas.

Bone tissue is uniquely organized into osteons or columns of bone tissue which are distinct when viewed microscopically (see Figure 9.7). Osteocytes occur in the margins between concentric lamellae. Canaliculi branch through the solid matrix and connect with the central canal. Bone functions in support of the body, attachment of skeletal muscles, and protection of internal organs. Bone tissue will be studied in more detail in a later laboratory exercise.

LABORATORY ACTIVITY	MICROSCOPIC OBSERVATIONS OF SUPPORTIVE CONNECTIVE TISSUES

MATERIALS

compound microscope
prepared microscope slides:
   hyaline cartilage
   elastic cartilage
   fibrocartilage
   bone

PROCEDURES

1. Examine each tissue under the microscope and locate each structure listed below.
2. Draw each tissue in the spaces provided in Figure 9.8. Label each drawing.

   a. Hyaline cartilage
   Scan the slide of hyaline cartilage at low and medium magnifications. Describe the location of the perichondrium and chondroblasts.

   _____

   _____

   Examine the deeper middle region of the cartilage. How do these cartilage cells differ from the cells in the perichondrium?

   _____

   _____

   Draw a section of hyaline cartilage and label the perichondrium, chondroblasts, chondrocytes, and lacunae.

   b. Elastic cartilage
   Identify the perichondrium with many small chondroblasts and the chondrocytes trapped in lacunae deeper in the tissue. What is visible in the matrix of this cartilage?

   _____

   _____

   Draw a section of elastic cartilage and label the perichondrium, chondroblasts, chondrocytes, lacunae, and elastic fibers in the matrix.

   c. Fibrocartilage
   Fibrocartilage is composed of groups of chondrocytes stacked up in lacunae. Collagen fibers occur in the matrix, yet may be difficult to see microscopically. How does the arrangement of cells in this cartilage differ from the cell distribution in hyaline and elastic cartilages?

   _____

   _____

   Draw a section of fibrocartilage and label the chondrocytes.

   d. Bone (osseous) tissue
   Scan the bone slide and observe the pattern of osteons. Increase the magnification and observe the detailed structure of a single osteon. How is this tissue similar and dissimilar to cartilage tissues?

   _____

   _____

Draw several osteons and label the central canal, canaliculi, lamellae, and osteocytes.

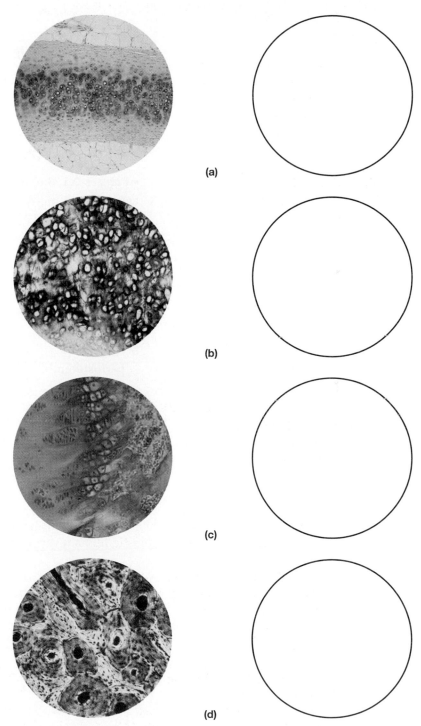

Figure 9.8
Supportive Connective Tissue.

(a) Hyaline cartilage (LM × 100) and sketch. **(b)** Elastic cartilage (LM × 100) and sketch.
(c) Fibrocartilage (LM × 100) and sketch. **(d)** Bone tissue (LM × 100) and sketch.

CONNECTIVE TISSUES CHECKLIST

This is a list of bold terms presented in Exercise 9. Use this list for review purposes after you have completed the laboratory activities.

INTRODUCTION TO CONNECTIVE TISSUE

- Figure 9.1 and Table 9.1

matrix
ground substance
collagen fiber
elastic fiber
reticular fiber
mesenchyme
connective tissue proper
fluid connective tissue
supportive connective tissue

CONNECTIVE TISSUE PROPER

- Figures 9.2–9.5 and Table 9.1

fibroblast
macrophage
mast cells
adipocytes
loose (areolar) connective tissue
adipose
dense connective tissue
reticular tissue

FLUID CONNECTIVE TISSUES

- Figure 9.1 and Table 9.1

blood
lymph

SUPPORTIVE CONNECTIVE TISSUES

- Figures 9.1, 9.6–9.8, Table 9.1

cartilage
   perichondrium
   chondroblast
   chondrocyte
   lacunae
hyaline cartilage
elastic cartilage
fibrocartilage
bone
   periosteum
   osteoblast
   osteocyte
   lacunae
   osteon
      central canal
      lamellae
      canaliculi
osteoclast

CONNECTIVE TISSUES

Laboratory Report Exercise 9

Complete the following sections after your microscopic observations of all connective tissues.

A. Identify the connective tissue described in each statement:

_____ Cells with nucleus and cytoplasm pushed against the cell membrane

_____ Cell types in the tissue include mast cells, fibroblasts, and macrophages

_____ Solid matrix, lamellae surrounding central canals

_____ Found in intervertebral disks, chondrocytes stacked inside lacunae

_____ Parallel bundles of collagen fibers with fibroblasts between fibers

_____ Gelatinous matrix, chondrocytes in lacunae, elastic fibers in matrix

_____ Numerous cells among network of reticular fibers

B. List the cell type(s) found in the following connective tissues:

hyaline cartilage _____

bone tissue _____

adipose tissue _____

loose connective tissue _____

elastic cartilage _____

dense connective tissue _____

reticular tissue _____

C. Short Answer Questions

1. How do connective tissues differ from epithelia?

2. What is the matrix composed of in loose connective tissue?

3. List the three major groups of connective tissues, and give an example of each.

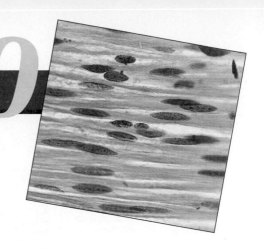

Muscle Tissue

OBJECTIVES

On completion of this exercise, you should be able to:

- List the three types of muscle tissue and a function of each.
- Describe the histological appearance of each type of muscle tissue.
- Identify each muscle tissue in microscope preparations.

WORD POWER

sarcolemma (sarco—flesh)
striated (stria—streaked)
myofilament (myo—muscle)
cardiac (cardi—heart)
visceral (viscera—internal organ)

MATERIALS

compound microscope
prepared slides:
  skeletal muscle
  cardiac muscle
  smooth muscle

INTRODUCTION

Muscle tissue specializes in contraction. Muscle cells shorten during contraction and produce force or tension that causes a variety of movements. Muscular valves control the movement of materials through the digestive system. Rapid skeletal muscle contractions generate body heat. The heart beats and pushes blood through the vascular system.

There are three types of muscle tissue, each named for its location in the body. **Skeletal** muscle is attached to bone and provides movement such as walking and moving the head. **Cardiac** muscle forms the walls of the heart. During the contraction phase of a heart beat, blood is pumped into blood vessels. The pressure generated by the heart's contraction forces blood to flow through the vascular system to supply cells with oxygen, nutrients, and other essential materials. **Smooth** or **visceral** muscle tissue is found inside hollow organs such as the stomach, intestines, blood vessels, and uterus. Smooth muscle controls the movement of materials through the digestive system, the diameter of blood vessels, and the contraction of the uterus during labor.

A. Skeletal Muscle

Skeletal muscle tissue is attached to bones of the skeletal system by tendons of strong, dense connective tissue. When skeletal muscle tissue contracts, it pulls on a tendon which, in turn, pulls and moves a bone. The functions of skeletal muscle tissue include movement for locomotion, facial expressions, and speech, maintenance of body posture and tone, and heat production during shivering.

Figure 10.1a highlights skeletal muscle tissue. The tissue is composed of large, long, cellular structures called **muscle fibers**. Each fiber is a composite of many fused myoblasts, creating a **multinucleated** fiber that is a syncytium, a cell with many nuclei. The nuclei are clustered under the **sarcolemma**, the muscle fiber's cell membrane.

SKELETAL MUSCLE TISSUE

Cells are long, cylindrical, striated, and multinucleate.

LOCATIONS: Combined with connective tissues and nervous tissue in skeletal muscles

FUNCTIONS: Moves or stabilizes the position of the skeleton; guards entrances and exits to the digestive, respiratory, and urinary tracts; generates heat; protects internal organs

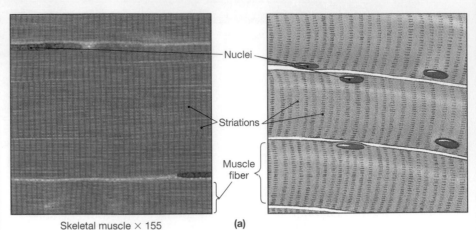

Skeletal muscle × 155

Nuclei

Striations

Muscle fiber

(a)

— gap junction

CARDIAC MUSCLE TISSUE

Cells are short, branched, and striated, usually with a single nucleus; cells are interconnected by intercalated discs.

LOCATION: Heart

FUNCTIONS: Circulates blood; maintains blood (hydrostatic) pressure

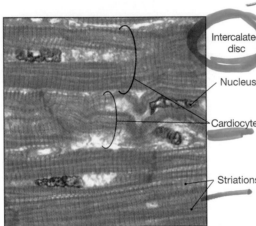

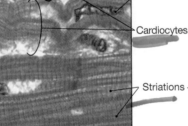

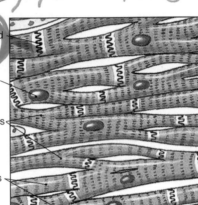

Intercalated disc

Nucleus

Cardiocytes

Striations

Cardiac muscle × 425

(b)

involuntary

SMOOTH MUSCLE TISSUE

Cells are short, spindle-shaped, and nonstriated, with a single, central nucleus.

LOCATIONS: Encircles blood vessels; in the walls of digestive, respiratory, urinary, and reproductive organs

FUNCTIONS: Moves food, urine, and reproductive tract secretions; controls diameters of respiratory passageways and blood vessels

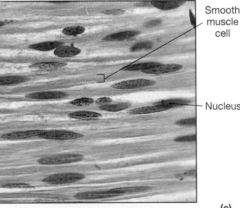

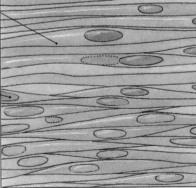

Smooth muscle cell

Nucleus

Smooth muscle × 330

(c)

involuntary   *1 nucleous NO striction*

**Figure 10.1
Muscle Tissue.**

(a) Skeletal muscle fibers are large, have multiple, peripherally located nuclei, and exhibit prominent striations (banding) and an unbranched arrangement. **(b)** Cardiac muscle cells differ from skeletal muscle fibers in three major ways: size (cardiac muscle cells are smaller), organization (cardiac muscle cells branch), and number and location of nuclei (a typical cardiac muscle cell has one centrally placed nucleus). Both skeletal and cardiac muscle cells contain actin and myosin filaments in an organized array that produces striations. **(c)** Smooth muscle cells are small and spindle-shaped, with a central nucleus. They do not branch or have striations.

Muscle fibers are **striated** with a distinct banded pattern due to the repeating organization of internal contractile proteins called **myofilaments**. Skeletal muscle tissue may be stimulated to contract consciously and is therefore under **voluntary** control.

The terms "voluntary" and "involuntary" muscle control are used more for convenience than for description. Skeletal muscle is said to be voluntary, yet you cannot stop your muscles from shivering when you are cold. Additionally, once you voluntarily start a muscle contraction, the brain assumes control of the muscle activity. The heart muscle is involuntary, yet some individuals can control their heart rate. Generally, the term "voluntary" is associated with skeletal muscle, and "involuntary" refers to cardiac and smooth muscle tissues.

LABORATORY ACTIVITY   MICROSCOPIC OBSERVATIONS OF SKELETAL MUSCLE TISSUE

MATERIALS

compound microscope
skeletal muscle slide (striated or voluntary muscle)

PROCEDURE

1. Use Figures 10.1 and 10.2 for reference and examine skeletal muscle tissue under the microscope at various magnifications.
2. Compare the appearance of the tissue in transverse and longitudinal sections.
3. Draw and label the microscopic structure of skeletal muscle tissue in the space provided in Figure 10.2.

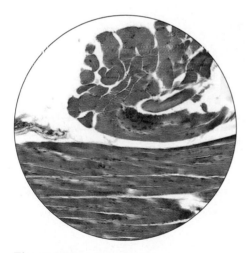
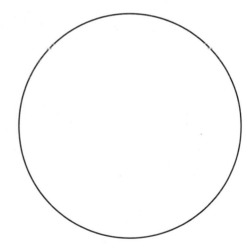

Figure 10.2
Skeletal Muscle Tissue.
(LM × 100)

B. Cardiac Muscle

Cardiac muscle tissue occurs only in the walls of the heart. Contraction of this muscle tissue pumps blood out of the heart and into blood vessels. The resulting blood pressure from the heart's contraction circulates blood through body tissues and back to the heart.

Compare the cardiac and skeletal muscle tissues in Figure 10.1a and b. Cardiac muscle tissue is striated like skeletal muscle tissue. Each cardiac muscle fiber, also called a **cardiocyte**, has a single nucleus (**uninucleated**) and branches. Cardiocytes are connected by **intercalated discs** which conduct the stimulus of contraction to neighboring cells. Unlike the control of skeletal muscles, that of cardiac muscle is **involuntary**. For example, when you exercise, autonomic nerves increase the heart rate to deliver more blood to the active tissues. When you relax or sleep, other autonomic nerves lower the heart rate.

LABORATORY ACTIVITY   MICROSCOPIC OBSERVATIONS OF CARDIAC MUSCLE

MATERIALS

    compound microscope

    cardiac muscle slide

PROCEDURE

1. Use Figures 10.1 and 10.3 for reference and examine the heart muscle under the microscope at various magnifications.

2. Compare the appearance of the tissue in transverse and longitudinal sections.

3. Examine individual cardiocytes at a high magnification. Do you see striations and branching? What structure connects adjacent cells?

4. Draw and label the microscopic structure of cardiac muscle tissue in the space provided.

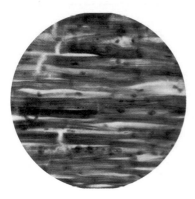 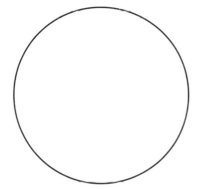

Figure 10.3
Cardiac Muscle Tissue.
(LM × 100)

C. Smooth (Visceral) Muscle

Smooth, or visceral, muscle tissue is found in the walls of internal hollow organs such as the stomach and intestines, urinary bladder, blood vessels, and uterus. Contraction of this tissue moves food through the digestive system, opens and closes valves, expels contents of bladders, and regulates the diameter of blood vessels. During labor, the smooth muscle tissue of the uterus contracts forcibly to deliver the fetus.

    Figures 10.1c and 10.4 show smooth muscle tissue. The muscle fibers are **nonstriated**, and lack the bands found in skeletal and cardiac muscle tissue. Each fiber is

uninucleated and spindle-shaped, thick in the middle, and tapered at the ends. The tissue usually occurs in double sheets of muscle organized at right angles. This arrangement enables the tissue to shorten structures and reduce their diameter. Smooth muscle is under **involuntary** control.

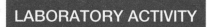

LABORATORY ACTIVITY   MICROSCOPIC OBSERVATIONS OF VISCERAL MUSCLE

MATERIALS

compound microscope
smooth muscle slide (visceral muscle, tissues may be teased apart)

PROCEDURE

1. Use Figures 10.1c and 10.4 for reference and examine smooth muscle tissue under the microscope at various magnifications.

2. Examine individual fibers at a high magnification. Where is the nucleus located in the cell? Do you see striations?

_____

3. Draw and label the microscopic structure of smooth muscle tissue in the space provided.

_____

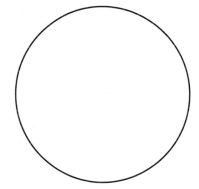

Figure 10.4
Smooth Muscle Tissue teased.
(LM × 100)

MUSCLE TISSUE CHECKLIST

This is a list of bold terms presented in Exercise 10. Use this list for review purposes after you have completed the laboratory activities.

SKELETAL MUSCLE TISSUE

• Figures 10.1a, 10.2
skeletal muscle
muscle fiber
multinucleated
sarcolemma
striated
myofilament
voluntary muscle

CARDIAC MUSCLE TISSUE

• Figures 10.1b, 10.3
cardiac muscle tissue
cardiocyte
uninucleated
intercalated disc
involuntary muscle

SMOOTH MUSCLE TISSUE

• Figures 10.1c, 10.4
smooth (visceral) muscle tissue
nonstriated
uninucleated
involuntary

MUSCLE TISSUE

Laboratory Report Exercise 10

A. Matching

Match the structure on the left with the correct muscle description on the right. Each choice may be used more than once.

1. _____ muscle fiber membrane   A. sarcolemma
2. _____ electrical conduction in heart B. intercalated disk
3. _____ striated, uninucleated   C. cardiac muscle
4. _____ muscle of the tip of tongue   D. skeletal muscle
5. _____ muscle tissue of an artery   E. smooth muscle
6. _____ voluntary muscle
7. _____ nonstriated
8. _____ involuntary, striated

B. Short-Answer Questions

1. How are skeletal and cardiac muscle tissues similar? How are they dissimilar?

2. Why are skeletal muscle fibers multinucleated?

3. Which types of muscle tissue are striated?

4. Where in the body does smooth muscle occur?

5. What is the function of intercalated disks in cardiac muscle?

6. Which muscle tissues are controlled involuntarily?

Nerve Tissue

Neuron cell Body

OBJECTIVES

On completion of this exercise, you should be able to:

- List the basic functions of neural tissue.
- Describe the two basic types of cells found in neural tissue.
- Identify a neuron and its basic structure under the microscope.
- Describe how neurons communicate with other cells across a synapse.
- List several functions of neuroglial cells.

WORD POWER

neuron (neuro—nerve)   axon (axo—axle)
glial (glia—glue)   synapse (synap—union)
soma (soma—body)
dendrite (dendro—tree)

MATERIALS

compound microscope
prepared slides:
   multipolar neuron
   astrocyte slide

INTRODUCTION

To maintain homeostasis, the body must constantly evaluate internal and external conditions and respond quickly and appropriately to environmental changes. The nervous system processes sensory information from sensory organs and responds with motor instructions to the body's effectors, the muscles and glands. Cells responsible for retrieving, interpreting, and sending the electrical signals of the nervous system are called **neurons**. There are other cells in the nervous system, including a diverse group called **neuroglia**. Neurons and neuroglia are collectively referred to as neural tissue.

Neurons are excitable; that is, they can respond to environmental changes by processing stimuli into electrical impulses called **action potentials**. These impulses may be conducted to adjacent neurons or to muscle and glandular tissues for homeostatic adjustments.

No messages

CNS – central Nervi. System

A. Neuron Structure

A typical neuron has distinct cellular regions. Examine Figure 11.1 and locate the **soma** surrounding the **nucleus**. This area also contains most of the organelles of the cell. Many fine extensions called **dendrites** conduct impulses toward the soma. The signal is then conducted into a single **axon** which carries information away from the soma, toward another neuron or to a muscle or gland. At the end of the axon is an enlarged **synaptic knob** which contains membranous **synaptic vesicles**. These vesicles contain **neurotransmitter** molecules, a chemical messenger used to excite or inhibit other cells. The axon communicates with an adjacent neuron, muscle fiber, or gland across a specialized junction called the **synapse**. Notice in Figure 11.1 the three axons on the left synapsing with the neuron shown in full on the right. At the synapse, cells do not touch; they are separated by a small **synaptic cleft**. When an action potential reaches

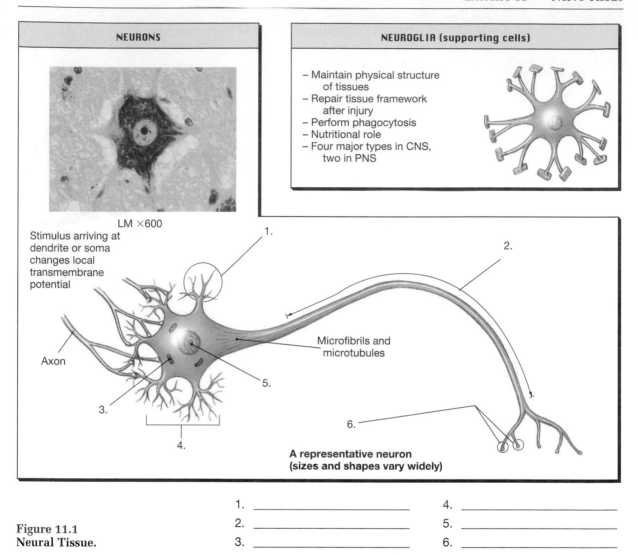

Figure 11.1
Neural Tissue.

1. _____   4. _____
2. _____   5. _____
3. _____   6. _____

the end of the axon, synaptic vesicles release neurotransmitter molecules which diffuse across the synaptic cleft and excite or inhibit the postsynaptic cell.

STUDY TIP

On most neuron slides it is difficult to distinguish the axon from the dendrites. Locate a neuron that is isolated from the others and examine the soma for a large extension. This is most likely the axon.

LABORATORY ACTIVITY MICROSCOPIC OBSERVATIONS OF THE NEURON

MATERIALS

compound microscope
prepared microscope slide of multipolar neurons

PROCEDURE

1. Start at low magnification and select a neuron to examine more closely. Center the neuron in the field and increase the magnification. Adjust the light setting of the microscope if necessary.

2. Identify the following structures of a neuron: soma, nucleus, dendrites (thin extensions), axon (thicker extension).

3. Draw and label several neurons in the space provided.

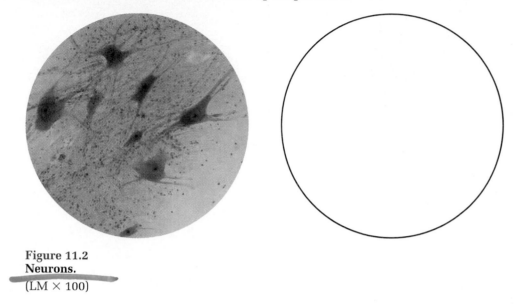

**Figure 11.2
Neurons.**
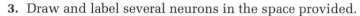
(LM × 100)

B. Neuroglial Cells

Neuroglial cells, or simply glial cells, are supportive cells of the nervous system. There are six types of glial cells, each with a specific function. **Astrocytes** attach blood vessels to neurons or anchor neurons in place. Phagocytic glial cells are responsible for housekeeping chores in the nervous system. Other glial cells protect neurons by wrapping around them to isolate them from chemicals present in the interstitial fluid. Schwann cells wrap around peripheral neurons to increase the speed at which they transmit action potentials. Repair of peripheral nerve tissue, nerves outside the brain and spinal cord, is made possible by special neuroglial cells that build a "repair tube" to reconnect the severed axons. In this exercise you will examine the most common glial cell in the nervous system, the astrocyte, shown in Figure 11.1.

LABORATORY ACTIVITY   MICROSCOPIC OBSERVATIONS OF NEUROGLIA

MATERIALS

compound microscope

prepared microscope slide of astrocytes

PROCEDURE

1. Adjust the magnification of the microscope to low power. Set the slide on the stage and slowly turn the coarse focus knob until you can clearly see the specimen. Now use the fine focus adjustment as you examine the tissue.

2. Scan the slide and locate a star-shaped astrocyte. Center the cell in your field of view and then increase the magnification for a more detailed observation. Notice the numerous feet extending from the cell.

...QUICKCHECK

NERVE TISSUE CHECKLIST

This is a list of bold terms presented in Exercise 11. Use this list for review purposes after you have completed the laboratory activities.

NERVE TISSUE

- Figure 11.1

neuroglia
  astrocyte
neuron
  soma
  nucleus
  dendrite
  axon
  synaptic knob
    synaptic vesicle
      neurotransmitter
synapse
synaptic cleft
action potential

NERVE TISSUE

Laboratory Report Exercise 11

A. Matching

Match each cell structure listed on the left with the correct description on the right.

1. _____ soma   A. membranous organelle containing neurotransmitters
2. _____ synaptic vesicles   B. chemical message released at synapse
3. _____ dendrite   C. cell body surrounding the nucleus
4. _____ neurotransmitter   D. space between two communicating neurons
5. _____ axon   E. propagates action potentials
6. _____ synaptic cleft   F. receiving branch of neuron

B. Short-Answer Questions

1. What are the basic functions of neural tissue?

2. How do neurons communicate with other cells?

3. Which part of a neuron conducts an impulse toward the soma?

4. In which direction does an action potential travel in an axon?

5. Arrange in order the following structures that an impulse would pass through, starting at the dendrite:

   synapse, soma, synaptic knob, axon, dendrite

121

6. List three functions of neuroglial cells.

C. **Drawing**

   1. Draw and label a multipolar neuron in the space provided.

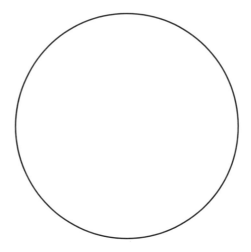

   2. Draw an astrocyte in the space provided.

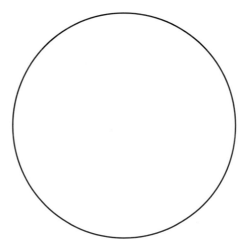

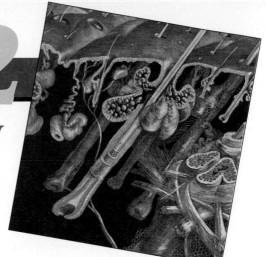

EXERCISE 12

The Integumentary System

OBJECTIVES

On completion of this exercise, you should be able to:

- Identify the two main layers of the skin.
- Identify the layers of the epidermis.
- Distinguish between the papillary and reticular layers of the dermis.
- Identify a hair follicle, the parts of a hair, and an arrector pili muscle.
- Distinguish between sebaceous and sudoriferous glands.
- Describe three sensory organs of the integument.

WORD POWER

epidermis (epi—over, upon)
stratum (strat—layer)
melanin (melan—black)
spinosum (spin—thorn)
granulosum (gran—grain)
lucidum (luci—clear)
corneum (corn—horn)

dermis (derm—the skin)
papillary (papill—nipple)
sebaceous (seb—grease)
apocrine (apo—off, from)
eccrine (ec—out, away)
pili (pil—hair)

MATERIALS

skin models
skin charts
compound microscope
prepared microscope slides:
  scalp cross section
  palm or sole cross section

INTRODUCTION

The integumentary system is the most visible organ system of the human body. Most individuals spend more time grooming their skin and skin derivatives such as hair and nails than exercising their muscular system. Our impressions of others are sometimes influenced by their skin. If their skin is wrinkled, we may think of them as old. Tanned skin is associated with health even though the sun may damage the skin. Skin seals your body in a protective barrier that is flexible yet resistant to abrasion and evaporative water loss. We contact the external environment with the skin. Caressing a baby's head, feeling the texture of granite, and testing the temperature of your bath water all involve sensory organs in the skin. Sweat glands in the skin cool the body to regulate body temperature. Vitamin D, an essential compound in calcium and phosphorus balance, is synthesized by the skin on exposure to sunlight. In this exercise, you will study the structure of the skin.

The skin is organized into two tissue layers, a superficial layer of epithelium, called the **epidermis**, and a deeper layer of connective tissue, the **dermis**. Locate these layers in Figure 12.1. Isolating the dermis from underlying structures is a fatty layer of connective tissue, the **hypodermis** or **subcutaneous** layer.

A. Epidermis

The epidermis consists of stratified squamous epithelium organized into many distinct cell bands or strata as shown in Figure 12.2. Thick-skinned areas such as the palms of

123

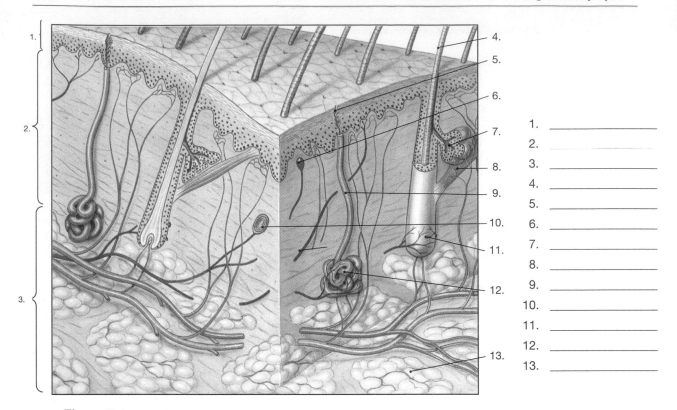

1. _____
2. _____
3. _____
4. _____
5. _____
6. _____
7. _____
8. _____
9. _____
10. _____
11. _____
12. _____
13. _____

Figure 12.1
Components of the Integumentary System.

Relationships among the major components of the integumentary system (with the exception of nails, shown in Figure 12.6).

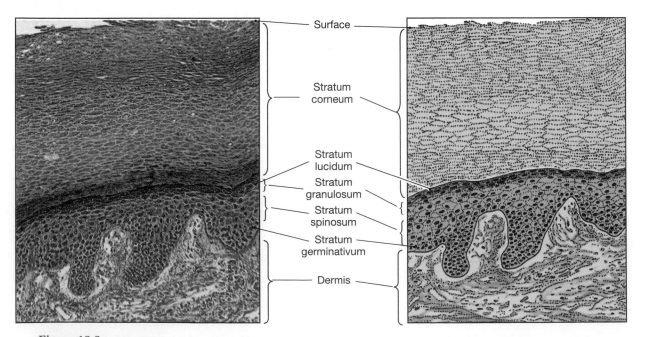

Surface

Stratum corneum

Stratum lucidum

Stratum granulosum

Stratum spinosum

Stratum germinativum

Dermis

Figure 12.2
Layers of the Epidermis.

A light micrograph through a portion of the epidermis, showing the major stratified layers of epidermal cells. (LM ×90)

the hands and soles of the feet have five layers, and thin-skinned areas have four. Cells are produced in the basal region of the epidermis and pushed upward toward the surface of the skin. During this migration, they synthesize and accumulate the protein keratin which reduces water loss across their cell membrane. The uppermost layer of the epidermis consists of dead, dry, scalelike cells.

The **stratum germinativum (STRA-tum jer-mi-na-TĒ-vum)**, or **stratum basale**, is a single cell layer between the base of the epidermis and the upper dermis. Locate this layer in Figure 12.2. The cells of this stratum are in a constant state of mitosis, replacing cells that have rubbed off the epidermal surface. As new cells are produced, they push previously formed cells toward the surface. It takes from 15 to 30 days for a cell to migrate from the stratum basale to the top of the stratum corneum.

Many different cell types populate the stratum basale. **Melanocytes** produce a pigment called **melanin (ME-la-nin)** which protects deeper cells from the harmful effects of ultraviolet (UV) radiation from the sun. Prolonged exposure to UV light causes an increase in melanin synthesis, resulting in a darkening or tanning of the skin. Other epidermal cells produce a yellow-orange pigment called **carotene**. This pigment is common in light-skinned individuals.

Superficial to the stratum basale is the **stratum spinosum**, which consists of five to seven rows of cells interconnected by thickened cell membranes called **desmosomes**. During the slide preparation process, cells in this layer often shrink, whereas the desmosome bridges between cells remain intact. This results in cells with a spiny outline, hence, the name "spinosum."

Superior to the stratum spinosum is a layer of darker cells that make up the **stratum granulosum**. As cells from the stratum basale are pushed upward, they synthesize the protein **keratohyalin (ker-a-tō-HĪ-a-lin)** which increases durability and reduces water loss from the skin. Keratohyalin granules stain dark and give this layer its color.

The uppermost layer of the epidermis is the **stratum corneum**, which has many layers of flattened, dead cells. As cells from the stratum granulosum move upward, keratohyalin granules are converted to the fibrous protein **keratin** which kills the cells and waterproofs them with a hard covering. If the skin is located in an area of high abrasion, like the thick-skinned palms and soles, a thin transparent layer of cells called the **stratum lucidum** lies between the stratum granulosum and the stratum corneum. In the stratum lucidum, keratohyalin is converted to a intermediate protein called **eleidin (a-lē-a-din)** which eventually becomes keratin. Cells of the stratum corneum are constantly being shed or worn off and replaced by new cells produced in and pushed up from deeper layers.

B. Dermis

Below the stratum germinativum is the **dermis**, a band of connective tissues that anchors the epidermis in place (see Figure 12.1). The dermis is divided into two regions, the **papillary** and **reticular** layers. Although no distinct boundary occurs between these layers, the upper fifth of the dermis is designated the papillary region. This layer consists of loose connective tissue with numerous collagen and elastic fibers. The tissue projects folds, called dermal papillae, into the epidermis, which result in the swirls of fingerprints. These patterns on the skin are genetically based and do not change during an individual's life. Within the projections of dermal tissue are small sensory receptors for light touch, **Meissner's corpuscles**.

Deep to the papillary layer is the reticular region, distinguished by widely scattered cells and dense irregular connective tissue with interlaced collagen fibers. By comparison, collagen fibers in the papillary layer are less distinct. Large blood vessels, sweat glands, and adipose tissue are less visible in the reticular region. Sensory receptors in this layer, called **Pacinian corpuscles**, detect deep pressure. Attaching the dermis to underlying structures is the **subcutaneous layer** or **hypodermis**, which is profuse with adipose tissue and loose connective tissue.

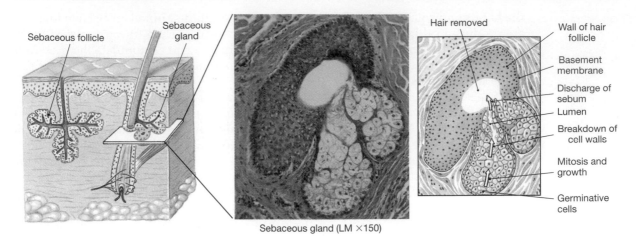

Sebaceous gland (LM ×150)

Figure 12.3
Sebaceous Glands and Follicles.
The structure of sebaceous glands and sebaceous follicles in the skin.

C. Accessory Structures of the Skin

During embryonic development, the epidermis produces accessory structures called *epidermal derivatives*, which include oil and sweat glands, hair, and nails. These structures are exposed on the surface of the skin and project deep into the dermis.

Sebaceous glands

Sebaceous (se-BĀ-shus) glands secret an oily substance, called **sebum**, which coats hair shafts and the epidermal surface. Observe in Figure 12.3 the infolding of epithelium to form the hair follicle and sebaceous gland. Notice how the glandular cells empty onto the follicle. The gland is a **holocrine** exocrine gland. The cells fill with sebum and burst to transport it into a duct. In sebaceous glands, oil-filled sebaceous cells are pushed toward the lumen of the duct. Once on the free surface of the lumen, their cell membranes rupture and sebum is discharged into the duct. Sebum coats the hair shaft to reduce brittleness and prevents excessive drying of the scalp.

Sebaceous follicles secrete sebum onto the surface of the skin. These follicles are not associated with hair and are distributed on the face, most of the trunk, and the male reproductive organs. Secretions from sebaceous follicles lubricate the skin and provide limited antibacterial action.

Most teenagers have dealt with skin blemishes called acne. When sebaceous ducts become blocked, sebum becomes trapped in the ducts. As the sebum accumulates, it causes the skin to raise, resulting in a pimple. The white head of the pimple shows the duct is infected with bacteria that are feeding on the sebum.

Sudoriferous (sweat) glands

Sudoriferous (sū-dor-IF-er-us) glands are scattered throughout the dermis of most of the skin. Sweat glands are exocrine glands that secrete their fluid into sweat ducts leading directly to the surface of the skin or into ducts that empty into hair follicles, as shown in Figure 12.4.

Two types of sweat glands occur in the dermis, apocrine and merocrine glands. **Apocrine glands** are found in the groin, nipples, and axillae. Apocrine sweat glands secrete a thick **sweat** into ducts associated with hair follicles. Bacteria on the hair metabolize the sweat and produce the characteristic body odor of, for example, axillary sweat. When body temperature increases, **merocrine (MER-ō-krin) glands** secrete sweat containing electrolytes, proteins, urea, and other compounds onto the body surface. The sweat absorbs body heat and evaporates from the skin, cooling the body. It also con-

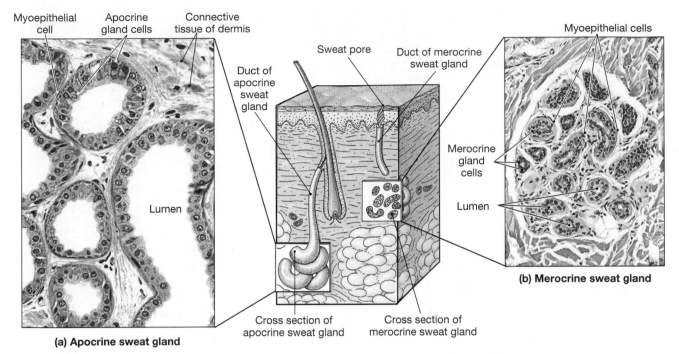

Figure 12.4
Sweat Glands.

(a) Apocrine sweat glands, located in the axillae, groin, and nipples, produce a thick, odorous fluid by apocrine secretion. (LM × 369) **(b)** Merocrine sweat glands produce a watery fluid by merocrine secretion. (LM × 194)

tributes to body odor because of the presence of urea and other wastes. Merocrine glands are not associated with hair follicles and are distributed throughout most of the skin.

Hair

Most of the skin is covered with hair. Only the lips, the nipples, portions of the external genitalia, and the palmar and plantar surfaces of the fingers and toes are without hair. Three major types of hair occur across the skin. **Terminal hairs** are the thick and heavy hairs on the scalp, eyebrows, and eyelashes. **Vellus** hairs are lightly pigmented and distributed over much of the skin as fine "peach fuzz." **Intermediate** hairs are the heavier hairs on the arms and legs. Hair generally serves a protective function. It cushions the scalp and prevents foreign objects from entering the eyes, ears, and nose. Hair also serves as a sensory receptor. Wrapped around the base of each hair is a **root hair plexus**, a sensory neuron sensitive to movement of the hair.

Examine Figure 12.5 and note that each hair is embedded in a **hair follicle**, a pocket of epidermis extending deep into the dermis. Deep in the hair follicle is the **hair root**. At the root is a **papilla** containing nerves and blood vessels and the living, proliferative part of the hair, the **matrix**. Cells in the matrix undergo mitotic divisions that cause elongation and growth of the hair. Above the matrix, keratinization of the hair cells causes them to harden and die. The resulting **hair shaft** contains an outer **cortex** and an inner **medulla**. Figure 12.5 also depicts an **arrector pili (a-REK-tor PI-lē) muscle** attached to each hair follicle. When furry animals such as dogs and cats are cold, these muscles contract to raise the hair and trap a layer of warm air next to the skin. In humans, the muscle has no known thermoregulatory use; humans do not have enough hair to gain an insulation benefit. These muscles contract when we are cold and produce "gooseflesh." The arrector pili muscles also respond to emotional stimuli such as beautiful music, touch, and chalk scratching on a chalk board. The arrector pili muscles in animals also have an emotional response. Animals raise their fur to look bigger when they feel threatened.

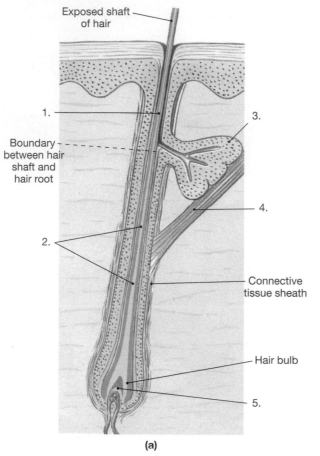

Exposed shaft
of hair

1.

Boundary
between hair
shaft and
hair root

2.

3.

4.

Connective
tissue sheath

Hair bulb

5.

(a)

1. _____

2. _____

3. _____

4. _____

5. _____

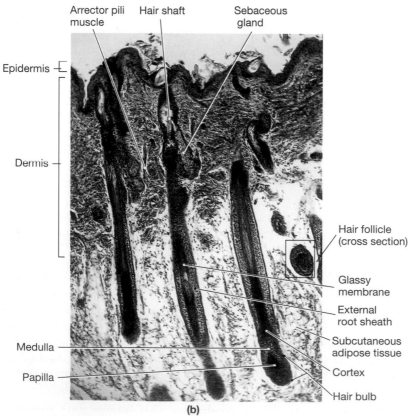

Arrector pili   Hair shaft   Sebaceous
muscle   gland

Epidermis

Dermis

Hair follicle
(cross section)

Glassy
membrane

External
root sheath

Subcutaneous
adipose tissue

Medulla

Cortex

Papilla

Hair bulb

(b)

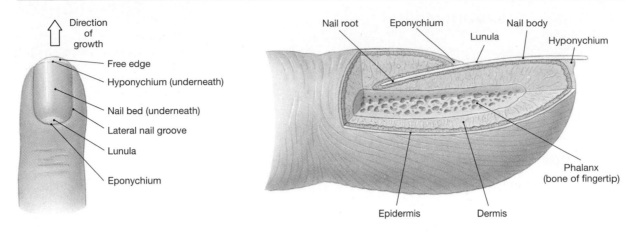

Figure 12.6
Structure of a Nail.
These drawings illustrate the prominent features of a typical fingernail as viewed from the surface and in section.

Nails

To protect the tips of the fingers and toes, hard **nails** cover their dorsal surface. Nails consist of tightly packed keratinized cells. Figure 12.6 illustrates the parts of a typical nail. The visible elongated body of the nail, called the **nail body**, protects the underlying **nail bed** of the skin. Blood vessels underneath the nail body give the nail its pinkish coloration. The **free edge** of the nail body extends past the end of the digit. The **nail root** is at the base of the nail where new growth occurs. The **lunula (LOO-nū-la)** is a whitish portion of the proximal nail body where blood vessels do not show through. The cuticle around the nail is called the **eponychium (ep-ō-NIK-ē-um)**. It is composed of a band of epidermis which seals the **nail groove** to the epidermis. Under the free edge of the nail is the **hyponychium (hī-pō-NIK-ē-um)**, a thicker region of the epidermis.

LABORATORY ACTIVITY   MICROSCOPIC OBSERVATION OF THE SKIN

MATERIALS

> compound microscope
> prepared slide of the scalp (cross section)

PROCEDURE

1. Label Figures 12.1, 12.3, 12.4, and 12.5.
2. Scan the scalp slide at low magnification and identify the epidermis, dermis, and hypodermis.
3. Increase the magnification and examine the epidermis. Locate the epidermal layers beginning with the deepest layer, the stratum germinativum.
4. Study the dermis and identify the papillary and reticular regions. In the papillary region, look for Meissner's corpuscles positioned where the dermis folds upward along the epidermis.
5. Locate a hair follicle. Identify the hair shaft, cortex, and medulla. Identify a sebaceous gland associated with the hair follicle.
6. Scan the dermis of the slide for a sudoriferous gland. Trace the duct from the gland to the surface of the skin.
7. Study the models, charts, and other material available in the lab.

INTEGUMENTARY SYSTEM CHECKLIST

This is a list of bold terms presented in Exercise 12. Use this list for review purposes after you have completed the laboratory activities.

EPIDERMIS

• Figures 12.1 and 12.2
epidermis
stratum germinativum (stratum basale)
 melanocytes
 melanin
 carotene
stratum spinosum
 desmosome
stratum granulosum
 keratohyalin
stratum lucidum
 eleidin
stratum corneum
 keratin

DERMIS

• Figure 12.1
dermis
papillary layer
reticular layer
Meissner's corpuscles
Pacinian corpuscles
subcutaneous layer (hypodermis)

SEBACEOUS AND SUDORIFEROUS GLANDS

• Figures 12.3 and 12.4
holocrine gland
sebaceous glands
 sebum
 sebaceous follicle

sudoriferous gland
 sweat
apocrine gland
merocrine gland

HAIR

• Figure 12.5
hair follicle
hair
root hair plexus
 hair root
 papilla
 matrix
 hair shaft
 cortex
 medulla
arrector pili muscle

NAILS

• Figure 12.6
 nail
 nail bed
 nail body
 free edge
 nail root
 lunula
 eponychium
 nail groove
 hyponychium

THE INTEGUMENTARY SYSTEM
Laboratory Report Exercise 12

A. Matching
    Match each skin structure listed on the left with the correct description on the right.

1. _____ sebaceous gland	**A.** layer in thickened areas of epidermis
2. _____ apocrine sweat gland	**B.** deep to the dermis, contains adipose
3. _____ keratin	**C.** sweat gland associated with hair follicle
4. _____ arrector pili	**D.** produces new epidermal cells
5. _____ stratum corneum	**E.** deep layer of the dermis
6. _____ papillary layer	**F.** protein that reduces water loss
7. _____ stratum germinativum	**G.** functions in thermoregulation
8. _____ reticular layer	**H.** surface layer of epidermis
9. _____ subcutaneous layer	**I.** muscle attached to hair follicle
10. _____ stratum lucidum	**J.** folded layer of dermis next to epidermis
11. _____ eccrine sweat gland	**K.** produces sebum

B. Short-Answer Questions
    1. Describe the layers of epidermis in an area where the skin is thick.

    2. Why is the epidermis keratinized?

    3. How does the skin tan when exposed to sunlight?

    4. List the types of sweat glands associated with the skin.

    5. How are cells replaced in the epidermis?

Body Membranes

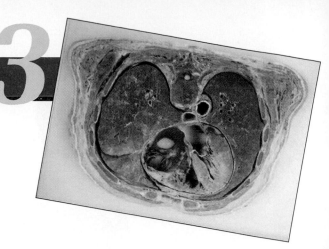

OBJECTIVES

On completion of this exercise, you should be able to:

- List and provide examples of the four types of body membranes.
- Discuss the tissue components of each body membrane.
- Describe the histological organization of each membrane.

WORD POWER

lamina propria (lamin—a sheet layer)
serous (sero—serum)
mesothelium (meso—middle)
pericardium (peri—around)

pleura (pleur—rib, side)
peritoneum (peritone—intestinal membrane)
cutaneous (cut—skin)
keratinized (kera—horn)

MATERIALS

compound microscope
prepared slides:
   ileum
   serous membrane
   skin section
fresh sectioned cow joint

INTRODUCTION

The term "membrane" is used to refer to a variety of anatomical structures. Cells possess a plasma membrane that defines their boundaries and regulates the passage of materials between the cell and the surrounding environment. In this exercise "membranes" refers to sheets of tissue that wrap around organs. Some body membranes cover and isolate structures from the surrounding anatomy; other membranes act as a barrier to prevent an infection from spreading from one organ to another. Most body membranes produce fluid and keep the cells on the exposed or free surface of the membrane moist. Absorption may occur across these moist membranes.

Body membranes are composed of epithelia and connective tissues. Epithelium occurs on the exposed surface of the membrane and is supported by underlying connective tissue. It is usually kept moist by secretions from the membrane layers. There are four major types of membranes, classified by their location: **mucous membranes** occur where an opening is exposed to the external environment, **serous membranes** wrap around organs in the ventral body cavity, the **cutaneous membrane** is the skin, and **synovial (si-NŌ-vē-ul) membranes** line the cavities of movable joints.

A. Mucous Membranes

Mucous membranes line organs and are exposed to the external environment. The digestive, respiratory, urinary, and reproductive systems are protected with a layer called the mucosa, the mucous membrane layer lining the wall of a tube or duct (see Figure 13.2). The **lining epithelium** of mucous membranes may be simple columnar, stratified squamous, or transitional. It is always attached to the **lamina propria (PRO-prē-uh)**, a sheet of loose connective tissue. **Mucus** is secreted by **goblet cells** in

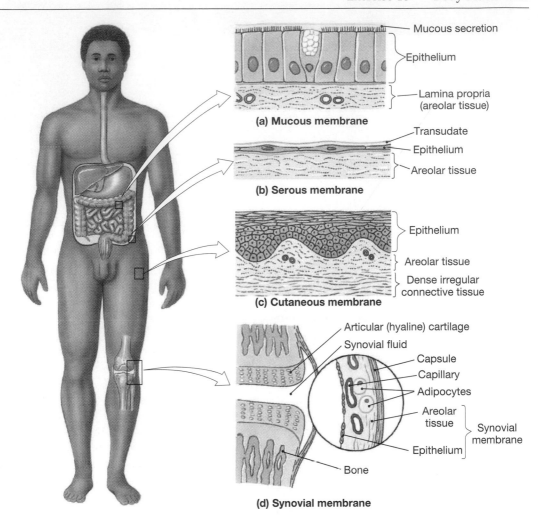

Figure 13.1
Membranes.

(a) Mucous membranes are coated with the secretions of mucous glands. These membranes line the digestive, respiratory, urinary, and reproductive tracts. **(b)** Serous membranes line the ventral body cavities (the peritoneal, pleural, and pericardial cavities). **(c)** The cutaneous membrane, or skin, covers the outer surface of the body. **(d)** Synovial membranes line joint cavities and produce the fluid within the joint.

the epithelium or by glands to protect the epithelium. Mucus is a viscous fluid containing a glycoprotein called mucin, salts, water, epithelial cells, and white blood cells. **Mucin** gives mucus its slippery and sticky attributes. Examine Figure 13.1a and locate the mucus coating the surface epithelium.

The mucous membrane of the digestive system has regional specializations. Where digestive contents are liquid, such as in the stomach and intestines, the epithelium is **simple columnar** with goblet cells. This mucous lining functions in absorption and secretion. The mouth, pharynx, and rectum process solid food or waste materials that are abrasive to the epithelial lining. **Stratified squamous epithelium** protects underlying structures in these areas.

Most mucous membranes are constantly replacing old cells. As materials move through a lumen, the exposed epithelial cells are scraped off the mucous membrane.

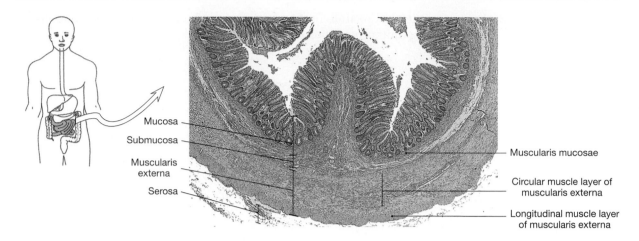

Figure 13.2
Mucous Membrane of Digestive Tract.
The mucosa of the digestive tract is a mucous membrane.

Simple columnar cells in the small intestine live for approximately 48 hr before they are replaced. The old cells are shed into the intestinal lumen and added to the feces.

The mucosa of the respiratory system is similar to that of the stomach and intestines. The nasal cavity, trachea (windpipe), and large bronchi of the lungs are lined with epithelium populated with goblet cells that secrete mucus to trap particulate matter in inhaled air. The epithelium is a **ciliated columnar** type that transports the dust-trapping mucus upward for removal from the respiratory system. The respiratory membrane of the lung is the moist epithelial surface where inhaled air exchanges gases with the blood. Small pouches called alveoli increase the respiratory membrane area for the diffusion of gases. Portions of the male and female reproductive systems are also lined with a ciliated mucous membrane to help in the transport of sperm or eggs.

The mucosa of the urinary bladder is a **transitional epithelium** that enables the bladder wall to stretch and recoil as the bladder fills and empties. Urine keeps the epithelium moist and prevents dehydration.

LABORATORY ACTIVITY   OBSERVATION OF A MUCOUS MEMBRANE

MATERIALS

compound microscope
prepared ileum slide (transverse section)

PROCEDURES

1. Examine a slide of the ileum, the last segment of the small intestine. Move the slide to the center of the tissue section and locate the lumen, the open space of the intestine. Lining the folds of the wall is the mucous membrane.

2. Exposed on the surface is a layer of simple columnar epithelium. Large goblet cells are scattered among the columnar cells and secrete mucus to protect the lining epithelium. Deep to the epithelium is a layer of connective tissue, the lamina propria, which serves to anchor the epithelium in place. The lamina propria is mostly loose connective tissue with numerous blood and lymphatic vessels.

3. Draw a portion of the slide in the space provided. Label the lumen, mucous membrane, lining epithelium, and lamina propria.

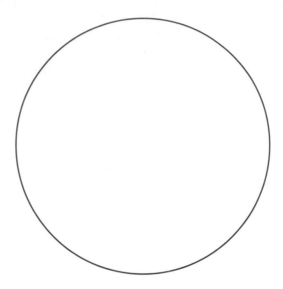

B. Serous Membranes

Serous membranes are double-layered membranes that cover internal organs and line the ventral body cavity to reduce friction between organs. The **visceral** layer of the membrane is in direct contact with the particular organ, and the **parietal** layer lines the wall of the cavity. Between the two layers of the membrane is a minute space filled with **serous fluid**, a slippery lubricant. Serous fluids are different from the mucus secreted by mucous membranes. Serous fluids are thin, watery secretions, similar to blood plasma. A thin covering of specialized epithelium called **mesothelium** faces the small cavity between the layers. Interstitial fluid from the underlying connective tissue transudes or passes through the mesothelium to form the serous fluid.

Three serous membranes occur in the body: the **pericardium** of the heart, the **pleurae (ploo-rē)** of the lungs, and the **peritoneum (per-i-tō-NĒ-um)** of the abdominal organs. Each of these was discussed in Exercise 3. Figure 13.3 details the peritoneum.

LABORATORY ACTIVITY   OBSERVATION OF A SEROUS MEMBRANE

MATERIALS

compound microscope

prepared serous membrane slide (pericardium, pleura, or peritoneum)

PROCEDURES

1. Examine a slide of serous membrane. Identify the epithelial and connective tissue components.

2. Draw a portion of the slide in the space provided on the next page. Label the visceral layer, the serous cavity, and the mesothelium.

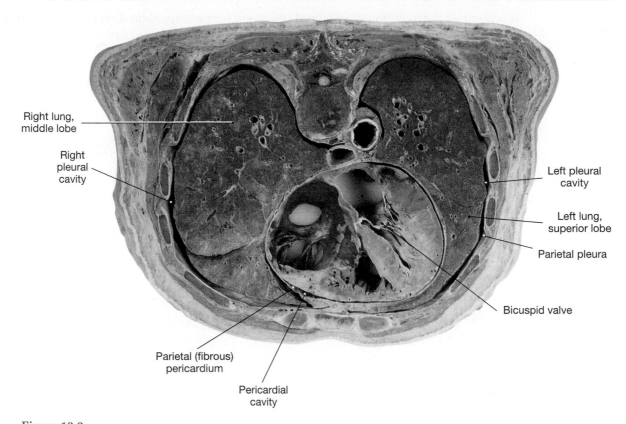

Figure 13.3
Serous Membranes of Thoracic Cavity.
The heart and lungs are encased in serous membranes.

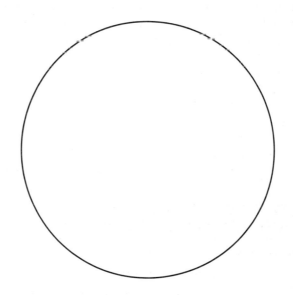

C. Cutaneous Membrane

The **cutaneous membrane** is the **skin**, the epidermis and dermis that covers the exterior surface of the body. The epidermis consists of **stratified squamous epithelium,** and the dermis is made up of a variety of connective tissues including loose connective and adipose tissues. Protection against abrasion and the entrance of microbes are major functions of the membrane. Unlike all mucous, serous, and synovial membranes which are kept moist, the cutaneous membrane is a dry membrane. A process called **keratinization** waterproofs the skin's surface. Epithelial cells synthesize the protein keratin, which kills the cells, leaving a dry protective layer of scaly cells on the surface of the skin. The cells are eventually shed and are constantly replaced. Exercise 12 examines the skin and the cutaneous membrane in greater detail.

LABORATORY ACTIVITY   OBSERVATION OF THE CUTANEOUS MEMBRANE

MATERIALS

compound microscope
prepared slide of the skin (transverse section)

PROCEDURES

1. Examine a slide of the skin and locate the cutaneous membrane. Distinguish between the epithelium of the epidermis and connective tissue of the dermis.
2. Draw a portion of the slide in the space provided. Label the epidermis, stratified squamous epithelium, and dermis.

D. Synovial Membrane

Freely movable joints of the skeletal system, such as the knee and elbow, are lined with a **synovial (sin-Ō-vē-ul) membrane**, see Figure 13.4. To allow free movement, the bones of the joint are separated by a small space, the **synovial cavity**. Within this cavity, **articular cartilage** lines the articular surfaces of the bones. Connective and epithelial tissues encapsulate the joint cavity and bones, forming the synovial membrane. Unlike other body membranes, synovial epithelium is not a continuous sheet of cells. Instead, epithelial cells are clustered, and gaps between the epithelia expose the underlying loose connective tissue. Fluids seep from the connective tissue and fill the synovial cavity. This **synovial fluid** is a clear, lubricating solution containing mucin, salts, and albumins.

Joints such as the shoulder have additional synovial membranes called **bursae** which cushion structures such as tendons and ligaments. Bursae also occur over bones where the skin is thin, such as the elbow, to reduce abrasion to the skin.

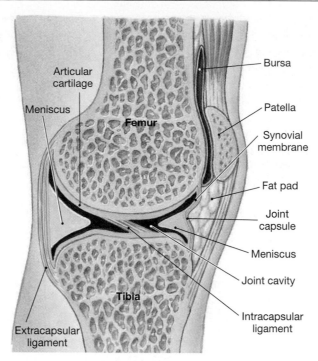

Figure 13.4
The Structure of a Synovial Joint.

A simplified sectional view of the knee joint. The synovial membrane secretes fluid into the joint cavity.

LABORATORY ACTIVITY   OBSERVATION OF A SYNOVIAL MEMBRANE

MATERIALS

   sectioned cow joint

PROCEDURES

   1. If possible, examine the gross anatomy of a fresh cow joint that has been sectioned along the longitudinal plane.
   2. Locate the articular cartilage at the ends of the articulating bones. Examine the joint closely for the synovial membrane and bursae.

BODY MEMBRANES CHECKLIST

This is a list of bold terms presented in Exercise 13. Use this list for review purposes after you have completed the laboratory activities.

MUCOUS MEMBRANES

• Figures 13.1 and 13.2
lining epithelium
lamina propria
goblet cells
  mucus
  mucin
simple columnar epithelium
stratified squamous epithelium
ciliated columnar epithelium
transitional epithelium

SEROUS MEMBRANES

• Figures 13.1 and 13.3
parietal layer
  serous fluid
visceral layer
mesothelium
pericardium
pleurae
peritoneum

CUTANEOUS MEMBRANES

• Figures 13.1 and 13.4
skin
  epidermis
  dermis
stratified squamous epithelium
keratin

SYNOVIAL MEMBRANES

• Figures 13.1 and 13.5
synovial membrane
  synovial cavity
  synovial fluid
articular cartilage
bursae

BODY MEMBRANES

Laboratory Report Exercise 13

A. **Short-Answer Questions**

   1. Where in your body do serous membranes occur?

   2. What type of tissue lines all body membranes?

   3. How are serous and synovial fluids produced?

   4. What is the function of goblet cells?

   5. What layers of the skin comprise your cutaneous membrane?

   6. A lamina propria occurs in what membrane?

B. **Matching**

Match the structure on the left with the correct description on the right.

1. _____ serous fluid
2. _____ lamina propria
3. _____ goblet cell
4. _____ visceral layer
5. _____ cilia
6. _____ parietal layer

A. deep serous membrane layer
B. produces mucus
C. superficial serous layer
D. propels mucus
E. functions as a lubricant
F. connective tissue of mucosa

14

Skeletal System Overview

OBJECTIVES

On completion of this exercise, you should be able to:

- List the components of the axial skeleton.
- Describe the bones of the appendicular skeleton.
- Describe the gross anatomy of a long bone.
- Describe the histological organization of compact and spongy bone.
- List the five shapes of bones and give an example of each type.
- Describe the bone markings visible on the skeleton.

WORD POWER

osteoblast (blast—immature)
medullary (medull—marrow)
metaphysis (meta—middle)
osteon (osteo—bone)
lamellae (lame—thin plate)
lacunae (lacun—space)
trabeculae (trab—beam)

MATERIALS

articulated skeleton
preserved long bone or fresh long bone
gloves
safety glasses
blunt probe
compound microscope
prepared slide:
  bone tissue (cross section)

INTRODUCTION

The skeletal system serves many functions. Bones support soft tissues of the body and protect vital internal organs. Calcium, lipid, and other materials are stored in the bones, and blood cells are manufactured in the red marrow of bones. Bones serve as levers for the muscular system to pull on and produce movement or maintain posture. In this exercise, you will study the gross structure of bone and the individual bones of the skeletal system.

A. Histological Organization of Bone

Two types of bone tissue are found in the skeleton: **compact bone** and **spongy bone**. Compact or dense bone seals the outer surface of bones and is found where stress arrives from one direction on the bone, as in the diaphysis of a long bone. Spongy bone is found in the epiphyses and the lining of the medullary cavity of long bones.

Compact bone is characterized by supportive columns called **osteons** or Haversian systems (see Figure 14.1). Each osteon consists of many rings of calcified matrix called **concentric lamellae (lah-MEL-lē)**. Between the lamellae are mature bone cells called **osteocytes** which are in small spaces in the matrix called **lacunae**.

Bone requires a substantial supply of nutrients and oxygen. Nerves and blood and lymphatic vessels pierce the periosteum and enter the bone in a **perforating canal**

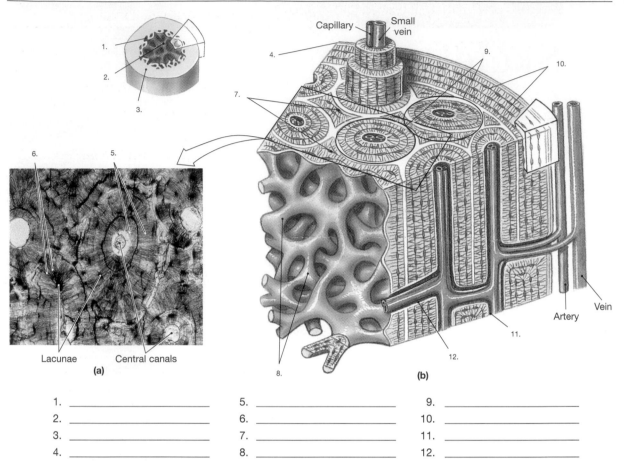

Lacunae   Central canals
(a)

(b)

1. _____	5. _____	9. _____
2. _____	6. _____	10. _____
3. _____	7. _____	11. _____
4. _____	8. _____	12. _____

Figure 14.1
Structure of a Typical Bone.

(a) A thin section through compact bone; in this procedure the intact matrix and central canals appear white, and the lacunae and canaliculi are shown in black. (LM ×272) **(b)** Diagrammatic view of the structure of a representative bone.

oriented perpendicular to the osteons. This canal interconnects with **central canals** positioned in the center of osteons. Radiating outward from a central canal are small diffusion channels called **canaliculi (kan-a-LIK-ū-lī)** which facilitate nutrient, gas, and waste exchange with the blood. Locate these structures in Figure 14.1a.

To maintain its strength and weight-bearing ability, bone tissue is continuously being remodeled. This remodeling process leaves distinct structural features in compact bone. Old osteons are partially removed, and the concentric rings of lamellae are fragmented, resulting in **interstitial lamellae** between intact osteons. Typically, the distal end of a long bone is extensively remodeled throughout life, whereas areas of the diaphysis may never be remodeled. Other lamellae occur underneath the periosteum and wrap around the entire bone. These **circumferential lamellae** are added during bone growth in diameter.

Spongy bone or cancellous tissue is found in the epiphyses and the medullary cavity of long bones and inside other types of bones (see Figure 14.1b). Unlike compact bone, spongy bone is not organized into osteons, rather it forms a lattice or meshwork of bony struts called **trabeculae (tre-BEK-ū-lē)**. Each trabecula is composed of layers of lamellae which are intersected with canaliculi. Filling the spaces between the trabeculae is **red marrow**, the tissue that produces most blood cells. Spongy bone is always sealed with a thin outer layer of compact bone.

LABORATORY ACTIVITY   HISTOLOGICAL STUDY OF BONE

MATERIALS

compound microscope

prepared slide of bone tissue (transverse section, cross section)

PROCEDURES

1. Label and review the structures in Figure 14.1.
2. Examine a bone model and locate each structure shown in Figure 14.1.
3. Obtain a prepared microscope slide of bone tissue. Most slides are a transverse section through bone that is ground very thin. This preparation process removes the bone cells but leaves the bone matrix intact for detailed studies.
4. At low magnification observe the overall organization of the bone tissue. How many osteons can you locate?

   _____

   _____

5. Select an osteon and observe it at a higher magnification. Identify the central canal, canaliculi, and lacunae.
6. Move the slide and locate an area of interstitial lamellae. How do these lamellae differ from the concentric lamellae?

   _____

   _____

B. Bone Structure

Bones are encapsulated in a tough, fibrous membrane called the **periosteum**. This membrane appears shiny and glossy in the living condition. Histologically, the periosteum is composed of two distinct layers: an outer **fibrous layer** where muscle tendons and bone ligaments attach and an inner **cellular layer** which produces cells called **osteoblasts (OS-tē-ō-blasts)** for bone growth and repair.

A long bone has a shaft, called the **diaphysis (dī-AF-i-sis),** with an **epiphysis (ē-PIF-i-sis)** on each end (see Figure 14.2). The proximal epiphysis is on the superior end where the bone is attached, and the distal epiphysis is on the inferior end. The diaphysis consists mainly of strong, compact bone, and the epiphysis consists of spongy bone. A layer of hyaline cartilage, the **articular cartilage**, covers the epiphysis where it articulates with another bone. The interior of the diaphysis is hollow, forming the **medullary** or **marrow cavity** where lipid is stored as **yellow marrow**. An inner membrane called the **endosteum** lines the bone tissue facing the medullary cavity. **Osteoclast** cells in the endosteum secrete a weak acid which dissolves bone matrix to tear down bone for remodeling or repair and the release of stored minerals into the blood.

Between the diaphysis and an epiphysis is the **metaphysis (me-TAF-i-sis)**. In a juvenile's bone, the metaphysis is called the **epiphyseal plate** and consists of a plate of hyaline cartilage which allows the bone to grow in length. By early adulthood, the rate of mitosis in the cartilage plate slows, and ossification fuses the epiphysis to the diaphysis. Once bone stops growing in length, all that remains of the metaphysis is a bony remnant of the growth plate, an **epiphyseal line**.

Flat bones are thin bones with no marrow cavity, such as the frontal and parietal bones of the skull. Flat bones have a layer of spongy bone sandwiched between layers of compact bone (see Figure 14.2b). The compact bone layers are called the external and internal tables and are thick to provide strength for the bone. The spongy bone layer is filled with marrow and is called the **diploë**.

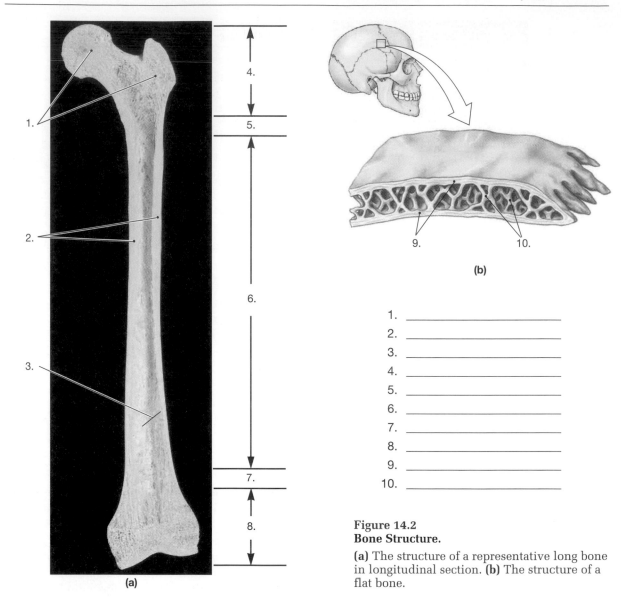

(a)

(b)

1. _____
2. _____
3. _____
4. _____
5. _____
6. _____
7. _____
8. _____
9. _____
10. _____

Figure 14.2
Bone Structure.

(a) The structure of a representative long bone in longitudinal section. **(b)** The structure of a flat bone.

LABORATORY ACTIVITY   EXAMINATION OF A FRESH LONG BONE

MATERIALS

preserved long bone or a fresh long bone from a butcher shop
gloves and safety glasses
blunt probe

PROCEDURES

1. Examine the long bone and locate the periosteum. Does it appear shiny? Are any tendons or ligaments attached to it?

_____

_____

2. If the bone has been sectioned, observe the internal bone tissue of the diaphysis. Is the bone tissue similar in all regions of the sectioned bone?

_____

_____

3. Locate an epiphysis and its articular cartilage. What is the function of the cartilage?

_____

_____

4. Look closely and locate the metaphysis. Most likely the bone has an epiphyseal line rather than a epiphyseal plate. Why?

_____

_____

_____

C. Classification of Bones

Bones may be grouped according to their shape, as illustrated in Figure 14.3. Bones of the limbs are **long bones** which are greater in length than in width. Bones of the wrist and ankle are **short bones** and are almost as wide as they are long. The skull has thin, platelike **flat bones**. The vertebrae of the spine are **irregular** bones and are not in any of the previous categories. **Sesamoid** bones form inside tendons. The patella or kneecap develops inside the tendons anterior to the knee and is the largest sesamoid bone. **Sutural** or **Wormian** bones occur along suture lines of the skull where the suture has branched and surrounded a small piece of bone. The number of sutural bones is variable among individuals and is not included when counting the number of bones in the skeletal system.

LABORATORY ACTIVITY   CLASSIFICATION OF BONES

MATERIALS

articulated skeleton

PROCEDURES

1. Locate the long bones of the skeleton. To which division of the skeletal system do they belong?

_____

_____

2. Examine the wrists and ankles of the articulated skeleton. What type of bones occur there?

_____

_____

3. Examine the bones of the skull. How are these bones classified? Do you see any sutural bones in the skull?

_____

_____

4. Find a sesamoid bone on the skeleton. Inside what structure does this bone develop?

_____

_____

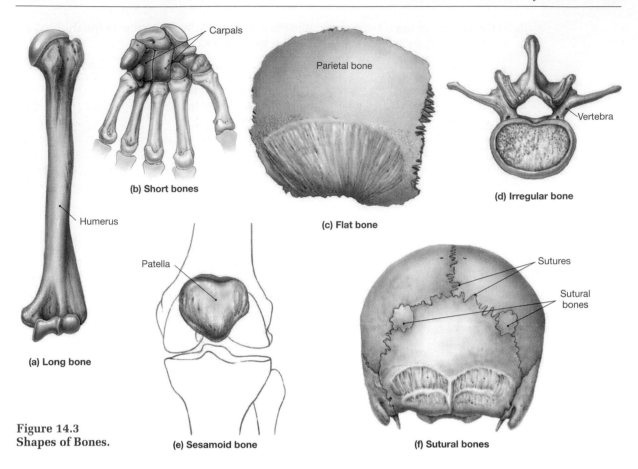

Figure 14.3
Shapes of Bones.

(a) Long bone
(b) Short bones
(c) Flat bone
(d) Irregular bone
(e) Sesamoid bone
(f) Sutural bones

Humerus
Carpals
Parietal bone
Vertebra
Patella
Sutures
Sutural bones

5. Observe the bones of the spine. What are these bones collectively called and how are they classified?

_____

_____

D. The Skeleton

The adult skeletal system consists of 206 bones. Each bone is an organ and includes osseous tissue, cartilage, and other connective tissues. The skeleton is organized into the axial and appendicular divisions. The **axial division** includes the **skull**, the **verte-brae**, the **sternum**, 12 pairs of **ribs**, and the **hyoid** bone, for a total of 80 bones. Locate these bones in Figure 14.4.

The **appendicular division** consists of the pectoral girdles plus the upper limbs and the pelvic girdle plus the lower limbs. Each of the girdles articulates with the axial division and provides attachment and mobility of the limbs. A total of 126 bones are found in the appendicular division of the adult skeleton.

Each **pectoral girdle** includes the shoulder blade or **scapula** and the collar bone or **clavicle**. Each **upper limb** consists of the arm, wrist, and hand. The **humerus** is the upper arm bone, and the **ulna** and **radius** comprise the forearm. The hand is composed of eight wrist bones collectively called the **carpals**, which articulate with the elongated

metacarpal bones of the palm. The individual bones of the fingers are the **phalanges**. Locate the bones of the pectoral girdle and upper limb in Figure 14.4.

The **pelvic girdle** is fashioned from two **coxal** bones, each an aggregate of three bones. The **ilium** is the superior bone and is the hip area of the torso. The **ischium** is inferior to the ilium and is used when sitting. The **pubis** is in the anterior pelvis. Each **lower limb** is comprised of the leg, ankle, and foot. The **femur** is the thighbone and is the largest bone in the body. It articulates with the coxal bone. The lower leg is comprised of a medial **tibia**, which bears most of the body weight, and a thin, lateral **fibula**. The kneecap, called the **patella**, occurs at the articulation between the femur and tibia. The eight ankle bones of the foot are collectively called the **tarsals**. **Metatarsals** form the arch of the foot, and **phalanges** form the toes. Locate the bones of the pelvic girdle and lower limb in Figure 14.4.

LABORATORY ACTIVITY	THE AXIAL AND APPENDICULAR SKELETAL DIVISIONS

MATERIALS

articulated skeleton

PROCEDURES

1. Use Figure 14.4 as a guide and locate the bones of the axial division of the skeleton. List the major components of the axial division.

_____

_____

2. Use Figure 14.4 as a guide and locate the major components of the appendicular division of the skeleton.

3. What bones are found in the shoulder and upper limb?

_____

_____

4. What three bones fuse to form a coxal bone?

_____

_____

5. List the bones of the lower limb.

_____

_____

E. Skeletal Terminology

Each bone has certain anatomical features on the surface called **bone markings** or surface markings. A particular bone marking may be unique to a single bone or may occur throughout the skeleton. Table 14.1 illustrates examples of bone markings and organizes the markings into five groups. The first group includes general anatomical structures, and the second group lists bony structures for tendon and ligament attachment. The third group contains structures that occur at sites of articulation with other bones. The last two groups include depressions and openings.

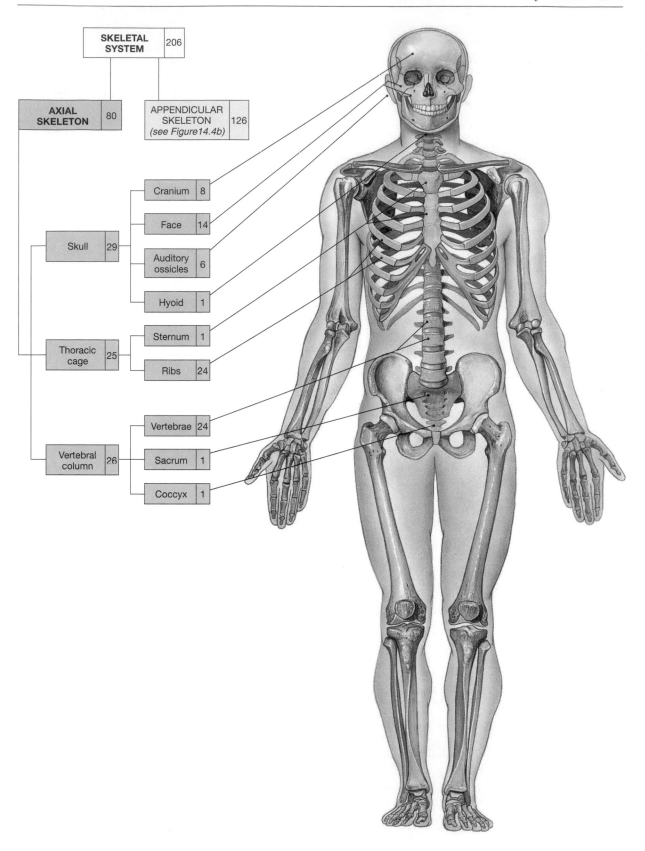

Figure 14.4a
The Axial Skeleton.

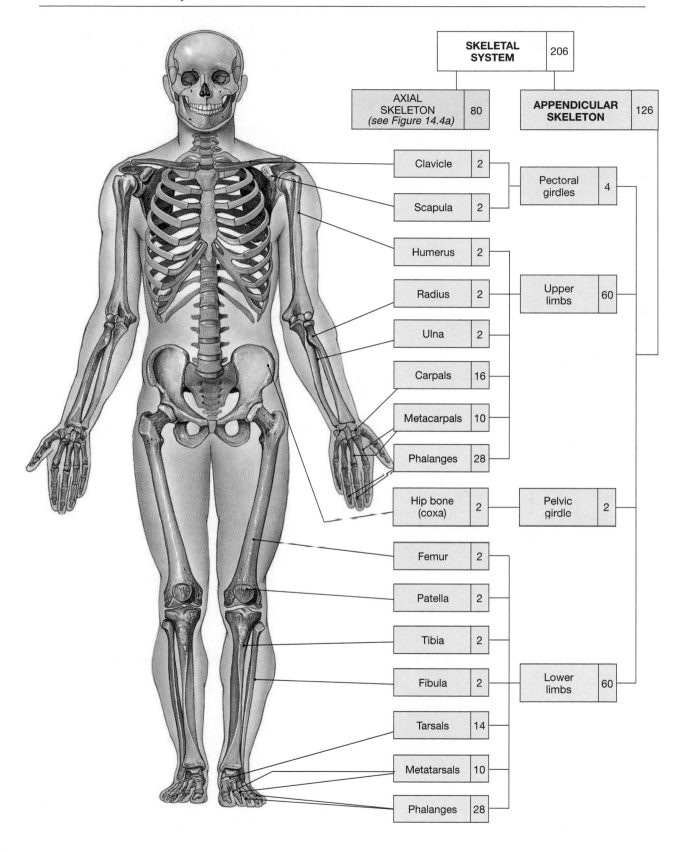

Figure 14.4b
The Appendicular Skeleton.

TABLE 14.1   An Introduction to Skeletal Terminology

General Description	Anatomical Term	Definition
Elevations and projections (general)	Process	Any projection or bump
	Ramus	An extension of a bone making an angle to the rest of the structure
Processes formed where tendons or ligaments attach	Trochanter	A large, rough projection
	Tuberosity	A smaller, rough projection
	Tubercle	A small, rounded projection
	Crest	A prominent ridge
	Line	A low ridge
	Spine	A pointed process
Processes formed for articulation with adjacent bones	Head	The expanded articular end of an epiphysis, separated from the shaft by the neck
	Neck	A narrow connection between the epiphysis and the diaphysis
	Condyle	A smooth, rounded articular process
	Trochlea	A smooth, grooved articular process shaped like a pulley
	Facet	A small, flat articular surface
Depressions	Fossa	A shallow depression
	Sulcus	A narrow groove
Openings	Foramen	A rounded passageway for blood vessels and/or nerves
	Canal	A large-diameter passageway through the substance of a bone
	Fissure	An elongated cleft
	Sinus or antrum	A chamber within a bone, normally filled with air

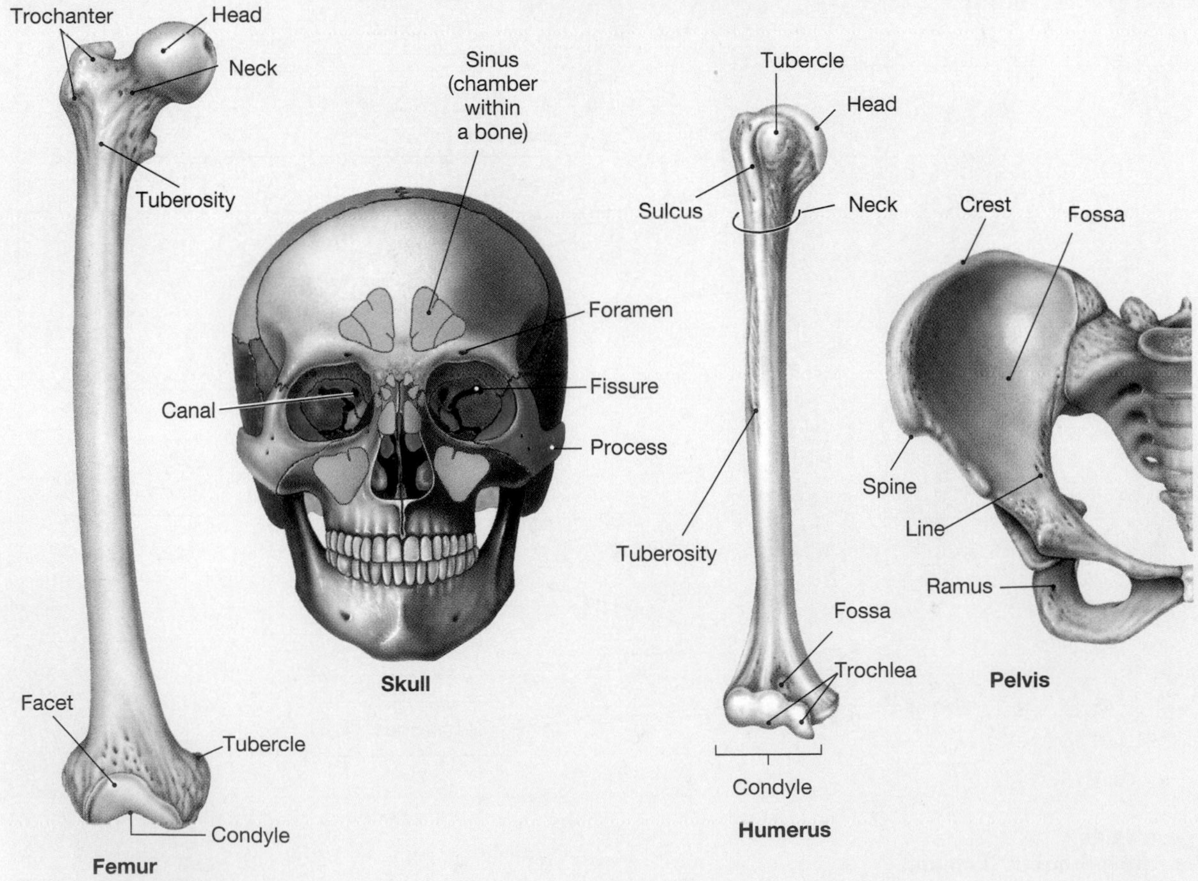

Femur

Skull

Humerus

Pelvis

LABORATORY ACTIVITY   BONE MARKINGS

MATERIALS

articulated skeleton

PROCEDURES

1. On the articulated skeleton locate each bone marking illustrated in Table 14.1.
2. In the space provided, list and define each bone marking you observe.

_____

_____

_____

_____

_____

_____

SKELETAL SYSTEM OVERVIEW CHECKLIST

This is a list of bold terms presented in Exercise 14. Use this list for review purposes after you have completed the laboratory activities.

HISTOLOGICAL ORGANIZATION OF BONE

• Figure 14.1
compact bone
osteon
  concentric lamellae
  lacunae
  canaliculi
  central canal
perforating canal
interstitial lamellae
spongy bone
trabeculae
red marrow

ANATOMY OF LONG BONE

• Figure 14.2
periosteum
  fibrous layer
  cellular layer
osteoblast
diaphysis
  medullary cavity
  yellow marrow
  endosteum
  osteoclast
metaphysis
  epiphyseal plate
  epiphyseal line
epiphysis
articular cartilage

CLASSIFICATION OF BONES

• Figure 14.3
long bones, short bones, flat bones
irregular bones, sesamoid bones
sutural bones (Wormian) bones

THE SKELETON

• Figure 14.4
axial division
  skull
  vertebrae
  sternum
  ribs
  hyoid
appendicular division
pectoral girdle
  scapula
  clavicle
upper limb
  humerus
  ulna
  radius
  carpals
  metacarpals
  phalanges
pelvic girdle
  coxal bone
   ilium
   ischium
   pubis
lower limb
  femur
  patella
  tibia
  fibula
  tarsals
  metatarsals
  phalanges

SKELETAL SYSTEM OVERVIEW
Laboratory Report Exercise 14

A. **Short-Answer Questions**
   1. List the components of the axial skeleton.

   2. List the components of the appendicular skeleton.

   3. Give an example of five different types of surface markings on bones.

   4. List the different shapes of bones.

B. **Matching**
   Match the structure on the left with the correct description on the right.

   1. _____ lacuna   A. bone shaft
   2. _____ trabecula   B. found in juvenile bones
   3. _____ articular cartilage   C. bony column
   4. _____ diaphysis   D. contains yellow marrow
   5. _____ epiphyseal plate   E. bone tip
   6. _____ osteon   F. lines marrow cavity
   7. _____ periosteum   G. found in adult bones
   8. _____ interstitial lamellae   H. forms osteon
   9. _____ epiphysis   I. bony projection
   10. _____ endosteum   J. produces bone matrix
   11. _____ medullary cavity   K. cellular space in bone matrix
   12. _____ epiphyseal line   L. remodeled osteons
   13. _____ concentric lamellae   M. outer membrane of bone
   14. _____ osteoblast   N. cartilage on epiphysis

C. **Fill in the Blanks**

Match each bone with the correct division and part of the skeleton. Each question may have more than one answer, and each choice can be used more than once.

  1. _____ scapula   **A.** axial division
  2. _____ coxal bone   **B.** appendicular division
  3. _____ patella   **C.** pectoral girdle
  4. _____ hyoid   **D.** upper limb
  5. _____ radius   **E.** pelvic girdle
  6. _____ metacarpal   **F.** lower limb
  7. _____ vertebra
  8. _____ clavicle
  9. _____ rib
 10. _____ femur
 11. _____ sternum
 12. _____ carpal

D. **Drawing**

   1. Draw and label a cross section through a long bone.

   2. Draw and label an osteon.

The Axial Skeleton

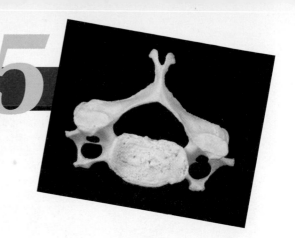

OBJECTIVES

On completion of this exercise, you should be able to:

- Identify the components of the axial skeleton.
- Identify the cranial and facial bones of the skull.
- Identify the surface features of the cranial and facial bones.
- Describe the skull of a fetus.
- Describe the four regions of the vertebral column and distinguish among the vertebrae of each region.
- Identify the features of a typical vertebra.
- Discuss the articulation of the ribs with the thoracic vertebrae.
- Identify the components of the sternum.

WORD POWER

foramen (foram—opening)
condyle (condyl—knuckle)
styloid (stylo—point)
petrous (petro—rock)
sella turcica (sella—saddle)
pterygoid (ptery—wing)
cribriform plate (cribr—sieve)
crista galli (crist—crest; galli—rooster)
alveolar (alve—socket)
ramus (ram—branch)

mental foramen (ment—chin)
sinus (sinu—hollow)
pedicle (pedic—little foot)
lamina (lamin—plate)
demifacet (demi—half; facette—small face)
manubrium (manu—hand)
xiphoid (xiph—sword)
dens (dens—tooth)
tubercle (tuber—knot)

MATERIALS

articulated skeleton
disarticulated skeleton
skull—unsectioned
skull—midsagittal section
skull—horizontal section
fetal skull
articulated vertebral column
disarticulated vertebral column

INTRODUCTION

The human skeleton consists of 206 bones, 80 of which are in the axial division. The axial skeleton includes the skull, hyoid bone, vertebral column, rib cage, and sternum. The axial skeleton provides a central framework for attachment of the appendicular skeleton (the shoulders, arms, hands, hips, legs, and feet). Internal organs such as the brain, heart, and lungs are protected by the bones of the axial skeleton.

A. The Skull

The skull has 22 bones organized into 14 facial bones and 8 cranial bones encasing the brain. Eight skull bones are paired with right and left bones, and the other 6 are single, unpaired bones located in the midline of the skull. Seven additional bones are associated with the skull, 6 bones of the middle ear (3 auditory ossicles per ear) and the hyoid bone. Therefore, the actual total number of skull bones is 29. The auditory ossicles are described in Exercise 34.

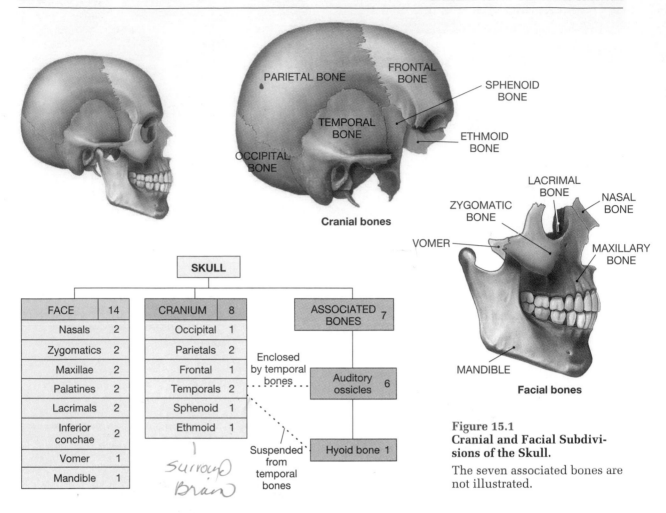

Cranial bones

Facial bones

Figure 15.1
Cranial and Facial Subdivisions of the Skull.

The seven associated bones are not illustrated.

FACE	14
Nasals	2
Zygomatics	2
Maxillae	2
Palatines	2
Lacrimals	2
Inferior conchae	2
Vomer	1
Mandible	1

CRANIUM	8
Occipital	1
Parietals	2
Frontal	1
Temporals	2
Sphenoid	1
Ethmoid	1

surroud Brain

SKULL

ASSOCIATED BONES	7

Enclosed by temporal bones

Auditory ossicles	6

Suspended from temporal bones

Hyoid bone	1

Figure 15.1 shows the subdivisions of the skull with the cranial bones disarticulated from the facial bones. The figure displays the skull in a lateral position. Looking directly into the face and eye sockets is the anterior view; the back of the head is the posterior view. Looking down on the skull from above is the superior view; the base underneath the skull is the inferior surface. Most study skulls have been sectioned horizontally to permit viewing of the interior cranial floor.

The cranium

The cranium consists of eight bones as follows: two temporal, two parietal, one occipital, one frontal, one sphenoid, and one ethmoid. Examine a skull from the superior view and locate the four bones that span the roof of the cranium. The **frontal** bone extends from the forehead posteriorly to the **coronal suture** and articulates with a pair of **parietal** bones. The parietal bones are joined superiorly by the **sagittal suture**. A single **occipital** bone meets the parietals at the **lambdoidal (lam-DOYD-ul) (occipitoparietal)** suture, completing the posterior wall of the cranium.

In the lateral view of the skull, represented in Figure 15.2, the **temporal** bone articulates with the parietal bone at the **squamosal suture**. The squamosal and coronal sutures are joined by the **sphenoparietal suture**. Below this suture, the **sphenoid** bone is partly visible as a vertical rectangle of bone. The sphenoid and parts of the frontal, temporal, and occipital bones form the floor of the cranium.

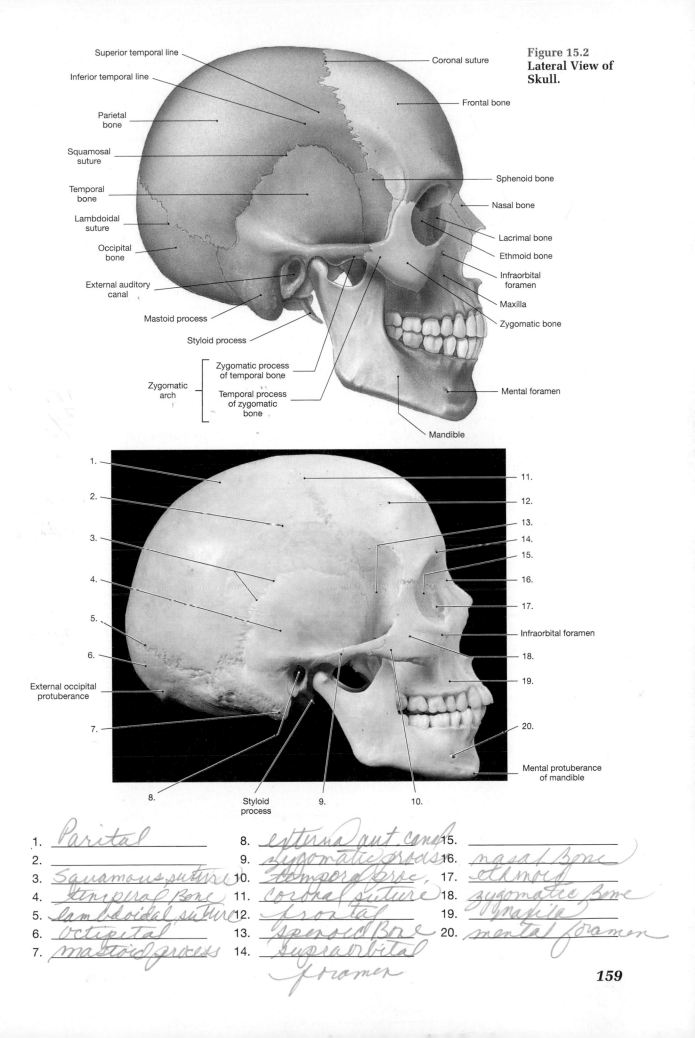

Superior temporal line

Inferior temporal line

Parietal bone

Squamosal suture

Temporal bone

Lambdoidal suture

Occipital bone

External auditory canal

Mastoid process

Styloid process

Zygomatic arch

Zygomatic process of temporal bone

Temporal process of zygomatic bone

Coronal suture

Frontal bone

Sphenoid bone

Nasal bone

Lacrimal bone

Ethmoid bone

Infraorbital foramen

Maxilla

Zygomatic bone

Mental foramen

Mandible

Figure 15.2
Lateral View of Skull.

External occipital protuberance

Infraorbital foramen

Styloid process

Mental protuberance of mandible

1. <u>Parietal</u>
2. _____
3. <u>Squamous suture</u>
4. <u>temporal Bone</u>
5. <u>lambdoidal suture</u>
6. <u>octipital</u>
7. <u>mastoid process</u>
8. <u>external aut. canal</u>
9. <u>zygomatic prods</u>
10. <u>temporal Bone</u>
11. <u>coronal suture</u>
12. <u>frontal</u>
13. <u>spenoid Bone</u>
14. <u>Supraorbital foramen</u>
15. _____
16. <u>nasal Bone</u>
17. <u>ethmoid</u>
18. <u>zygomatic Bone</u>
19. <u>maxila</u>
20. <u>mental foramen</u>

159

The **ethmoid** is a small, rectangular bone deep in the eye orbit, behind the bridge of the nose. Locate the ethmoid in Figure 15.3, an anterior view of the skull. On the study skull, insert your thumb and forefinger halfway into separate eye orbits and gently pinch the bone that is between them. This is the general location of the ethmoid. Remove your fingers and examine the medial wall of the orbit. The ethmoid is directly posterior to the small lacrimal bone, the bone with a small canal.

The face

The face is constructed of 14 bones: 2 nasal, 2 maxillae, 2 lacrimal, 2 zygomatic, 2 palatine, 2 inferior nasal conchae, 1 vomer, and 1 mandible. Refer to Figure 15.3, an anterior view of the skull, and locate each facial bone. The **nasal** bones are small bones that form the bridge of the nose. Lateral to the nasals are the **maxillary** bones, or the maxillae. These bones form the floor of the eye orbits and extend inferiorly to comprise the upper jaw. Below the eye orbits are the **zygomatic** bones, commonly called the cheek bones. At the bridge of the nose, lateral to each maxilla, are the small **lacrimal** bones of the medial eye orbitals. A small canal passes through each lacrimal bone for drainage of tears into the nasal cavity. The **inferior nasal conchae (KONG-kē)** are the lower shelves of bone in the nasal cavity. The other conchae in the nasal cavity are part of the maxillae. The lower jaw bone is the **mandible**.

Turn the skull over to view the inferior surface as shown in Figure 15.4. Locate the **palatine** bones of the posterior roof of the mouth next to the last molar tooth. Notice that the nasal cavity is divided by a thin wall of bone, the **vomer**.

LABORATORY ACTIVITY   IDENTIFICATION OF SKULL BONES

MATERIALS

skull (unsectioned)

PROCEDURES

1. On a study skull locate the 22 bones of the skull from each view of the skull.
2. Determine the general location of each skull bone on your head.

B. The Cranial Bones

The floor of the cranium has three depressions or fossae, illustrated in Figure 15.5. The **anterior cranial fossa** is mostly the base of the frontal bone. Small portions of the ethmoid and sphenoid also contribute to the floor of this area. The **middle cranial fossa** is a composite of the sphenoid, temporal, and occipital bones. The **posterior cranial fossa** is composed of the occipital bone.

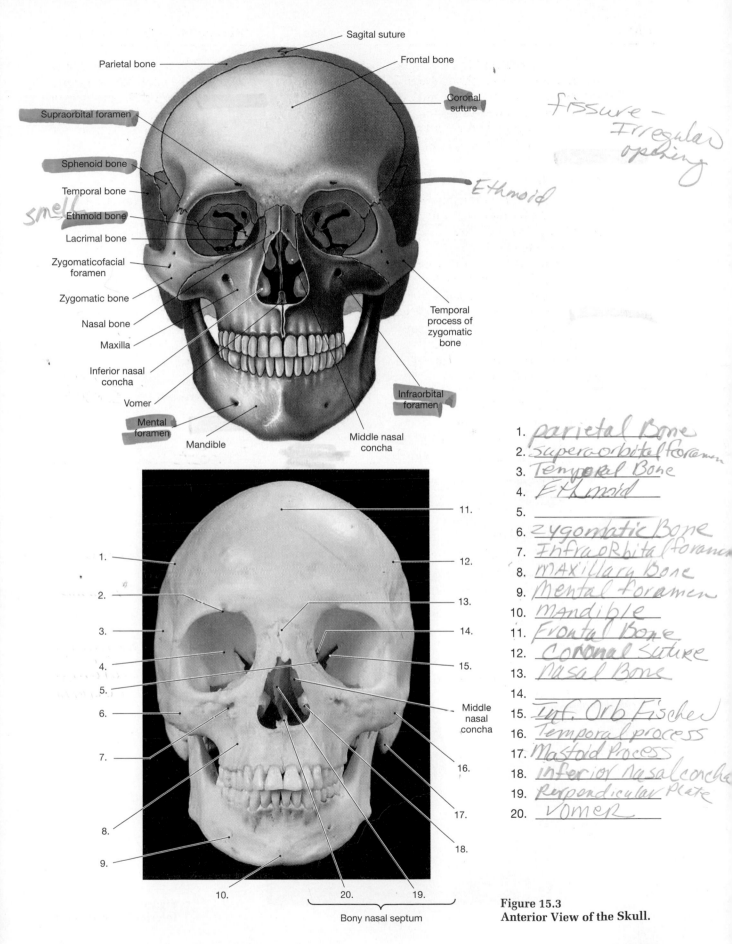

Figure 15.3
Anterior View of the Skull.

Sagital suture
Parietal bone
Frontal bone
Coronal suture
Supraorbital foramen
Sphenoid bone
Temporal bone
Ethmoid bone
Lacrimal bone
Zygomaticofacial foramen
Zygomatic bone
Nasal bone
Maxilla
Inferior nasal concha
Vomer
Mental foramen
Mandible
Middle nasal concha
Temporal process of zygomatic bone
Infraorbital foramen

Middle nasal concha
Bony nasal septum

Handwritten annotations:
fissure – Irregular opening
Ethmoid
smell

1. parietal Bone
2. superaorbital foramen
3. Temporal Bone
4. Ethmoid
5.
6. zygomatic Bone
7. InfraoRbital foramen
8. MAXillary Bone
9. Mental foramen
10. Mandible
11. Frontal Bone
12. Coronal suture
13. Nasal Bone
14.
15. Inf. Orb Fischer
16. Temporal process
17. Mastoid Process
18. inferior nasalconcha
19. Perpendicular Plate
20. VoMeR

161

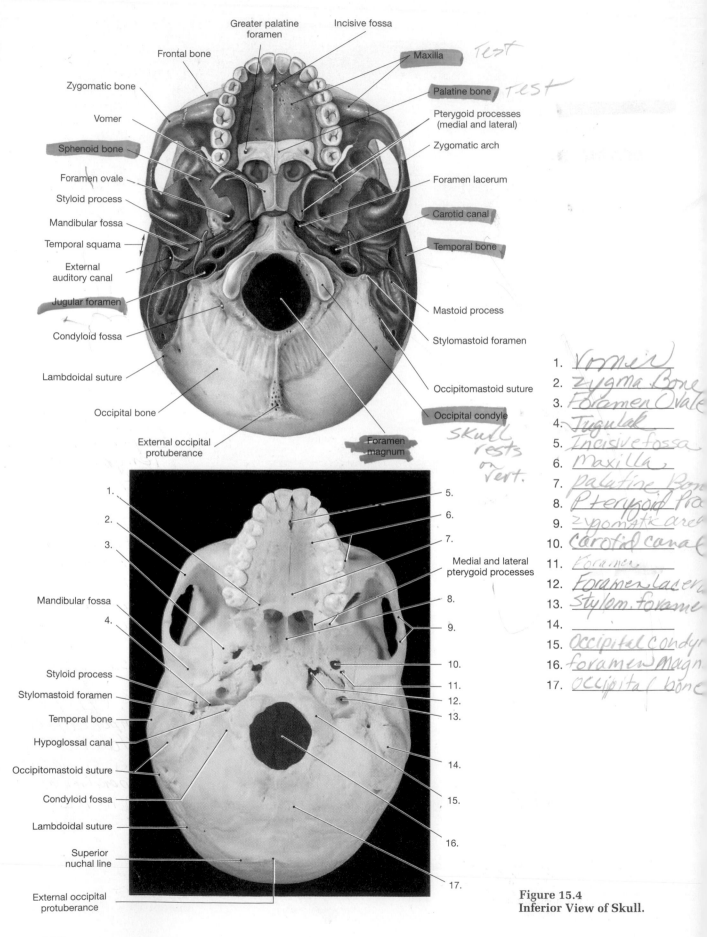

Greater palatine foramen

Incisive fossa

Frontal bone

Maxilla — *Test*

Zygomatic bone

Palatine bone — *Test*

Vomer

Pterygoid processes (medial and lateral)

Sphenoid bone

Zygomatic arch

Foramen ovale

Foramen lacerum

Styloid process

Carotid canal

Mandibular fossa

Temporal bone

Temporal squama

External auditory canal

Jugular foramen

Mastoid process

Condyloid fossa

Stylomastoid foramen

Lambdoidal suture

Occipital bone

Occipitomastoid suture

External occipital protuberance

Occipital condyle

Foramen magnum

Skull rests on Vert.

Mandibular fossa

Medial and lateral pterygoid processes

Styloid process

Stylomastoid foramen

Temporal bone

Hypoglossal canal

Occipitomastoid suture

Condyloid fossa

Lambdoidal suture

Superior nuchal line

External occipital protuberance

1. *Vomer*
2. *Zygma Bone*
3. *Foramen Ovale*
4. *Jugular*
5. *Incisive fossa*
6. *Maxilla*
7. *palatine Bone*
8. *Pterygoid Pro*
9. *zygomatic arch*
10. *Carotid canal*
11. *Foramen*
12. *Foramen Lacer*
13. *Stylom. forame*
14. _____
15. *Occipital condy*
16. *foramen Magn*
17. *Occipital bone*

Figure 15.4
Inferior View of Skull.

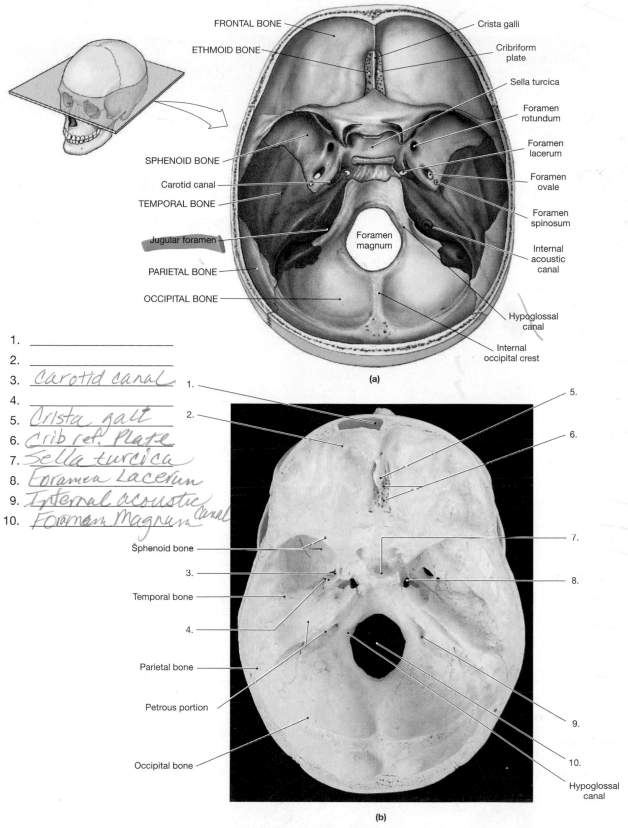

(a)

FRONTAL BONE

ETHMOID BONE

SPHENOID BONE

Carotid canal

TEMPORAL BONE

Jugular foramen

PARIETAL BONE

OCCIPITAL BONE

Crista galli

Cribriform plate

Sella turcica

Foramen rotundum

Foramen lacerum

Foramen ovale

Foramen spinosum

Internal acoustic canal

Foramen magnum

Hypoglossal canal

Internal occipital crest

1. _____

2. _____

3. _Carotid canal_

4. _____

5. _Crista galt_

6. _Crib ref. Plate_

7. _Sella turcica_

8. _Foramen Lacerum_

9. _Internal acoustic canal_

10. _Foramen Magnum canal_

Sphenoid bone

3.

Temporal bone

4.

Parietal bone

Petrous portion

Occipital bone

1.

2.

5.

6.

7.

8.

9.

10.

Hypoglossal canal

(b)

Figure 15.5
Sectional Anatomy of the Skull.

(a) Horizontal section through the skull, showing the floor of the cranial cavity. **(b)** Human skull, interior view, horizontal section.

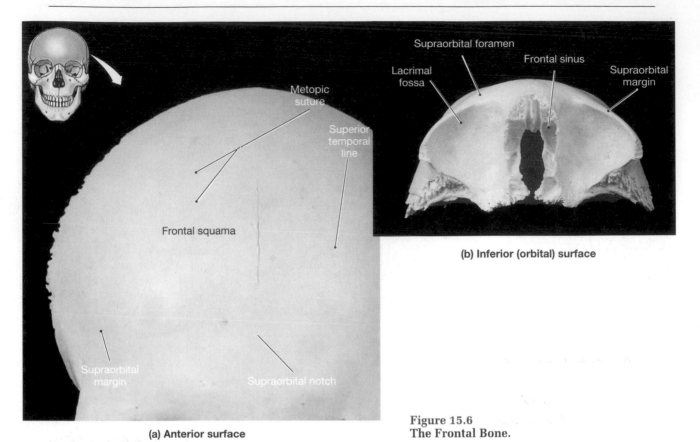

Metopic suture

Superior temporal line

Frontal squama

Supraorbital margin

Supraorbital notch

(a) Anterior surface

Supraorbital foramen

Lacrimal fossa

Frontal sinus

Supraorbital margin

(b) Inferior (orbital) surface

Figure 15.6
The Frontal Bone.

Frontal bone

The frontal bone forms the roof, walls, and floor of the anterior cranium (see Figure 15.6). The **frontal squama** is the flattened expanse commonly called the forehead. In the mid-sagittal plane of the squama is the **metopic (frontal) suture** where the two frontal bones fused in early childhood. This suture may not be visible on the study skull and is not illustrated in the figures. The frontal bone forms the superior portion of the eye orbit. Superior to the orbit is the **supraorbital foramen** which may occur as a small notch on some skulls rather than a complete hole.

Occipital bone

The occipital bone comprises the posterior floor and wall of the skull. Position your study skull as in Figure 15.7, an inferior view. The most conspicuous structure of the occipital bone is the **foramen magnum**, the large hole where the spinal cord enters the skull and joins the brain. Along the lateral margins of the foramen magnum are by flattened **occipital condyles** which articulate with the first vertebra of the spine. Passing under each occipital condyle is the **hypoglossal canal,** a passageway for the hypoglossal nerve that controls muscles of the tongue and throat.

The occipital bone has many external surface marks that show where muscles and ligaments attach. The **occipital crest** is a ridge that extends posteriorly from the foramen magnum to a small bump, the **external occipital protuberance.** Wrapping around the occipital bone lateral from the crest and protuberance are the **superior** and **inferior nuchal (NOO-kul) line**, surface marks indicating where muscles of the neck attach to the skull.

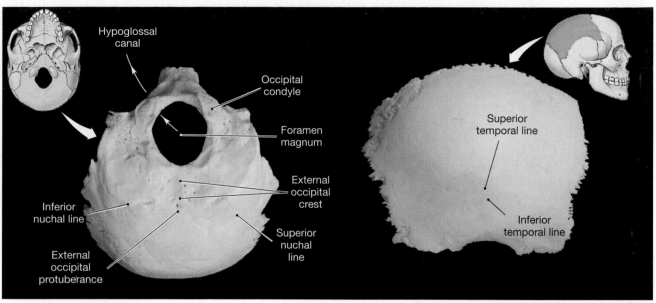

(a) Occipital bone, inferior view

(b) Right parietal bone, lateral view

Figure 15.7
The Occipital and Parietal Bones.

Parietal bones

The parietal bones form the posterior crest of the skull and are joined by the sagittal suture. The bones are smooth and have few surface features. The faint **superior temporal line** above the squamosal suture is a site of muscle attachment (see Figure 15.7). No major foramina pass through the parietal bones.

Temporal bones

The temporal bones form the lower lateral walls and part of the floor of the middle cranial fossa. Each temporal bone is a complex bone with many surface features. Figure 15.8a details the lateral aspect of a temporal bone. Locate the **external auditory canal**, a tube for conducting sound waves toward the eardrum. The large, flattened area above the canal is the **squamous portion**. Directly posterior to the external auditory canal is the **mastoid process**, an attachment site for muscles that move the head. Feel the mastoid process behind one of your ears. Anterior to the external auditory canal is a thin extension of bone, the **zygomatic process** of the temporal bone. The name of this structure refers to the bone it articulates with, not the bone it occurs on.

Examine the floor of the temporal bones as shown in Figure 15.8b. The large crest arising from the floor is the **petrous portion** of the temporal bone. Inside this bony mass are the organs for hearing and equilibrium and the tiny auditory ossicle bones. Identify the **internal acoustic canal** on the posterior medial surface of the petrous portion. The internal acoustic canal is not continuous with the external auditory canal. Below (inferior to) the canal is the jugular foramen.

Position the study skull as in Figure 15.4 to view the inferior aspect of the temporal bones. Remove the mandible from your specimen if allowed by your lab instructor. A good anatomical landmark in this view is the mastoid process. The **styloid process** is medial and anterior to the mastoid process. This delicate process serves as a muscle attachment site and is often broken off in study skulls. Between the styloid and the mastoid processes is a small foramen, the **stylomastoid foramen,** where the facial nerve exits the cranium. Anterior to the external auditory canal is the **mandibular fossa**, the shallow depression where the mandible articulates with the temporal. Between the mastoid

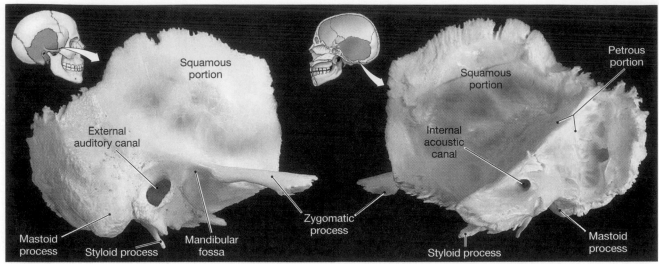

(a) **Lateral view**   (b) **Medial view**

(c) **Mastoid air cells**

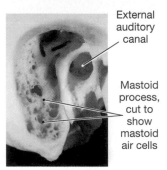

External
auditory
canal

Mastoid
process,
cut to
show
mastoid
air cells

Figure 15.8
The Temporal Bones.

(a,b) The right temporal bone. **(c)** A cutaway
view of the mastoid air cells.

process and the occipital condyle is the **jugular foramen**. This foramen is an exit hole
for cranial nerves and the jugular vein, the large vein that drains blood from the brain.
Directly anterior to the jugular foramen is the **carotid canal**, a passageway for the carotid
artery, the artery that supplies oxygenated blood to the head. This canal burrows medially through the temporal and sphenoid bones and exits on the floor of the skull.

STUDY TIP

Notice how close the jugular foramen and carotid canal are to each other on the inferior view of
the skull. When observing the floor, however, the temporal bone's petrous portion separates the
jugular foramen posteriorly from the anterior carotid canal. Imagine the petrous portion as a
mountain with a passageway on each side.

Sphenoid bone

The sphenoid bone is the hub of the cranium, and all cranial bones articulate with the
sphenoid. It can be seen from all views of the skull. Use Figures 15.2–15.5 and study
the location of the sphenoid bone.

The easiest aspect of the sphenoid to work with is the floor, the inside of the skull,
as shown in Figure 15.4. Face your study skull in the same direction as the figure. This
anatomy is also detailed in Figure 15.9, an isolated view of the sphenoid bone. The
anterior margin of the sphenoid is the bat-shaped **lesser wing**. The **greater wing** is posterior to the lesser wing and contributes to the floor of the middle cranial fossa. The
greater wing is also visible on the lateral surface of the skull, just inferior to the

(a) Superior surface

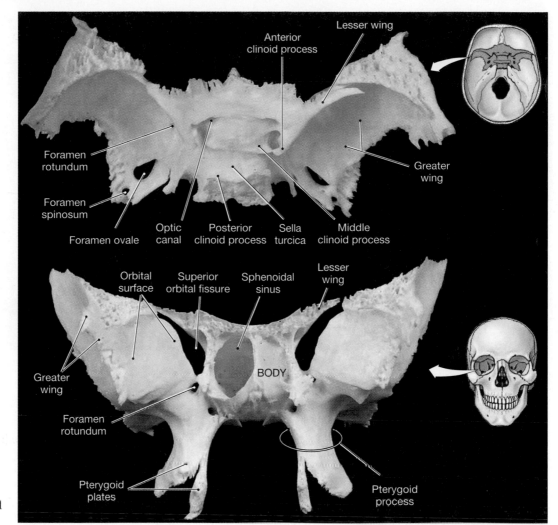

(b) Anterior surface

Figure 15.9
The Sphenoid
Bone.

sphenoparietal suture. In the middle of the sphenoid is a raised platform, the **sella turcica (TUR-si-kuh)**, or Turk's saddle, where the pituitary gland sits. The upper margins of the sella turcica are bordered by two pairs of processes, the **anterior clinoid processes** and the **posterior clinoid processes**.

Three foramina are aligned lateral to each side of the sella turcica and serve as passageways for blood vessels and nerves. The **foramen ovale** is the oval hole. Posterior to the foramen ovale is a small **foramen spinosum**. These foramina are passageways for parts of the trigeminal nerve. The round foramen anterior to the foramen ovale is the **foramen rotundum**. Notice the tubular entrance into the foramen rotundum. The **foramen lacerum** is medial to the carotid canal. On many skulls, the carotid canal merges with the foramen lacerum to form a single foramen. Anterior to the foramen rotundum is a cleft in the sphenoid, the **superior orbital fissure**, where nerves to the ocular muscles pass. At the base of the anterior clinoid process is the **optic canal** where the optic nerve exits the eye socket to carry visual signals to the brain.

The pterygoid processes extend from the inferior sphenoid. The **lateral** and **medial pterygoid (TER-i-goyd) processes** are attachment sites for muscles involving the mouth.

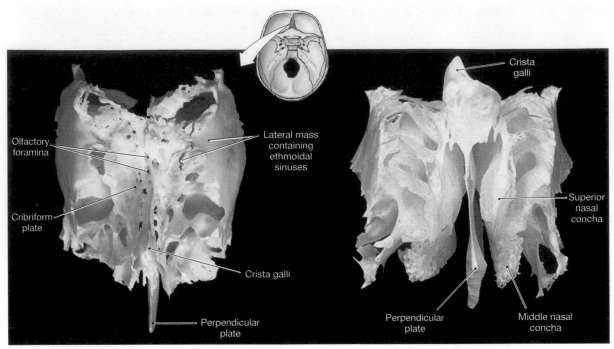

(a) **Superior surface**   (b) **Posterior surface**

Figure 15.10
The Ethmoid Bone.

Ethmoid bone

The **ethmoid** is a single rectangular bone deep in the eye orbit behind the bridge of the nose. Locate the ethmoid in Figure 15.3, an anterior view of the skull. The ethmoid is directly posterior to the small lacrimal bone, the bone with a small foramen in the medial "corner" of the eye orbit. Observe in Figure 15.10 the vertical crest of bone called the **crista galli** and the surrounding screen-like **cribriform plate**. Membranes that protect and support the brain attach to the crista galli. Fine extensions of the olfactory nerves pass through many foramina that perforate the cribriform plate. The inferior ethmoid has a thin sheet of bone, the **perpendicular plate**, which divides the nasal cavity. Thin shelves of bone extend from the lateral walls of the ethmoid as the **superior** and **middle conchae**.

LABORATORY ACTIVITY	BONES OF THE CRANIUM

MATERIALS

skull (horizontal section)

PROCEDURES

1. Label and review the cranial structures in Figures 15.2–15.5.
2. On a study skull locate all the bones of the cranium. Be able to identify each bone from all views of the skull.
3. Identify each surface feature on the cranial bones.

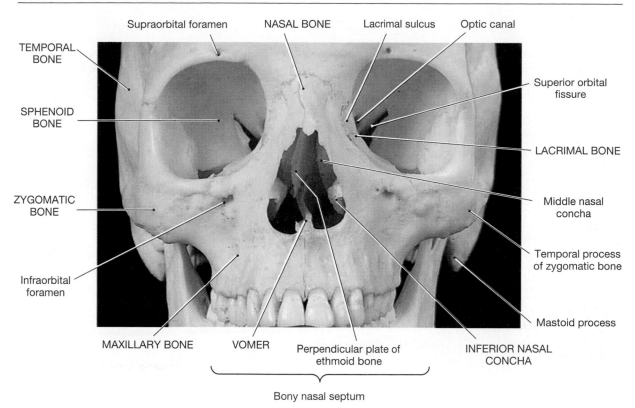

TEMPORAL BONE

SPHENOID BONE

ZYGOMATIC BONE

Infraorbital foramen

Supraorbital foramen   NASAL BONE   Lacrimal sulcus   Optic canal

Superior orbital fissure

LACRIMAL BONE

Middle nasal concha

Temporal process of zygomatic bone

Mastoid process

MAXILLARY BONE   VOMER   Perpendicular plate of ethmoid bone   INFERIOR NASAL CONCHA

Bony nasal septum

Figure 15.11
The Smaller Bones of the Face.

C. The Facial Bones

Maxillae

The paired maxillae are the foundation of the face (see Figure 15.11). Inferior to the orbit is the **infraorbital foramen**. The **alveolar process** consists of the U-shaped processes where the upper teeth are embedded in the maxilla. From the inferior aspect, the **palatine process** of the maxilla is visible (see Figure 15.4). This bony shelf forms the anterior hard palate of the mouth. At the anterior margin of the palatine process is the incisive fossa. Feel your hard palate by placing your tongue on the roof of your mouth just behind your upper teeth.

Palatine bones

The palatine bones are small bones in the roof of the mouth posterior to the palatine processes of the maxillae. These bones, along with the maxillae, form the roof of the mouth and separate the oral cavity from the nasal cavity. This separation of cavities allows us to chew and breathe at the same time. Each palatine bone has a **greater palatine foramen** on the lateral margin as detailed in Figure 15.4. Only the inferior portion of the palatine bone is completely visible. The superior surface forms the floor of the nasal cavity and supports the vomer.

Zygomatic bones

The check bones are the zygomatic bones. This bone also contributes to the floor and lateral wall of the orbit (see Figure 15.11). Lateral and slightly inferior to the orbit is the small **zygomaticofacial foramen**. The posterior margin of the zygomatic bone narrows inferiorly to the **temporal process** of the zygomatic. This process joins the temporal bone's zygomatic process to complete the **zygomatic arch**.

Lacrimal bones

The lacrimal bones are the anterior portions of the medial orbital wall, illustrated in Figure 15.11. Each lacrimal bone is named after the lacrimal glands that produce tears. Tears flow medially across the eye and drain into the inferior **lacrimal fossa** which transports them to the nasal cavity.

Nasal bones

The nasal bones are the bridge of the nose. These bones articulate superiorly with the frontal bone, laterally with the maxilla, and posteriorly (internally) with the ethmoid bone. Review the nasal bones in Figure 15.11.

Vomer

The vomer separates the nasal chamber into right and left cavities. This thin sheet of bone is best viewed from the inferior aspect of the skull looking into the nasal cavities, as in Figure 15.11.

Inferior nasal conchae

The inferior nasal conchae are bony shelves that extend medially from the lower lateral portion of the nasal wall. They cause inspired air to swirl in the nasal cavity so that the moist mucous membrane lining can warm, cleanse, and moisten it. Similar shelves of bone occur on the lateral walls of the ethmoid. Locate the inferior nasal conchae in Figure 15.11.

Mandible

The mandible is the U-shaped bone of the lower jaw, detailed in Figure 15.12. The horizontal **body** bends posteriorly at the **angle** to a raised projection, the **ramus**, which terminates at a U-shaped **mandibular notch**. Two processes extend upward from the notch, the anterior **coronoid (ko-RŌ-noyd) process** and a posterior **condylar process**. The smooth condylar process, also called the mandibular condyle, articulates in the mandibular fossa on the temporal bone to form the temporomandibular joint (TMJ). Open and close your mouth to feel this articulation. The **alveolar process** is the crest of bone where the lower teeth articulate with the mandible. Lateral to the chin or **mental protuberance** is the **mental foramen**. "Mental" is the Latin word for chin. The medial mandibular surface, shown in Figure 15.12b, features the **mandibular groove** where the submandibular salivary gland rests against the bone. At the posterior end of the groove is the **mandibular foramen**, a passageway for a sensory nerve from the lower teeth and gums.

LABORATORY ACTIVITY   THE FACIAL BONES

MATERIALS

skull (unsectioned)

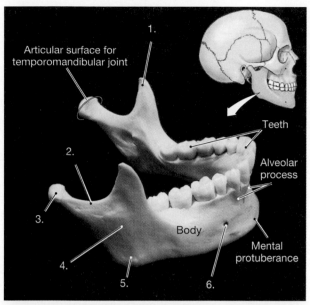

Articular surface for
temporomandibular joint

Teeth

Alveolar
process

Body

Mental
protuberance

(a) Lateral view

1. *Coranoid Process*
2. *Manidiblar notch*
3. *Head*
4. *Ramus*
5. *Angle*
6. *Mental Foramen*
7. *Mandibular Foramen*
8. *alveolar process*
9. *Depression*
   submandibular
   salivtory gland

Figure 15.12
The Mandible and
Hyoid Bone.

(a) A lateral view of the
mandible. **(b)** A medial
view of the right mandible.
(c) An anterior view of the
hyoid bone.

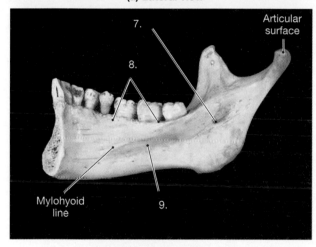

Articular
surface

Mylohyoid
line

(b) Medial view

ramous - Branch

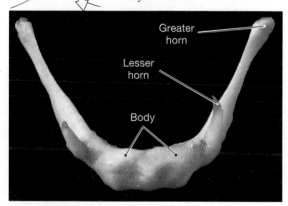

Greater
horn

Lesser
horn

Body

(c) Anterior view

PROCEDURES

1. Label and review the skeletal features of the face in Figures 15.2–15.12.
2. On a study skull, locate all the bones of the face from each view of the skull.
3. Identify all surface features located on the facial bones.

D. Hyoid Bone

The hyoid bone is a U-shaped bone inferior to the mandible. The bone is unique
because it does not articulate with other bones. You cannot feel your hyoid because it
is surrounded by the ligaments and muscles of the throat and neck. Figure 15.12 details
the anatomy of the hyoid bone. Two peglike processes for muscle attachment occur on
the bone, an anterior pair of **lesser cornua** and a larger pair of posterior **greater cornua**.
Muscles that move the tongue and larynx attach to these pegs.

LABORATORY ACTIVITY   HYOID BONE

MATERIALS

articulated skeleton

PROCEDURES

1. Locate the hyoid bone on an articulated skeleton.
2. Identify the surface feature of the hyoid bone.

E. Sinuses of the Skull

The skull contains four cavities called **paranasal sinuses** which interconnect with the nasal cavity and reduce the weight of the skull (see Figure 15.13). The sinuses are called paranasal sinuses because they are near the nasal cavity. The nasal cavity and the sinuses are lined with a mucous membrane. The **frontal sinus** is located above the articulation with the nasal bones. This sinus extends laterally over the orbit of the eyes. The **sphenoid sinus** is located in the sphenoid bone directly inferior to the sella turcica. The ethmoid bone is full of **air cells** which collectively comprise the **ethmoid sinus**. Each maxilla contains a large **maxillary sinus** situated lateral to the nasal cavity.

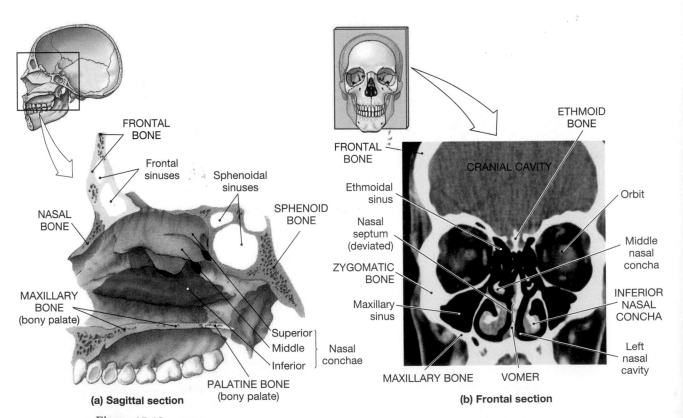

Figure 15.13
The Nasal Complex.

(a) A sagittal section through the skull, with the nasal septum removed to show major features of the wall of the right nasal cavity. **(b)** An MRI scan showing a frontal section through the ethmoidal and maxillary sinuses, part of the paranasal sinuses. The sphenoidal sinuses are visible in (a).

LABORATORY ACTIVITY   SINUSES OF THE SKULL

MATERIALS

skull (midsagittal section)

PROCEDURES

1. Locate the four sinuses on a midsagittally sectioned skull.

F. Fetal Skull

The fetal skull has many bones that are unfused and incompletely ossified. These bones develop as unfused patches in membranes of fibrous connective tissue. These soft spots, called **fontanels (fon-tuh-NELZ)**, allow the skull to expand as the brain increases in size. At birth the skull is still not completely formed, and the fontanels enable it to flex and squeeze through the birth canal during delivery. By the age of 4 years, most of the bones have ossified across their fontanels. Identify each fontanel in Figure 15.14.

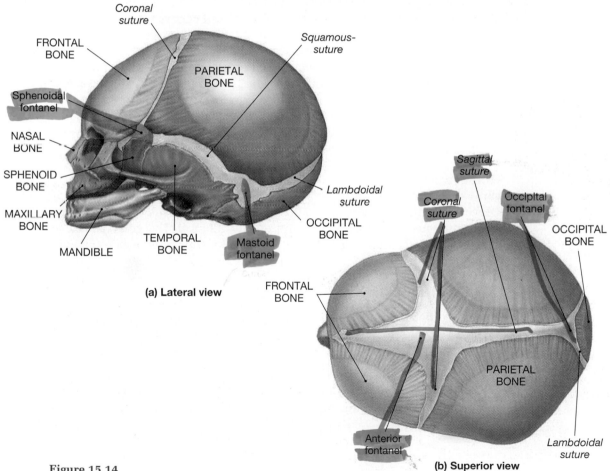

Figure 15.14
The Skull of an Infant.

(a) A lateral view. The skull of an infant contains more individual bones than that of an adult. Many of the bones eventually fuse; thus the adult skull has fewer bones. The flat bones of the skull are separated by areas of fibrous connective tissue, allowing for cranial expansion and the distortion of the skull during birth. The large fibrous areas are called fontanels. By about age 4, these areas will disappear. **(b)** A superior view.

LABORATORY ACTIVITY FETAL SKULL

MATERIALS

   skull (unsectioned)
   fetal skull

PROCEDURES

1. Identify each fontanel on a fetal skull using Figure 15.14 as a guide.
2. Compare a fetal skull and the skull of an adult. Which skull has more bones?

G. Vertebral Column

The vertebral column, or spine, is a flexible chain of 33 vertebrae. It articulates with the skull superiorly, the pelvic girdle inferiorly, and the ribs laterally. Vertebrae are grouped into four regions based on their location in the spine and their anatomical features (see Figure 15.15). The first 7 vertebrae are the **cervical** vertebrae of the neck. Twelve **thoracic** vertebrae articulate with the ribs. The lower back has 5 **lumbar** vertebrae, and a **sacrum** joining the hips is fashioned from 5 fused vertebrae. The **coccyx (KOK-siks)** is the tailbone and consists of 3 to 5 fused coccygeal vertebrae.

Notice that the vertebral column is not straight but curved. These **spinal curves** are necessary to balance the body weight for standing. At birth, the spinal column is generally rounded posteriorly, like a bow. The posterior curvature of the thoracic and sacral regions are called **primary curvatures**. A few months after birth, **secondary curvatures** begin to develop. The cervical vertebrae start to curve anteriorly to support the head, which the baby can now hold up. The lumbar region begins to bend anteriorly to balance the body weight for standing. Once the child is approximately 10 years old, the spinal curves are established and the fully developed column has alternating secondary and primary curvatures.

From the cervical to the lumbar region are cushions of fibrocartilage called **intervertebral discs** between the bodies of articulating vertebrae. Each disc consists of an outer layer of strong fibrocartilage called the **annulus fibrosus** that surrounds an inner mass, the **nucleus pulposus.** Water and elastic fibers in the gelatinous mass of the nucleus pulposus absorb stress between the vertebrae during movement and support of the body.

Vertebral anatomy

Figure 15.16 illustrates anatomical features of a typical vertebra from each region of the spine. Each vertebra has a large anterior disc and elongated processes posteriorly.

Although there is regional specialization of the vertebrae, they share many surface features. Each vertebra has the following structures:

- **body**: The thick, disk-shaped anterior portion; also called the **centrum**.
- **spinous process**: A long, single extension of the posterior vertebral wall.
- **transverse process**: A pair of extensions lateral to the spinous process.

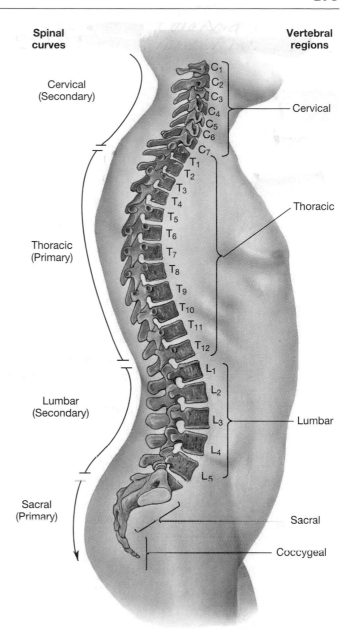

Spinal curves

Cervical (Secondary)

Thoracic (Primary)

Lumbar (Secondary)

Sacral (Primary)

Vertebral regions

C_1
C_2
C_3
C_4
C_5
C_6
C_7

T_1
T_2
T_3
T_4
T_5
T_6
T_7
T_8
T_9
T_{10}
T_{11}
T_{12}

L_1
L_2
L_3
L_4
L_5

Cervical

Thoracic

Lumbar

Sacral

Coccygeal

Figure 15.15
The Vertebral Column.

The major divisions of the vertebral column, showing the four spinal curves.

- **vertebral foramen**: The large hole posterior to the centrum; the anterior wall is formed by the centrum, the lateral walls by the pedicles, and the roof by the lamina.
- **pedicle (PE-di-kulz)**: The strut of bone extending posteriorly from the body to a transverse process.
- **lamina (LA-mi-na)**: A flat plate of bone between the transverse and spinous processes.

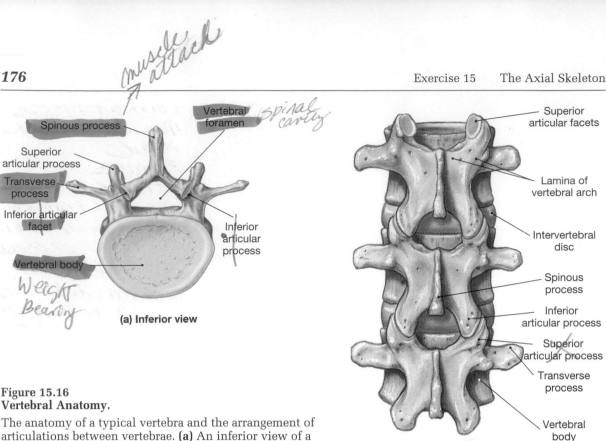

muscle attack (handwritten)

spinal cavity (handwritten)

Weight Bearing (handwritten)

Spinous process

Superior articular process

Transverse process

Inferior articular facet

Vertebral body

Vertebral foramen

Inferior articular process

(a) Inferior view

Superior articular facets

Lamina of vertebral arch

Intervertebral disc

Spinous process

Inferior articular process

Superior articular process

Transverse process

Vertebral body

(b) Posterior view

**Figure 15.16
Vertebral Anatomy.**

The anatomy of a typical vertebra and the arrangement of articulations between vertebrae. **(a)** An inferior view of a vertebra. **(b)** A posterior view of three articulated vertebrae.

- **superior articular process**: A projection on the superior surface of the posterior pedicle.
- **superior articular facet**: A smooth articular surface on the posterior tip of the superior articular process.
- **inferior articular process**: A downward projection of the inferior lamina wall.
- **inferior articular facet**: A smooth articular surface on the anterior tip of the inferior articular process.
- **inferior vertebral notch**: An inverted "U" on the inferior side of the pedicle.

Vertebral articulations

The vertebral column moves much like a flexible gooseneck lamp, each joint moves slightly, but collectively they permit a wide range of motion. The greatest flexibility is in the cervical region for head movement. Closely examine the articulations between the vertebrae in Figure 15.16. Observe how the inferior articular process of the *upper* vertebra articulates with the superior articular process of the *lower* vertebra. Also locate the **intervertebral foramen**, the hole created by the inferior vertebral notch of the upper vertebra and the pedicle of the lower vertebra.

Cervical vertebrae

Seven cervical vertebrae are located in the neck. Figure 15.17 illustrates the distinguishing feature of cervical vertebrae, the **transverse foramen** located on each transverse process. Only cervical vertebrae have transverse foramina.

The first two cervical vertebrae are modified for special articulations. The first cervical vertebra, C_1, is called the **atlas**, named after the Greek character who carried the world on his shoulders. The atlas is the only vertebra that articulates with the skull. Notice in Figure 15.17b that the superior articular facets of the atlas are greatly enlarged. The occipital condyles of the occipital bone fit in the facet like two spoons

1. <u>Vertebral arch</u>
2. <u>Vertebral foramen</u>
3. <u>Pedicle</u>
4. <u>Transverse Process</u>
5. <u>Spinous process</u>
6. <u>Lamina</u>
7. <u>Transverse foramen</u>
8. <u>Vertebral Body</u>

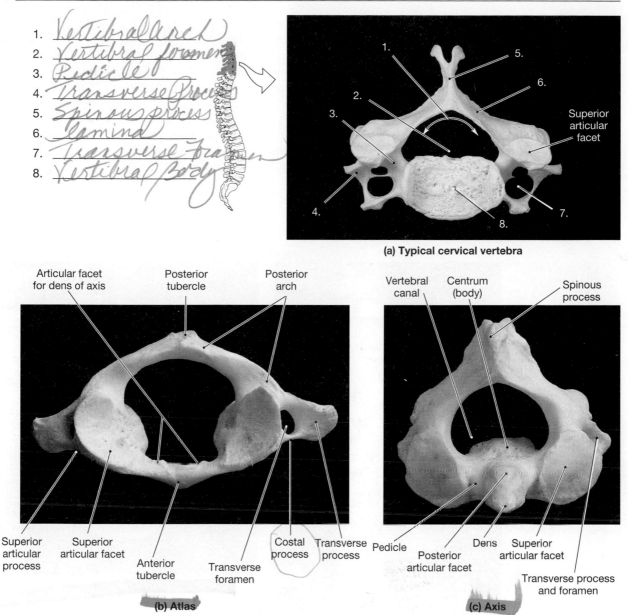

(a) Typical cervical vertebra

Figure 15.17
The Cervical Vertebrae.

(a) Features of a typical cervical vertebra highlighting the transverse foramen, superior view.
(b) The Atlas (C_1) in a superior view. **(c)** Superior view of the Axis (C_2).

nested together. When you nod your head "yes," the atlas remains stationary while the occipital condyles glide in the facets.

The atlas is unusual in that it lacks a centrum and a spinous process. A small, rough structure, the **posterior tubercle**, occurs where the spinous process normally resides. A long spinous process would interfere with the occipitoatlas articulation. Without a body, the atlas has a very large vertebral foramen.

The **axis** is the second cervical vertebra. It is specialized to articulate with the atlas. A peglike **dens (DENZ)** or **odontoid process** arises superiorly from the body of the axis (see Figure 15.17c). It fits against the anterior wall of the vertebral foramen and provides the atlas with a pivot point for when the head moves laterally and medially. A transverse ligament secures the atlas around the dens.

Thoracic vertebrae

The 12 thoracic vertebrae are larger than the cervical vertebrae and increase in size as they approach the lumbar region. The thoracic vertebrae articulate with the 12 pairs of ribs. Most ribs attach at two sites on **articular facets** (transverse costal facets) at the tip of the transverse process, and on a **demifacet** located on the posterior of the body. Two demifacets usually are present on the same body, a **superior** and an **inferior demifacet**. Locate the facet and demifacets in Figures 15.18. Notice the variation in the occurrence of facets and demifacets along the thoracic region. Facets and demifacets are found only on thoracic vertebrae. The articulations between the ribs and the thoracic vertebrae are examined in more detail in the upcoming section on the ribs.

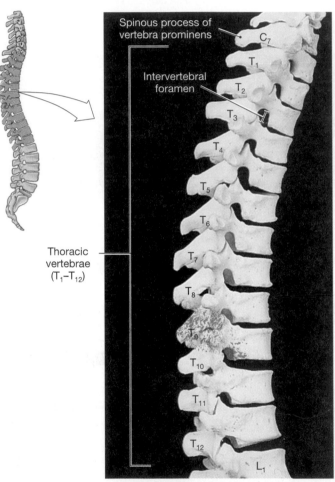

(a) Thoracic vertebrae, lateral view

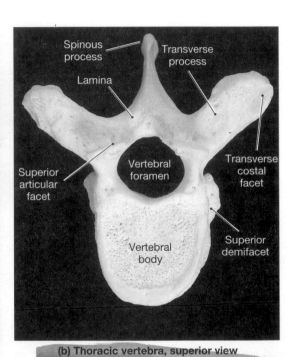

(b) Thoracic vertebra, superior view

(c) Thoracic vertebra, lateral view

Figure 15.18
The Thoracic Vertebrae.

(a) A lateral view of the thoracic region of the vertebral column. The vertebra prominens (C7) resembles T1 but lacks facets for rib articulation. Vertebra T12 resembles the first lumbar vertebra (L1) but has a facet for rib articulation. **(b)** Thoracic vertebra, superior view. **(c)** Thoracic vertebra, lateral view.

Lumbar vertebrae

The five lumbar vertebrae are large and heavy so as to support the weight of the head, neck, and trunk. Compared to thoracic vertebrae, lumbar vertebrae have a wider body, a blunt and horizontal spinous process, and shorter transverse processes, see Figure 15.19. The vertebral foramen is smaller than that of thoracic vertebrae. To prevent twisting of the back while lifting and carrying objects, the superior articular process is turned medially and the inferior articular processes are oriented laterally to interlock the lumbar vertebrae. No facets or transverse foramina occur on the lumbar region.

Sacral and coccygeal vertebrae

The **sacrum** is a single bony element composed of five fused sacral vertebrae (see Figure 15.20). It articulates with the ilium of the pelvic girdle to form the posterior wall of the pelvis. Fusion of the sacral bones before birth consolidates the vertebral canal into the **sacral canal**. **Sacral foramina** occur along the lateral margin of the fused vertebral bodies. The spinous processes fuse to form an elevation called the **median sacral crest**. A **lateral sacral crest** extends from the lateral margin of the sacrum.

The coccyx articulates with the fifth fused sacral vertebra at the **coccygeal cornua (KORN-ū-uh)**. Although they are variable in number, in most individuals there are four coccygeal bones.

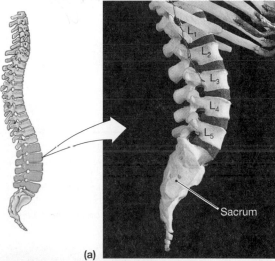

(a)

Figure 15.19
The Lumbar Vertebrae.
(a) A lateral view of the lumbar vertebrae and sacrum. **(b)** A lateral view of a typical lumbar vertebra. **(c)** A superior view of the same vertebra.

(b) Lateral view

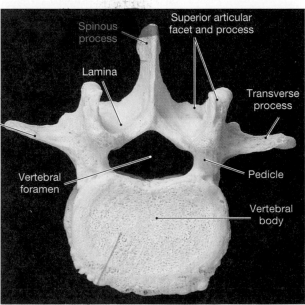

(c) Superior view

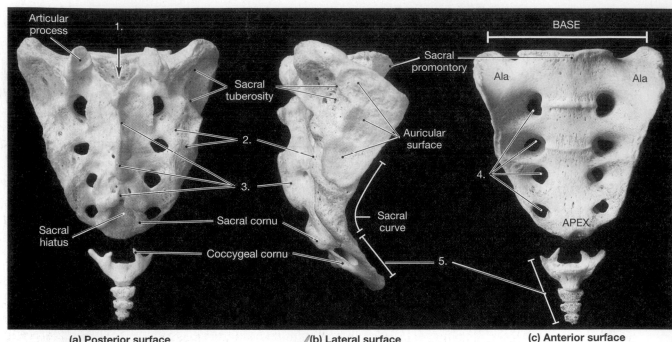

(a) Posterior surface   (b) Lateral surface   (c) Anterior surface

1. *Entrance to sacral canal*   3. *Median Sacral Crest*   5. *Coccyx*
2. *Lateral Sacral crest*   4. *Sacral foramina*

Figure 15.20
The Sacrum and Coccyx.
(a) A posterior view. **(b)** A lateral view from the right side. **(c)** An anterior view.

LABORATORY ACTIVITY   THE VERTEBRAL COLUMN

MATERIALS

articulated skeleton
articulated vertebral column
disarticulated vertebral column

PROCEDURES

1. Label and review the vertebral anatomy presented in Figures 15.15–15.20.
2. Identify the four regions of the vertebral column on an articulated skeleton.
3. Describe the anatomy of a typical vertebra. Locate each feature on a vertebra.
4. Distinguish the anatomical differences among cervical, thoracic, and lumbar vertebrae.
5. Identify the unique features of the atlas and the axis. How do these two vertebrae articulate with the skull and with one another?
6. Describe the articular features for rib attachment on a thoracic vertebra.
7. Discuss how a lumbar vertebra differs from a thoracic vertebra.
8. Describe the anatomy of the sacrum and the coccyx.

H. **Thoracic Cage**

There are 12 pairs of ribs in both males and females. They articulate with the thoracic vertebrae posteriorly and the sternum anteriorly to enclose the thoracic organs in a pro-

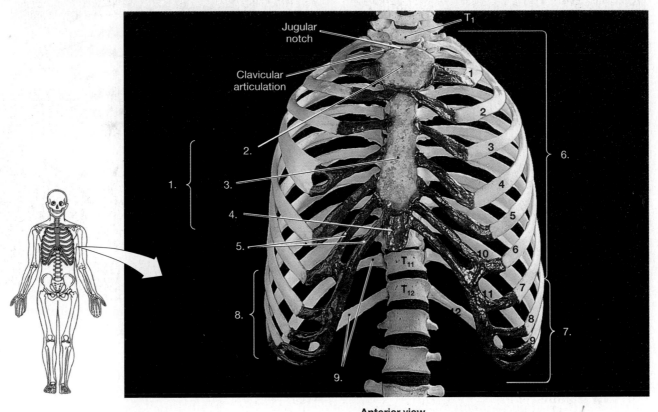

Anterior view

1. *sterum*
2. *manubrium*
3. *body*

4. *xiphoid process*
5. *costal cartilages*
6. *true, vertebrosternal ribs*

7. *false ribs*
8. *vertebrochondral rib*
9. *floating rib (ribs)*

Figure 15.21
The Thoracic Cage.
An anterior view of the thoracic cage and sternum.

tective rib cage. In breathing, muscles move the ribs to increase or decrease the size of the thoracic cavity and cause air to move in or out of the lungs.

Sternum

The **sternum** is composed of three bony elements, a superior **manubrium (ma-NOO-brē-um)**, a middle **body**, and an inferior **xiphoid (ZĪ-foyd) process**, all shown in Figure 15.21. The manubrium is triangular and articulates with the first pair of ribs and the clavicle. Muscles that move the head and neck attach to the manubrium. The body is elongated and receives the costal cartilage of ribs 2 through 7. The xiphoid process is shaped like an arrowhead and projects inferiorly off the sternal body. This process is cartilaginous until late in adulthood when it completely ossifies.

Ribs

Ribs, also called **costae**, are classified according to how they articulate with the sternum. Examine Figure 15.22, an anterior view of the rib cage. The first seven pairs are called **vertebrosternal** or **true ribs** because their cartilage, the costal cartilage, attaches directly to the sternum. The next five pairs are called **false ribs** because their costal

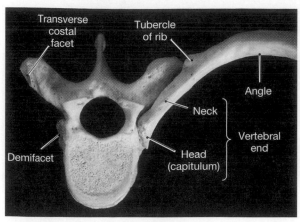

(a) Superior view

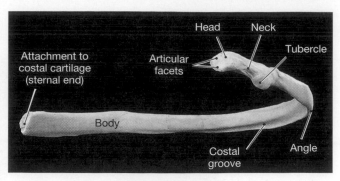

(b) Posterior view

Figure 15.22
The Ribs.

(a) Details of rib structure and the articulations between the ribs and thoracic vertebrae. **(b)** A posterior view of the head of a representative rib from the right side (ribs 2–9).

cartilages do not connect directly with the sternum. Ribs 8–10 are **vertebrochondral ribs** and their cartilages fuse with the cartilage of rib 7. Pairs 11 and 12 are called **floating (vertebral) ribs** and do not articulate with the sternum.

Figure 15.22b presents the gross anatomy of a rib. A rib has a **head** or **capitulum (ka-PIT-ū-lum)** with two **facets** for articulating with the demifacets of thoracic vertebrae, and a **tubercle** that articulates with the costal facet of the transverse process. Between the head and tubercle is a slender **neck**.

Differences in the articulation of ribs with the thoracic vertebrae are reflected in the variation of the demifacets. Vertebrae T_1 through T_8 all have paired demifacets with a superior demifacet and an inferior demifacet. The first rib articulates with an articular facet of T_1. The second rib articulates with the inferior demifacet of T_1 and the superior demifacet of T_2. Ribs 3 through 9 continue this pattern of articulating with two demifacets. Vertebrae T_9 through T_{12} have a single facet on the centrum, and the corresponding ribs articulate entirely on the one facet. After each rib articulates on a demifacet, the rib bends laterally and articulates with the costal facet on the transverse process. The last two pairs of ribs do not articulate on costal facets.

LABORATORY ACTIVITY   THE THORACIC CAGE

MATERIALS

articulated skeleton
articulated vertebral column with ribs
disarticulated vertebral column and ribs

PROCEDURES

1. Label Figure 15.21.
2. Discuss the anatomy of a typical rib. How many pairs of ribs do males and females have?

_____

3. Describe the anatomical features involved in the articulation of a rib on a thoracic vertebra.

_____

4. Identify the differences of articular facets along the thoracic region and relate this to how each rib articulates with the vertebrae.
5. Identify the bony elements of the sternum.

AXIAL SKELETON CHECKLIST

This is a list of bold terms presented in Exercise 15. Use this list for review purposes after you have completed the laboratory activities.

SUTURES AND FOSSAE OF THE SKULL

• Figures 15.2–15.5

coronal suture

sagittal suture

lambdoidal (occipitoparietal) suture

squamosal suture

sphenoparietal suture

anterior cranial fossa

middle cranial fossa

posterior cranial fossa

CRANIAL BONES

• Figures 15.1–15.10

frontal bone (1)

frontal squama

supraorbital foramen

metopic (frontal) suture

occipital bone (1)

foramen magnum

occipital condyles

hypoglossal canal

occipital crest

external occipital protuberance

superior nuchal line

inferior nuchal line

parietal bones (2)

superior temporal line

temporal bones (2)

external auditory canal

squamous portion

mastoid process

zygomatic process

styloid process

stylomastoid foramen

mandibular fossa

jugular foramen

carotid canal

petrous portion

internal acoustic canal

Sphenoid bone (1)

lesser wing

greater wing

sella turcica

anterior clinoid process

posterior clinoid process

foramen ovale

foramen lacerum

foramen spinosum

foramen rotundum

superior orbital fissure

optic canal

lateral and medial pterygoid processes

Ethmoid bone (1)

crista galli

cribriform plate

perpendicular plate

superior and middle conchae

FACIAL BONES OF THE SKULL

• Figures 15.1–15.4, 15.11

Maxillae (2)

infraorbital foramen

alveolar process

palatine process

Palatine bones (2)

greater palatine foramen

Zygomatic bones (2)

zygomaticofacial foramen

temporal process

zygomatic arch

Lacrimal bones (2)

lacrimal fossa

Nasal bones (2)

Vomer (1)

Inferior nasal conchae (2)

Mandible (1)
   ramus
   coronoid process
   condylar (mandibular) process
   mandibular notch
   angle
   body
   mental protuberance
   mental foramen
   mandibular groove
   mandibular foramen

HYOID BONE (1)

- Figure 15.12
   greater cornu
   lesser cornu

SINUSES OF THE SKULL

- Figure 15.13
paranasal sinuses
frontal sinus
sphenoid sinus
ethmoid sinus
air cells
maxillary sinus

FETAL SKULL

- Figure 15.14
fontanels

VERTEBRAL COLUMN

- Figure 15.15–15.20
intervertebral discs
spinal curves
primary curvatures
secondary curvatures

Gross vertebral anatomy
   centrum (body)
   spinous process
   transverse process
   pedicle
   lamina

vertebral foramen
superior articular process
superior articular facet
inferior articular process
inferior articular facet
inferior vertebral notch
intervertebral foramen (between adjacent vertebrae)

Cervical vertebrae (7)
   transverse foramen
   atlas
      posterior tubercle
   axis
      dens (odontoid process)

Thoracic vertebrae (12)
   articular facet
   superior demifacet
   inferior demifacet

Lumbar vertebrae (5)

Sacral vertebrae (5)
   sacrum
   sacral canal
   median sacral crest
   lateral sacral crest

Coccygeal vertebrae (usually 4)
coccygeal cornua

THORACIC CAGE

STERNUM (1)

- Figure 15.21
   manubrium
   body
   xiphoid process

RIBS (24)

- Figure 15.22
   head
   capitulum
   facet
   tubercle, neck
vertebrosternal (true) ribs (pairs 1–7)
false ribs (pairs 8–12)
vertebrochondral ribs (pairs 8–10)
floating (vertebral) ribs (pairs 11–12)

AXIAL SKELETON

Laboratory Report Exercise 15

A. **Short-Answer Questions**

1. List the components of the axial skeleton.

2. How many bones are found in the cranium and the face?

3. Describe the three cranial fossae and the bones that form the floor of each.

4. List the sutures of the skull and the bones that articulate at each.

5. Describe the four regions of the vertebral column.

B. **Matching**

Match each structure of the skull with the correct bone. Each choice may be used more than once.

1. _A_ sella turcica	**A.** sphenoid
2. _E_ crista galli	**B.** maxilla
3. _J_ external auditory meatus	**C.** frontal
4. _E_ foramen magnum	**D.** parietal
5. _J_ zygomatic process	**E.** occipital
6. _I_ condylar process	**F.** ethmoid
7. _J_ petrous portion	**G.** nasal
8. _A_ lesser wing	**H.** zygomatic
9. _J_ mandibular fossa	**I.** mandible
10. _J_ styloid process	**J.** temporal
11. _I_ coronoid process	
12. _J_ jugular foramen	
13. _e_ superior nuchal line	
14. ___ superior temporal line	
15. _C_ supraorbital foramen	

C. **Matching**

Match each structure of the vertebral column and rib cage with the correct description. Each choice can be used more than once.

1. _____ spinous process
2. _____ transverse foramen
3. _____ manubrium
4. _____ capitulum
5. _____ vertebrosternal rib
6. _____ pedicle
7. _____ tubercle
8. _____ xiphoid process
9. _____ centrum
10. _____ vertebral foramen
11. _____ vertebrocostal rib
12. _____ axis
13. _____ dens
14. _____ vertebral ribs
15. _____ facet

A. all vertebrae
B. second cervical vertebra
C. thoracic vertebrae
D. head of rib
E. true rib
F. body of vertebra
G. articulates with facet
H. false rib
I. ribs 11 through 12
J. sternum

D. **Labeling**

Label Figure 15.23, a typical lumbar vertebra.

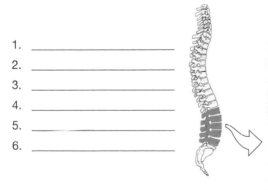

1. _____
2. _____
3. _____
4. _____
5. _____
6. _____

Figure 15.23
Typical Vertebra of the
Lumbar Region.

Vertebra is shown in superior view.

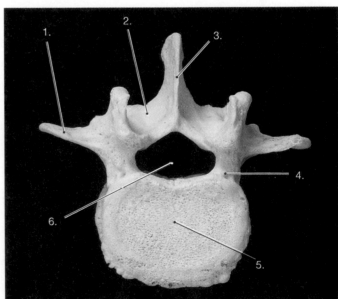

Typical lumber vertebra

The Appendicular Skeleton

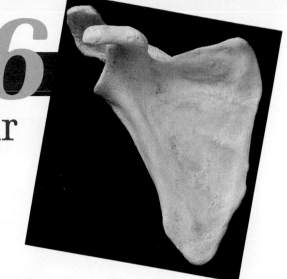

OBJECTIVES

On completion of this exercise, you should be able to:

- Identify the bones and surface features of the pectoral girdle and upper limb.
- Articulate the clavicle with the scapula.
- Articulate the humerus, radius, and ulna.
- Identify the bones of the hand.
- Identify the bones and surface features of the pelvic girdle and lower limb.
- Articulate the coxal bones with the sacrum to form the pelvis.
- Identify the bones of the foot.

WORD POWER

scapula (scapul—shoulder blade)
glenoid (glen—a pit)
coracoid (cora—a crow)
acromion (acromi—shoulder
blade point)
tubercle (tuber—a knot)
capitulum (capit—the head)
trochlea (trochle—a pulley)
coronoid (coron—a crown)
olecranon (olecran—the elbow)

carpus (carp—the wrist)
pisiform (pisi—a pea, form—shape)
hamate (hamat—a hook)
coxa (cox—the hip)
acetabulum (acetabul—a vinegar
cup)
trochanter (trochanter—a runner)
malleolus (malle—a hammer)
talus (talus—the ankle)
cuneiform (cuni—a wedge)

MATERIALS

articulated skeleton
disarticulated skeleton

INTRODUCTION

The appendicular division of the skeletal system hangs off the bones of the axial division. It provides the bony structure to the limbs, permitting us to move around and to interact with the surrounding environment. The shoulder or pectoral girdle is loosely attached to the sternum to allow the shoulder and arm a wide range of movement. The arm or upper limb is used to handle and carry objects. The pelvic girdle is securely attached to the sacrum of the spine. The lower limb of the leg supports the body and is moved for locomotion.

THE PECTORAL GIRDLE AND UPPER LIMB

The **pectoral girdle** of each shoulder has a shoulder blade or **scapula (SKAP-ū-la)** and a collarbone or **clavicle (KLAV-i-kul)**.

Each upper limb includes the bones of the arm, wrist, and hand. A total of 30 bones make up each limb, and all but 3 of them are in the wrist and hand. The upper arm bone is the **humerus**, and inferior to the humerus are the lateral **radius** and the medial **ulna**. The ulna and radius articulate with the humerus at the elbow. Eight **carpal** bones

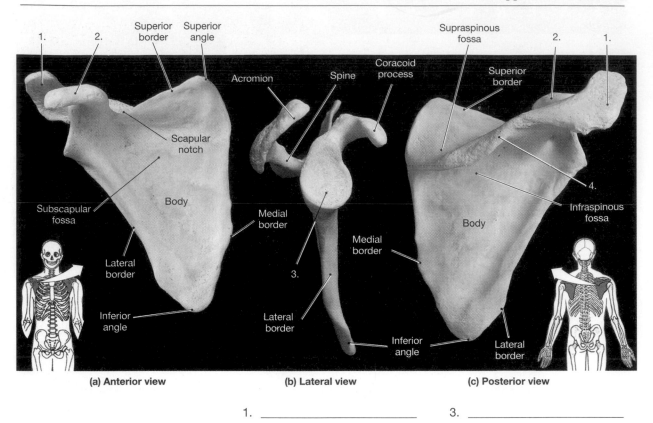

(a) Anterior view   **(b) Lateral view**   **(c) Posterior view**

1. _____   3. _____

2. _____   4. _____

Figure 16.1
The Scapula.
(a) Anterior, **(b)** lateral, and **(c)** posterior views of the right scapula.

comprise the wrist. The hand contains 5 long, slender bones in the palm called **metacarpals** and 14 **phalanges** of the fingers.

A. The Pectoral Girdle

The scapula

Locate the scapula on your disarticulated study skeleton. Orient the bone as in Figure 16.1c to view the posterior surface.

Most conspicuous is a prominent ridge of bone, the **spine**, extending across the scapula. Superior and inferior along the length of the spine are depressions, the **supraspinous** and **infraspinous fossae**. The **body** of the scapula is below the spine. At the lateral tip of the spine is the **acromion (a-KRŌ-mē-on)** which hangs over the **glenoid cavity** where the humerus articulates.

The edges of the scapula are the **superior**, **medial**, and **lateral borders** and the **superior** and **inferior angles**. The **subscapular fossa** is the smooth, triangular surface where the anterior of the scapula articulates with the ribs. Superior to the glenoid cavity is the beak-shaped **coracoid (KOR-uh-koyd) process**. At the base of the coracoid process, locate the **scapular notch**, an indentation in the superior border. The scapular **neck** is the ring of bone around the base of the coracoid process and the glenoid fossa. Rotate the scapula to the lateral view and notice how the coracoid process extends over the anterior surface of the scapula. Can you feel the spine and acromion on your scapula?

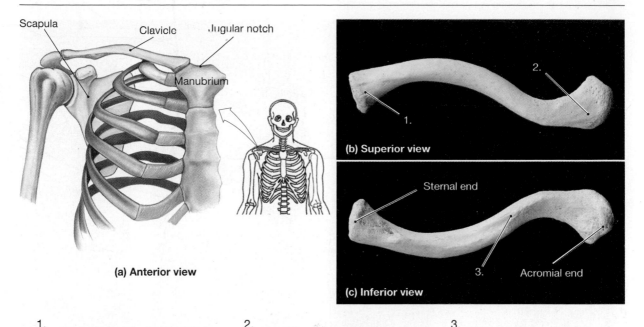

Figure 16.2
The Clavicle.

(a) The position of the clavicle, anterior view. **(b)** Superior and **(c)** inferior views of the right clavicle.

The clavicle

The S-shaped **clavicle** is the only connection between the pectoral girdle and the axial skeleton. The round **sternal end** articulates with the sternum, and the **acromial end** joins the scapula (see Figure 16.2). Inferiorly, toward the acromion where the clavicle bends, is the **conoid tubercle**, an attachment site for shoulder muscles.

LABORATORY ACTIVITY   THE PECTORAL GIRDLE

MATERIALS

> articulated skeleton
> disarticulated skeleton

PROCEDURES

1. Locate the scapula and clavicle on your body.
2. Identify the surface features of the scapula and label Figure 16.1.
3. Identify the surface features of the clavicle and label Figure 16.2.
4. Place the clavicle on your shoulder and determine how it would articulate with your scapula.

B. The Upper Limb

The humerus

Refer to Figure 16.3a and examine the proximal end of a humerus. The **head** of the humerus articulates with the glenoid cavity of the scapula. Lateral to the head is the

1. _____

2. _____

3. _____

4. _____

5. _____

6. _____

7. _____

8. _____

9. _____

10. _____

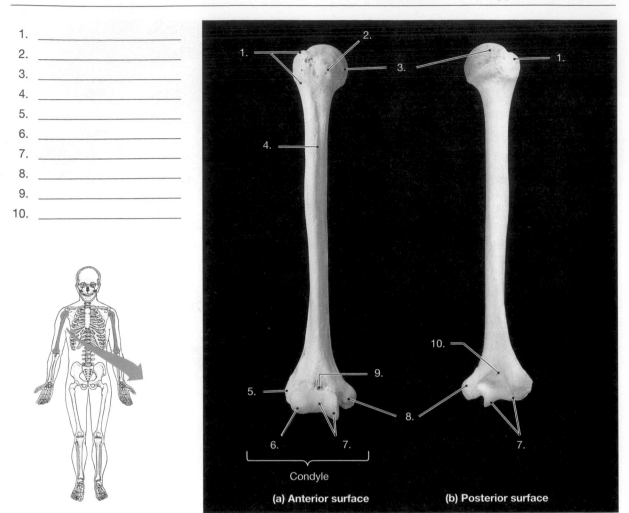

Figure 16.3
The Humerus.
(a) Anterior view of the right humerus. **(b)** Posterior view of the right humerus.

greater tubercle, and medial to the head is the **lesser tubercle**, both sites for muscle attachment. The **intertubercular groove** separates the tubercles. Between the head and the tubercles is the **anatomical neck**, and inferior to the tubercles is the **surgical neck**. Trace a finger distally from the greater tubercle and feel the rough **deltoid tuberosity** where the large deltoid muscle of the shoulder attaches.

The distal end of the humerus is specialized to accommodate two joints, the hingelike elbow joint and a pivot joint of the forearm. Locate the **medial epicondyle** in Figure 16.3a. Notice how this epicondyle is larger than the **lateral epicondyle** on the opposite side. Below the epicondyles is a single **condyle** with two distinct features, the round **capitulum** on the lateral side and the cylindrical **trochlea** located medially. Above the trochlea are two depressions, the anterior **coronoid fossa** and the posterior triangular **olecranon (ō-LEK-ruh-non) fossa** (Figure 16.3b). Can you feel the epicondyles on your humerus?

The ulna

The forearm has two parallel bones, the ulna and the radius. The ulna is easy to identify by the conspicuous U-shaped **trochlear notch**, as shown in Figure 16.4. The notch

1. _____
2. _____
3. _____
4. _____
5. _____
6. _____
7. _____
8. _____
9. _____
10. _____

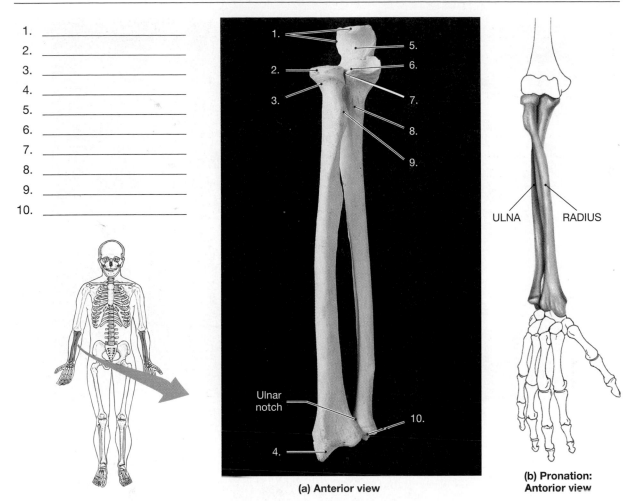

Ulnar notch

(a) Anterior view

ULNA   RADIUS

**(b) Pronation:
Anterior view**

**Figure 16.4
The Radius and Ulna.**

(a) The radius and ulna are shown in anterior view. **(b)** Note the changes that occur during pronation.

is like a C clamp with two processes that attach to the humerus, the superior **olecranon process** and the inferior **coronoid process**. The processes fit into their corresponding fossae on the humerus. On the lateral surface of the coronoid process is the flat **radial notch**. Below the notch is the rough **ulnar tuberosity**. The distal ulna has a pointed **styloid process**.

STUDY TIP

Notice that the terminology of the elbow is consistent in the humerus and ulna. The trochlear notch of the ulna fits into the trochlea of the humerus. The coronoid and olecranon processes fit into their respective fossae on the humerus.

The radius

The radius (see Figure 16.4) has a disk-shaped **head** that pivots in the radial notch of the ulna. Below the head is the **neck**, and inferior to the neck is the **radial tuberosity**. On the distal portion, the **ulnar notch** on the medial surface articulates with the ulna. The **styloid process** of the radius is larger and not as pointed as the styloid process of the ulna.

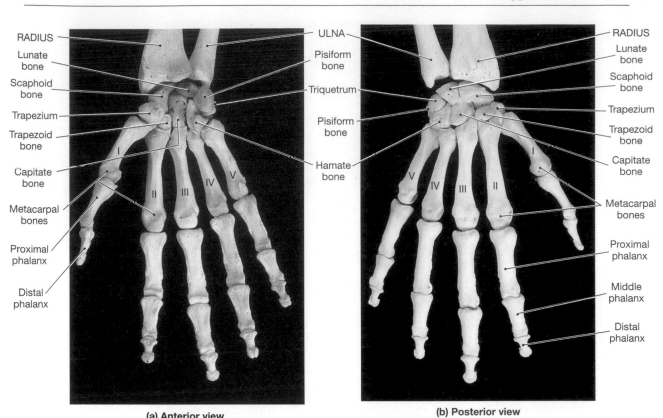

(a) Anterior view   (b) Posterior view

Figure 16.5
Bones of the Wrist and Hand.

(a) Anterior and **(b)** posterior views of the right hand.

C. The Wrist and Hand

Each wrist and hand (see Figure 16.5) contains 27 bones. The wrist has 8 **carpals (KAR-pulz)** arranged in two rows of 4, the proximal and the distal carpals. An easy method of identifying the carpals is to use the anterior wrist and start with the carpal next to the styloid process of the radius. From this reference point moving medially, proximal carpals are the **scaphoid**, **lunate**, **triangular (triquetral)**, and small **pisiform**. Returning on the lateral side, identify the four distal carpals: the **trapezium**, **trapezoid**, **capitate**, and **hamate**.

The 5 long bones of the palm are **metacarpals**. Each metacarpal is designated by a Roman numeral starting with the thumb metacarpus as I. The 14 bones of the fingers are called **phalanges**. Each finger has 3 individual bones, and the thumb, or *pollex*, has two phalanges.

LABORATORY ACTIVITY THE UPPER LIMB

MATERIALS

articulated skeleton
disarticulated skeleton

PROCEDURES

1. Locate the humerus, ulna, and radius on one side of your body.
2. Identify the surface features of the humerus. Label Figure 16.3.
3. Identify the bone markings of the ulna and radius and label Figure 16.4.
4. Identify the bones of the hand and wrist and study Figure 16.5.
5. Articulate the bones of the pectoral girdle and the upper limb.

THE PELVIC GIRDLE AND LOWER LIMB

The pelvic girdle forms the hips and is the anchoring point of the legs and spine. Each hip or **coxal** bone is formed by the fusion of three bones, the **ilium (IL-ē-um)**, **ischium (IS-kē-um)**, and **pubis (PŪ-bis)**. The **pelvis** is the region that includes the sacrum of the axial skeleton and the pelvic girdle. The lower limb consists of the **femur** in the thigh, a kneecap or **patella**, and the **tibia (TI-bē-uh)** and **fibula (FIB-ū-la)** of the lower leg. Each ankle contains 7 **tarsal** bones, and the foot has 5 **metatarsals** and 14 **phalanges**. Locate each bone of the lower limb in Figure 16.1 and on the study skeleton.

D. The Pelvic Girdle

The coxa

Locate a coxal bone on your study skeleton and refer to Figure 16.6a, a lateral view of the coxa. The **ilium**, **pubis**, and **ischium** fuse along a horizontal axis passing through the middle of the **acetabulum (a-se-TAB-ū-lum)**, the deep socket where the head of the femur articulates with the pelvic girdle. Usually, by 25 years of age the three bones have completely fused to create a single **coxa**. The superior ridge of the ilium is the **iliac crest** which is shaped like a shovel blade with an **anterior superior iliac spine** and a **posterior iliac spine**. The large indentation below the posterior iliac spine is the **greater sciatic (sī-A-tik) notch**. The notch terminates at a bony point, the **ischial spine**. Inferior to this spine is the **lesser sciatic notch**. The **ischial tuberosity** is in the most inferior portion of the ischium.

The pubis bone is on the anterior coxa. It joins the ischium just anterior to the ischial tuberosity, creating the **obturator (OB-tū-rā-tor) foramen**. The most anterior region of the pubis is the pointed **pubic tubercle**.

Examine the medial side of the coxa. Most conspicuous on the posterior iliac crest is the rough **auricular surface** where the **sacroiliac joint** attaches the pelvic girdle to the sacrum of the axial skeleton. Anteriorly, the pubis bones join at the **pubic symphysis**, a strong joint containing fibrocartilage. Locate these joints in Figure 16.6b.

The male and female pelvis are anatomically different (see Figure 16.7). The female pelvis has a wider pelvic outlet, the space between the ischial spines. The female pelvis also has a wider U-shaped pubic angle at the pubis symphysis than the V-shaped male pelvis. The wider female pelvis provides a larger passageway for childbirth.

LABORATORY ACTIVITY   THE PELVIC GIRDLE

MATERIALS

articulated skeleton
disarticulated skeleton

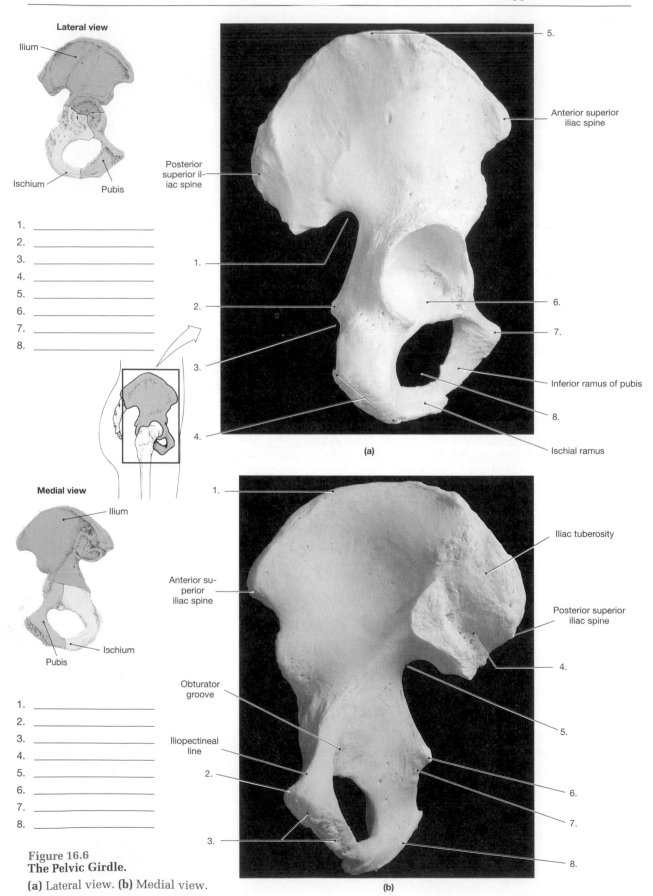

Lateral view

Ilium

Ischium   Pubis

1. _____
2. _____
3. _____
4. _____
5. _____
6. _____
7. _____
8. _____

Medial view

Ilium

Ischium

Pubis

1. _____
2. _____
3. _____
4. _____
5. _____
6. _____
7. _____
8. _____

5.

Anterior superior
iliac spine

Posterior
superior il-
iac spine

1.

2.

3.

6.

7.

Inferior ramus of pubis

4.

8.

Ischial ramus

(a)

1.

Iliac tuberosity

Anterior su-
perior
iliac spine

Posterior superior
iliac spine

4.

Obturator
groove

5.

Iliopectineal
line

2.

6.

7.

3.

8.

Figure 16.6
The Pelvic Girdle.

(a) Lateral view. **(b)** Medial view.

(b)

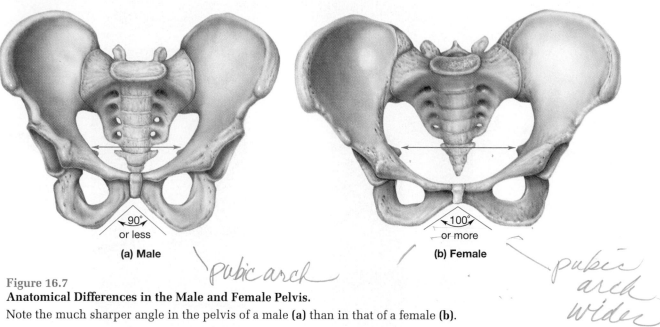

Figure 16.7
Anatomical Differences in the Male and Female Pelvis.
Note the much sharper angle in the pelvis of a male **(a)** than in that of a female **(b)**.

pubic arch (handwritten)   *pubic arch wider* (handwritten)

PROCEDURES

1. Locate a coxal bone on your body.
2. Identify the ilium, ischium, and pubis on a coxal bone on your study skeleton.
3. Identify the surface features of the coxa and label Figure 16.6.
4. Articulate both coxae and the sacrum to form the pelvis.

E. The Lower Limb

The femur

Locate a femur on your study skeleton. The femur is the largest bone of the skeleton. It supports the body's weight and bears the stress from the legs. Use Figure 16.8 of the femur as a guide to locate the smooth, round **head** of the femur. The head fits into the acetabulum of the coxa and permits the femur a wide range of movement. Below the head is a narrow **neck** that joins the head to the diaphysis. Lateral to the head is a large stump, the **greater trochanter (trō-KAN-ter)**; the **lesser trochanter** is on the medial surface. These large processes are attachment sites for powerful hip and thigh muscles. Along the middle of the posterior diaphysis is the **linea aspera**, a rough line for thigh muscle attachment. At the distal end of the femur are the **lateral** and **medial condyles** which articulate with the tibia. The condyles are separated posteriorly by the **intercondylar fossa**. A smooth **patellar surface** spans the lateral and medial condyles and serves as a gliding platform for the patella. Superior to the condyles are the **lateral** and **medial epicondyles**.

The patella

The patella is encased within the tendons of the anterior thigh muscles. The superior border of the bone is the flat **base**; the **apex** is at the inferior tip. Tendons attach to the rough anterior surface, and the smooth posterior **facets** glide over the condyles of the femur. The **medial facet** is narrower than the **lateral facet** (see Figure 16.9).

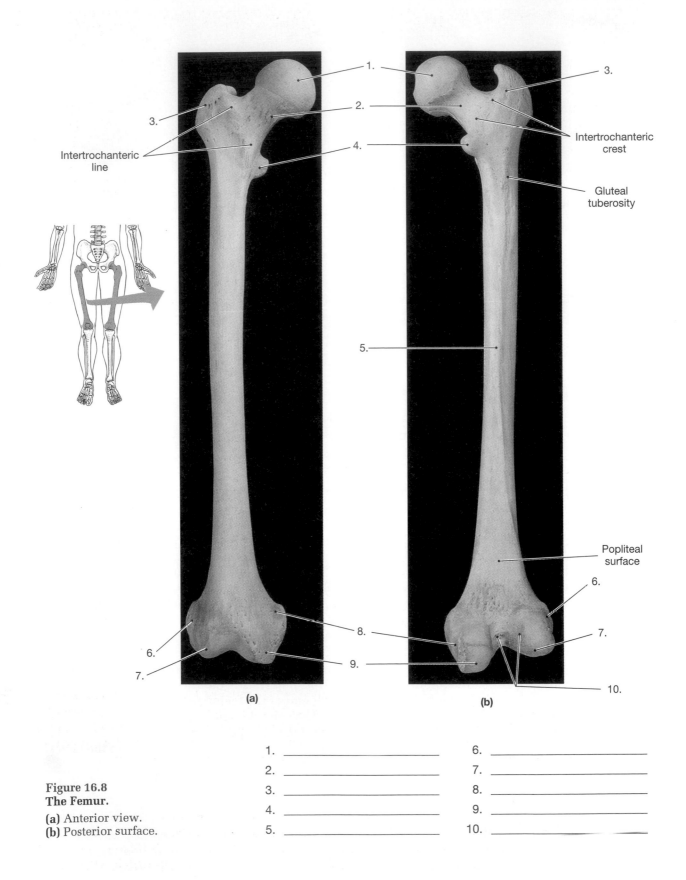

Figure 16.8
The Femur.

(a) Anterior view.
(b) Posterior surface.

1. _____   6. _____
2. _____   7. _____
3. _____   8. _____
4. _____   9. _____
5. _____   10. _____

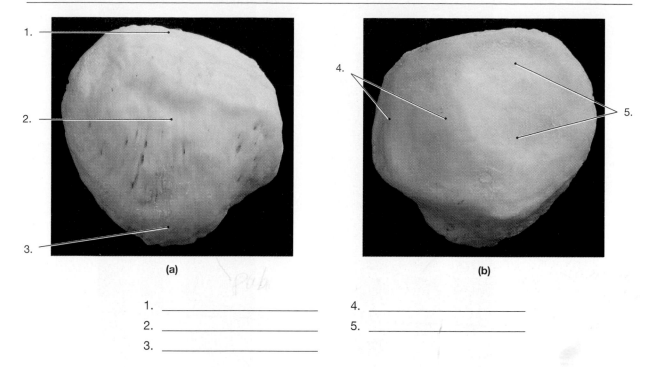

(a)   (b)

1. _____   4. _____
2. _____   5. _____
3. _____

Figure 16.9
The Patella.
(a) Anterior surface. **(b)** Posterior surface.

The tibia

The lower leg bones are illustrated in Figure 16.10. The tibia is the large medial bone of the lower leg. The proximal portion of the tibia flares to develop the **lateral** and **medial condyles** which articulate with the corresponding femoral condyles. Separating the tibial condyles is a ridge of bone, the **intercondylar eminence**. This eminence fits into the intercondylar fossa of the femur. Anteriorly below the condyles is the large **tibial tuberosity,** where thigh muscles attach. The anterior diaphysis forms a ridge, the **anterior crest**, along most of the shaft's length. The distal tibia is constructed to articulate with the foot. A large wedge, the **medial malleolus (ma-LĒ-ō-lus)**, helps to stabilize the ankle joint. The **inferior articular surface** is smooth so that it can slide over the talus of the ankle.

The fibula

The fibula is the slender bone lateral to the tibia (see Figure 16.10). The proximal and distal regions of the fibula appear very similar at first, but closer examination reveals the proximal **head** is rounded and less pointed than the distal **lateral malleolus**. The head of the fibula articulates below the lateral condyle of the tibia at the **superior tibiofibular joint**. The distal articulation creates the **inferior tibiofibular joint**.

F.   The Ankle and Foot

The ankle is composed of seven **tarsal** bones. The **talus** sits on top of the heel bone, the **calcaneus (kal-KĀ-nē-us)**, and articulates with the tibia and the lateral malleolus of the fibula. Anterior to the talus is the **navicular** which articulates with three **cuneiform** bones. Lateral to the navicular and the third cuneiform is the **cuboid** which articulates

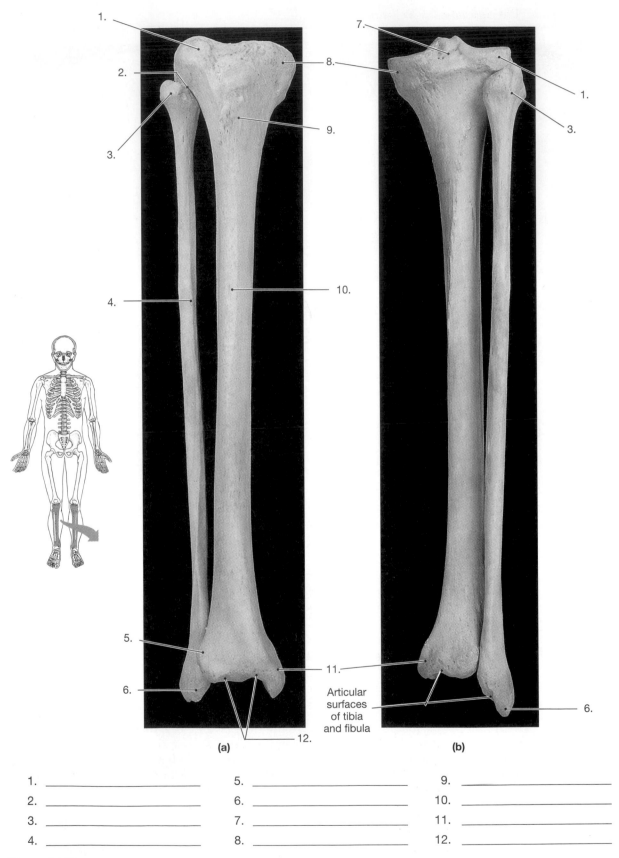

(a) **(b)**

Articular
surfaces
of tibia
and fibula

1. _____   5. _____   9. _____
2. _____   6. _____   10. _____
3. _____   7. _____   11. _____
4. _____   8. _____   12. _____

Figure 16.10
The Tibia and Fibula.

(a) Anterior. **(b)** Posterior.

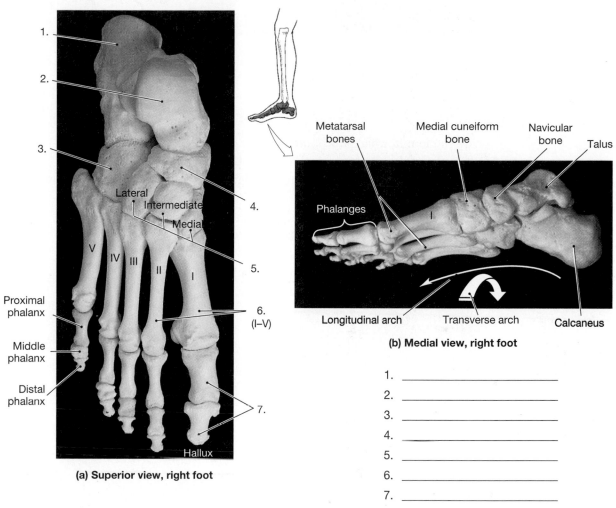

Figure 16.11
Bones of the Ankle and Foot.

posteriorly with the calcaneus. The arch of the foot is formed by five **metatarsals** spanning the arch. They are named with Roman numerals I through V, from medial to lateral. Like the fingers of the hand, each toe has three **phalanges** except the *hallux* or big toe which has two bones. Identify each bone of the foot in Figure 16.11.

STUDY TIP

It's easy to confuse the carpals and metacarpals of the hand with the tarsals and metatarsals of the foot. Just remember, when you listen to music you clap your carpals and tap your tarsals!

LABORATORY ACTIVITY   THE LOWER LIMB

MATERIALS

articulated skeleton

disarticulated skeleton

PROCEDURES

1. Locate the femur, tibia, and fibula on your body.
2. Identify the surface features of the femur. Label Figure 16.8.
3. Identify the surface features of the patella, tibia, and fibula. Label Figures 16.9 and 16.10.
4. Identify the bones of the ankle and foot. Label Figure 16.11.
5. Articulate the pelvis and the bones of the lower limb.

APPENDICULAR SKELETON CHECKLIST

This is a list of bold terms presented in Exercise 16. Use this list for review purposes after you have completed the laboratory activities.

PECTORAL GIRDLE

Scapula (2)
- Figure 16.1
  superior borders
  medial borders
  lateral borders
  superior angle
  inferior angle
  subscapular fossa
  glenoid cavity
  coracoid process
  scapular notch
  neck
  acromion process
  spine
  supraspinous fossa
  infraspinous fossa

Clavicle (2)
- Figure 16.2
  sternal end
  acromial end
  conoid tubercle

ARM

Humerus (2)
- Figure 16.3
  head
  greater tubercle
  lesser tubercle
  intertubercular groove
  anatomical neck
  surgical neck
  deltoid tuberosity
  medial epicondyle
  lateral epicondyle
  condyle
  capitulum

trochlea
coronoid fossa
olecranon fossa

Ulna (2)
- Figure 16.4
  trochlear notch (semilunar notch)
  olecranon process
  coronoid process
  radial notch
  ulnar tuberosity
  styloid process

Radius (2)
- Figure 16.4
  head
  neck
  radial tuberosity
  ulnar notch
  styloid process

WRIST AND HAND

- Figure 16.5
Carpals (16 total, 8 per hand)
  scaphoid
  lunate
  triangular (triquetral)
  pisiform
  trapezium
  trapezoid
  capitate
  hamate
Metacarpals (10 total, 5 per hand)
Phalanges (28 total, 14 per hand)

PELVIC GIRDLE

• Figures 16.6 and 16.7

Coxa (os coxa) (2)
    acetabulum
    obturator foramen
    auricular surface
    sacroiliac joint

Ilium (2)
    iliac crest,
    greater sciatic notch
    anterior superior iliac spine
    posterior superior iliac spine

Ischium (2)
    ischial tuberosity
    lesser sciatic notch
    ischial spine

Pubis (2)
    pubic tubercle
    pubic symphysis

LEG

Femur (2)
• Figure 16.8
head
neck
linea aspera
greater trochanter
lesser trochanter
lateral condyle
medial condyle
intercondylar fossa
patellar surface
lateral epicondyle
medial epicondyle

Patella (2)
• Figure 16.9
base
apex
lateral facet
medial facet

Tibia (2)
• Figure 16.10
lateral condyle
medial condyle
intercondylar eminence
tibial tuberosity
anterior crest
medial malleolus
inferior articular surface

Fibula (2)
• Figure 16.10
head
lateral malleolus
superior tibiofibular joint
inferior tibiofibular joint

FOOT

• Figure 16.11

Tarsals (14 total, 7 per foot)
    calcaneus
    talus
    navicular
    cuneiforms **(6 total, 3 per foot)**
Metatarsals (10 total, 5 per foot)
Phalanges (28 total, 14 per foot)

APPENDICULAR SKELETON

Laboratory Report Exercise 16

A. Matching

Match the surface feature on the left with the correct bone on the right. Each choice on the right can be used more than once.

1. _____ acromion
2. _____ intercondylar fossa
3. _____ trochlea
4. _____ glenoid fossa
5. _____ ulnar notch
6. _____ deltoid tuberosity
7. _____ greater trochanter
8. _____ sternal extremity
9. _____ lateral malleolus
10. _____ linea aspera
11. _____ capitulum
12. _____ medial malleolus
13. _____ intercondylar eminence
14. _____ base

A. clavicle
B. patella
C. fibula
D. humerus
E. femur
F. scapula
G. tibia
H. radius

B. Fill in the Blanks

Complete each statement by filling in the blank with the correct directional term.

1. The humerus is _____ to the radius.
2. The fibula is _____ to the tibia.
3. The talus is _____ to the calcaneus.
4. The clavicle is _____ to the scapula.
5. The patella is _____ to the femur.
6. The ilium is _____ to the ischium.
7. The ulna is _____ to the radius.
8. The metacarpals are _____ to the carpals.

C. **Short-Answer Questions**

1. List the bones of the pectoral girdle and the upper limb.

2. Discuss the features on the distal extremity of the humerus where the ulna and radius articulate.

3. Compare the bones of the hands and feet.

4. Describe the coxal bone.

5. List the bones of the lower limb.

6. Compare the pelvis in males and females.

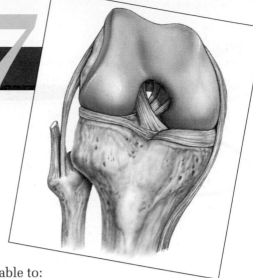

EXERCISE 17

Articulations

OBJECTIVES

On completion of this exercise, you should be able to:

- List the three groups of functional joints and give an example of each.
- List the four types of structural joints and give an example of each.
- Describe the types of diarthroses and the movement each produces.
- Describe the anatomy of a typical synovial joint.
- Describe and demonstrate the various movements of synovial joints.

WORD POWER

arthrology (arthro—joint)
synarthrosis (syn—together)
amphiarthrosis (amphi—semi)
diarthrosis (di—apart)
syndesmosis (desmo—band)
symphysis (symphysis—growing together)

gomphosis (gompho—peg, nail)
abduct (ab—away)
adduct (ad—toward)
pronate (pron—bent forward)
supinate (supin—lying on the back)

MATERIALS

articulated skeleton
fresh sectioned beef joint

INTRODUCTION

Arthrology is the study of the structure and function of joints. An articulation occurs where two or more bones fit together and form a joint. Identify several joints on your body. You most likely located joints that allow a free range of movement, such as your knee or hip joints. Many joints, however, allow little movement or no movement. Joints where the bones are closely held together are strong joints with limited movement between the bones. Joints contain a cavity between the bones that permits free movement. Some individuals have more movement in a particular joint than most people and are said to be "double jointed." In reality, they do not have two joints, and the additional movement allowed is due to the anatomy of the articulating bones, the structure of the joint, or the position of tendons and ligaments around the joint.

A. Classification of Joints

Two systems of classification are commonly used for articulations. The **functional** classification scheme groups joints by the amount of movement permitted. The **structural** system classifies joints by the type of connective tissue found between the articulating bones. Each classification method is useful when describing joints anatomically or functionally.

Functional classification of joints

The functional classification scheme has three groups of joints: immovable joints, the **synarthroses**; semimovable joints, the **amphiarthroses**; and freely movable joints, the **diarthroses**. Table 17.1 summarizes each group of functional joints.

TABLE 17.1	A Functional Classification of Articulations	*Kind of Movement*	
Functional Category	**Structural Category**	**Description**	**Example**
Synarthrosis (no movement)	**Fibrous**		
	Suture	Fibrous connections plus interlocking projections	Between the bones of the skull
	Gomphosis	Fibrous connections plus insertion in alveolar process	Between the teeth and jaws
	Cartilaginous		
	Synchondrosis	Interposition of cartilage plate	Epiphyseal plates
	Bony fusion		
	Synostosis	Conversion of other articular form to solid mass of bone	Portions of the skull, epiphyseal lines
Amphiarthrosis (little movement)	**Fibrous**		
	Syndesmosis	Ligamentous connection	Between the tibia and fibula
	Cartilaginous		
	Symphysis	Connection by a fibrocartilage pad	Between right and left halves of pelvis; between adjacent vertebral bodies along vertebral column
Diarthrosis (free movement)	**Synovial**	Complex joint bounded by joint capsule and containing synovial fluid	Numerous; subdivided by range of movement
	Monaxial	Permits movement in one plane	Elbow, ankle
	Biaxial	Permits movement in two planes	Ribs, wrist
	Triaxial	Permits movement in all three planes	Shoulder, hip

Synarthroses (sin-ar-THRŌ-sēz) have bones that are closely fitted together or have a strong ligament surrounding the bones. Three types of synarthroses are found in the skeleton. **Sutures** occur where the bones of the skull interdigitate or interlace together. This forms a very strong but immovable joint. The joint between the teeth and the alveolar bone of the jaw is a **gomphosis (gom-FŌ-sis)**. This joint is lined with a strong periodontal ligament which holds the teeth in place and permits no movement. Other synarthroses include joints between the diaphysis and epiphyses of long bones. In an immature bone, a plate of cartilage called the **epiphyseal plate** is the site of bone growth. Because this plate separates the bone's shaft from its end, it is considered an articulation and is called a **synchondrosis** (together with cartilage). On maturity, the plate of cartilage is replaced by bony tissue, and the epiphyses fuse with the diaphysis. This synarthrotic joint is then called a **synostosis (sin-os-TŌ-sis)**, a bony fusion across a joint.

The second category of functional joints are the **amphiarthroses (am-fē-ar-THRŌ-sēz)**. These joints are held together by strong connective tissues and exhibit minimal movement. Between parallel bones of the forearm and lower leg a ligament of fibrous connective tissue forms and wraps around the bones. This strong band results in a **syndesmosis (sin-dez-MŌ-sis)** joint (syn—together; desmo—band). The syndesmosis prevents excessive movement in these joints.

The **symphyses** are amphiarthroses characterized by the presence of **fibrocartilage** between the articulating bones. The **intervertebral disks** between the vertebrae are composed of fibrocartilage and construct a symphysis between the articulating vertebrae. Where the two coxal bones unite at the pubis is another symphysis, the **symphysis pubis**.

This strong joint limits flexion of the pelvis. During childbirth, a hormone softens the fibrocartilage to widen the pelvic bowl. Locate each symphysis on an articulated skeleton.

Diarthroses (dī-ar-THRŌ-sēz) are freely movable joints. Bones in these joints are separated by a small **joint cavity** lined with a **synovial (sin-NŌ-vē-ul) membrane**. The anatomy of a diarthrotic joint will be examined in more detail in a later section in this exercise.

STUDY TIP

The terminology used to describe articulations is often difficult to master without an understanding of the meaning of the terms. Refer to the Word Power list at the beginning of the exercise frequently and learn the prefixes used for each group of joints.

The presence of a joint cavity in diarthroses allows a wide range of motion. Movements are classified according to the number of planes the bones move through. **Monaxial (mon-AKS-ē-ul)** joints, like the elbow, move in one plane, and a **biaxial (bī-AKS-ē-ul)** joint allows movement in two planes. Move your wrist up and down and side to side to demonstrate biaxial movement. **Triaxial (trī-AKS-ē-ul)** joints occur in the ball-and-socket joints of the shoulder and hip and permit movement in three planes.

Structural classification of joints

The structural classification of joints groups articulations according to the type of connective tissue present between bones. This method is important when discussing joint anatomy, rather than movement. Each group of structural joints has anatomical features that distinguish them from other groups. Table 17.2 summarizes the structural classification of joints and relates each type of joint to its functional category.

Four types of structural joints occur in the skeleton. The first group is the **bony fusion**, joints where bones have fused together and permit no movement. A good example of a bony fusion is the frontal bone. Humans are born with two frontal bones, and by the age of 8 years they fuse into a single frontal bone. The old articulation site is then occupied by bony tissue. This joint is also called a synostosis (together with bone).

Fibrous joints have a strong fibrous connective tissue between the articulating bones. Little to no movement occurs in fibrous joints. There are several types of fibrous joints; for example, the **sutures** of the skull are lined with fibrous connective tissue. The fibrous tissue is removed during the preparation of the skull and is not present on the study skulls in the laboratory. The **syndesmosis** is a fibrous ligament between the tibia and fibula that prevents excessive movement. A syndesmosis also occurs between the ulna and radius.

Cartilaginous joints, as their name implies, have cartilage between the bones. The type of cartilage, hyaline or fibrocartilage, determines the type of cartilaginous joint. Between the ribs and the sternum is a **synchondrosis (sin-kon-DRŌ-sis)**, a cartilaginous joint with hyaline cartilage. The **symphysis** joint of the pubis bone and intervertebral discs has fibrocartilage between the bones.

An interesting synchondrosis is the temporary joint in a juvenile long bone, the growth plate or **epiphyseal plate**. While the bone is immature, the diaphysis and epiphysis are separated by a plate of cartilage, the tissue of the immovable synchondrosis. A long bone stops growing in length once the cartilage plate has been replaced by bone tissue. All that remains of the joint is a remnant called the **epiphyseal line**, and the former synchondrosis is then classified as a synostosis.

Synovial joints have a joint cavity lined by a synovial membrane. All the free-moving joints, the diarthroses, are also synovial joints. In the next section you will learn the anatomy of a synovial joint.

what holds together)

TABLE 17.2	A Structural Classification of Articulations		
Structure	*Type*	*Functional Category*	*Example*
Bony fusion	Synostosis (illustrated)	Synarthrosis	**Frontal bone**
Fibrous joint	Suture (illustrated) Gomphosis Syndesmosis	Synarthrosis Synarthrosis Amphiarthrosis	**Skull**
Cartilaginous joint	Synchrondrosis Symphysis (illustrated)	Synarthrosis Amphiarthrosis	**Pelvis**
Synovial joint	Monaxial Biaxial Triaxial (illustrated)	Diarthroses	**Synovial joint**

CLINICAL APPLICATION—ARTHRITIS

Arthritis is a disease that destroys a synovial joint by damaging the articular cartilage. Two types of arthritis occur: rheumatoid arthritis and osteoarthritis. **Rheumatoid arthritis** is an autoimmune disease that occurs when the body's immune system attacks the cartilage and synovial membrane of the joint. As the disease progresses, the joint cavity is eliminated and the articulating bones fuse, resulting in painful disfiguration of the joint and loss of joint function. **Osteoarthritis** is a degenerative joint disease that often occurs due to age and wearing of the joint tissues. The articular cartilage is damaged, and bone spurs may project into the joint cavity. Osteoarthritis tends to occur in the knee and hip joints, while rheumatoid arthritis is more common in the smaller joints of the hand.

LABORATORY ACTIVITY   CLASSIFICATION OF JOINTS

MATERIALS

PROCEDURES

1. Locate on your body a joint from each functional and structural group. Complete Table 17.3 as you make your observations.

TABLE 17.3	Classification of Joints		
Joint	*Functional Group*	*Structural Group*	*Example*
Syndesmosis	_____	_____	_____
Synchondrosis	_____	_____	_____
Suture	_____	_____	_____
Synovial	_____	_____	_____
Symphysis	_____	_____	_____

2. Identify two monaxial joints, two biaxial joints, and two triaxial joints on your body. Fill in Table 17.4 with your observations.

TABLE 17.4	Representative Synovial Joints	
Plane of Motion	*Example 1*	*Example 2*
Monaxial joint	_____	_____
Biaxial joint	_____	_____
Triaxial joint	_____	_____

B. Anatomical Structure of Synovial (Diarthrotic) Joints

As mentioned previously, all diarthrotic joints are capable of free movement. This range of motion is due to the anatomical organization of the joint. Between the bones of all diarthrotic joints is a cavity lined with a synovial membrane. Because of this, diarthrotic joints are structurally classified as synovial joints.

Figure 17.1a highlights the important features of a typical synovial joint. First, identify the **joint cavity** between the bones and the **articular cartilage** on the end of each bone. This cartilage is hyaline cartilage and provides a slippery, gelatinous surface should the two bones make contact across the cavity. Locate the **synovial membrane** that produces synovial fluid. How is this membrane positioned in the joint? Injury to a joint may cause inflammation of this membrane and lead to excessive fluid production. Finally, notice how the periosteum of each bone is continuous with the strong **articular capsule** that encases the joint.

Figure 17.1b illustrates the general features of the knee joint. Notice the large **bursa** between the patella and the femur. Bursae are similar to synovial membranes, except instead of lining joint cavities they provide padding between bones and other structures.

LABORATORY ACTIVITY   SYNOVIAL JOINT STRUCTURE

MATERIALS

fresh beef joint

PROCEDURES

1. Complete the labeling of Figure 17.1.

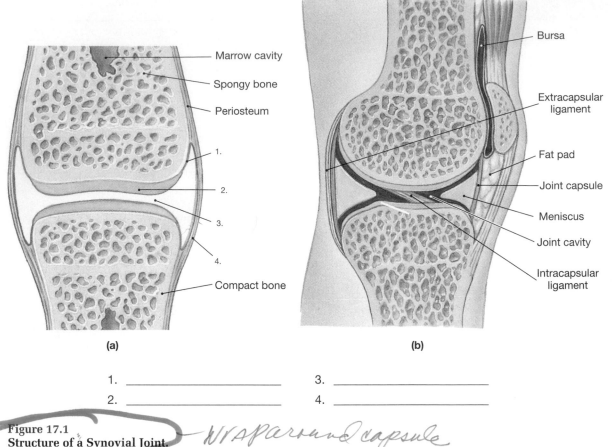

(a)   (b)

1. _____   3. _____

2. _____   4. _____

Figure 17.1
Structure of a Synovial Joint.   *Wraparound capsule*

(a) Basic structure of a typical synovial joint. (b) Sectional view of the knee joint.

2. If one is available in the laboratory, examine a fresh beef joint. Locate and
   describe the following structures:
   a. synovial cavity
   b. articular cartilage
   c. articular capsule

C. Types of Diarthroses

Six types of diarthroses or synovial joints occur in the skeleton. Each type permits a
certain amount of movement owing to the joining surfaces of the articulating bones.
Figure 17.2 details each type of joint and includes a mechanical representation of each
joint to show planes of motion.

Hinge joints are monaxial joints that permit movement in one plane, like a door
hinge. Hinge joints are located in the elbow, the fingers and toes, and the knee. Imagine
standing in anatomical position and then bending your legs and arms. This movement
is possible because of your hinge joints.

Pivot joints are monaxial joints that permit one bone to rotate around a relatively stationary bone. Although rotation occurs, the bones move within only one plane. Shake
your head "no" to demonstrate the pivot joint between your first two cervical vertebrae.
The first cervical vertebra, the atlas, pivots around the second cervical vertebra, the axis.

Gliding joints are common joints where flat articular surfaces, such as in the wrist,
slide by neighboring bones. The movement is typically monaxial. Other glide joints
occur between bones of the sternum and between tarsals of the foot. Flatted your hand
on your desk and observe the gliding of your wrist bones.

extension +f
into Another Open Close
Fingers
Knee

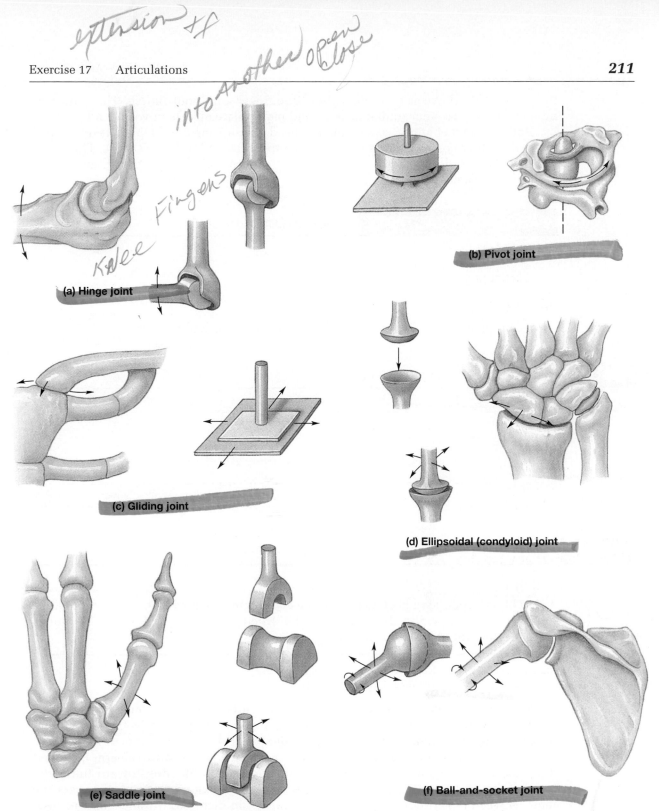

(a) Hinge joint

(b) Pivot joint

(c) Gliding joint

(d) Ellipsoidal (condyloid) joint

(e) Saddle joint

(f) Ball-and-socket joint

Figure 17.2
A Functional Classification of Synovial Joints.

Ellipsoidal joints are also called **condyloid** joints. A convex surface of one bone articulates in a depression of another bone. This concave-to-convex spooning of articulating surfaces permits biaxial movement in two planes. The articulation between the bones of the forearm and wrist is an ellipsoidal joint.

The **saddle joint** is a special biaxial joint found only at the junction of the thumb metacarpus and the triquetrum of the wrist. Place a finger on your lateral wrist and feel the saddle joint move as you touch your little finger with your thumb. This joint permits you to oppose your thumb to grasp and manipulate objects in your hand.

Ball-and-socket joints have a rounded head of one bone that fits into a cup-shaped fossa of another bone, as in the joint between the humerus and the scapula. This triaxial joint permits dynamic movement in many planes. Locate another ball-and-socket joint in your skeleton.

LABORATORY ACTIVITY   TYPES OF DIARTHROSES

MATERIALS

articulated skeleton

PROCEDURES

1. Complete Table 17.5 by locating each joint on your body.

TABLE 17.5   **Examples of Diarthrotic Joints**

Type of Diarthrosis	Example
Pivot	
Hinge	
Ball and socket	
Ellipsoidal	
Gliding	
Saddle	

2. On an articulated skeleton, examine each type of joint located in the previous procedure. Specifically, notice how the structure of the joining bones determines the amount of joint movement.

D. Skeletal Movement at Diarthrotic Joints

The diversity of bone shapes and joint types permits a variety of movements in the skeleton. Examine Figure 17.3 which illustrates angular movements. Notice that the figure includes a small dot at the joint where a demonstrated movement is described.

Figure 17.3a demonstrates side-to-side movements. Notice how the arm is moved at the shoulder for **abduction** away from the body and **adduction** toward the body. Practice this movement with your arms and fingers. **Flexion** of a joint decreases the angle between the articulating bones, while **extension** increases the angle between the bones (see Figure 17.3b). Hang your arm down at your side and move it into anatomical position with your palm forward. Now flex your arm by moving the elbow joint. Your hand should be up by your shoulder. Notice how close the antebrachium is to the brachium and how the angle between them has decreased. Is your flexed arm still in anatomical position?

Now extend your arm to return it to anatomical position. How has the angle changed? **Hyperextension** moves the body beyond anatomical position. Follow Figure 17.3b and flex, extend, and hyperextend your head. How are flexion and hyperextension different?

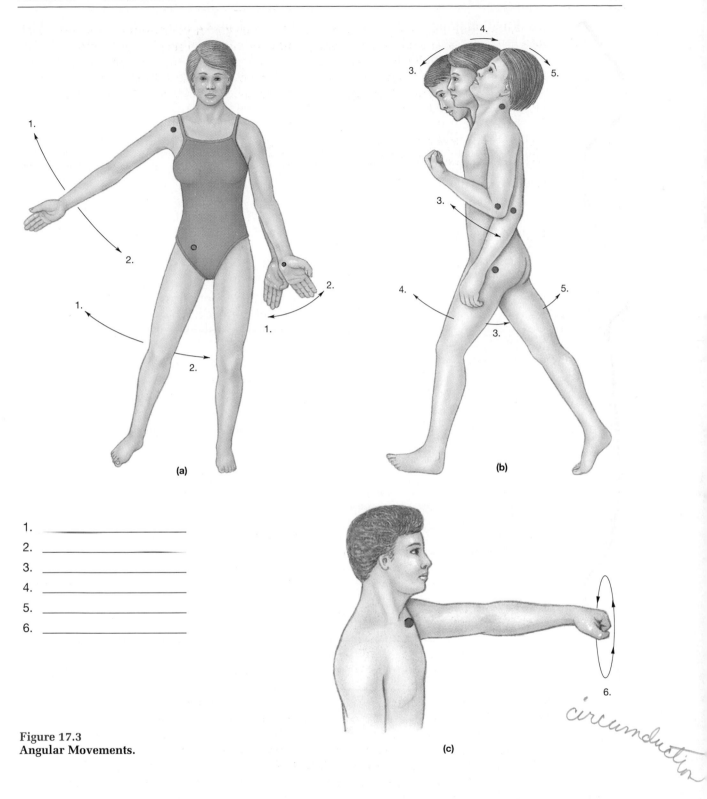

1. _____
2. _____
3. _____
4. _____
5. _____
6. _____

Figure 17.3
Angular Movements.

circumduction

   Circumduction is another movement at a ball-and-socket joint. Swing your arm over your head as if it were a fan blade. Notice how the proximal region of the arm is relatively stationary while the distal portion traces a wide circle in the air. Then stand up and circumduct your leg. Circumduction is a great movement for working the muscles of the thigh.

Rotational movement occurs at the ball-and-socket joints located in the shoulder and hip (see Figure 17.4). This movement turns the rounded head of one bone in a cup-shaped socket of another bone. The resulting motion is **lateral** or **medial rotation**. With your arm in anatomical position, rotate it to move the palm of your hand posteriorly. Your elbow joint should not move during the rotation. All movement is at the ball-and-socket joint of the pectoral girdle. When might you use this type of movement?

Figure 17.5 represents special movements that occur only at specific diarthrotic joints. Place your arms in anatomical position. Now, without causing rotation at the shoulder, turn your palms so they face posteriorly. This movement out of anatomical position is called **pronation (prō-NĀ-shun)**. **Supination (soo-pi-NĀ-shun)** of the palm returns the hand to the anterior position, the anatomical position. Pronation and supination occur in the elbow as the radius pivots around the ulna. The humerus functions as a foundation for muscles to pull from. What type of daily activities require you to pronate and supinate? Touch your thumb pad with the pad of your little finger. This movement is called **opposition**.

Lateral movement of the ankle to move the sole outward is **eversion (ē-VER-zhun)**. Positioning the sole medially is **inversion**; the foot moves "in." Terms that are somewhat confusing are those related to moving the ankle. **Dorsiflexion** permits you to walk on your heels; the foot is raised, and the angle of the ankle is decreased. **Plantar flexion** moves the foot to walk on your tiptoes. Remember to "plant" your toes for plantar flexion. The body may be thrust forward or pulled back.

Retraction, which means to take back, moves structures posteriorly out of anatomical position. Move your mandible back, as if to demonstrate an overbite. Now **protract** your mandible by jutting your chin out as if it were a pointer. Practice retracting and protracting your shoulders, too.

The mandible also moves up and down. Lower your mandible as if to take a bite of food. This movement is called **depression**. Closing your mouth **elevates** the mandible. Examine these movements in Figure 17.5. Name another joint that permits depression and elevation.

LABORATORY ACTIVITY   MOVEMENT AT DIARTHROTIC JOINTS

MATERIALS

PROCEDURES

1. Demonstrate the following movements and complete Table 17.6.

TABLE 17.6   **Movements at Selected Synovial Joints**

Movement	Description	Example
Supination	_____	_____
Hyperextension	_____	_____
Abduction	_____	_____
Rotation	_____	_____
Flexion	_____	_____

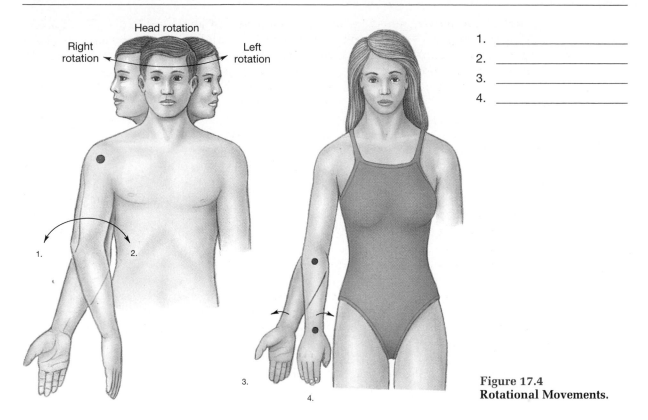

1. _____
2. _____
3. _____
4. _____

Figure 17.4
Rotational Movements.

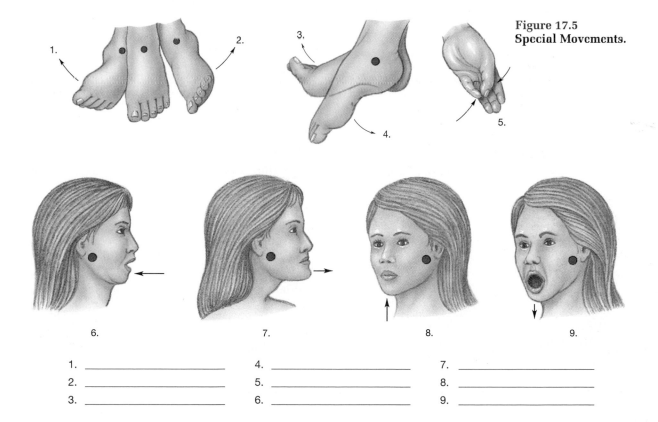

Figure 17.5
Special Movements.

1. _____   4. _____   7. _____
2. _____   5. _____   8. _____
3. _____   6. _____   9. _____

ARTICULATIONS CHECKLIST

This is a list of bold terms presented in Exercise 17. Use this list for review purposes after you have completed the laboratory activities.

CLASSIFICATION OF JOINTS

• Tables 17.1 and 17.2

Functional classification

synarthroses

   sutures

   gomphosis

   synchondrosis

   synostosis

amphiarthroses

   syndesmosis

   symphysis

   intervertebral discs

   symphysis pubis

diarthroses

   monaxial joint

   biaxial joint

   triaxial joint

structural classification

bony fusion

fibrous joints

   sutures

   syndesmosis

cartilaginous joints

   synchondrosis

   epiphyseal plate

   symphysis

synovial joints

ANATOMY OF A SYNOVIAL JOINT

• Figure 17.1

joint cavity

articular cartilage

synovial membrane

articular capsule

bursa

TYPES OF DIARTHROSES

• Figure 17.2

hinge joints (monaxial)

pivot joints (monaxial)

gliding joints (monaxial)

ellipsoidal or condyloid joints (triaxial)

saddle joint (biaxial)

ball-and-socket joints (triaxial)

MOVEMENT AT DIARTHROTIC JOINTS

• Figures 17.3, 17.4, and 17.5

flexion

extension

hyperextension

rotation (lateral and medial)

circumduction

abduction

adduction

opposition

eversion

inversion

dorsiflexion

plantar flexion

pronation

supination

retraction

protraction

depression

elevation

ARTICULATIONS

Laboratory Report Exercise 17

A. Matching

Match the joint on the left with the correct description on the right.

1. _____ pivot
2. _____ symphysis
3. _____ ball and socket
4 _____ gomphosis
5. _____ hinge
6. _____ suture
7. _____ synostosis
8. _____ syndesmosis
9. _____ condyloid
10. _____ synchrondrosis

A. forearm and wrist joint
B. joint between parietal bones
C. rib and sternum joint
D. joint between vertebral bodies
E. femur and coxa joint
F. phalangeal joint
G. distal tibio-fibular joint
H. atlas and axis joint
I. fused frontal bones
J. tooth in socket

B. Matching

Match the movement on the left with the correct description on the right.

1. _____ retraction
2. _____ dorsifloxion
3. _____ eversion
4 _____ inversion
5. _____ pronation
6. _____ plantar flexion
7. _____ protraction
8. _____ supination
9. _____ adduction
10. _____ abduction

A. movement away from midline
B. foot turned outward
C. palm faces posteriorly
D. palm faces anteriorly
E. movement to posterior plane
F. standing on tiptoes
G. movement in anterior plane
H. foot turned inward
I. standing on heels
J. movement toward midline

C. Describe the joints and movements involved in:

a. walking _____
b. throwing a ball _____
c. turning a doorknob _____
d. crossing your legs while sitting _____
e. shaking your head "no" _____
f. chewing food _____

D. **Short-Answer Questions**

   1. Describe the three types of functional joints.

   2. What factors limit the movement of a joint?

   3. Outline the various types of structural joints.

E. **Drawing**

   Sketch the basic structure of a synovial joint.

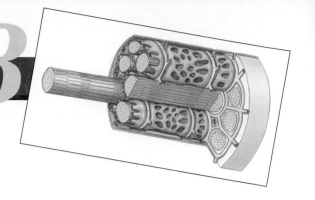

Muscular System Overview

OBJECTIVES

On completion of this exercise, you should be able to:

- Describe the basic functions of the muscular system.
- Describe the organization of a skeletal muscle.
- Describe the microanatomy of a muscle fiber.
- Discuss and provide examples of a lever system.
- Understand some of the rules that determine names of muscles.

WORD POWER

epimysium (epi—over; mys—muscle)
perimysium (peri—around)
fascicle (fasci—a bundle)

sarcolemma (sacro—flesh; lemma—a sheath)
synergist (synerg—work together)
synapse (synap—a union)

MATERIALS

muscle cell model
heavy textbook
pencils
teased skeletal muscles
dissecting microscope

INTRODUCTION

Every time a part of the body is moved, either consciously or unconsciously, muscles are used. Skeletal muscles are primarily responsible for **locomotion**, movement of the body. Actions such as rolling your eyes, writing your name, and speaking are the result of highly coordinated muscle contractions. Other functions of skeletal muscle include maintenance of **posture** and body **temperature** and **support** of soft tissues, as with the muscles of the abdomen. Smooth muscle is found in blood vessels, in the respiratory and digestive tracts, and in the reproductive system. Heart muscle comprises the walls of the heart and pushes blood through the cardiovascular system.

In addition to muscle tissue's ability to contract, it has several other unique characteristics. Muscles, along with nerve tissue, are **excitable** and produce electrical impulses in response to a stimulus. Muscle tissue is **extensible** and can be stretched. When the ends of a stretched muscle are released, it recoils to its original size, like a rubber band. This property is called **elasticity**.

A. Skeletal Muscle Organization

Connective tissue coverings

A collagenous connective tissue layer called the **epimysium (ep-i-MĪZ-ē-um)** covers the entire muscle. As shown in Figure 18.1, the epimysium is continuous with the fascia and deep fascia described below and separates the muscle from neighboring structures. The epimysium folds into the belly of the muscle as the **perimysium (per-i-MĪZ-ē-um)** and separates the muscle into bundles of muscle cells called **fascicles (FA-sik-ulz)** or **fasciculae**. Fascicles can be easily seen as parallel threadlike fibers when a muscle is

219

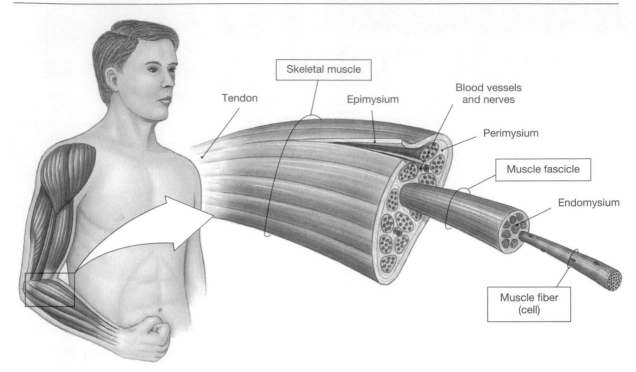

Figure 18.1
Organization of Skeletal Muscles.
Gross organization of skeletal muscle.

teased apart with a probe or a scalpel blade (see the Laboratory Activity below). An extension of the perimysium, called the **endomysium (en-dō-MĪZ-ē-um)**, plunges deep into each fascicle and surrounds each individual muscle fiber.

The epimysium extends as a strong cord called the **tendon** which blends into the periosteum of the attached bone. When the muscle fibers contract and generate tension, they transmit this force through the connective tissue layers to the tendons, which pulls on the associated bone and produces movement.

Structure of a skeletal muscle fiber

Skeletal muscles are composed of cellular structures called **muscle fibers** (see Figure 18.2). Each muscle fiber is a composite of many cells that fused during embryonic development. The cell membrane of a muscle fiber is called the **sarcolemma (sar-cō-LEM-uh)**. Connecting the sarcolemma to the interior of the muscle fiber are many **transverse tubules** (T tubules) which function in passing contraction stimuli to deeper regions of the muscle fiber.

Inside the muscle fiber are proteins arranged in thousands of rods called **myofibrils** which extend the length of the fiber. Each myofibril is surrounded by membranes of the **sarcoplasmic reticulum**, a site of calcium storage. The sarcoplasmic reticulum (SR) surrounds each myofibril within the **sarcoplasm**. The branches of the SR fuse to form large, expanded chambers called **terminal cisternae (sis-TUR-nē)**, which lie adjacent to the transverse tubules. The terminal cisternae and lumen of the SR serve as an intracellular storage site for calcium. The membranes of the SR contain specialized proteins for transporting calcium ions into the cisternae. Transverse tubules stimulate the release of calcium ions from the sarcoplasmic reticulum during muscle contraction.

Each myofibril consists of several different kinds of proteins arranged in about 3,000 **thin** and 1,500 **thick** filaments (see Figure 18.2b). During contraction, thick and thin protein molecules interact to produce tension and shorten the muscle. The

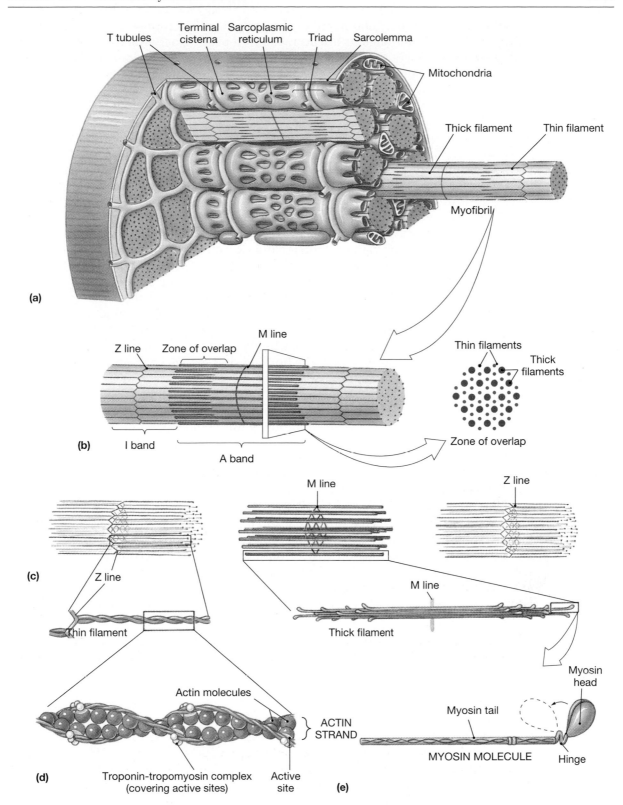

Figure 18.2
Organization of Skeletal Muscles.

(a) Structure of a skeletal muscle fiber. **(b)** Diagrammatic organization of a sarcomere, part of a single myofibril. **(c)** The sacromere in part (c), stretched apart to the point where thick and thin filaments no longer overlap. (This cannot happen in an intact muscle fiber.) **(d)** Structure of a thin filament. **(e)** Structure of a thick filament.

thin filaments are mostly composed of the protein **actin**, and the thick filaments are made of **myosin**. The filaments are arranged in repeating patterns called **sarcomeres (SAR-kō-mērz)** along a myofibril. Distinct regions of the sarcomere are distinguishable. The thin filaments connect at the **Z lines** on each end of the sarcomere. Areas near the Z line that contain only thin filaments are **I bands**. Between the I bands of a sarcomere is the **A band**, the area within the length of thick filaments. The edges of the A band are the **zone of overlap** where the thick and thin filaments bind during muscle contraction. The middle region of the A band is the **H band** (**H zone**) and contains only thick filaments. A dense **M line** in the center of the A band attaches the thick filaments. Because the filaments do not overlap completely, some areas of the sarcomere appear lighter than others. This organization results in the striated or striped appearance of skeletal muscle tissue.

A thin filament consists of two intertwining strands of the protein actin (Figure 18.2d). Three protein components make up the actin strands. The first, called **G actin,** forms the actin filament (often called **F actin**), which looks something like a twisted pearl necklace. Located on each G actin subunit is an **active site** where myosin molecules bind. Associated with the actin filaments are two other proteins, **tropomyosin (trō-pō-MĪ-ō-sin)** and **troponin (TRŌ-pō-nin)**. Tropomyosin follows the twisted actin filaments and blocks active sites to regulate muscle contraction. Troponin holds tropomyosin in position and has binding sites for calcium ions. When calcium ions are released into the sarcoplasm, they bind to and cause troponin to change shape. This moves tropomyosin away from the binding sites, exposing them to myosin heads for the interactions of contraction.

A thick filament is made of approximately 200 subunits of the protein **myosin** (Figure 18.2e). Each subunit consists of two strands: two intertwined **tail regions** and two **globular heads**. Bundles of myosin subunits, with the tails parallel to each other and the heads projecting outward, comprise a thick filament. Each globular myosin head contains a binding site for actin and a region that functions as an ATPase enzyme. This portion of the head splits an ATP molecule and absorbs the released energy to pivot and slide the thin filament inward.

When the muscle is stimulated, calcium is released from the SR and binds to troponin, which in turn moves tropomyosin and exposes the active sites on the G actins. The myosin heads attach to the active sites and ratchet the thin filaments inward, much like a tug-a-war team pulling on a rope. As thin filaments slide into the H band, the sarcomere shortens. The additive effect of the shortening of many sarcomeres along the myofibril results in a decrease in the length of the myofibril and contraction of the muscle. Compare the relaxed and contracted sarcomeres in Figure 18.3.

LABORATORY ACTIVITY   ORGANIZATION OF SKELETAL MUSCLE

MATERIALS

    muscle model
    muscle fiber model
    round steak or similar cut of meat

PROCEDURES

1. Review the organization of muscles in Figures 18.1 and 18.2.
2. Review the histology of skeletal muscle fibers in Exercise 10.
3. Identify the connective tissue coverings of muscles on the laboratory models. Your laboratory instructor may have prepared a muscle demonstration from a round steak or similar cut of meat. Examine the meat for the various connective tissues. Are fascicles visible on the specimen? _____

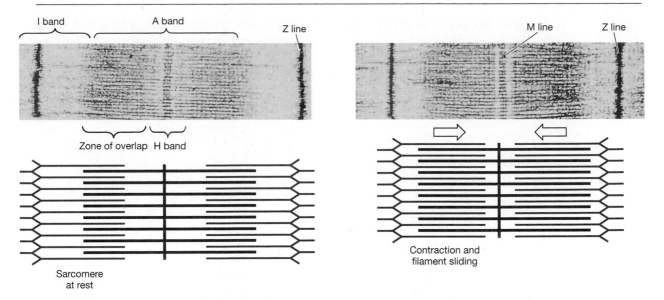

Figure 18.3
Changes in the Appearance of a Sarcomere during Contraction of a Skeletal Muscle Fiber.
During a contraction the A band stays the same width, but the Z lines move closer together and the I band gets smaller.

4. Examine the skeletal muscle fiber models in the laboratory and identify each feature. Describe the location of the sarcoplasmic reticulum, myofibrils, sarcomeres, and filaments. _____

_____

5. Draw the sarcomere and label the following structures:
   thick and thin filaments
   sarcomere
   Z line
   I band
   A band
   H band
   zone of overlap
   M line

6. Examine a specimen of preserved muscle tissue. Place the tissue in a dish with saline solution. Using tweezers and a probe, tease apart the muscle. Notice the fascicles appear as strands of muscle tissue. Examine the fascicles under a dissecting microscope. How are they arranged in the muscle?

_____

_____

B. The Neuromuscular Junction

Every skeletal muscle fiber is controlled by a motor neuron at a single **neuromuscular junction** (see Figure 18.4). The end of the neuron, the axon, expands to form a bulbous **synaptic knob** containing chemical messenger molecules called **acetylcholine (as-ē-til-KŌ-lēn)** (ACh). A small gap, the **synaptic cleft**, separates the synaptic knob from a folded area of the sarcolemma, the **motor end plate**. When a nerve impulse, an action potential, travels down a neuron and reaches the synaptic knob, membranous structures called **synaptic vesicles** release ACh into the synaptic cleft. Receptor proteins in the sarcolemma fill with chemical stimulus, and the sarcolemma generates an **action potential**. This muscle impulse spreads across the sarcolemma, down transverse tubules, and causes the release of calcium ions (Ca^{2+}) into the sarcoplasm of the muscle fiber. Calcium concentrations within the sarcoplasm of muscles cells at rest are very low. When the action potential reaches the SR, calcium channels open across the SR membrane and calcium flows down its concentration gradient into the sarcoplasm.

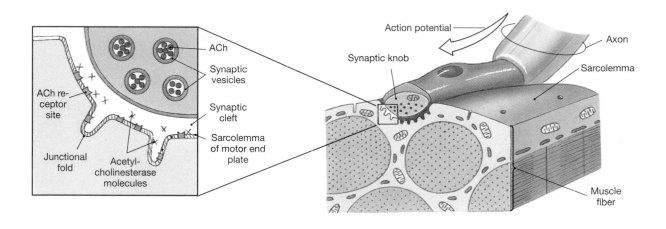

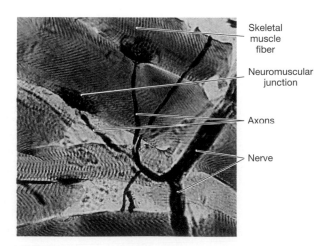

Figure 18.4
Neuromuscular Junction.

Because each myofibril is surrounded by the SR, calcium is quickly and efficiently released into the myofibril. The release of calcium ions into the sarcoplasm triggers events for the interaction between thick and thin filaments that result in contraction of the muscle fiber.

LABORATORY ACTIVITY   NEUROMUSCULAR JUNCTION

MATERIALS

compound microscope
neuromuscular junction slide

PROCEDURES

1. Review the structure of the neuromuscular junction in Figure 18.4.

2. Examine the slide of the neuromuscular junction at low and medium powers. Identify the long, dark, threadlike structures and the oval disks. Describe the appearance of the muscle fibers. _____

_____

3. In the space below, sketch several muscle fibers and their neuromuscular junctions.

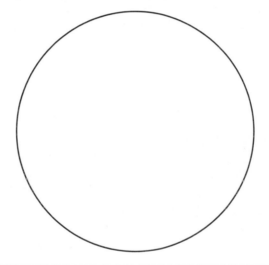

C. Muscles as Lever Systems

The property of contractility enables the body to move all its various parts, particularly where muscles attach to bones. This combination of muscles, bones, and joints forms a **lever system**, which is an efficient way of using force to move an object (see Figure 18.5). By applying a force at a certain point in a lever system, you can increase the amount of work achieved by the same amount of effort. A **lever** system consists of a rigid rod, a lever (L), which moves around a point called the **fulcrum** (F). The force required to move the lever is the **effort** (E). The **resistance** (R) of the system is the weight of the lever or the mass resting on the lever. In a muscle-bone lever system, the body weight is the resistance, the bone is the rod, the joint is the fulcrum, and the muscle supplies the effort. A common application of the lever is a child's seesaw. In principle, two children on a seesaw are of similar weight and the board or lever pivots at the fulcrum. Imagine, however, a child sitting alone on a seesaw. The child is the resistance to be moved by the lever, the board or pole under the resistance. Pulling down on the unoccupied end of the seesaw easily moves the weight of the child as the lever pivots around the fulcrum of the seesaw.

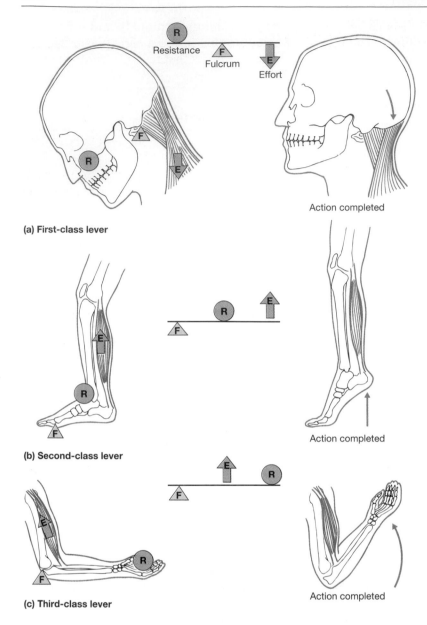

Figure 18.5
The Three Classes of Levers.

(a) In a first-class lever, the applied force and the resistance are on opposite sides of the fulcrum. First-class levers can change the amount of force transmitted to the resistance and alter the direction and speed of movement. (b) In a second-class lever, the resistance lies between the applied force and the fulcrum. This arrangement magnifies force at the expense of distance and speed; the direction of movement remains unchanged. (c) In a third-class lever, the force is applied between the resistance and the fulcrum. This arrangement increases speed and distance moved but requires a larger applied force. The example shows the force and distance relationships involved in the contraction of the biceps brachii muscle that flexes the forearm.

The position of the fulcrum along the lever arm can dramatically change the range of motion and amount of work produced. For example, if the insertion of a muscle is close to a joint, the muscle will be able to produce a great range of motion. However, this comes at a loss in the work done by that muscle. Conversely, if another muscle of the same strength has its insertion further from the joint, this second muscle can generate more work, but with a limited range of motion. This principle is called **leverage.** Most lever systems in the human body are of the type where the effort is applied between the fulcrum and the resistance; for example, the biceps moves the arm by applying a force between the resistance (the mass of the hand) and the elbow joint.

LABORATORY ACTIVITY   LEVERAGE

MATERIALS

textbook, two pencils

PROCEDURE

1. Review the organization of lever systems in Figure 18.5. You are going to construct a simple first-class lever system to demonstrate leverage.

2. Lay the book down on the top of a desk and place a pencil parallel to the book spine for use as a fulcrum. Put the sharpened end of the other pencil under the spine and on top of the fulcrum. This pencil is the lever.

3. Move the fulcrum away from the book until it is close to the eraser of the lever pencil.

4. Using care to prevent breaking the pencil, push down on the lever. Notice how difficult it is to move the resistance (the book) with the fulcrum at the far end of the lever.

5. Move the fulcrum pencil toward the spine of the book and push down on the lever as before. This time the book is easily raised. What you have demonstrated is the principle of leverage. By moving the fulcrum (the point of contact between the two pencils) closer to the weight (the book), it is easier for you to apply enough force to raise the book. When the fulcrum is further away from the weight, it is far more difficult to elevate the book.

6. Next, place the pencils as they were in the first example, with the fulcrum far from the weight, and then remove the book. Now press down on the end of the pencil and note that, without the book, the tip moves easily.

7. Shift the fulcrum toward the tip of the pencil. Now press down on the lever pencil. Again the tip moves easily, but this time the distance the tip moves is shorter. The force on the lever is used to move either a heavy weight over a short distance or a small weight over a greater distance. Strength is sacrificed for range of movement. Muscles systems work in a similar manner.

D. **Action, Origin, and Insertion**

Skeletal muscles attach to bones and other tissues by tendons. Each muscle causes a movement or **action** that is dependent upon many factors, especially the shape of the attached bones. For the muscle to produce a smooth, coordinated action, one end of it must serve as an attachment site while the other end moves the intended bone. The less movable end of the muscle is called the **origin**. The opposite end that moves the bone is called the **insertion**. During the action the insertion moves toward the origin to generate a pulling force. Usually, when a muscle pulls in one direction, an antagonistic muscle pulls in the opposite direction to produce resistance and promote a smooth movement.

Muscles can generate only a pulling force; they do not push. Muscles operate in **antagonistic** sets. When a muscle produces movement in one direction, another muscle pulls in the opposite direction. For example, when the arm is flexed, the biceps brachii muscle is the **prime mover** while the triceps brachii is the antagonist. Each muscle's role reverses when the arm is extended.

Muscles that help the prime mover are called **synergists**. For example, the supinator muscle of the arm turns the hand so that the palm faces up. The biceps brachii acts as a synergist for this action since this muscle secondarily supinates the forearm. **Fixators** are muscles that anchor bones in place to allow other muscles to perform their actions more efficiently. For example, many muscles surrounding the scapula help fix that bone in place if the arm is required to apply a greater force. If greater flexibility is needed, the fixators relax and allow the scapula to shift its position. In this way one can have both great strength and great flexibility in the arm, although not at the same time.

E. Naming Muscles

A variety of methods are used to name muscles. Examine Table 18.1 and notice that muscles are named after the bones they attach to, body regions, muscle shapes, and muscle actions, origins, and insertions. The names are often compound words. For example, the temporalis muscle is found on the temporal bone, and the occipitalis sits on the occipital bone of the skull. Another easily identifiable muscle is the sternoclei-domastoid which originates on the sternum (sterno-) and the clavicle (-cleido) and inserts on the mastoid process of the temporal bone.

The size of the muscle is often reflected in its name. *Maximus* refers to large muscles, as in gluteus maximus, and *minimus* refers to small muscles, as in gluteus minimus. The name adductor longus implies the muscle is long. *Brevis* refers to a short muscle.

The direction of muscle fascicles is also used to name muscles. The term *rectus* is used if the fibers of the muscle are parallel to the body or limb. For example, in the abdominus rectus, the muscle fascicles are parallel to the body. *Transverse* refers to fascicles that are perpendicular to the midline, as in transversus abdominis. In the internal and external *obliques,* the fascicles cut diagonally across the body.

Many muscles have multiple origins. The triceps, for example, has three origins or heads, and they all insert on the olecranon process of the elbow. Look for the prefixes *bi-* for two, *tri-* for three, and *quad-* for four origins. Anatomists in antiquity often devised names based on the shape of the muscle. The *deltoid* has a broad origin and inserts on a very narrow region of the humerus. This gives this muscle a triangular or deltoid shape, hence the name. The *trapezius* is a very large muscle covering the middle back. When both muscles are viewed together, they form a trapezoid shape.

Many muscles are named based on their actions. The term *flexor carpi ulnaris* appears complex, but it is really quite easy to understand if you examine the name step by step. *Flexor* means the muscle flexes something. What does it flex? Look at the next two terms. *Carpi* refers to carpals, which are the bones of the wrist. The term *ulnaris* suggests that the muscle flexes the carpi on the medial side of the wrist. Therefore, the flexor carpi ulnaris is a muscle that flexes the wrist laterally. Once you have an idea of the action, you can then determine where on the body the muscle is located. In this case, it is the forearm. The origin and insertion can sometimes be immediately determined. This muscle inserts on the carpi. The origin is a little more difficult and may require some memorization. Try using this analysis procedure with the muscle *levator scapulae.*

TABLE 18.1   Muscle Terminology

Terms Indicating Direction Relative to the Longitudinal Axis of the Body	Terms Indicating Actions	Terms Indicating Specific Regions of the Body	Terms Indicating Structural Characteristics of the Muscle
Lateralis (lateral)	**General**	Abdominis (abdomen)	**Origin/Insertion**
Medialis, medius (medial, middle)	Abductor	Anconeus (elbow)	Biceps (two heads)
Obliquus (oblique)	Adductor	Auricularis (auricle of ear)	Triceps (three heads)
Rectus (straight, parallel)	Depressor	Brachialis (brachium)	Quadriceps (four heads)
Transversus (transverse)	Extensor	Capitis (head)	
	Flexor	Carpi (wrist)	**Shape**
	Levator	Cervicis (neck)	Deltoid (triangle)
	Pronator	Cleido/clavius (clavicle)	Orbicularis (circle)
	Rotator	Coccygeus (coccyx)	Pectinate (comblike)
	Supinator	Costalis (ribs)	Piriformis (pear-shaped)
	Tensor	Cutaneous (skin)	Platy- (flat)
		Femoris (femur)	Pyramidal (pyramid)
	Specific	Genio- (chin)	Rhomboideus (rhomboid)
	Buccinator (trumpeter)	Glossal (tongue)	Serratus (serrated)
	Risorius (laughter)	Hallucis (great toe)	Splenius (bandage)
	Sartorius (like a tailor)	Ilio- (ilium)	Tendinosus (tendinous)
		Inguinal (groin)	Trapezius (diamond)
		Lumborum (lumbar region)	
		Nasalis (nose)	**Other Striking Features**
		Nuchal (back of neck)	Alba (white)
		Oculo- (eye)	Anterior
		Oris (mouth)	Brevis (short)
		Palpebrae (eyelid)	Externus (superficial)
		Pollicis (thumb)	Extrinsic (outside)
		Popliteus (behind knee)	Gracilis (slender)
		Psoas (loin)	Inferioris (inferior)
		Radialis (radius)	Internus (deep, internal)
		Scapularis (scapula)	Intrinsic (inside)
		Temporalis (temples)	Lateralis (lateral)
		Thoracis (thoracic region)	Latissimus (widest)
		Tibialis (tibia)	Longus (long)
		Ulnaris (ulna)	Longissimus (longest)
		Uro- (urinary)	Magnus (large)
			Major (larger)
			Maximus (largest)
			Medialis (medial)
			Minimus (smallest)
			Minor (smaller)
			Posterior
			Profundus (deep)
			Superficialis (superficial)
			Superioris (superior)
			Teres (long and round)
			Vastus (great)

MUSCULAR SYSTEM OVERVIEW CHECKLIST

This is a list of bold terms presented in Exercise 18. Use this list for review purposes after you have completed the laboratory activities.

FUNCTIONS OF THE MUSCULAR SYSTEM

movement
posture
body temperature
support
contractility, excitability
extensibility, elasticity

ORGANIZATION

• Figure 18.1
connective tissue
epimysium
  perimysium
  fasciculae
    endomysium
    muscle fiber
      myofibril

MICROANATOMY OF MUSCLE FIBERS

• Figures 18.2, 18.3, 18.4
sarcolemma
  sarcoplasm
  transverse tubules (T tubules)
  sarcoplasmic reticulum (SR)
    terminal cisternae
      calcium
  myofibrils
    filaments
    thin filaments
      actin
        G actin
        F actin
      tropomyosin
      troponin
    thick filaments
      myosin
        globular heads
        tail regions

sarcomere
  Z line
  I band
  A band
    zone of overlap
    H band
    M line

NEUROMUSCULAR JUNCTION

• Figure 18.4
motor neuron
  synaptic knob
    synaptic vesicles
      acetylcholine (ACh)
synaptic cleft
muscle fiber
  motor end plate
    ACh receptors
action potential

MUSCLES AS LEVER SYSTEMS

• Figure 18.5
lever system
lever
fulcrum
effort
resistance
leverage

ACTION, ORIGIN, AND INSERTION

tendon
origin, insertion, action
antagonistic pairs
prime mover
synergist
fixator

NAMING MUSCLES

• Table 18.1
maximus, minimus
longus, brevis
rectus, transverse, oblique

Name _____   Date _____

Section _____   Number _____

MUSCULAR SYSTEM OVERVIEW
Laboratory Report Exercise 18

A. Matching
Match each term listed on the left with the correct description on the right.

1. _____ sarcomere	A. boundary of sarcomere
2. _____ extensibility	B. banding patterns in muscle tissue
3. _____ elasticity	C. storage site for calcium
4. _____ epimysium	D. thin filament
5. _____ perimysium	E. between two Z lines
6. _____ endomysium	F. attaches muscle to bone
7. _____ tendon	G. rods of filaments
8. _____ Z line	H. muscle cell
9. _____ muscle fiber	I. thick filament
10. _____ myofibril	J. carries action potential to SR
11. _____ striations	K. muscles can stretch
12. _____ myosin	L. muscle fiber cell membrane
13. _____ sarcolemma	M. stretched muscle springs back
14. _____ sarcoplasm	N. muscle fiber cytoplasm
15. _____ neuromuscular junction	O. connective tissue covering fascicles
16. _____ transverse tubule	P. connective tissue covering fibers
17. _____ sarcoplasmic reticulum	Q. connective tissue covering muscles
18. _____ actin	R. interface between nerve and muscle

B. Short-Answer Questions
1. List the connective tissue layers surrounding a muscle, working from the out-side inward.

2. The gastrocnemius muscle attaches to the calcaneus (heel) bone of the foot. Describe this organization in terms of a lever system. Where are the fulcrum and lever located? What is the resistance and what applies the force?

C. **Matching**
     Match each term listed on the left with the correct description on the right.

1.	_____ abductor	**A.**	tongue
2.	_____ depressor	**B.**	wrist
3.	_____ flexor	**C.**	great
4.	_____ levator	**D.**	trapezoid shape
5.	_____ pronator	**E.**	clavicle
6.	_____ supinator	**F.**	short
7.	_____ tensor	**G.**	circular
8.	_____ cleido	**H.**	slender
9.	_____ glossal	**I.**	rib
10.	_____ anconeus	**J.**	moves away
11.	_____ capitis	**K.**	lowers
12	_____ cervicis	**L.**	thoracic region
13.	_____ coccygeus	**M.**	elbow
14.	_____ oris	**N.**	widest part
15.	_____ oculi	**O.**	scapula
16.	_____ lumborum	**P.**	tenses
17.	_____ scapularis	**Q.**	turns palms anteriorly
18.	_____ thoracis	**R.**	turns palms posteriorly
19.	_____ orbicularis	**S.**	mouth
20.	_____ trapezius	**T.**	raises
21.	_____ brevis	**U.**	eye
22.	_____ gracilis	**V.**	coccyx
23.	_____ vastus	**W.**	lumbar region
24.	_____ latissimus	**X.**	neck region
25.	_____ costalis	**Y.**	flexes
26.	_____ carpi	**Z.**	head

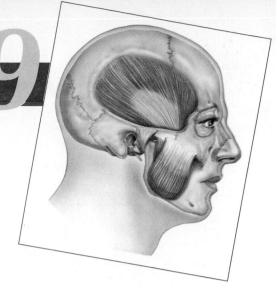

Muscles of the Head and Neck

OBJECTIVES

On completion of this exercise, you should be able to:

- Identify the origin, insertion, and action of the muscles of facial expression and mastication.
- Identify the origin, insertion, and action of the muscles that move the eye.
- Identify the origin, insertion, and action of the muscles that move the head and neck.

WORD POWER

orbicularis oculi (orb—a circle; oculi—the eye)
levator (leva—lift up)
rectus (rect—straight)
styloglossus (stylo—pointed instrument; glossus—tongue

buccinator (buccin—a trumpet)
platysma (platy—broad, flat)
masseter (masseter—a chewer)
temporalis (tempor—the temples)
sternocleidomastoid (sterno—breastbone; cleido—clavicle)

MATERIALS

head model
torso model
eye model

INTRODUCTION

The muscles of the head and neck produce a wide range of motions for facial expressions, processing food, producing speech, and positioning the head. The names of these muscles usually indicate the bone they attach to or the structure they surround. In this exercise you will identify the major muscles of facial expression and mastication (chewing), the muscles that move the eyes, and the muscles that position the head and neck. Attempt to find the general location of each muscle on your body. Contract the muscle and observe the movement you generate.

A. Muscles of Facial Expression

Epicranius

The **epicranius (ep-i-KRĀ-ne-us)**, also called the **occipitofrontalis**, consists of two smaller muscles: the **frontalis** and the **occipitalis**, see Figure 19.1. The frontalis is the broad anterior muscle on the forehead that covers the frontal bone. It originates on the superior margin of the eye orbit, near the eyebrows. The muscle fibers blend into the sheet of connective tissue called the **galea aponeurotica (GĀ-lē-uh ap-o-nū-RO-ti-kuh)** which covers the cap of the skull. The actions of the frontalis include wrinkling the forehead, raising the eyebrows, and pulling the scalp forward.

The **occipitalis** covers the back of the skull. It arises on the occipital bone and the mastoid process of the temporal bone, extends superiorly, and inserts on the galea aponeurotica, completing the posterior part of the epicranius. This muscle draws the scalp backward, an action difficult for most people to isolate and perform.

1. _____

2. _____

3. _____

4. _____

5. _____

6. _____

7. _____

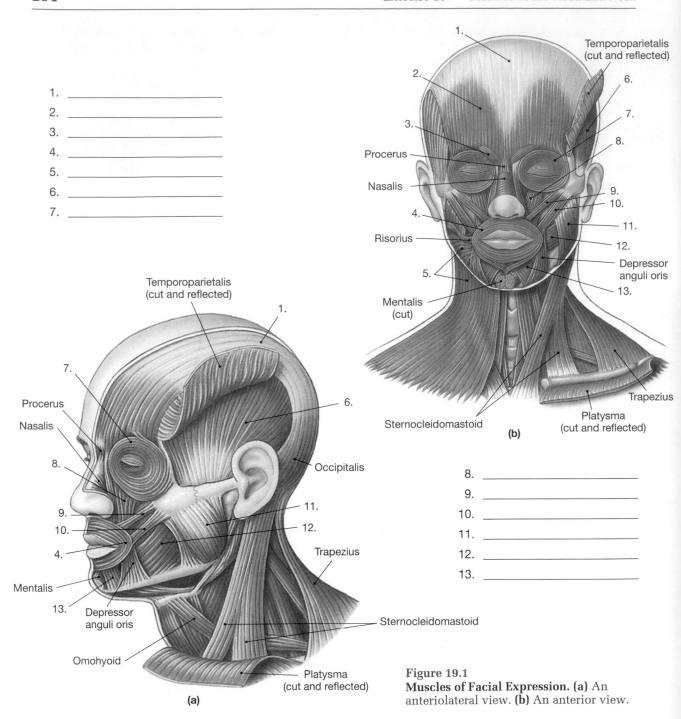

Temporoparietalis
(cut and reflected)

Procerus

Nasalis

Risorius

Mentalis
(cut)

Sternocleidomastoid

Depressor
anguli oris

Trapezius

Platysma
(cut and reflected)

(b)

Temporoparietalis
(cut and reflected)

Procerus

Nasalis

Occipitalis

Trapezius

Mentalis

Depressor
anguli oris

Omohyoid

Sternocleidomastoid

Platysma
(cut and reflected)

(a)

8. _____

9. _____

10. _____

11. _____

12. _____

13. _____

Figure 19.1
Muscles of Facial Expression. (a) An
anteriolateral view. **(b)** An anterior view.

Orbicularis oculi

The sphincter muscle of the eye is the **orbicularis oculi (or-bik-ū-LA-ris ok-ū-lī)**. Locate
this muscle on Figure 19.1. It arises from the medial wall of the orbit, and its fibers
form a band of muscle that passes around the circumference of the eye, which serves as
the insertion. The muscle acts to close the eye, as during an exaggerated blink.

Corrugator supercilii

The **corrugator supercilii** begins on the ridge between the eyebrows and the bridge of
the nose and angles laterally to insert into the skin of the forehead. It acts to pull the

skin downward and wrinkles the forehead into a frown. Think of a corrugated tin roof when you study this muscle. The corrugator supercilii bunches the skin of the forehead into folds or corrugations.

Orbicularis oris

The **orbicularis oris** is a sphincter muscle that surrounds the mouth. It originates on the bones surrounding the mouth and inserts on the lips. This muscle shapes the lips for a variety of functions including speech, food manipulation, and facial expressions and purses the lips together for a kiss. Identify this muscle in Figure 19.1 and on your face.

Buccinator

The **buccinator (BUK-si-nā-tor)** is the horizontal muscle spanning between the jaws (see Figure 19.1). It originates on the alveolar processes of the maxilla and mandible and inserts on the fibers of the orbicularis oris. While chewing, the buccinator moves food across the teeth from the cheeks. The name "buccinator" refers to trumpet players because they use the buccinator to pressurize their exhalations. You use this muscle when you contract and compress your cheeks during eating, when you tense your cheeks when blowing out, and when you suck on a straw.

Depressor labii and levator labii

The labii muscles insert on the lips and move them according to their name. The **depressor labii** arises from the region between the symphysis and mental foramen of the mandible (see Figure 19.1). It inserts on the skin of the lower lip and blends into the orbicularis oris muscle. The depressor labii pulls the lower lip down and slightly outward. The origin of the **levator labii** is immediately superior to the infraorbital foramen of maxilla. It inserts on the skin at the angle of the mouth and blends into the orbicularis oris. It raises the upper lip, much as when we pretend to whinny like a horse. Move your lips and practice the actions of these labii muscles.

Risorius

The **risorius** is a narrow muscle that begins on the fascia over the parotid salivary gland and ends on the angle of the mouth (see Figure 19.1). When it contracts, the risorius pulls and produces a grimacelike tensing of the mouth. Although the term "risorius" refers to a smile, the muscle is probably more associated with the expression of pain rather than pleasure. In the disease tetanus, the risorius is involved in the painful contractions that pull the corners of the mouth backward into "lockjaw." Contract your risorius and notice your facial expression.

Zygomaticus major and minor

The **zygomaticus major** arises from the temporal process of the zygomatic bone (Figure 19.1). The origin of the **zygomaticus minor** lies just medial to the origin of the zygomaticus major, near the maxillary suture. They both insert on the skin and corners of the mouth. These muscles pull the skin of the mouth upward and laterally and produce a smile. Place your fingers on your zygomatic arches and smile. Do you feel your zygomaticus muscles contract?

Platysma

The **platysma (pla-TIZ-muh)** is a thin, broad muscle covering the sides of the neck (see Figure 19.1). It originates on the fascia covering the pectoralis and deltoid muscles and extends upward to insert on the inferior edge of the mandible. Some of the fibers of the platysma also extend into the fascia and muscles of the lower face. The platysma

P.217

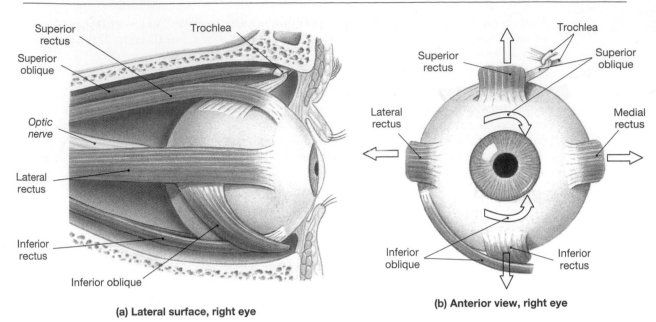

(a) Lateral surface, right eye

(b) Anterior view, right eye

Figure 19.2
Extrinsic Eye Muscles.

depresses the mandible and the soft structures of the lower face, resulting in an expression of horror and disgust. Compare this action with that of the risorius.

LABORATORY ACTIVITY   **MUSCLES OF FACIAL EXPRESSION**

MATERIALS

> torso model
> head model
> muscle chart

PROCEDURES

1. Label and review the muscles of the head in Figure 19.1.
2. Examine the head of the torso model and locate each muscle described.
3. Find the general location of the muscles of facial expression on your face. Practice the action of each muscle and observe how your facial expression changes.

B. Muscles of the Eye

Examine a model of the eye with the extrinsic muscles attached. These oculomotor muscles are located on the outside of the eye, as shown in Figure 19.2, and insert on the sclera, the white, fibrous covering of the eye. Intrinsic muscles are the internal muscles of the eye and are involved in focusing the eye for vision. These muscles are discussed in Exercise 36.

Six extrinsic (external) eye muscles control the movements of the eye (see Figure 19.2). The **superior rectus**, **inferior rectus**, **medial rectus**, and **lateral rectus** are straight muscles that move the eyeball up and down and side to side. They originate around the optic foramen within the orbit, insert on the sclera, and roll the eyeball as indicated by their names. The **superior** and **inferior oblique** muscles attach diagonally on the eyeball. The superior oblique has a tendon passing through a trochlea (pulley) located on the upper orbit. It rolls the eye downward, and the inferior oblique rolls the eye upward.

LABORATORY ACTIVITY   MUSCLES OF THE EYE

MATERIALS

> torso model
> head model
> eye model
> eye muscle chart

PROCEDURES

1. Review the muscles of the eye in Figure 19.2.
2. Examine an eye model and locate each extrinsic muscle.
3. Practice the action of each eye muscle by moving your eyeballs.

C. Muscles of Mastication

Masseter and temporalis

The muscles involved in mastication or chewing are shown in Figure 19.3. The **masseter (MAS-se-tur)** is a short, thick muscle originating on the maxilla and the anterior portion of the zygomatic arch. It inserts on the angle and the ramus of the mandible. The action is described below. You can feel the **temporalis (tem-pō-RA-lis)** covering almost the entire temporal fossa, which serves as the origin. It inserts on the coronoid process of the mandible. When the masseter and the temporalis contract, they elevate your jaw against the teeth, producing the tremendous force used in chewing. The masseter protrudes the jaw slightly when it acts alone. The temporalis retracts your jaw during contraction.

STUDY TIP

Put your fingertips at the angle of your jaw and clench your teeth. You should feel the masseter bunch up as it forces the teeth together.

LABORATORY ACTIVITY   MUSCLES OF MASTICATION

MATERIALS

> torso model
> head model
> muscle chart

PROCEDURES

1. Label and review the muscles of the head in Figures 19.1 and 19.3.
2. Examine the head of the torso model and locate each muscle described.
3. Find the general location of the muscles of mastication on your face. Practice the action of each muscle and observe how your mandible moves.

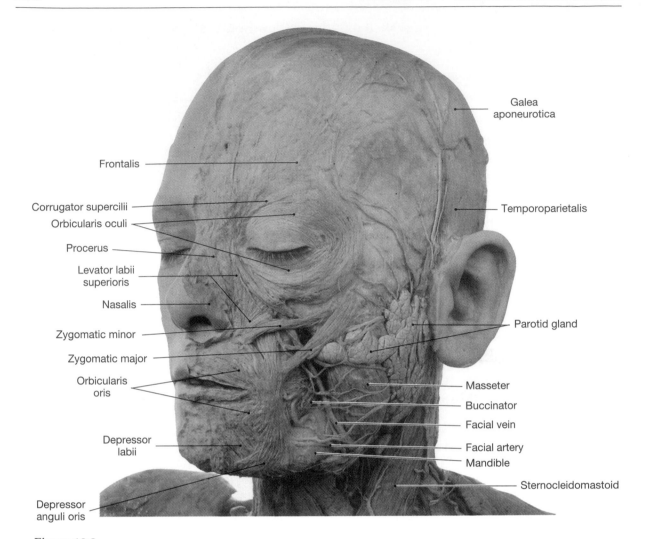

Figure 19.3
Cadaver Head and Neck, Lateral View.
Note the superficial arteries, veins, facial nerve, parotid gland, and muscles.

D. Muscles of the Tongue

The muscles of the tongue form the floor of the oral cavity (see Figure 19.4). The root word for this muscle group is "glossus," Greek for tongue. Each prefix indicates the muscle's origin. The **palatoglossus** arises from the soft palate and inserts on the lateral tongue. During swallowing, the palatoglossus elevates the tongue and depresses the soft palate.

Antagonistic muscles protract and retract the tongue. Anteriorly, the **glenioglossus** originates on the posterolateral area of the mandible. It depresses and protracts the tongue, as in initiating the licking of an ice cream cone. Posterior, the **styloglossus** and **hyoglossus** muscles arise on the styloid process and the hyoid bone, respectively. Both muscles retract the tongue. The styloglossus also raises the sides of the tongue.

1. _____
2. _____
3. _____
4. _____
5. _____

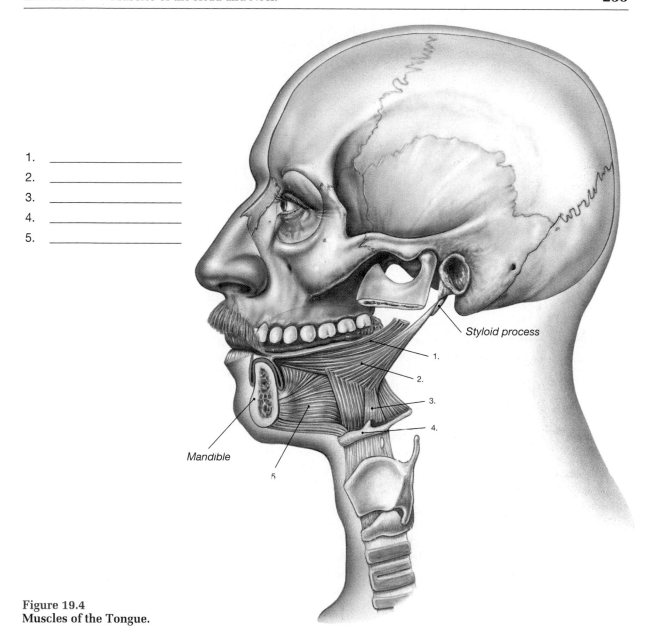

Figure 19.4
Muscles of the Tongue.

LABORATORY ACTIVITY   MUSCLES OF THE TONGUE

MATERIALS

    torso model
    head model
    muscle chart

PROCEDURES

1. Label and review the muscles of the tongue in Figure 19.4.
2. Examine the muscle models and identify each muscle of the tongue.

3. Practice the action of each tongue muscle. The ability to curl your tongue with the styloglossus is genetically controlled by a single gene. Individuals with the dominant gene are "rollers" and can curl the tongue. Those with the recessive form of the gene are "nonrollers." Is it possible for nonrollers to learn how to roll their tongue? Are you, your parents, or your children, rollers or nonrollers?

E. **Muscles of the Anterior Neck**

All the muscles described below with the suffix -hyoid, insert on the hyoid bone. This bone is suspended by muscles of the inferior and posterior mandible. It is also connected to the larynx and serves as a foundation for the muscles of the tongue. The muscles that insert on the hyoid are the **suprahyoid** and **infrahyoid** muscle groups. As the names imply, they are located either above or below the hyoid bone. The origin is different in each case, as indicated by the prefix; for example, genio- refers to the chin and stylo- refers to the styloid process of the temporal bone.

Superior muscles of the hyoid

Examine Figure 19.5 and locate the **geniohyoid** which originates on the inside of the mandible near the symphysis and acts to depress the mandible and elevate the larynx. The **stylohyoid** originates on the styloid process. It pulls on the styloid process of the skull and raises the hyoid and the larynx. The **mylohyoid** originates along a broad origin on the inner border of the mandible. Along with the diagastric muscle, it elevates the hyoid and depresses the lower jaw. The **digastric** has two parts or bellies: the **anterior** belly originates on the inside of the mandible near the chin, and the **posterior** belly arises on the mastoid process. The bellies insert on the hyoid and form a muscular swing that acts to elevate the hyoid or lower the mandible to open the jaw.

Inferior muscles of the hyoid

The infrahyoid muscles arise below the hyoid and act to depress the hyoid and larynx (see Figure 19.5). The **omohyoid (ō-mā-HĪ-oyd)** has its origin on the clavicle and 1st rib. The **sternohyoid** is a thin, straplike muscle that originates on the sternal end of the clavicle. The **sternothyroid** originates on the manubrium of the sternum and terminates on the thyroid cartilage of the larynx. These muscles act to depress the hyoid bone. The **thyrohyoid** originates on the thyroid cartilage of the larynx, sweeps upward, and inserts on the hyoid to the larynx.

LABORATORY ACTIVITY   MUSCLES OF THE ANTERIOR NECK

MATERIALS

torso model
head model
muscle chart

PROCEDURES

1. Label and review the anterior neck muscles in Figure 19.5.
2. Locate each muscle on the head and neck models.
3. Produce the actions of your suprahyoid and infrahyoid muscles and observe the movement of your larynx.

Figure 19.5
Muscles of the Anterior Neck.

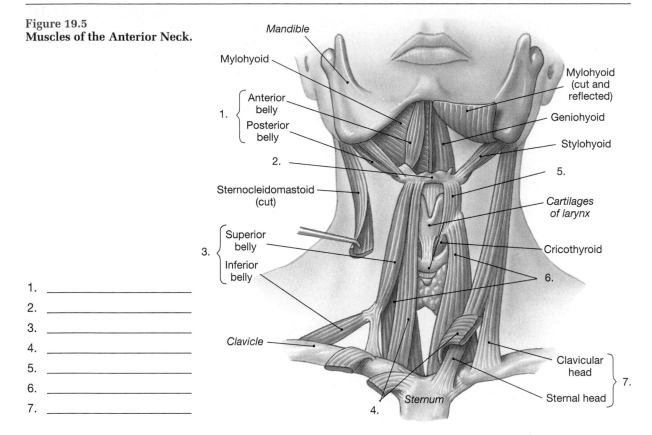

Mandible

Mylohyoid

1. { Anterior belly / Posterior belly

2.

Sternocleidomastoid (cut)

3. { Superior belly / Inferior belly

Clavicle

Sternum

4.

Mylohyoid (cut and reflected)

Geniohyoid

Stylohyoid

5.

Cartilages of larynx

Cricothyroid

6.

Clavicular head
Sternal head
} 7.

1. _____

2. _____

3. _____

4. _____

5. _____

6. _____

7. _____

F. Muscles that Move the Head

The muscles that move the neck are long, slender muscles. Each consists of several divisions arising from different points. The sternocleidomastoid muscle is on the anterior neck and acts to flex the head. The posterior muscles of the neck include the **scalenus**, the **semispinalis**, and the **splenius** muscles. Because the posterior neck muscles also move the chest, they are discussed in Exercise 20.

Sternocleidomastoid

The **sternocleidomastoid (ster-nō-klī-dō-MAS-toyd)** muscle occurs on both sides of the neck and is named after its points of attachment on the sternum, clavicle, and mastoid process of the temporal bone (see Figure 19.1). Because this muscle spans the head and sternum, when both sides contract, the neck flexes and the head is tucked down toward the sternum. If only one side contracts, the head is pulled toward the shoulder of that side. Rotate your head until your chin almost touches your right shoulder and locate your left sternocleidomastoid just above the manubrium of the sternum.

LABORATORY ACTIVITY   MUSCLES THAT MOVE THE HEAD

---

1. Examine the torso model and locate the sternocleidomastoid.
2. Contract the muscle on one side and observe your head movement. Next, contract both sides and note that flexion of the head occurs.

MUSCLES OF THE HEAD AND NECK CHECKLIST

This is a list of bold terms presented in Exercise 19. Use this list for review purposes after you have completed the laboratory activities.

MUSCLES OF FACIAL EXPRESSION

• Figures 19.1 and 19.3
epicranius (occipitofrontalis)
frontalis
occipitalis
galea aponeurotica
orbicularis oculi
corrugator supercilii
orbicularis oris
buccinator
depressor labii
levator labii
risorius
zygomaticus major
zygomaticus minor
platysma

MUSCLES OF MASTICATION

• Figures 19.1 and 19.3
masseter
temporalis

MUSCLES OF THE EYE

• Figure 19.2
rectus muscles of the eye
    medial rectus
    lateral rectus
    superior rectus
    inferior rectus
oblique muscles of the eye
    superior oblique
    inferior oblique

MUSCLES OF THE TONGUE

• Figure 19.4
palatoglossus
glenioglossus
styloglossus
hyoglossus

MUSCLES OF THE NECK

• Figure 19.5
suprahyoid muscles of the neck
    geniohyoid
    stylohyoid
    mylohyoid
    digastric (anterior and posterior bellies)
infrahyoid muscles of the neck
    omohyoid
    sternohyoid
    sternothyroid
    thyrohyoid

MUSCLES THAT MOVE THE HEAD

• Figures 19.1 and 19.3
sternocleidomastoid

MUSCLES OF THE HEAD AND NECK
Laboratory Report Exercise 19

A. Matching

Match each term listed on the left with the correct description on the right. O means origin, I means insertion, and A means action.

1. _____ orbicularis oculi
2. _____ buccinator
3. _____ zygomaticus minor
4. _____ masseter
5. _____ frontalis
6. _____ depressor labii
7. _____ temporalis
8. _____ medial rectus
9. _____ occipitalis
10. _____ levator labii
11. _____ superior oblique
12. _____ geniohyoid
13. _____ sternohyoid
14. _____ platysma
15. _____ corrugator supercilii
16. _____ risorius
17. _____ mylohyoid
18. _____ orbicularis oris
19. _____ zygomaticus major
20. _____ digastric

A. moves scalp backward
B. elevates upper lip
C. rotates eye along main axis
D. closes mouth
E. small muscle used in smiling
F. large muscles used in smiling
G. A: elevates jaw; O: angle of the jaw
H. neck muscle, consists of two parts
I. A: elevates larynx; O: chin; I: hyoid
J. tenses cheeks
K. moves scalp forward
L. A: elevates jaw; O: temporal bone
M. A: depresses hyoid; O: sternum; I: hyoid
N. depresses lower lip
O. rotates eye medially
P. closes eye
Q. wrinkles forehead
R. tenses angle of mouth laterally
S. A: elevates hyoid; O: inner border of mandible
T. thin muscle, covers sides of neck, depresses jaw

B. **Short-Answer Questions**

1. Describe the position and action of the muscles of mastication. Which muscles oppose the action of these muscles?

2. Describe the muscles involved in smiling and grimacing.

3. Explain how the muscles of the tongue and anterior neck are named.

4. Describe the movement produced by each extrinsic eye muscle.

Muscles of the Chest, Abdomen, and Spine

OBJECTIVES

On completion of this exercise, you should be able to:

- Identify the origin, insertion, and action of the major muscles of the anterior chest.
- Identify the origin, insertion, and action of the major muscles of the posterior chest.
- Identify the origin, insertion, and action of the major muscles of the thoracic and abdominal regions.

WORD POWER

pectoralis (pector—the chest)
serratus (serrat—a saw)
nuchae (nuch—the nape)

latissimus (lati—wide)
longissimus (longi—long)
intercostal (costa—rib)

MATERIALS

torso model
arm model
muscle chart

INTRODUCTION

The torso of the body supports many muscles spread over a large area. These muscles are not as densely packed as those of the head and neck, but because they act on the limbs and the head, their actions are complex. The muscles are generally grouped based on the region of the torso that they cover or the limb they act upon. For example, the muscles of the chest (including the posterior surface of the chest) act on the arms and the head. The primary function of the abdominal muscles is to support the abdomen, viscera, and lower back and to move the legs.

A. Anterior Muscles of the Chest

Pectoralis major

The largest muscle of the chest is the **pectoralis (pek-to-RA-lis) major** muscle (see Figure 20.1), which covers most of the upper rib cage on each side of the chest. Locate this muscle on your chest. In females, the lower part of the muscle mass is covered by the breast. You may be able to feel a portion of the pectoralis major near its origin on the sternum or clavicle while pressing your arm against your side. The pectoralis major muscle has several parts. The **sternal portion** arises from the costal cartilage, the sternum, and the ribs. The **clavicular portion** originates along the anterior side of the clavicle near the sternum. Both portions insert along the crest of the greater tubercle of the humerus and form the anterior wall of the axilla. The pectoralis major acts to adduct the arm and flex and rotate the humerus at the shoulder.

STUDY TIP

Your hands can be used to simulate a muscle's origin, insertion, and action. For example, place your right hand over your left pectoralis major muscle. Your palm represents the stationary origin of the muscle at the sternum, while your fingers touching the humerus act as the insertion. Flex your fingers and pull the humerus medially, the major action of the muscle.

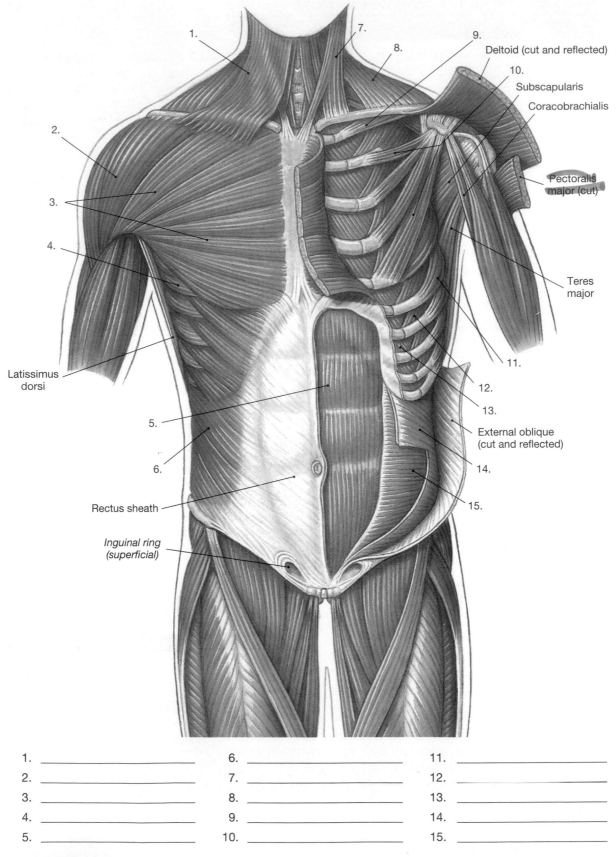

Deltoid (cut and reflected)

10.

Subscapularis

Coracobrachialis

Pectoralis
major (cut)

Teres
major

11.

Latissimus
dorsi

12.

13.

External oblique
(cut and reflected)

14.

Rectus sheath

15.

Inguinal ring
(superficial)

1. _____   6. _____   11. _____
2. _____   7. _____   12. _____
3. _____   8. _____   13. _____
4. _____   9. _____   14. _____
5. _____   10. _____   15. _____

Figure 20.1
Muscles of the Anterior Trunk.

Pectoralis minor

Tucked under the pectoralis major is the **pectoralis minor**. This smaller, triangle-shaped muscle has a much different function than its larger cousin. It originates along the edges of the third, fourth, and fifth ribs and inserts into the coracoid process of the scapula. It acts to pull the top of the scapula forward and depress the shoulders. It also elevates the ribs during forced inspiration, as during strenuous exercise.

Subclavius

The **subclavius (sub-KLĀ-vē-us)**, as the name implies, is under the clavicle (see Figure 20.1). It arises from the first rib, inserts on the underside of the clavicle, and acts to depress and protract the clavicle.

Serratus anterior

The **serratus anterior** appears as wedges on the side of the chest (see Figure 20.1). This gives the muscle a sawtooth appearance similar to that of a bread knife. The muscle also looks like a fan with the blades resting on the ribs and the handle on the scapula. Straps of muscle arise on the upper edge of the first eight ribs, project back, and insert on the costal surface of the scapula near the medial border. The serratus pulls the scapula forward and rotates it. It is also a synergist to a variety of muscles that move the arm. For example, when the deltoid has abducted the arms, it acts as a synergist to the trapezius in moving the arm into a vertical position, such as in raising your arm above your head.

Muscles of respiration

Intercostal Muscles. The **intercostal** muscles are found between the ribs and assist the diaphragm during breathing, see Figure 20.1. These muscles are difficult to palpate because they are deep to other chest muscles. The **external intercostal** spans the region between each rib, with the origin on the lower portion of the rib. Because the insertion is on the superior region of the rib below the origin, when the external intercostal contracts, it elevates the ribs and increases the volume of the thoracic cavity, much like pulling open a bellows to fan a fire. This increase in volume is accompanied by a corresponding decrease in pulmonary pressure and air flows into the lungs for inspiration. The intercostal muscles of beef and pork are the barbeque "ribs" that you might enjoy.

The **internal intercostals** lie deep to the external intercostals. They act to depress the rib cage for forced expiration during exercise. Respiratory volumes and rates increase during exercise. The internal intercostals become active at this time to quickly depress the rib cage and force air out of the lungs. Sometimes during exercise you may notice a pain in the side of your chest. This is most likely caused by the intercostal muscles cramping due to overexertion. The more the intercostal muscles are exercised, the less they cramp.

STUDY TIP

Notice the difference in the orientation of the muscle fibers between the external and internal intercostals. The external fibers flare laterally as they are traced from bottom to top, while the internal fibers are directed medially. This tip is also useful in examining the external and interal oblique muscles of the abdomen.

Diaphragm

The **diaphragm** is a sheet of muscle that forms the thoracic floor and separates the thoracic and abdominopelvic cavities (see Figure 20.2). It originates at many points along

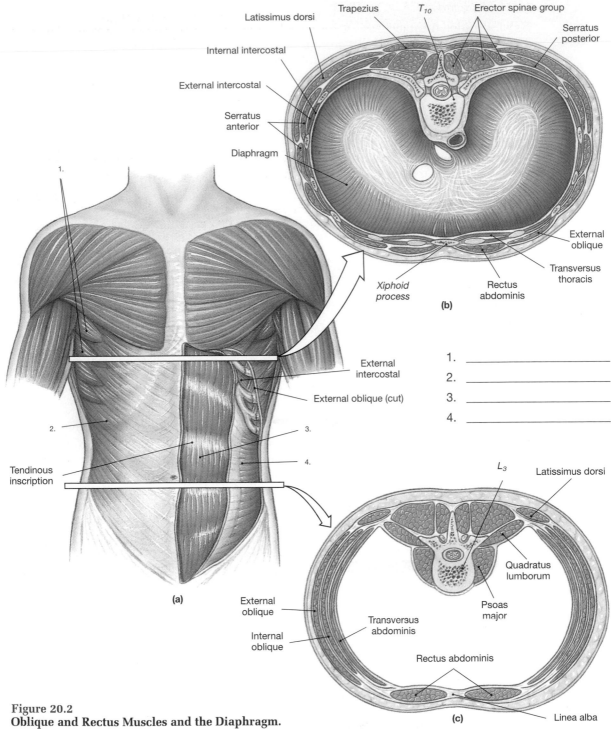

Figure 20.2
Oblique and Rectus Muscles and the Diaphragm.

(a) An anterior view. **(b)** A sectional view at the level of the diaphragm. **(c)** A sectional view at the level of the umbilicus.

its edges, and the muscle fibers meet at a central tendon. When the muscle contracts, it pulls down upon the central tendon and lowers the thoracic floor. This increases the volume of the thoracic cavity and results in a pressure decrease in the lungs. Contracting the diaphragm to expand the thoracic cavity is the muscular process by which air is inhaled into the lungs.

LABORATORY ACTIVITY   MUSCLES OF THE ANTERIOR CHEST

MATERIALS

torso models
arm models
muscle charts

PROCEDURES

1. Label and review the muscles of the anterior chest in Figures 20.1–20.2.
2. Identify each muscle of the anterior chest on the laboratory models.
3. Use Figure 20.3 as a guide to locate the position of these muscles on your trunk. Practice the action of each chest muscle on your body.

B. Abdominal Muscles

Rectus abdominis

The **rectus abdominis** (Figure 20.2) is the straight muscle along the abdomen between the pubic symphysis and the xiphoid process of the sternum. This muscle is divided by a midsagittal fibrous line called the **linea alba**. A well-developed rectus abdominis muscle has a "washboard" appearance because transverse bands of tendons divide it into several segments. When the rectus abdominis contracts, it flexes the pubic symphysis and the xiphoid process toward each other, the movement that occurs in situps.

External and internal oblique muscles

Lateral to the rectus abdominis is the **external oblique**, a thin membranous muscle that covers the sides of the abdomen. The external oblique originates on the inferior borders of the lower eight ribs and inserts on the anterior portion of the iliac crest and the linea alba. The **internal oblique** lies deep to the external oblique. The muscle fibers sweep upward from the lower abdomen and insert on the lower ribs, the xiphoid process, and the linea alba. Both the external and internal oblique muscles act with the rectus abdominis to flex the vertebral column to compress the abdomen. They also increase the pressure in the abdomen during defecation, urination, and childbirth. Locate these muscles in Figure 20.2.

Transverse abdominis

The **transverse abdominis** is deep to the internal oblique, see Figure 20.2. It originates on the lower ribs, the iliac crest, and the lumbar vertebrae and inserts on the linea alba and the pubic symphysis. The transverse abdominis muscle contracts with the other abdominal muscles to compress the abdomen.

LABORATORY ACTIVITY   MUSCLES OF THE ABDOMEN

MATERIALS

torso model
muscle chart

PROCEDURES

1. Label and review the muscles of the abdomen in Figures 20.1 and 20.2.

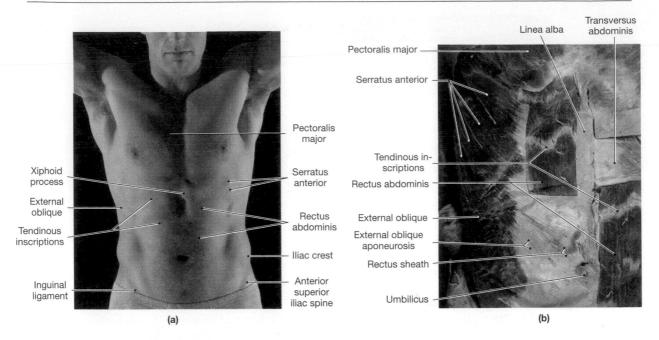

Figure 20.3
Oblique and Rectus Muscles.

(a) Surface anatomy of the abdominal wall, anterior view. (b) Cadaver, anterior superficial view of the abdominal wall.

2. Identify each abdominal muscle on the laboratory models.
3. Use Figure 20.3 as a guide to locate the position of these muscles on your abdomen.

C. Posterior Muscles of the Chest

Trapezius

The large, diamond-shaped muscle of the upper back is the **trapezius**, (Figure 20.4). It spans the gap between the scapulae and extends from the lower thoracic vertebrae to the back of the head. Locate this muscle on your body. The trapezius has numerous origins. The upper portion arises from the **ligamentum nuchae**, a mass of fibers extending from the cervical vertebrae to the occipital bone. From the ligamentum nuchae, the upper part of the trapezius inserts posteriorly on the distal third of the clavicle. The middle and lower muscle masses originate on the spinous processes of the thoracic vertebrae and insert on the acromion and spine of the scapula. Since each part of the trapezius has a unique origin and insertion, each part also produces a unique movement. The upper trapezius elevates the clavicle and scapula by pulling in the direction of the occipital bone. The middle region adducts the scapula, and the lower fibers depress and medially rotate the scapula. Perform each of these actions separately and observe the resulting motion.

STUDY TIP

In the cat and other animals, the trapezius is divided into three distinct muscles, each having its own origin and action. For your own study you may wish to think of the trapezius as three muscles (upper, middle, and lower) united into one and learn the action of each "portion" separately.

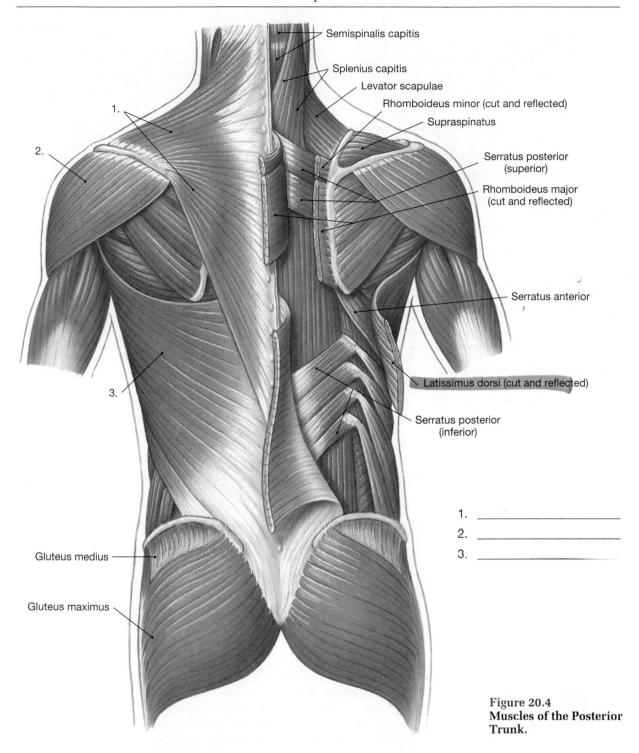

Semispinalis capitis

Splenius capitis

Levator scapulae

Rhomboideus minor (cut and reflected)

Supraspinatus

Serratus posterior
(superior)

Rhomboideus major
(cut and reflected)

Serratus anterior

Latissimus dorsi (cut and reflected)

Serratus posterior
(inferior)

Gluteus medius

Gluteus maximus

1.
2.
3.

1. _____
2. _____
3. _____

**Figure 20.4
Muscles of the Posterior
Trunk.**

Latissimus dorsi

The **latissimus dorsi (la-TIS-i-mus DOR-sē)** is the large muscle wrapping around the lower back. Find this muscle in Figure 20.4 and on your body. The latissimus dorsi has a broad origin from the sacral and lumbar vertebrae up to the sixth thoracic vertebra. This muscle sweeps up and inserts on the humerus, close to the attachment of the pectoralis major. It acts to adduct and extend the humerus, as when pulling a rope toward you.

STUDY TIP

Always notice whether a muscle passes in front of or behind the bone it inserts onto. This will determine how the bone moves. For example, the latissimus dorsi passes in front of the humerus before it inserts into the intertubercular groove. This allows medial rotation of the humerus. If it passed behind the humerus, the resulting action would be lateral rotation.

LABORATORY ACTIVITY   MUSCLES OF THE POSTERIOR CHEST

MATERIALS

    torso model
    muscle chart

PROCEDURES

1. Review the muscles of the posterior chest in Figure 20.4.
2. Identify each abdominal muscle on the laboratory models.
3. Locate the position of these muscles on your chest and practice their actions.

D. **Muscles of the Posterior Spine**

The muscles of the spine are deep to the trapezius, latissimus dorsi, and other back muscles. The **erector spinae** muscles include the longissiumus, iliocostalis, and other muscle groups. These muscles act to extend the spine.

Longissimus

The **longissimus** group consists of three muscles: the longissimus capitis, cervicis, and thoracis. Locate each of these muscles on Figure 20.5. All originate on the transverse processes of the vertebral column. The origins of the **longissimus capitis** lie on the transverse processes of the lower cervical and upper thoracic vertebrae, and it inserts on the mastoid process. The **longissimus cervicis** arises from the processes of the thoracic vertebrae and inserts on the processes of the lower cervical vertebrae. The **longissimus thoracis** originates on the processes of the lower thoracic and lumbar vertebrae and inserts on the transverse processes of the vertebrae above and on the ribs. The longissimus capitis and cervicis extend the head, and the thoracis extends the spine. If one side of any of the set contracts, it tilts the head or bends the spine to that side.

Iliocostalis

The iliocostalis group consists of three sets of muscles that arise on the posterior surface of the ribs and the iliac crest (see Figure 20.5). This muscle group acts to extend the spine and stabilize the thoracic vertebrae. Notice in Figure 20.5 that the longissimus thoracis and the iliocostalis muscles are difficult to differentiate in the lower back. In this region they are collectively called the **sacrospinalis** muscle.

Semispinalis capitis and splenius capitis

The **semispinalis capitis** extends from the spinous processes of the upper thoracic vertebrae and the seventh cervical vertebra to insert on the occipital bone (see Figure 20.5). It is medial to the splenius and the longissimus capitis and rotates the head to the side. If both contract, the head extends.

    The **splenius capitis** is medial to the longissimus capitis. It originates from the spines of vertebrae C_1 through T_4 and inserts on the occipital bone. Locate this muscle

**Figure 20.5
Muscles of the Spine.**

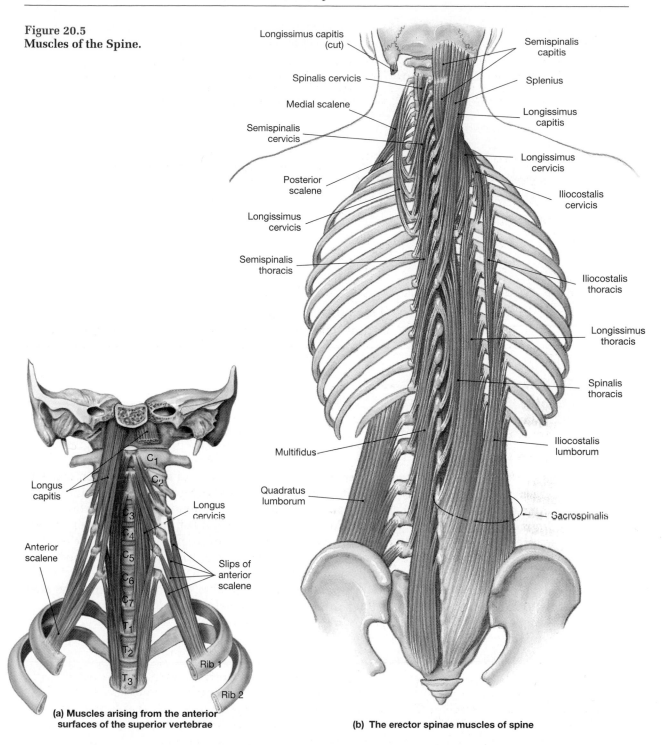

(a) **Muscles arising from the anterior
surfaces of the superior vertebrae**

(b) **The erector spinae muscles of spine**

in Figure 20.5. The splenius capitis is a synergist with the semispinalis capitis and longissimus capitis and extends the head.

Scalenes

The **scalenes** consists of an anterior, medial, and posterior group of muscles that all originate on the transverse processes of the cervical vertebrae and insert on the first or second rib. Locate the anterior scalenes in Figure 20.5. When the ribs are fixed, the scalenes flex the neck. If the neck is stationary, they act to raise the ribs during inspiration.

Quadratus lumborum

The **quadratus lumborum** (Figure 20.5) arises on the middle portion of the iliac crest and inserts on the inferior border of the twelfth rib and the transverse processes of the lumber vertebrae. It flexes the spine or, acting alone, bends the spine laterally toward the side of contraction.

LABORATORY ACTIVITY   MUSCLES OF THE POSTERIOR SPINE

MATERIALS

> torso model
> muscle chart

PROCEDURES

1. Review the muscles of the posterior spine in Figure 20.5.
2. Examine the posterior spine of the torso model and locate each muscle.

E. Muscles of the Pelvic Region

Muscles of the pelvic floor

The pelvic floor and wall form a bowl that holds the pelvic organs. The floor of the pelvis mainly consists of the **coccygeus** and the **levator ani** muscles (see Figure 20.6). The coccygeus is the most posterior of the two; it originates on the ischial spine, passes posteriorly, and inserts on the lateral and inferior borders of the sacrum. The levator ani has its origins on the inside edge of the pubis and the ischial spine and inserts on the coccyx. The pelvic diaphragm acts to flex the coccyx and tense the pelvic floor. Following the depression and protrusion of the external anal sphincter during defecation, the levator ani acts to elevate and retract the anus. During pregnancy, the expanding uterus bears down on the pelvic floor, and the coccygeus and levator ani support the weight of the fetus.

The **external anal sphincter** originates on the coccyx and inserts around the anal opening. This muscle closes the anus and is consciously relaxed for defecation.

LABORATORY ACTIVITY   MUSCLES OF THE PELVIC REGION

MATERIALS

> torso model
> muscle chart

PROCEDURES

1. Review the muscles of the pelvis in Figure 20.6.
2. Examine the pelvic region of the torso model and locate each muscle.

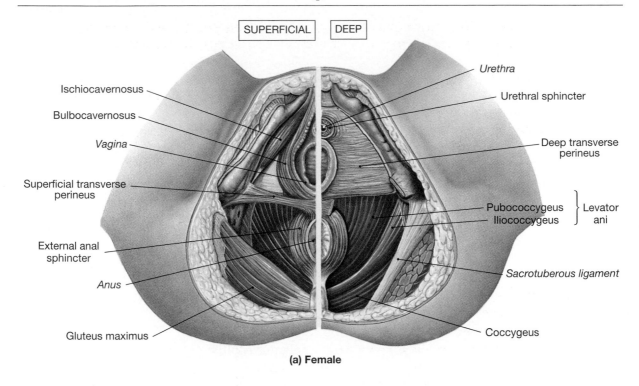

(a) Female

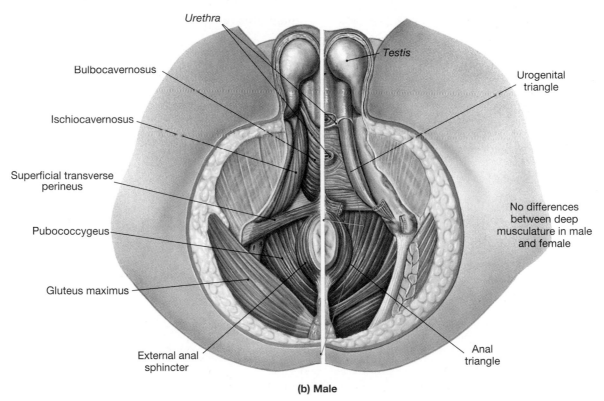

(b) Male

Figure 20.6
Muscles of the Pelvic Floor.

MUSCLES OF THE CHEST, ABDOMEN, AND SPINE CHECKLIST

This is a list of bold terms presented in Exercise 20. Use this list for review purposes after you have completed the laboratory activities.

MUSCLES OF THE ANTERIOR CHEST

• Figures 20.1, 20.2, and 20.3
pectoralis major
    sternal portion
    clavicular portion
pectoralis minor
subclavius
serratus anterior
external intercostals
internal intercostals
diaphragm

ABDOMINAL MUSCLES

• Figures 20.1, 20.2, and 20.3
rectus abdominis
linea alba
external oblique
internal oblique
transversus abdominis

POSTERIOR MUSCLES OF THE CHEST

• Figure 20.4
trapezius
ligamentum nuchae
latissimus dorsi

MUSCLES OF THE POSTERIOR SPINE

• Figures 20.4 and 20.5
erector spinae
    longissimus capitis
    longissimus cervicis
    longissimus thoracis
    iliocostalis
semispinalis capitis
splenius capitis
scalenes
sacrospinalis
quadratus lumborum

MUSCLES OF THE PELVIC REGION

• Figure 20.6
coccygeus
levator ani
external anal sphincter

MUSCLES OF THE CHEST, ABDOMEN, AND SPINE
Laboratory Report Exercise 20

A. Matching
Match each term listed on the left with the correct description on the right.

1. _____ pectoralis minor	**A.** diamond-shaped upper back muscle
2. _____ trapezius	**B.** outermost lateral muscle of the abdomen
3. _____ coccygeus	**C.** dorsal muscle that flexes the lower spine
4. _____ rectus abdominis	**D.** elevates the anal sphincter
5. _____ transversus abdominis	**E.** middle lateral muscle layer of the abdomen
6. _____ external oblique	**F.** posterior muscle of the pelvis diaphragm
7. _____ quadratus lumborum	**G.** dorsal neck muscles that extend the head
8. _____ external intercostals	**H.** small chest muscle, depresses scapula
9. _____ serratus anterior	**I.** separates the perineum from the pelvic viscera
10. _____ ligamentum nuchae	**J.** major muscle of inhalation
11. _____ levator ani	**K.** abdominal muscle with horizontal fibers
12. _____ latissimus dorsi	**L.** fibrous line located in the midline of the trunk
13. _____ pectoralis major	**M.** on midline of the trunk, compresses the abdomen
14. _____ pelvic diaphragm	**N.** major muscle used during a bench press
15. _____ diaphragm	**O.** found between the ribs, elevates the rib cage
16. _____ longissimus	**P.** large muscle wrapping around lower back
17. _____ internal intercostals	**Q.** fan-shaped muscle, inserts on scapula
18. _____ subclavius	**R.** sheet of fibers from ligaments of the cervical spines
19. _____ internal oblique	**S.** found inferior to clavicle, depresses clavicle
20. _____ linea alba	**T.** found between the ribs, depresses the rib cage

B. **Short-Answer Questions**

1. Which muscle is used when you shrug your shoulders? Describe how this action is accomplished.

2. When you blow hard through your mouth, in trying to inflate a balloon for example, your abdominal muscles contract. Why?

3. Describe the locations of the four abdominal muscles and their relationships to each other.

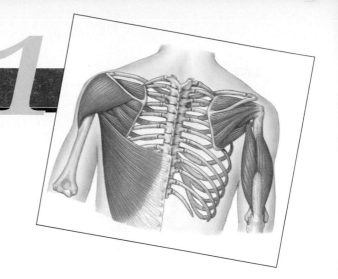

Muscles of the Shoulder, Arm, and Hand

OBJECTIVES

On completion of this exercise, you should be able to:

- Identify the origin, insertion, and action of major muscles of the shoulder region.
- Identify the origin, insertion, and action of major muscles of the upper arm.
- Identify the origin, insertion, and action of major muscles of the forearm.
- Identify the origin, insertion, and action of major muscles of the intrinsic hand.

WORD POWER

teres (teres—smooth)
rhomboideus (rhombo—a top)
levator (levator—a lifter)
deltoid (delt—triangular)

biceps (bi—two; ceps—heads)
retinaculum (retinacul—a holdfast)
thenar (thenar—palm of the hand)

MATERIALS

torso model
arm model

INTRODUCTION

The function of the muscles in this region of the body is to support and move the arm, hand, and fingers. The muscles of the shoulder help to anchor the humerus to the scapula and also to stabilize the scapula on the posterior chest. Muscles of the upper arm flex and extend the arm, muscles of the forearm move the wrist and hand.

A. Muscles of the Shoulder

Supraspinatus

Examine the posterior surface of the scapula on the torso and arm models. Locate the spine of the scapula on the model and on your body. You can use this bony protuberance as a landmark to identify several muscles. The **supraspinatus (soo-pra-spī-NA-tus)** originates on the supraspinous fossa, the depression located superior to the spine of the scapula (see Figure 21.1). The supraspinatus passes laterally under the acromion and in front of the distal end of the scapular spine to converge on a tendon that covers the upper capsule of the shoulder joint. The tendon eventually inserts on the upper part of the greater tubercle of the humerus. The supraspinatus acts to abduct the arm and assists the deltoid in this action. It also keeps the head of the humerus firmly seated in the glenoid fossa.

Infraspinatus

The **infraspinatus** is located below the scapular spine (see Figure 21.1). Again, use your scapular spine to help locate this muscle on your body. The infraspinatus originates on

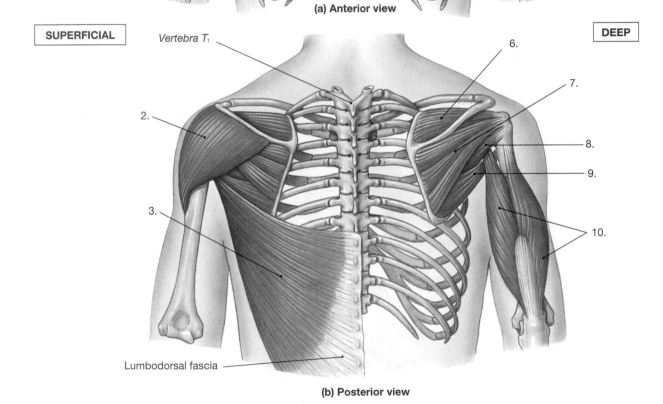

(a) Anterior view

SUPERFICIAL

DEEP

Vertebra T₁

Lumbodorsal fascia

(b) Posterior view

1. _____   6. _____
2. _____   7. _____
3. _____   8. _____
4. _____   9. _____
5. _____   10. _____

Figure 21.1
Muscles That Move the Arm.

the infraspinous fossa and inserts on the posterior surface of the greater tubercle, just inferior to the insertion of the supraspinatus. This muscle acts to laterally rotate the humerus and keeps the humoral head in the glenoid cavity by protecting the posterior side of the shoulder joint.

Subscapularis

The **subscapularis**, as the name implies, is found under the scapula next to the posterior surface of the rib cage (see Figure 21.1a). You cannot feel this muscle on your body. The subscapularis occupies most of the subscapular fossa. The muscle narrows laterally and passes in front of the humerus to insert on the upper part of the lesser tubercle. It medially rotates the humerus and so is an antagonist of the infraspinatus.

Teres minor

Originating along the lateral border of the scapula is the **teres minor**. It is a small, flat muscle, as shown in Figure 21.1, which passes laterally and superiorly to insert on the greater tubercle of the humerus near the insertion of the infraspinatus. The teres minor rotates the humerus laterally.

The four shoulder muscles described above, the supraspinatus, infraspinatus, subscapularis, and teres minor, all act to firmly seat the head of the humerus in the glenoid fossa. This fossa is a shallow depression, and without the protective function of these four muscles, the humerus would be easily dislocated from the shoulder. These muscles also make up a structure called the **rotator cuff**. Although the supraspinatus is part of the rotator cuff, it is not itself a rotator but an abductor muscle. You may be familiar with the rotator cuff if you are a baseball fan. The windup and throw of a pitcher involves circumduction of the humerus. This places much stress on the shoulder joint and the rotator cuff, which causes premature degeneration of the joint. To protect the shoulder joint and muscles, bursal sacs are interspersed between the tendons of the rotator cuff muscles and the neighboring bony structures. Repeated friction on the bursae may result in an inflammation called bursitis.

Teres major

The **teres major** is a thick muscle that arises on the inferior angle of the posterior surface of the scapula (see Figure 21.1). The muscle converges up and laterally into a flat tendon that ends on the anterior side of the humerus. This insertion point is just medial to the intertubercular groove on the crest of the lesser tubercle. Since the muscle passes to the anterior side of the humerus, it acts to medially rotate that bone and assist the latissimus dorsi in moving the arm downward.

Rhomboideus

The **rhomboideus** muscles extend between the upper thoracic vertebrae and the scapula (see Figure 21.2). They are deep to the trapezius, and you cannot feel them on your body. The **rhomboideus major** originates along the spinous processes of the upper throracic vertebrae and inserts on the lower medial border of the scapula. The **rhomboideus minor** arises from vertebrae T_7 and T_1 and attaches to the upper border of the scapula. The rhomboideus muscles act to adduct and laterally rotate the scapula.

Levator scapulae

The term **levator scapulae** implies that this muscle elevates the scapula. To do so it must originate superior to the scapula (see Figure 21.2). The muscle is deep to the trapezius and therefore difficult to feel. It arises from the transverse processes of vertebrae C_1 through C_4 and inserts on the upper medial border of the scapula. The levator scapulae also tilts the head toward the same side of contraction.

Figure 21.2
Muscles That Position the Shoulder Girdle.
Posterior view.

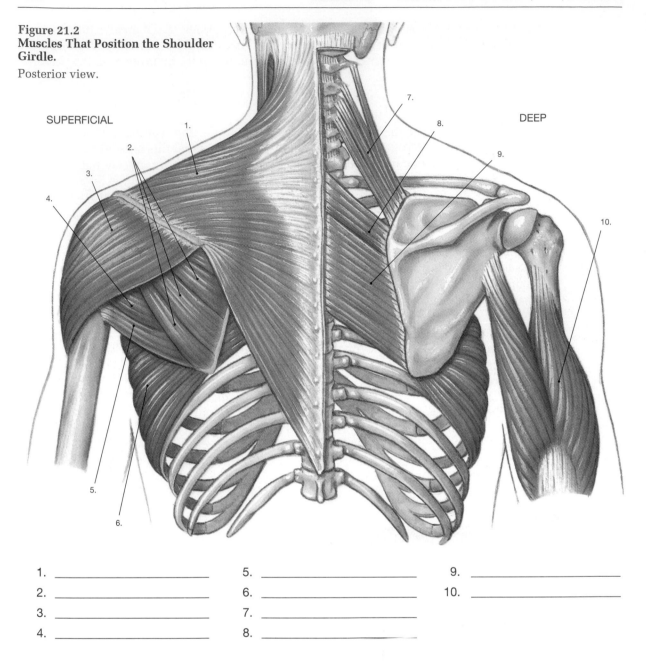

SUPERFICIAL

DEEP

1. _____	5. _____	9. _____
2. _____	6. _____	10. _____
3. _____	7. _____	
4. _____	8. _____	

Deltoid

The very large, easily identifiable muscle that covers the shoulder is the **deltoid**. Examine Figures 21.1 and 21.2 and locate this muscle on the upper part of your humerus, just below the point of the shoulder. It has a broad origin and several actions. As with the trapezius, it may help to consider the deltoid as having three parts. The posterior portion arises along the inferior margin of the scapular spine, the middle region originates on the acromion of the scapula, and the anterior portion originates on the distal end of the anterior surface of the clavicle. All parts of the muscle converge toward the humerus and insert on the large deltoid tuberosity. The three portions have different actions. The anterior portion extends the arm, while the posterior segment performs the opposite action and flexes the arm. The middle portion acts as the major abductor of the humerus.

LABORATORY ACTIVITY   MUSCLES OF THE SHOULDER

MATERIALS

torso model
arm model
muscle chart

PROCEDURES

1. Label and review the muscles in Figures 21.1 and 21.2.
2. Examine the shoulder of the torso model and locate each muscle of the shoulder.
3. Locate the general position of each shoulder muscle on your body. Contract each muscle and observe the action of your shoulder.

B. Muscles of the Upper Arm

Biceps brachii

The **biceps brachii** (Figure 21.3) makes up the fleshy mass of the anterior humerus when the arm is flexed. The term "biceps" refers to the presence of two origins or "heads." The first, called the **short head**, begins on the coracoid process of the scapula as a tendon that expands into the muscle belly. The second origin, the **long head**, begins on a roughened area on the superior lip of the glenoid fossa of the scapula, the supraglenoid tubercle. A tendon passes over the top of the humerus into the intertubercular groove and blends into the muscle. The tendon of the long head is enclosed within a protective covering called the intertubercular synovial sheath. The two heads of the muscle fuse and comprise most of the mass of the anterior brachium. The biceps brachii inserts onto the radial tuberosity. It is a principal flexor of the forearm. The biceps brachii is also a supinator. When the forearm is in a prone position, the radial tuberosity faces posteriorly and the tendon of the biceps passes between the radius and ulna to insert on the radius. The contraction of the biceps pulls the radial tuberosity anteriorly and rotates the radius laterally. This rotates (supinates) the palm into a forward-facing position.

Brachialis

The **brachialis (brā-kē-A-lis)** also flexes the forearm. It is located under the distal end of the biceps brachii (see Figure 21.3). You can feel a small part of it if you flex your arm and palpate the area just lateral to the tendon of the biceps. The brachialis originates on the distal humerus and inserts onto the ulnar tuberosity below the coronoid process. Since the ulna is limited to a hingelike movement, the action of the brachialis is flexion of the forearm.

Coracobrachialis

The **coracobrachialis (KOR-uh-kō-brā-kē-A-lis)** is a small muscle that originates on the coracoid process of the scapula and shares a common origin with the biceps (see Figure 21.3). The coracobrachialis passes inferiorly and inserts midway along the medial surface of the humerus between the origins of the brachialis and triceps muscles. It adducts and flexes the humerus.

Triceps brachii

The principal antagonist to the biceps brachii and brachialis is the **triceps brachii**. This muscle forms most of the mass of the upper posterior side of the arm. Examine Figure 21.3 and locate the triceps brachii on the back of the upper arm. The name indicates that it has three origins, two on the scapula and one on the humerus. The **long head**

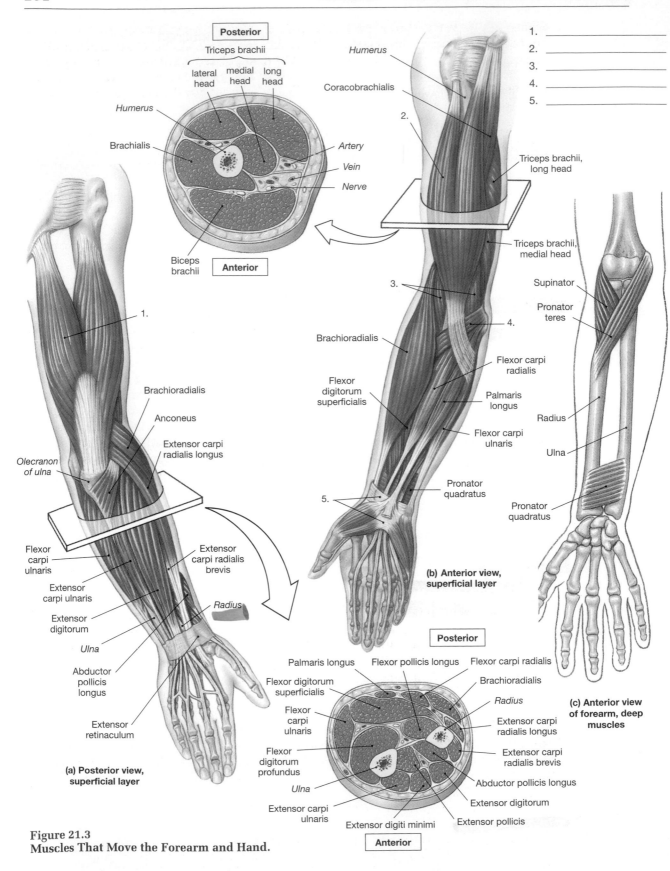

Posterior

Triceps brachii

lateral   medial   long
head   head   head

Humerus

Brachialis

Artery

Vein

Nerve

Biceps
brachii

Anterior

Humerus

Coracobrachialis

2.

Triceps brachii,
long head

Triceps brachii,
medial head

Supinator

Pronator
teres

1. _____
2. _____
3. _____
4. _____
5. _____

3.

4.

Brachioradialis

Flexor carpi
radialis

Flexor
digitorum
superficialis

Palmaris
longus

Flexor carpi
ulnaris

Radius

Ulna

Pronator
quadratus

5.

Pronator
quadratus

**(b) Anterior view,
superficial layer**

**(c) Anterior view
of forearm, deep
muscles**

1.

Brachioradialis

Anconeus

Extensor carpi
radialis longus

*Olecranon
of ulna*

Flexor
carpi
ulnaris

Extensor
carpi ulnaris

Extensor
digitorum

Ulna

Abductor
pollicis
longus

Extensor
retinaculum

Extensor
carpi radialis
brevis

Radius

**(a) Posterior view,
superficial layer**

Posterior

Palmaris longus   Flexor pollicis longus   Flexor carpi radialis

Flexor digitorum
superficialis

Flexor
carpi
ulnaris

Flexor
digitorum
profundus

Ulna

Extensor carpi
ulnaris

Extensor digiti minimi

Brachioradialis

Radius

Extensor carpi
radialis longus

Extensor carpi
radialis brevis

Abductor pollicis longus

Extensor digitorum

Extensor pollicis

Anterior

**Figure 21.3
Muscles That Move the Forearm and Hand.**

begins as a flattened tendon on the scapula at the roughened area below the glenoid cavity. The **lateral head** arises from a slender origin on the posterior surface of the humerus below the head of the humerus. The **medial head** originates on a broad region covering one-third of the distal posterior humerus. The three heads merge into a common tendon that begins at about the middle of the muscle and inserts on the olecranon process of the ulna. The triceps acts to extend the forearm.

LABORATORY ACTIVITY   Muscles of the Upper Arm

MATERIALS

    torso model
    arm model
    muscle chart

PROCEDURES

1. Label and review the muscles in Figure 21.3.
2. Examine the torso and arm models and locate each muscle of the upper arm.
3. Use Figure 21.4 as guides and locate the general position of each upper arm muscle on your body. Contract each muscle and observe the action of your arm.

C. Muscles of the Forearm

Supinator

The **supinator** is found on the lateral side of the forearm deep to several muscles. It arises on the lateral epicondyle of the humerus, crosses the antecubital region, and inserts on the lateral side of the radius distal to the radial head. When the supinator contracts, it rotates the radius into a position parallel to the ulna, resulting in supination of the forearm. Locate the supinator in Figure 21.3.

Pronator teres and pronator quadratus

The principal pronator muscle of the arm is the **pronator teres** (see Figure 21.3). Since pronation is the opposite action of supination, the origin and insertion of the pronator teres must be opposite that of the supinator. The pronator teres originates on the medial epicondyle of the humerus and on the proximal part of the ulna. It inserts on the lateral side of the radius about midway down the shaft of that bone. When the forearm is in a supine position, the pronator teres pulls the lateral edge of the radius toward the ulna and medially rotates the radius. This results in pronation of the forearm. Note how the pronator teres and supinator cross over each other for antagonistic actions.

The **pronator quadratus** is found just proximal to the wrist joint, on the anterior surface of the forearm. Examine Figure 21.3 to view this muscle. It originates on the anterior surface of the distal end of the ulna. The pronator quadratus passes laterally to insert on the shallow concavity on the distal end of the radius. It acts as a synergist to the pronator teres in pronating the forearm and can also cause medial rotation of the forearm as a whole.

STUDY TIP

When you study the flexors and extensors of the arm model, follow the tendons to their insertion points for clues about the names of the muscles. For example, the tendon of the flexor carpi ulnaris muscle inserts on the carpals on the ulnar side of the wrist and flexes the wrist (carpi). Since this muscle acts to flex the wrist, it must be located on the anterior surface of the forearm.

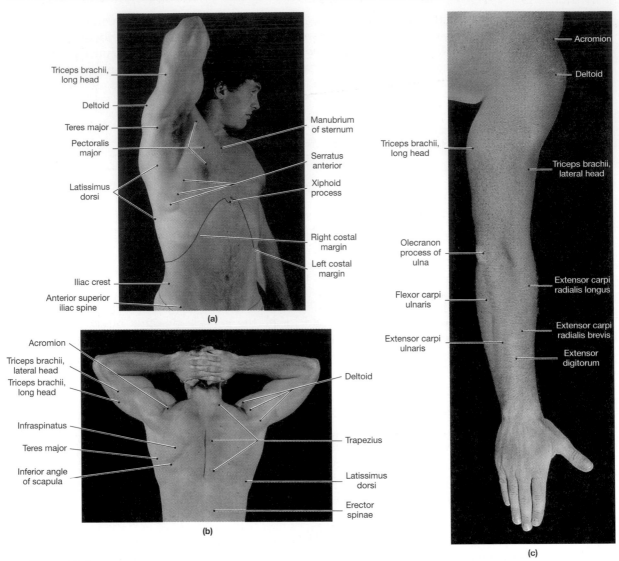

Figure 21.4
Surface Anatomy of the Neck, Shoulder, Arm, and Trunk, Male Torso.
(a) Lateral view. **(b)** Posterior view. **(c)** Right arm, posterior view.

Superficial flexors of the wrist

The superficial flexor muscles of the wrist and hand joint are located on the forearm; see Figure 21.3. They all have a similar origin, the common flexor tendon attached to the medial epicondyle of the humerus. The first is the **flexor carpi radialis** which lies just medial to the pronator teres and is the flexor muscle closest to the radius. The fibers of this muscle blend into a long tendon that inserts on the base of the second metacarpal. A smaller branch terminates on the base of the third metacarpal. Medial to the flexor carpi radialis is the **palmaris longus** which inserts on the palmer aponeurotica and is continuous with the palmer fascia. Medial to the palmaris is the **flexor carpi ulnaris**. This muscle rests on the ulnar side of the forearm and inserts on the pisiform and hamate bones of the carpals and the base of the third and fourth metacarpals. The radialis flexes and abducts the wrist. The ulnaris flexes and adducts the wrist. The palmaris longus tenses the palmer fascia and assists in wrist flexion.

Flexor digitorum superficialis

The **flexor digitorum superficialis** is located deep to the superficial flexors of the hand. Find this muscle in Figure 21.5. It arises from several points, one of which is located on the common flexor tendon described above. The others are on the lateral surface of the coronoid process of the ulna and the anterior surface of the radius. Four tendons exit the distal end of the muscle and pass under a transverse connective tissue sheath called the **flexor retinaculum (ret-i-NAK-ū-lum)**. The tendons then pass through a narrow valley bounded by carpal bones called the carpal tunnel. They are covered by a synovial sheath which protects and lubricates them in the tunnel. Repeated flexing of the hand and fingers, such as extended typing or piano playing, causes the sheath to swell and compress the median nerve. Pain and numbness occurs in the palm during flexion, a condition called carpal tunnel syndrome.

Extensors of the hand and fingers

The extensor muscles of the hand and fingers are located on the posterior forearm. They all originate from a common tendon attached to the lateral epicondyle of the humerus (except the extensor carpi radialis longus). Examine Figure 21.3 to find these muscles. The **extensor carpi ulnaris** lies next to the flexor carpi ulnaris near the ulna. The extensor carpi ulnaris also has an origin on the posterior surface of the ulna nearby. This muscle inserts on the base of the fifth metacarpal. As you work toward the radial side of the forearm, you will first see the extensor digitorum mini, which will not be considered here. Lateral to this muscle is the **extensor digitorum** (see Figure 21.5). It sends three or four tendons under a fibrous band that cuts transversely across the posterior aspect of the wrist, called the **extensor retinaculum**. The tendons insert on the dorsal surface of the base of the middle phalanges. The **extensor carpi radialis brevis** is a short, thick muscle located just lateral to the digitorum, which inserts on the base of the third metacarpal. The last extensor considered here is the **extensor carpi radialis longus**. It does not have a common origin, for the most part, with the other extensors. It arises from the humerus just proximal to the lateral epicondyle, although a few fibers do extend from the common tendon. The belly of this long muscle lies lateral to and slightly above the brevis. It inserts on the base of the second metacarpal. These muscles oppose the actions of the flexors of the forearm. The extensor carpi ulnaris and radialis (both longus and brevis) extend the wrist. The former adducts and the latter abducts the wrist as well. The extensor digitorum extends the fingers.

Brachioradialis

The superficial muscle located lateral to the extensor carpi radialis longus is the **brachioradialis**. This muscle is shown in Figure 21.3. You can feel the muscle on the lateral side of the anterior surface of your forearm. It originates on the humerus proximal to the lateral epicondyle and superior to the origin of the extensor carpi radialis longus. The brachioradialis flexes the forearm and assists the action of the biceps and brachialis. It also aids in lateral movement of the forearm.

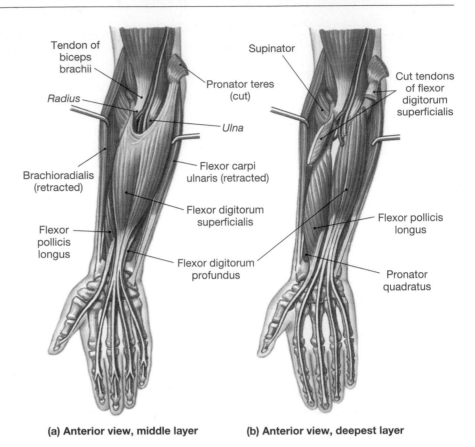

(a) Anterior view, middle layer

(b) Anterior view, deepest layer

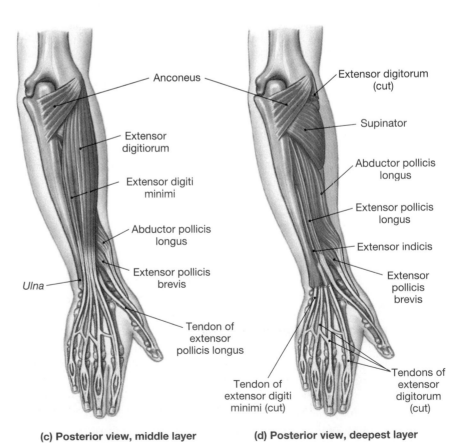

(c) Posterior view, middle layer

(d) Posterior view, deepest layer

Figure 21.5
Muscles That Move the Hand and Fingers.

Middle and deep muscle layers of the right forearm; for superficial muscles, see Figure 21.3.

LABORATORY ACTIVITY   MUSCLES OF THE FOREARM

MATERIALS

torso model
arm model
muscle chart

PROCEDURES

1. Label and review the muscles of the forearm in Figures 21.3 and 21.5.
2. Examine the forearm of the torso and arm models and locate each muscle.
3. Use Figure 21.4 and locate as many of the forearm muscles on your body as possible. Contract each muscle and observe the action of your forearm.

D. Intrinsic Muscles of the Hand

The intrinsic muscles of the hand are all located on the hand itself and usually control the movements of the fingers and thumb. You can locate the muscles described below in Figure 21.6.

Muscles of the thenar eminence

The mass of tissue at the base of the thumb is called the **thenar eminence** and consists of several muscles. If you examine your hand, you can see these muscles making up the fleshy region on the lateral side of the palm, near the thumb. All these muscles insert on the lateral side of the base of the proximal phalanx of the thumb and originate on the flexor retinaculum, but the origins are at slightly different positions along the retinaculum and include other bones of the wrist. These differences define the muscle's individual functions. The most medial of the thenar muscles is the **flexor pollicus brevis**. It acts to flex and adduct the thumb. Lateral to the flexor is the **abductor pollicus brevis**, which not only originates on the flexor retinaculum but also arises from the scaphoid and trapezium bones of the carpals. This muscle abducts your thumb. The most lateral thenar muscle is the **opponens pollicis**, and it partially originates on the trapezium of the wrist. It acts to bring the thumb across the palm toward the little finger. This movement is called opposition. Although the **adductor pollicis** is often not considered part of the thenar eminence, it is found just medial to the flexor pollicus brevis deep in the web of tissue between the thumb and palm. It originates on the metacarpals and carpals of the hand and inserts on the medial side of the proximal phalanx of the thumb. It serves to adduct the thumb and opposes the action of the abductor pollicus brevis.

Muscles of the hypothenar eminence

The **hypothenar eminence** is found at the base of the little finger and consists of three muscles. On your hand, it is the fleshy mass on the medial side of the palm. The most medial of hypothenar muscles is called the **opponens digiti minimi**, and it originates on the flexor retinaculum and the hamate of the carpals. It inserts on the fifth metacarpal. The opponens digiti acts in concert with the opponens pollicus to bring the little finger toward the thumb during opposition. The **flexor digiti minimi**, located medial to the opponens digiti, has the same origin as this muscle but inserts on the proximal, fifth phalanx. It acts to flex the little finger. The most medial of these muscles is the **abductor digiti minimi**. It arises from the pisiform bone, inserts on the proximal phalanx of the little finger, and abducts that finger. You should note that no muscles originate on the fingers. The phalanges serve as insertion points for muscles whose origins are more proximal.

Figure 21.6
Intrinsic Muscles of the Hand.

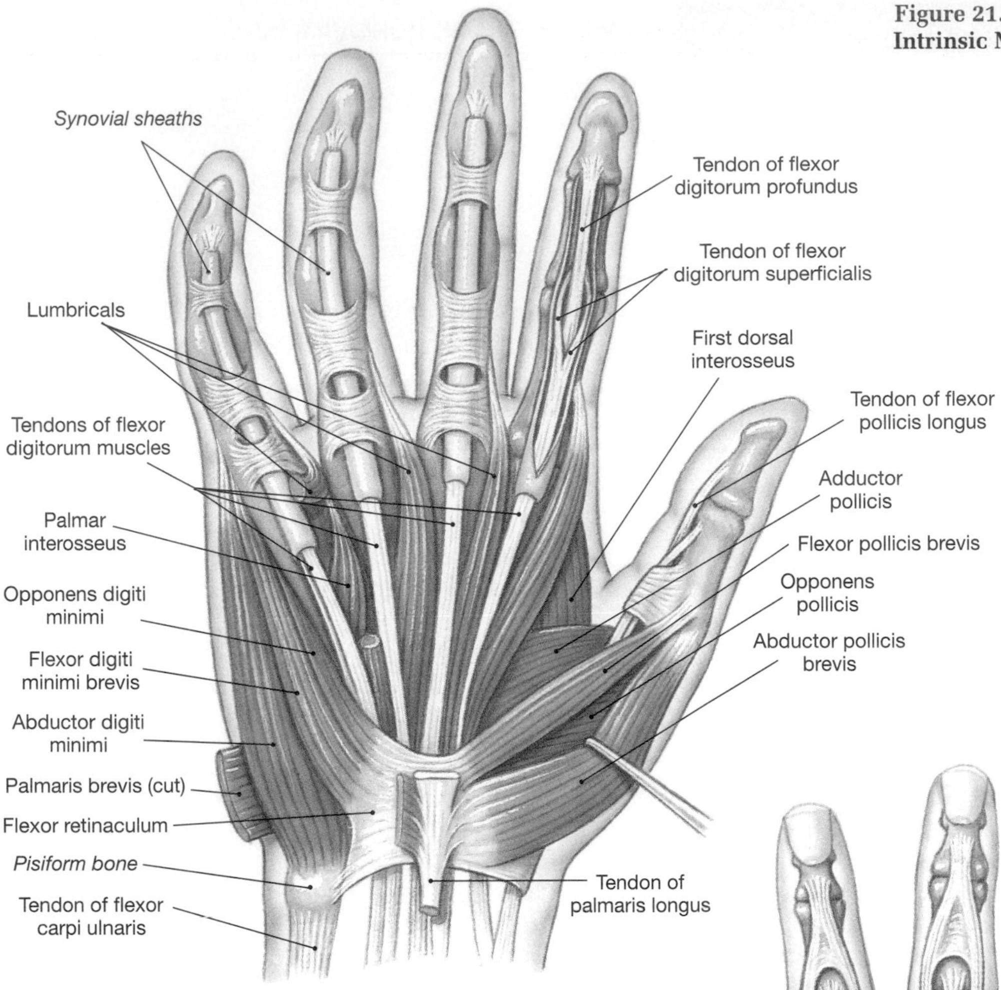

Synovial sheaths

Tendon of flexor
digitorum profundus

Tendon of flexor
digitorum superficialis

Lumbricals

First dorsal
interosseus

Tendon of flexor
pollicis longus

Tendons of flexor
digitorum muscles

Adductor
pollicis

Palmar
interosseus

Flexor pollicis brevis

Opponens digiti
minimi

Opponens
pollicis

Flexor digiti
minimi brevis

Abductor pollicis
brevis

Abductor digiti
minimi

Palmaris brevis (cut)

Flexor retinaculum

Pisiform bone

Tendon of
palmaris longus

Tendon of flexor
carpi ulnaris

**(a) Anterior view
of the right hand**

Tendon of
extensor
indicis

First dorsal
interosseus

Tendon of
extensor
digiti minimi

Tendon of
extensor
pollicis
longus

Abductor
digiti minimi

Tendon of extensor
pollicis brevis

Tendon of ex-
tensor
carpi ulnaris

Tendon of extensor
carpi radialis longus

Extensor
retinaculum

Tendon of extensor
carpi radialis brevis

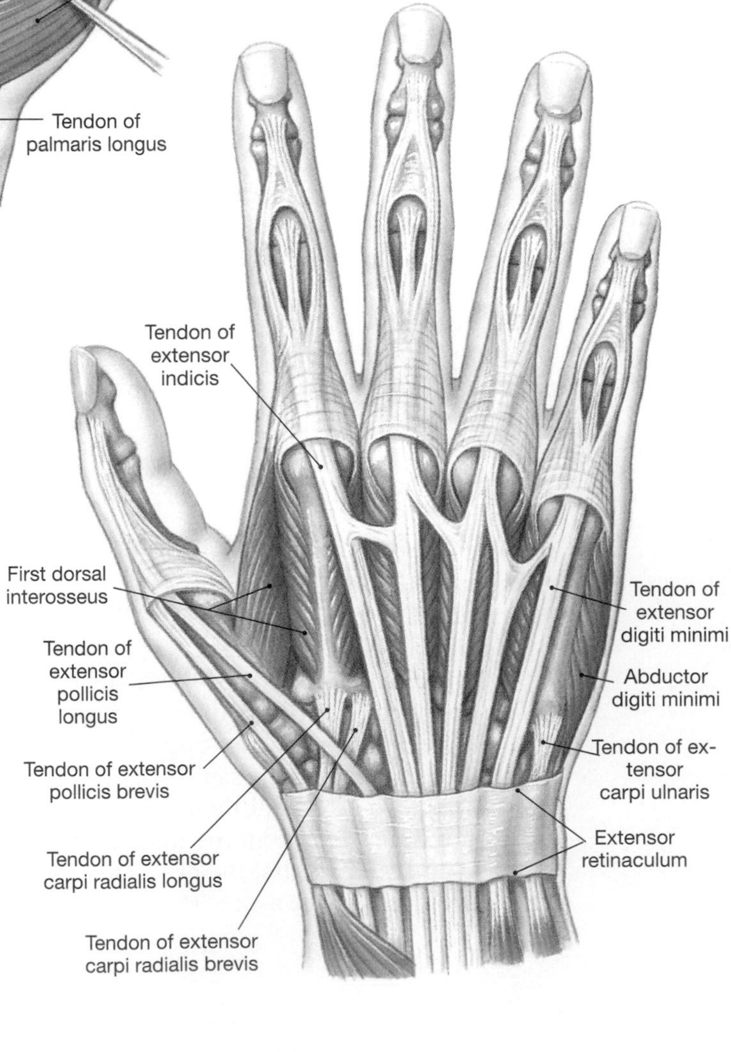

**(b) Posterior view
of the right hand**

LABORATORY ACTIVITY   INTRINSIC MUSCLES OF THE HAND

MATERIALS

torso model
arm model
muscle chart

PROCEDURES

1. Review the muscles of the hand in Figure 21.6.
2. Examine an arm or hand model and identify each muscle of the hand.
3. Use Figure 21.6 and locate as many muscles on your hand as possible. Contract each muscle and observe the action of your hand.

MUSCLES OF THE SHOULDER, ARM, AND HAND CHECKLIST

This is a list of bold terms presented in Exercise 21. Use this list for review purposes after you have completed the laboratory activities.

MUSCLES OF THE SHOULDER

• Figures 21.1, 21.2 and 21.4
rotator cuff
  supraspinatus
  infraspinatus
  subscapularis
  teres minor
  teres major
rhomboideus major
rhomboideus minor
levator scapulae
deltoid

MUSCLES OF THE UPPER ARM

• Figures 21.1, 21.2, 21.3, and 21.5
biceps brachii
  short head
  long head
brachialis
coracobrachialis
triceps brachii
  long head
  lateral head
  medial head

MUSCLES OF THE FOREARM

• Figures 21.3, 21.4 and 21.5
supinator
pronator teres
pronator quadratus

flexor carpi radialis
palmaris longus
flexor carpi ulnaris
flexor digitorum superficialis
flexor retinaculum
extensor carpi ulnaris
extensor digitorum
extensor retinaculum
extensor carpi radialis brevis
extensor carpi radialis longus
brachioradialis

INTRINSIC MUSCLES OF THE HAND

• Figure 21.6
thenar eminence
  flexor pollicis brevis
  abductor pollicis brevis
  adductor pollicis
  opponens pollicis
hypothenar eminence
  opponens digiti minimi
  flexor digiti minimi
  abductor digiti minimi

MUSCLES OF THE SHOULDER, ARM, AND HAND
Laboratory Report Exercise 21

A. Matching

Match each term listed on the left with the correct description on the right.

1. _____ triceps brachii		A. tenses palmer fascia and flexes wrist
2. _____ infraspinatus		B. major pronator of the arm
3. _____ flexor pollicis brevis		C. flexes and adducts the wrist
4. _____ teres minor		D. found between the metacarpals
5. _____ opponens digiti minimi		E. small muscle, has common origin with biceps
6. _____ flexor digiti minimi		F. small muscle that abducts the thumb
7. _____ biceps brachii		G. major supinator of the forearm
8. _____ brachialis		H. extends and adducts the wrist
9. _____ palmaris longus		I. part of rotator cuff, medially rotates humerus
10. _____ supraspinatus		J. band of connective tissue on the flexor tendons
11. _____ pronator teres		K. most medial thenar muscle, flexes and adducts thumb
12. _____ flexor carpi ulnaris		L. flexes little finger
13. _____ extensor carpi ulnaris		M. originates on inferior scapula, medially rotates humerus
14. _____ extensor digitorum		N. major flexor of the forearm
15. _____ extensor carpi radialis		O. brings little finger toward the thumb
16. _____ coracobrachialis		P. major flexor and supinator of the - forearm
17. _____ abductor pollicis brevis		Q. below the scapular spine, laterally rotates the humerus
18. _____ flexor digitorum superficialis		R. part of rotator cuff that does not rotate the humerus
19. _____ supinator		S. muscle of upper arm, flexes, extends and abducts arm
20. _____ teres major		T. extends the fingers
21. _____ flexor retinaculum		U. originates below scapula, a rotator cuff muscle
22. _____ deltoid		V. brings the thumb toward the little finger
23. _____ adductor pollicis		W. extends and abducts the wrist
24. _____ opponens pollicis		X. major extensor of the forearm
25. _____ subscapularis		Y. tendons pass through carpal tunnel, flexes the digits

B. **Short-Answer Questions**

1. A brace placed on your wrist to treat carpal tunnel syndrome would prevent what action of the wrist? What would you accomplish by limiting this action?

2. If you repeatedly grasped a tennis ball in your hand to develop grip strength, what part of your arm would develop in size?

3. Why would a dislocated shoulder also potentially result in a rotator cuff injury?

22

Muscles of the Pelvis, Leg, and Foot

OBJECTIVES

On completion of this exercise, you should be able to:

- Identify the muscles of the pelvis and gluteal regions.
- Identify the muscles of the anterior, medial, and posterior upper leg.
- Identify the muscles of the anterior and posterior lower leg.
- Identify the muscles of the foot.

WORD POWER

psoas (psoa—loin)
maximus (maxim—largest)
gracilis (gracil—slender)
tensor fasciae (tens—stretch; fasci—band)

latea (lat—wide)
sartoris (sartori—tailor)
vastus (vast—huge)

MATERIALS

torso model
leg model

INTRODUCTION

The muscles of the pelvis help support the mass of the body and stabilize the pelvic girdle. The leg is surrounded by muscles that move the thigh, knee, and foot. Flexors of the leg are on the posterior thigh, and extensor muscles are located anteriorly. Abductors are found on the lateral side of the thigh, and adductors are on the medial side.

A. Muscles of the Pelvis and Gluteal Region

Iliopsoas

The **iliopsoas (il-ē-ō-sō-us)** consists of two muscles, the psoas major and the iliacus (see Figure 22.1c). Although these muscles are located in the lower abdomen, they act on the femur and move the thigh. The **psoas (SŌ-us) major** originates on the body and transverse processes of vertebrae T_{12} through L_5. The muscle sweeps downward, passing between the femur and the ischial ramus, and inserts on the lesser trochanter of the femur. The **iliacus (il-Ē-uh-kus)** originates on the iliac fossa on the medial portion of the ilium and joins the tendon of the psoas. Collectively, the iliopsoas flexes the thigh, bringing the anterior surface of the upper leg toward the abdomen.

STUDY TIP

Abductors move a structure away from the midline of the body or limb and are generally found on the lateral side of the structure being moved. Adductors are found on the medial side.

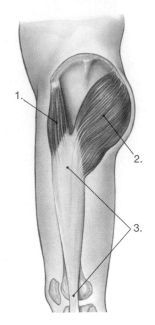

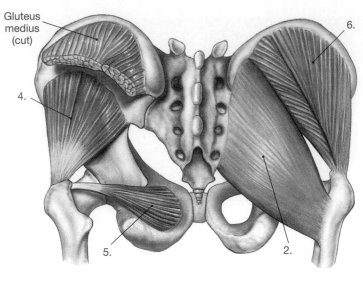

Gluteus
medius
(cut)

1.

2.

3.

4.

5.

6.

2.

(a) The gluteal muscle group

7.

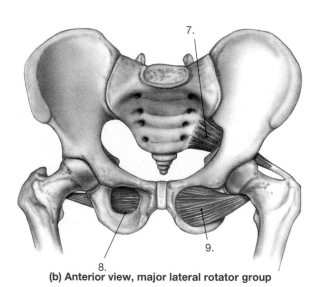

8.

9.

(b) Anterior view, major lateral rotator group

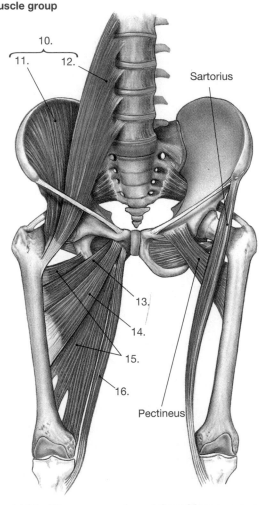

10.

11. 12.

Sartorius

13.

14.

15.

16.

Pectineus

(c) The Iliopsoas muscle and the adductor group

Figure 22.1
Muscles That Move the Thigh.

1. _____

2. _____

3. _____

4. _____

5. _____

6. _____

7. _____

8. _____

9. _____

10. _____

11. _____

12. _____

13. _____

14. _____

15. _____

16. _____

Deep leg rotators

The internal and external obturators and the piriformis are deep rotators of the thigh (see Figure 22.1). The **internal** and **external obturators** originate along the medial and lateral edges of the obturator foramen and insert on the trochanteric fossa, a shallow depression on the medial side of the greater trochanter. The **piriformis (pir-i-FOR-mis)** arises from the anterior and lateral surfaces of the sacrum and inserts on the greater trochanter of the femur. The piriform and obturator muscles rotate the thigh laterally. The piriformis also abducts the thigh.

Gluteal muscles

The posterior muscles of the pelvis are three gluteal muscles which constitute the buttocks (Figure 22.1). The most superficial and prominent is the **gluteus maximus**. It is a large, fleshy muscle and is easily located as the major muscle of the buttocks. It extends and laterally rotates the thigh. It originates along the posterior of the iliac crest and passes over the sacrum and down the side of the coccyx. From this broad origin, muscle fibers pass laterally and downward to converge in a thick tendon that inserts into the **fascia lata** of the thigh. The fascia lata is the deep fascia covering most of the thigh. It begins on the crest of the ilium and passes downward over the lateral surface of the upper thigh where it receives the insertion of the gluteus maximus and the tensor fasciae latae muscles. The fascia lata continues inferiorly as a thick band called the **iliotibial tract** which attaches to the lateral condyle of the tibia.

The **gluteus medius** is a thick muscle partially covered by the gluteus maximus. It originates on the lateral surface of the ilium and gathers laterally into a thick tendon that inserts posteriorly on the greater trochanter. The **gluteus minimus** begins on the lateral surface of the ilium, tucked under the origin of the medius. Its fibers also pass laterally to insert on the anterior surface of the greater trochanter. Both the gluteus medius and minimus abduct and medially rotate the thigh.

Tensor fasciae latae

The **tensor fasciae latae (FASH-ē-ē LĀ-tā)** originates on the outer surface of the anterior superior spinous process of the ilium. See Figure 22.1 for a view of this muscle. The muscle fibers project downward and join the gluteus maximus on the iliotibial tract. As the name implies, the tensor fasciae latae tenses the fascia of the thigh and helps to stabilize the pelvis on the femur. The muscle also acts to abduct and medially rotate the thigh.

LABORATORY ACTIVITY   MUSCLES OF THE PELVIS

MATERIALS

> torso model
> leg model

PROCEDURES

1. Label and review the muscles of the pelvis in Figure 22.1.
2. Examine the pelvis of the torso model and identify each muscle.

B. Muscles of the Upper Leg

Adductors of the thigh

The major adductor muscles of the thigh include the pectineus, the gracilis, and the adductor muscle group. These muscles arise on the pubis bone and insert on the medial femur to adduct the thigh. Additionally, these muscles act to flex and medially rotate the thigh.

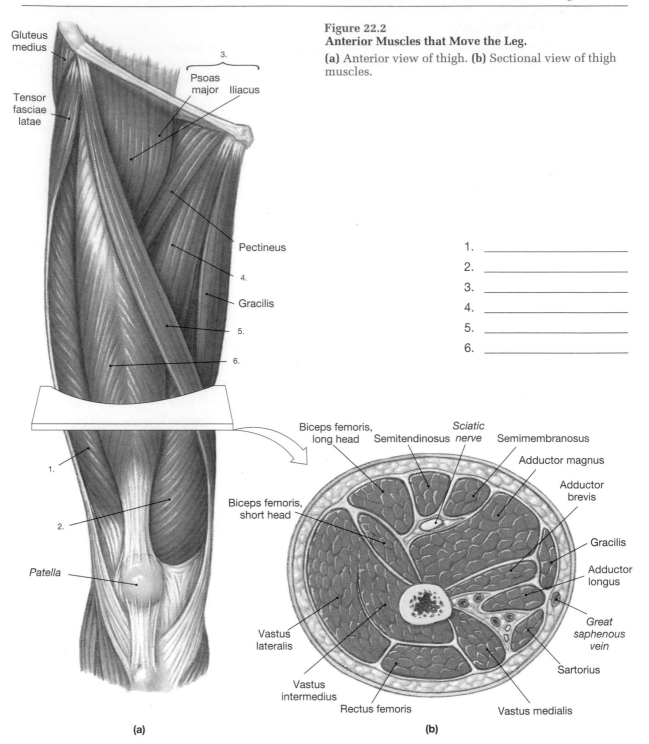

Figure 22.2
Anterior Muscles that Move the Leg.
(a) Anterior view of thigh. **(b)** Sectional view of thigh muscles.

1. _____
2. _____
3. _____
4. _____
5. _____
6. _____

(a)   (b)

Gracilis

The mass of tissue that can be felt on the inside of the thigh is comprised of the adductor muscles of the thigh. These muscles can be located on Figures 22.1 and 22.2. The **gracilis (GRAS-i-lis)** is the most superficial of the thigh adductors. It arises from the superior ramus of the pubic bone, near the symphysis, extends downward along the medial surface of the thigh, and inserts just medial to the insertion of the sartorius near the tibial tuberosity. Since it passes over the knee joint, it acts to flex the knee and adduct the thigh.

Pectineus

The **pectineus (pek-TIN-ē-us)** is another superficial adductor muscle of the medial thigh. It is located between the iliacus and adductor longus muscles. It originates along the superior ramus of the pubic bone and inserts on a small ridge, the pectineal line, inferior to the lesser trochanter of the femur. Locate this muscle in Figure 22.1.

Adductor muscle group

The adductor muscle group includes the adductor magnus, adductor longus, and the adductor brevis. The **adductor magnus** is the largest of the adductor muscles. It arises on the inferior ramus of the pubis and inserts along the length of the linea aspera of the femur. It is easily observed on the leg model if the superficial muscles are removed. Superficial to the adductor magnus is the **adductor longus.** The **adductor brevis** is superior and posterior to the adductor longus. The adductor muscles originate on the inferior pubis and insert on the posterior femur and are powerful adductors of the thigh. They also allow flexion and medial rotation of the thigh.

Sartorius

The **sartorius** is a thin, ribbonlike muscle originating on the anterior superior iliac spine and passing downward, cutting obliquely across the thigh. It is the longest muscle in the human body (see Figure 22.2). Tailors used to sit cross-legged while working, and because the sartorius muscle assists in maintaining this posture, it is commonly referred to as the "tailor's" muscle. The sartorius crosses the knee joint to insert on the medial surface of the tibia near the tibial tuberosity. This muscle is a flexor of the knee and thigh and a lateral rotator of the thigh.

The quadriceps femoris group

The extensors of the lower leg are collectively called the **quadriceps femoris group** and include the **rectus femoris, vastus medialis, vastus lateralis, and vastus intermedius**. They make up the bulk of the anterior mass of the thigh and are easy to locate on your leg.

The **rectus femoris** (see Figure 22.2) is located along the midline of the anterior surface of the thigh. It originates from the iliac spine and the acetabulum. This muscle blends into the tendon of the quadriceps group and inserts on the patella. All the muscles of the quadriceps group converge on this tendon which inserts on the tibial tuberosity. Because the rectus femoris crosses two joints, the hip and knee, it produces flexion of the hip joint and extension of the lower leg.

The vastus muscles are shown in Figure 22.2. Covering almost the entire medial surface of the femur is the **vastus medialis**. The **vastus lateralis** is located on the lateral side of the rectus femoris. Directly under the rectus femoris and between the vastus lateralis and medialis is the **vastus intermedius**. The vastus muscles originate on the anterior femur and posteriorly on the linea aspera of the femur. They converge on the tendon of the quadriceps muscle and act together to extend the knee.

Muscles of the posterior thigh

The Hamstring Group. The major muscles of the posterior upper leg are collectively called the **hamstrings**. They consist of the semitendinosus, semimembranosus and biceps femoris, as shown in Figure 22.3. They all have a common origin on the ischial tuberosity.

The **biceps femoris** is the lateral muscle of the posterior thigh. It has two heads and two origins, one on the ischial tuberosity and a second on the linea aspera of the femur. The two heads merge to form the belly of the muscle and insert on the lateral condyle of the tibia and the head of the fibula. Because this muscle spans the hip and the knee joints, it can flex the knee and extend the thigh.

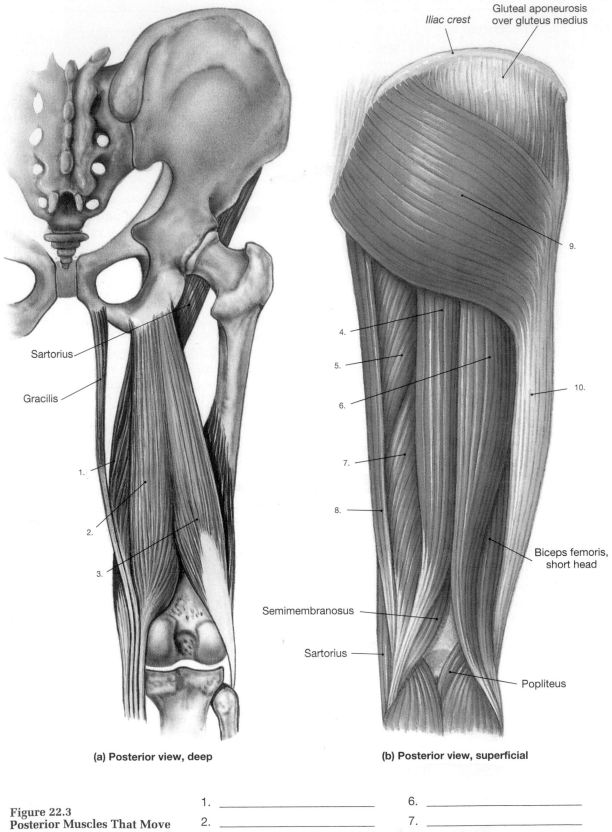

Iliac crest

Gluteal aponeurosis over gluteus medius

9.

10.

4.

5.

6.

7.

8.

Sartorius

Gracilis

1.

2.

3.

Biceps femoris, short head

Semimembranosus

Sartorius

Popliteus

(a) Posterior view, deep

(b) Posterior view, superficial

Figure 22.3
Posterior Muscles That Move the Leg.

(a) Diagrammatic view of leg flexors. **(b)** Posterior view of thigh.

1. _____

2. _____

3. _____

4. _____

5. _____

6. _____

7. _____

8. _____

9. _____

10. _____

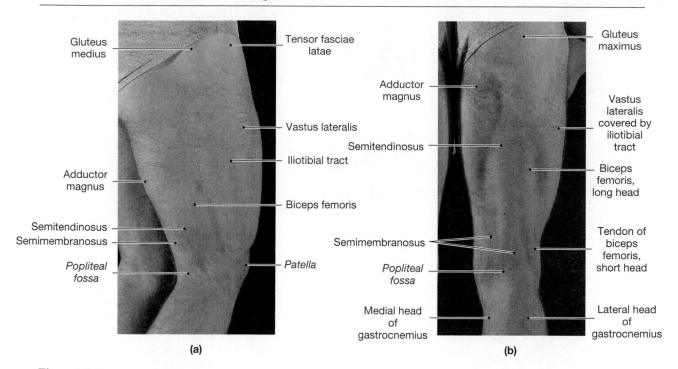

Figure 22.4
Surface Anatomy of Right Thigh.
(a) Lateral View. **(b)** Posterior View.

Medial to the biceps femoris is the **semitendinosus.** It is a long muscle that passes behind the knee to insert on the posteromedial surface of the tibia. The **semimembranosus** is medial to the semitendinosus and inserts on the medial tibia. These muscles cross both the hip and knee joints and act to extend the thigh and flex the knee. The hamstrings are therefore antagonists to the quadriceps. When the thigh is flexed and drawn up toward the pelvis, the hamstrings act to extend the thigh.

Flex your knee and feel the tendons of the semimembranosus and semitendinosus located just above the back of the knee on the medial side. Similarly, on the lateral side of the knee, just above the fibular head, the tendon of the biceps femoris can be palpated.

LABORATORY ACTIVITY   MUSCLES OF THE UPPER LEG

MATERIALS

torso model
leg model

PROCEDURES

1. Label and review the muscles of the thigh in Figures 22.1–22.3.
2. Examine the leg model and identify the flexors, extensors, adductors, and abductors of the leg.
3. Using Figure 22.4 as a guide, locate as many thigh muscles on your upper leg as possible. Practice the actions of the muscles and observe how your leg moves.

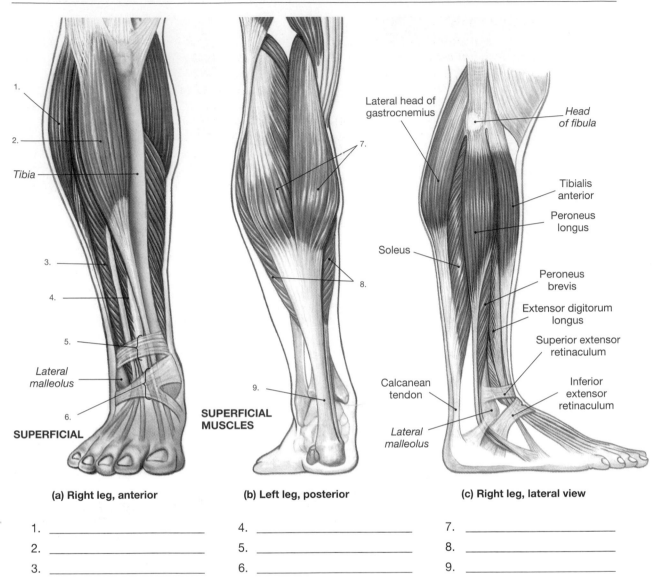

Tibia

1.
2.

3.

4.

5.

Lateral
malleolus

6.

SUPERFICIAL

(a) Right leg, anterior

7.

8.

9.

**SUPERFICIAL
MUSCLES**

(b) Left leg, posterior

Lateral head of
gastrocnemius

Soleus

Calcanean
tendon

Lateral
malleolus

*Head
of fibula*

Tibialis
anterior

Peroneus
longus

Peroneus
brevis

Extensor digitorum
longus

Superior extensor
retinaculum

Inferior
extensor
retinaculum

(c) Right leg, lateral view

1. _____	4. _____	7. _____
2. _____	5. _____	8. _____
3. _____	6. _____	9. _____

**Figure 22.5
Muscles That Move the Foot
and Toes.**

C.   Muscles of the Lower Leg

Tibialis anterior

The **tibialis (tib-ē-A-lis) anterior** (see Figure 22.5), is located on the anterior side of the lower leg and comprises the fleshy mass just lateral to the anterior border of the tibia (the shin). The muscle has two origins, one of which is found on the anterolateral side of the tibia and the other on a membrane between the tibia and fibula. The tendon passes over the dorsal surface of the foot and inserts on the medial cuneiform and the first metatarsal. This muscle dorsiflexes and inverts the foot. Place your fingers on your tibialis and dorsiflex your foot to feel the muscle contract.

Extensor muscles of the toes

Two extensor muscles arise on the anterior lower leg and insert on the phalanges. The **extensor hallucis longus** is lateral and deep to the tibialis anterior (see Figure 22.5). It arises from the middle of the fibula and the interosseous membrane between the tibia and fibula and inserts on the dorsal surface of the distal phalanx of the great toe. This muscle extends the big toe. The tendon of this muscle becomes very prominent on the dorsal surface of the foot when this action is performed.

Lateral to the extensor hallucis is the **extensor digitorum longus**. It arises on the lateral condyle of the tibia and the anterior surface of the fibula. The tendon spreads into four branches which insert on the dorsal surface of the phalanges of toes 2 through 5. The extensor digitorum acts to extend toes 2 through 5. When the toes are extended, these tendons are clearly visible on the top of the foot.

Muscles of the lateral lower leg

Peroneus muscles. Two muscles are located on the lateral side of the lower leg, the peroneus longus and the peroneus brevis (see Figure 22.5). The **peroneus longus** begins on the upper half of the shaft of the fibula, and the **peroneus brevis** originates inferior to the origin of the longus near the middle of the fibular body. The tendons pass behind the lateral malleolus and curve under the foot. The peroneus longus inserts on the base of the first metatarsal, and the peroneus brevis inserts on the dorsal surface of the base of the fifth metatarsal. These muscles evert or pull the lateral edge of the foot upward. The longus also plantar flexes the foot. Practice this action with your peroneal muscles.

Muscles of the posterior lower leg

Gastrocnemius. The posterior lower leg is covered by the prominent calf muscles, the gastrocnemius and the soleus (see Figure 22.5b). These muscles share the calcaneal (Achilles) tendon which inserts on the calcaneus of the foot. The **gastrocnemius (gas-trok-NĒ-mē-us)**, the most superficial calf muscle, originates on the lateral and medial epicondyles of the femur. The two heads cross over the back of the knee and form a fleshy mass consisting of two bellies before blending about halfway down the lower leg. The gastrocnemius acts to plantar flex, adduct, and invert the foot. Since the gastrocnemius arises on the femur it also flexes the knee.

Soleus. Deep to the gastrocnemius is the **soleus (SŌ-lē-us)** which has an origin on the upper third of the tibia and on the fibula (see Figure 22.5b). The belly of the muscle passes beneath the gastrocnemius and eventually converges into the calcaneal tendon. The soleus contracts with the gastrocnemius to plantar flex, adduct, and invert the foot.

Tibialis Posterior. The origin of the **tibialis posterior** is found midway along the shaft of the tibia and fibula (see Figure 22.5). The tendon passes medially to the calcaneus and inserts on the plantar surface of the navicular bone, the cuneiform bones, and the second, third, and fourth metatarsals. The tibialis posterior adducts, inverts, and plantar flexes the foot.

Flexor Hallucis Longus. The **flexor hallucis longus** begins lateral to the origin of the tibialis posterior on the fibular shaft. Its tendon runs parallel to that of the tibialis, passing medial to the calcaneus and on to insert on the plantar surface of the distal phalanx of the great toe (see Figure 22.5). It acts to flex the toe and opposes the action of the extensor hallucis longus.

LABORATORY ACTIVITY MUSCLES OF THE LOWER LEG

MATERIALS

leg model

PROCEDURES

1. Label and review the muscles of the lower leg in Figure 22.5.
2. Examine the pelvis of the torso model and identify each muscle.
3. Locate as many lower leg muscles on your leg as possible. Practice the actions of the muscles and observe how your leg moves.

D. Muscles of the Foot

Digitorum brevis

The muscles that move the toes arise from the calcaneus of the foot (see Figure 22.6a). The **extensor digitorum brevis** is located on the dorsal surface of the foot and originates on the lateral superior surface of the calcaneus. It passes obliquely across the foot and sends out four tendons which insert into the dorsal surface of the first through fourth proximal phalanges. Since this muscle is on the top of the foot, it acts to extend the

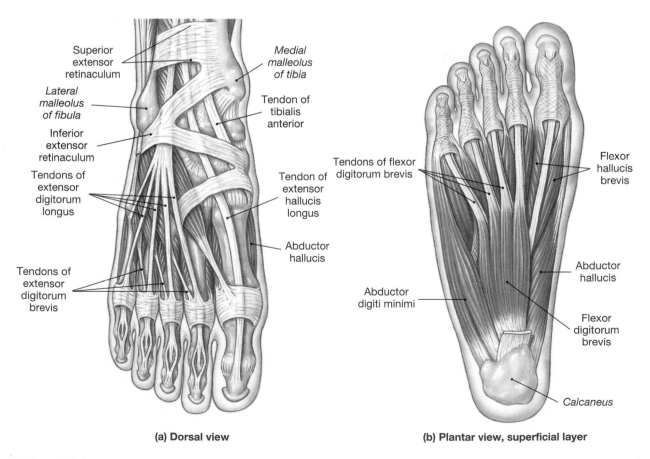

(a) Dorsal view

(b) Plantar view, superficial layer

Figure 22.6
Intrinsic Muscles of the Foot.

toes. One of the antagonists to this action is the **flexor digitorum brevis** which also origi-nates on the calcaneus, but on the plantar surface. It is situated in the middle of the sole of the foot. The tendons from this muscle insert into the base of the second through fifth toes. It acts to flex the toes.

Hallucis muscles of the foot

The **abductor hallucis** (see Figure 22.6) is found on the inner margin of the foot and originates on the plantar side of the calcaneus. It inserts on the medial side of the first phalanx of the great toe and acts to move that toe away from the others. The **flexor hallucis brevis** originates on the plantar surface of the cuneiform and cuboid bones of the foot and splits into two, a medial and lateral head. Each head sends a tendon onto the base of the first phalanx of the great toe, to either the lateral or the medial side. The muscle flexes the great toe and is an antagonist of the extensor hallucis longus.

Abductor digiti minimi

Examine the term **abductor digiti minimi** to define the action of this muscle. It abducts the little toe away from the others. The muscle is located on the outer margin of the foot and originates on the plantar and lateral surfaces of the calcaneus (see Figure 22.6). It inserts on the lateral side of the proximal phalanx of the little toe.

LABORATORY ACTIVITY   Muscles of the Foot

MATERIALS

leg model

PROCEDURES

1. Label and review the muscles of the foot in Figure 22.6.
2. Examine the calf and foot of the leg model and identify each muscle.
3. On your leg, locate muscles that move your foot. Practice the actions of the mus-cles and observe how your foot moves.

MUSCLES OF THE PELVIS, LEG, AND FOOT CHECKLIST

This is a list of bold terms presented in Exercise 22. Use this list for review purposes after you have completed the laboratory activities.

MUSCLES OF THE PELVIS AND GLUTEAL REGION

• Figures 22.1 and 22.2
iliacus
  iliopsoas
  psoas major
internal and external obturator
piriformis
gluteus maximus
gluteus medius
gluteus minimus
fascia lata
iliotibial tract
tensor fasciae latae

MUSCLES OF THE UPPER LEG

• Figures 22.1–22.4
anterior upper leg
sartorius
quadriceps femoris group
  rectus femoris
  vastus medialis
  vastus lateralis
  vastus intermedius
medial upper leg
  gracilis
  pectineus
  adductor brevis
  adductor longus
  adductor magnus

posterior upper leg
  hamstring group
    semitendinosus
    semimembranosus
    biceps femoris

MUSCLES OF THE LOWER LEG

• Figure 22.5
tibialis anterior
extensor hallucis longus
extensor digitorum longus
peroneus longus
peroneus brevis
gastrocnemius
soleus
tibialis posterior
flexor hallucis longus

MUSCLES OF THE FOOT

• Figure 22.6
extensor digitorum brevis
flexor digitorum brevis
abductor hallucis
flexor hallucis brevis
abductor digiti minimi

MUSCLES OF THE PELVIS, LEG, AND FOOT

Laboratory Report Exercise 22

A. Matching

Match each term listed on the left with the correct description on the right.

1. _____ sartorius
2. _____ semitendinosus
3. _____ peroneus longus
4. _____ gastrocnemius
5. _____ psoas major
6. _____ rectus femoris
7. _____ semimembranosus
8. _____ flexor hallucis longus
9. _____ abductor digiti minimi
10. _____ vastus lateralis
11. _____ biceps femoris
12. _____ gluteus maximus
13. _____ abductor hallucis
14. _____ vastus intermedius
15. _____ gracilis
16. _____ gluteus medius
17. _____ patellar ligament
18. _____ adductor brevis
19. _____ tibialis anterior
20. _____ tensor fasciae latae
21. _____ extensor hallucis longus
22. _____ extensor digitorum longus
23. _____ piriformis
24. _____ adductor magnus
25. _____ internal obturator

A. most anterior and superficial quadriceps muscle

B. abducts the great toe

C. quadriceps located deep to the rectus femoris

D. most medial of the leg adductors

E. abductor of the femur, originates on the iliac crest

F. connects the kneecap and the tibial tuberosity

G. short adductor of the femur

H. located on anterior lower leg, dorsiflexes the foot

I. The "tailor's" muscle

J. tenses the fasciae latae

K. originates on anterior lower leg, extends great toe

L. originates on sacrum; inserts on greater trochanter

M. largest adductor of the femur

N. long muscle that flexes the great toe

O. abducts the little toe

P. originates on the medial side of the obturator foramen

Q. hamstring, inserts with semimembranosus

R. muscle of lateral lower leg, plantar flexes the foot

S. most lateral of the quadriceps

T. hamstring, has two heads

U. large muscle used in climbing stairs

V. originates on anterior lower leg, extends toes

W. superficial calf muscle

X. hamstring, inserts on medial condyle of the tibia

Y. forms the iliopsoas muscle

B. **Short-Answer Questions**

1. Which muscles of the leg serve a similar function as do the rotator cuff muscles of the arm?

2. Which muscles help keep the head of the femur firmly seated in the acetabulum?

3. How are the muscles of the leg similar to the muscles of the arm? How are they different?

4. Describe how the hamstring muscle group moves the leg. Which muscle group is antagonist to the hamstring?

5. Describe the abductor and adductor muscles of the thigh.

D. **Completion**

Label each muscle in Figure 22.7.

1. _____

2. _____

3. _____

4. _____

5. _____

6. _____

7. _____

8. _____

9. _____

10. _____

11. _____

12. _____

13. _____

14. _____

15. _____

16. _____

17. _____

18. _____

19. _____

20. _____

21. _____

22. _____

23. _____

24. _____

25. _____

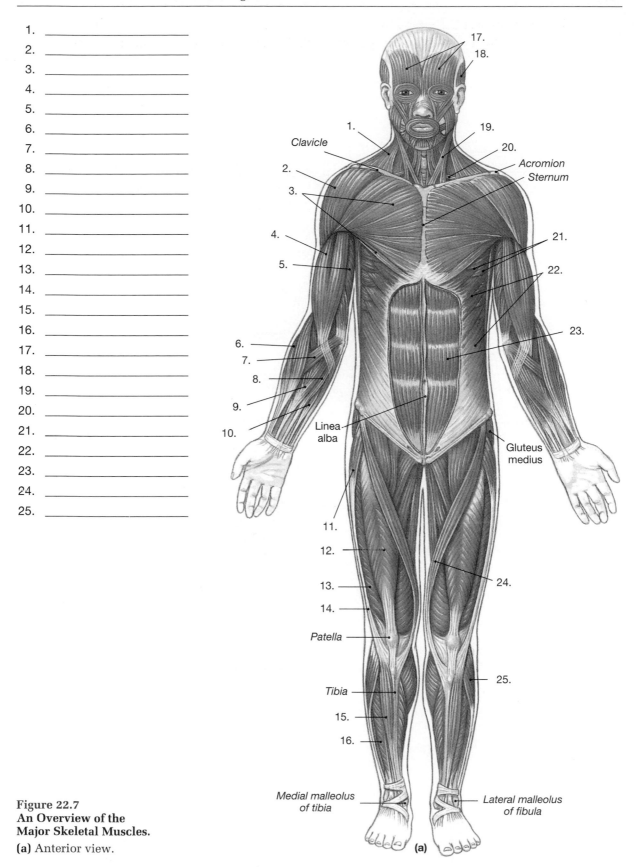

Figure 22.7
An Overview of the
Major Skeletal Muscles.

(a) Anterior view.

1. _____
2. _____
3. _____
4. _____
5. _____
6. _____
7. _____
8. _____
9. _____
10. _____
11. _____
12. _____
13. _____
14. _____
15. _____
16. _____
17. _____
18. _____

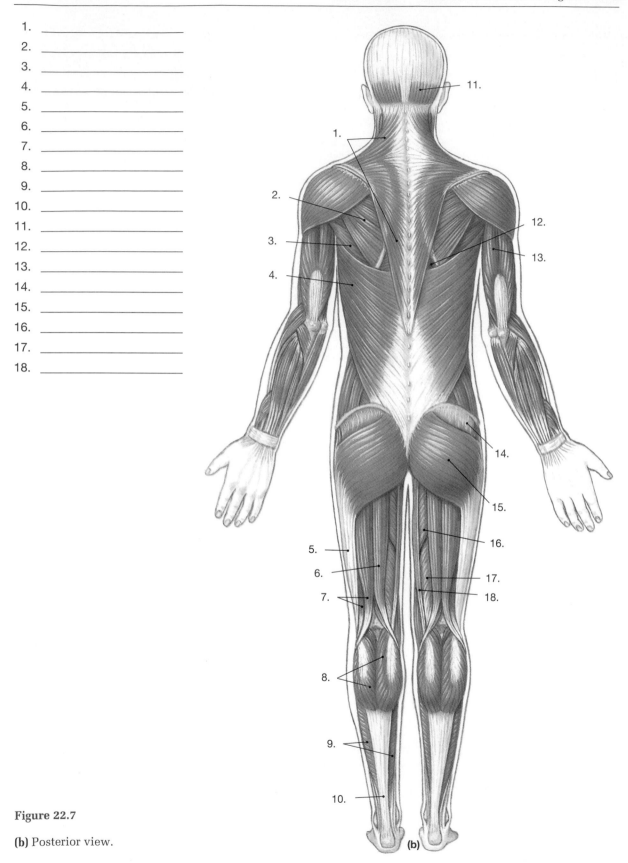

Figure 22.7

(b) Posterior view.

23

Muscles of the Cat

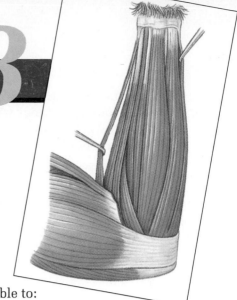

OBJECTIVES

On completion of this exercise, you should be able to:

- Compare and contrast the anatomy of human muscles and cat muscles.
- Identify the muscles of the back and the upper forelimb.
- Identify the muscles of the neck.
- Identify the muscles of the chest and abdomen.
- Identify the muscles of the thigh.
- Identify the muscles of the hindlimb.

WORD POWER

acromiodeltoid (acromio—tip of the shoulder)
clavodeltoid (clavo—clavicle)
xiphihumeralis (xiphi—a sword)

anconeus (anco—a bend, the elbow)
communis (communis—united)
tenuissimus (tenu—slender, thin)

MATERIALS

gloves
safety glasses
dissecting tools
dissecting pan
preserved cat

INTRODUCTION

This exercise is intended to complement the study of human muscles. Some muscles found in animals are lacking in humans. Other muscles that are fused in humans occur as separate muscles in animals. Also, the terminology used to describe the position and location of body parts for four-legged animals differs slightly from that used for the two-legged human animal since four-legged animals move forward head-first with the abdominal surface parallel to the ground. Anatomical position for a four-legged animal is all four limbs on the ground. "Superior" refers to the back or **dorsal** surface, and "inferior" relates to the belly or **ventral** surface. **Cephalic** means toward the front or anterior, and **caudal** refers to posterior structures.

PREPARING THE CAT FOR DISSECTION AND REMOVAL OF THE SKIN

All students must practice the highest levels of laboratory safety while handling and dissecting the cat. Keep the following guidelines in mind during the dissection:

1. Wear gloves and safety glasses to protect yourself from the fixatives used to preserve the specimen.
2. Do not dispose of the fixative from your specimen. You will use it to keep your specimen moist during storage, and it will prevent decay.
3. Be extremely careful when using a scalpel or other sharp instrument. Always direct cutting and scissor motions away from you to prevent an accident if the instrument slips on the moist tissues.

291

4. Before cutting, make sure that the tissue is freed from underlying or adjacent tissues so that they will not be accidentally severed.

5. Never discard tissue in the sink or trash. Your instructor will inform you of proper disposal procedures for tissues and specimens.

Read this entire section and familiarize yourself completely with the exercise before proceeding. You must exercise care to prevent muscle damage during skin removal. The attachment of the skin to the underlying hypodermis is variable; in some places it is loosely attached to the body with subcutaneous fat between the skin and the underlying muscle (e.g., the abdominopelvic region), whereas in other places it is held tightly to underlying muscle by tough fascia (e.g., the thigh). Additionally, the cat has some large, thin masses of muscle in the area of the neck and face that attach directly to and move the skin. The muscle is termed the platysma, but in humans it is smaller.

LABORATORY ACTIVITY   PREPARATION AND DISSECTION OF THE CAT

Remove the skin as a single intact piece. It should be saved to wrap the body for storage. The skin wrapping keeps the body moist and prevents excessive drying of the muscles. Carefully wrap the body with the skin prior to storage. Never rinse your cat with water, as this will remove the preservative and promote the growth of mold. You may need to supplement the skin wrap with some paper towels. Store the cat in the plastic bag provided and seal the bag. Attach a name tag with your name to the cat and the bag and place in the assigned storage area.

MATERIALS

gloves, safety glasses
dissecting tools
dissecting pan
preserved cat

PROCEDURES

I. Removal of the Skin

1. Place the cat dorsal side down on a dissecting tray before you begin removing the skin.

2. With your scalpel make a short, shallow incision in the midline just anterior to the tail (see Figure 23.1). Insert a blunt probe or your finger to separate the skin from underlying muscle and fascia (Figure 23.2). Extend your incision anteriorly by cutting the separated skin with scissors. Continue to separate with the probe and cut with the scissors until you have an incision that extends up to the neck. Remember to place the blunt-ended blade of the scissors under the skin.

3. On either side of the midventral incision, separate as much skin as possible from the underlying muscle and fascia.

4. Use bone forceps to cut off the tail and discard it as described by your instructor.

5. From the ventral surface, make several incision lines. *Always make sure to separate the skin from the underlying tissues using a blunt probe or your finger prior to cutting the skin.* The incisions include:
   a. complete encirclement of the neck
   b. on the lateral side of each forelimb from the dorsal incision to the wrist; completely cut the skin around the wrists,
   c. on the lateral side of each hindleg from the dorsal incision to the ankle; completely cut the skin around the ankle, and
   d. encircling the anus and genital organs. Loosen the skin as much as possible using your fingers. If the skin does not come off easily, it must be freed by cutting with the scalpel.

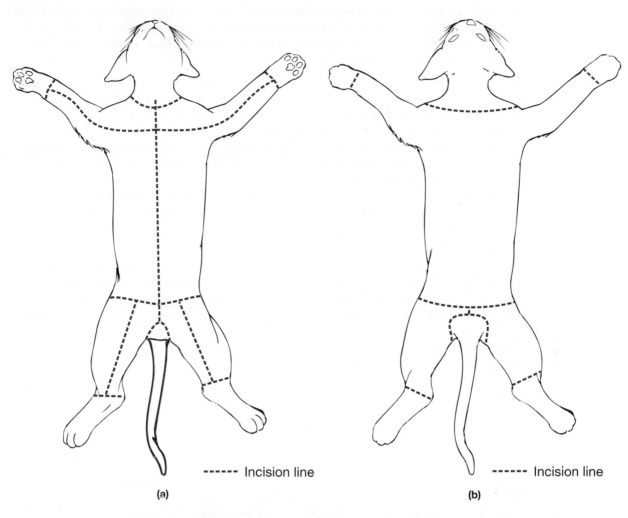

----- Incision line

(a)

----- Incision line

(b)

Figure 23.1
Cat Skinning Incision Lines.

(a) Ventral incisions. **(b)** Dorsal incisions.

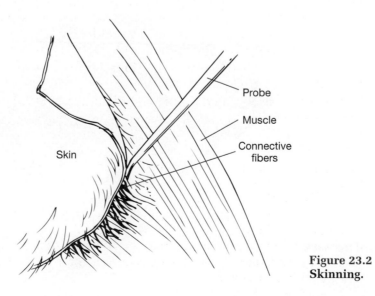

Probe

Muscle

Connective
fibers

Skin

Figure 23.2
Skinning.

6. Use your fingers to free the skin from loose connective tissue attaching the skin to the underlying structures. Work from the dorsal surface toward the ventral surface at the posterior of the cat and then work on the ventral surface from the posterior end of the cat toward its neck.

7. Depending upon your specimen, you may observe some or all of the following: thin, rubber band–like red or blue latex-injected blood vessels projecting between the muscles and the skin; in female cats, mammary glands between the skin and the underlying muscle; cutaneous nerves which are small, white, cordlike structures extending from the muscles to the skin. In male cats leave the skin associated with the male external genitalia. It will be removed at the time of dissection of the reproductive system.

8. Before beginning dissection of the muscles, remove as much of the extraneous fat, hair, platysma, and cutaneous maximus muscle and loose fascia as possible. If at any time you are confused about the nature of any of the material, check with your instructor before removing anything.

9. To store your specimen, wrap it in the skin and moisten it with fixative. Use paper towels if necessary to cover the entire specimen. Return it to the storage bag and seal the bag securely. Label the bag with your name and place it in the storage area as indicated by your instructor.

II. Dissection of the Cat

A. Superficial Muscles of the Back and Shoulder

Begin your dissection in the upper back region between the two forelimbs (see Figure 23.3). Be sure that the skin has been removed from the neck up to the ears.

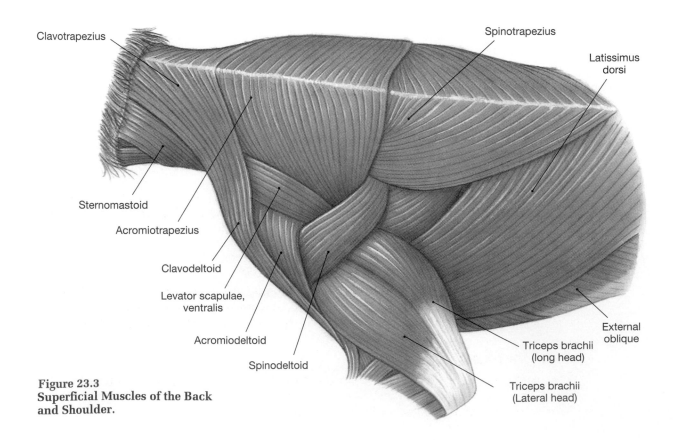

Figure 23.3
Superficial Muscles of the Back and Shoulder.

1. Trapezius group

Locate the **trapezius group** of muscles covering the dorsal surface of the neck and all of the scapula. The single trapezius in humans occurs as three distinct muscles in cats. Their prefix describes the insertion of each muscle.

The **spinotrapezius** is the most posterior of the trapezius muscles. This triangle-shaped muscle should be separated from the acromiotrapezius (anteriorly) and the latissimus dorsi (posteromedially). It originates from the spinous processes of posterior thoracic vertebrae, inserts on the scapular spine. It pulls the scapula dorsocaudad.

Anterior to the spinotrapezius is the **acromiotrapezius,** a large, almost square muscle originating from the spinous processes of cervical and anterior thoracic vertebrae and inserting on the scapular spine. It holds the scapula in place.

The **clavotrapezius** is a broad muscle anterior to the acromiotrapezius. It originates from the lambdoidal crest and axis and inserts on the clavicle. It draws the scapula craniad and dorsad (cranially and dorsally).

2. Latissimus dorsi

The large, flat muscle posterior to the trapezius group is the **latissimus dorsi**. Its origin is on the spines of thoracic and lumbar vertebrae, and it inserts on the medial aspect of the humerus. The latissimus dorsi acts to pull the forelimb posteriorly and dorsally.

3. Levator scapulae ventralis

A flat, straplike muscle, the **levator scapulae ventralis,** lies on the side of the neck between the clavotrapezius and the acromiotrapezius. The occipital bone and atlas are the origin, and the vertebral border of the scapula is the insertion. This muscle pulls the scapula forward. It is *not found in humans.*

4. Deltoid group

Locate the **deltoid group** of muscles, the shoulder muscles lateral to the trapezius group, see Figure 23.3. The following three cat muscles are homologous with the single deltoid muscle in the human.

The **spinodeltoid** is the most posterior of the deltoid group. It originates on the scapula ventral to the insertion of the acromiotrapezius and inserts on the proximal humerus. The action of this muscle is to flex the humerus and rotate it laterally.

The middle muscle of the deltoid group is the **acromiodeltoid.** It originates from the acromion process of the scapula deep to the levator scapulae ventralis and inserts on the proximal end of the humerus. The acromiodeltoid flexes the humerus and rotates it laterally.

The **clavodeltoid** (clavobrachialis) originates on the clavicle and inserts on the ulna. It is a continuation of the clavotrapezius below the clavicle and extends down the arm from the clavicle. It functions to flex the forearm.

B. Deep Muscles of the Back and Shoulder

To expose deeper muscles of the shoulder and back (see Figure 23.4), carefully bisect (divide into two equal parts) both the acromiotrapezius and spinotrapezius and then reflect (fold back) these muscles. Repeat the procedure with the latissimus dorsi.

1. Supraspinatus

Deep to the acromiotrapezius, the **supraspinatus** occupies the lateral surface of the scapula in the supraspinous fossa. It originates on the scapula and inserts on the humerus. It functions to extend the humerus.

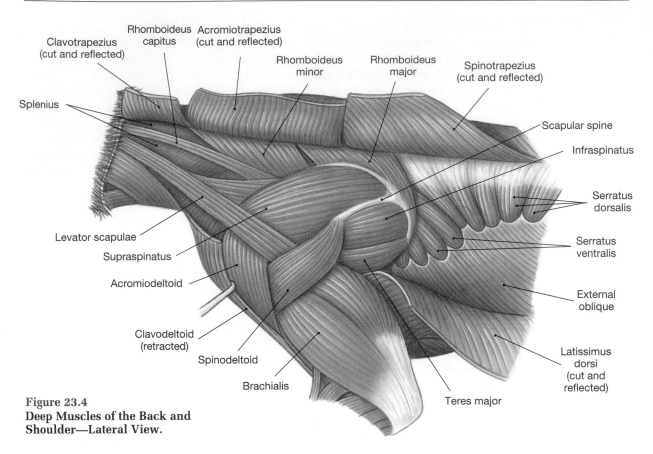

Figure 23.4
Deep Muscles of the Back and
Shoulder—Lateral View.

2. Infraspinatus

On the lateral surface of the scapula, the **infraspinatus** occupies the infraspinous fossa. It originates on the scapula, inserts on the humerus, and causes the humerus to rotate laterally.

3. Teres major

The **teres major** occupies the axillary border of the scapula where it has its origin. It inserts on the proximal end of the humerus and acts to rotate and draw the humerus posteriorly.

4. Rhomboideus group

The **rhomboideus group** connects the spinous processes of cervical and thoracic vertebrae with the vertebral border of the scapula. They hold the dorsal part of the scapula to the body.

The posterior muscle of this group is the **rhomboideus major**, see Figure 23.4. This fan-shaped muscle originates from the spinous processes and ligaments of posterior cervical and anterior thoracic vertebrae and inserts on dorsal posterior angle of the scapula. It draws the scapula dorsally and anteriorly.

The larger muscle anterior to the rhomboideus major is the **rhomboideus minor**. Its origin is on the spines of posterior cervical and anterior thoracic vertebrae. This muscle inserts along the vertebral border of the scapula and functions to draw the scapula forward and dorsally.

The most anterolateral muscle of the rhomboideus group is a narrow, thin, ribbonlike muscle, the **rhomboideus capitis**. It originates from the spinous nuchal line and inserts on the vertebral border of the scapula. It elevates and rotates the scapula. This muscle is *not found in humans.*

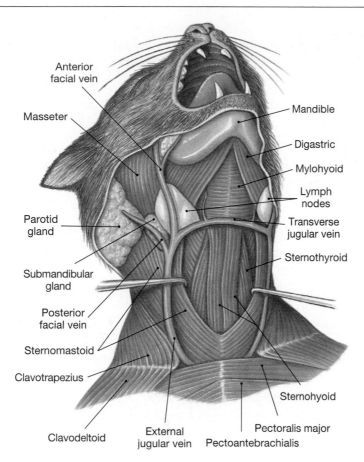

Figure 23.5
Superficial Muscles of the Neck—Ventral View.

5. Splenius

Deep to the rhomboideus capitus, a broad, flat, thin muscle covers most of the lateral surface of the cervical and thoracic vertebrae. This muscle, the **splenius**, has its origin on the spines of thoracic vertebrae and its insertion on the superior nuchal line of skull. It acts to turn or raise the head.

6. Subscapularis

To observe the muscle that occupies the subscapular fossa, the **subscapularis**, position the elbows to touch in the midline. It originates from the subscapular fossa and inserts on the humerus. It is an adductor of the humerus.

C. Superficial Muscles of the Neck

Remove all remaining skin from the side of the neck up to the ear. Be careful not to sever the external jugular vein, the large vein (blue-colored) lying on the ventral surface of the neck (see Figure 23.5). Free this vein from the underlying muscles and clean the connective tissue from the back of the shoulder and the ventral and lateral surfaces of the neck. Do not remove the fascia in the midline of the back since this is the origin of the trapezius muscle group.

1. Sternomastoid

The large V-shaped muscle between the sternum and head, the **sternomastoid,** originates from the manubrium of the sternum and passes obliquely around the neck to insert on the superior nuchal line and on the mastoid process. It turns and depresses the head (see Figure 23.5). This muscle is the sternocleidomastoid in humans.

2. Sternohyoids

The **sternohyoids** are a narrow pair of muscles (they may be fused in the midline) that lie over the larynx, along the midventral line of the neck. Their origin is the costal cartilage of the first rib, while their insertion is on the hyoid bone. They act to depress the hyoid bone.

3. Digastrics

Locate the pair of superficial muscles extending along the inner surface of the mandible, the **digastrics**. Each originates on the occipital bone and mastoid process and functions as a depressor of the mandible.

4. Mylohyoid

The **mylohyoid** is a superficial muscle running transversely in the midline and passing deep to the digastrics. It originates on the mandible and inserts on the hyoid bone. The mylohyoid raises the floor of the mouth.

5. Masseter

The large muscle mass anteroventral to the parotid gland at the angle of each jaw is the **masseter**. This cheek muscle originates from the zygomatic bone. It inserts on the posterolateral surface of the dentary bone and elevates the mandible.

D. Superficial Muscles of the Abdomen (Ventral View)

Three layers of muscle form the abdominal wall (see Figure 23.6). They are very thin, and care must be taken in separating them. Collectively, these muscles act to compress the abdomen.

1. External oblique

The **external oblique** is the most superficial of the lateral abdominal muscles. It originates on posterior ribs and lumbodorsal fascia and inserts on the linea alba from the sternum to the pubis. Its fibers run from anterodorsal to posteroventral, and it acts to compress the abdomen.

2. Internal oblique

Deep to the external oblique lies the **internal oblique**, with its fibers running perpendicular to those of the external oblique in a posterodorsal-to-anteroventral orientation. It originates from the pelvis and lumbodorsal fascia and inserts on the linea alba where it functions to compress the abdomen. Reflect the external oblique to view the internal oblique.

3. Transverse abdominis

The **transverse abdominis** is deep to the internal oblique, has fibers that run transversely across the abdomen, and forms the deepest layer of the abdominal wall. It originates from the posterior ribs, lumbar vertebrae, and the ilium and inserts on the linea alba. It acts to compress the abdomen.

4. Rectus abdominis

The **rectus abdominis** is a long, ribbonlike muscle in the midline of the ventral of the abdomen. It originates from the pubic symphysis and inserts on the costal cartilage. It compresses the internal organs of the abdomen.

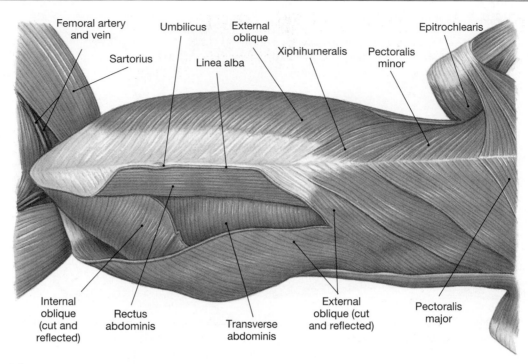

Figure 23.6
Superficial Muscles of the Abdomen—Ventral View.

E. Superficial Muscles of the Chest

1. Pectoralis group

The **pectoralis group** consists of the large muscles covering the ventral surface of the chest (see Figure 23.7). They arise from the sternum and attach, for the most part, to the humerus. There are four subdivisions in the cat, but only two in humans. In the cat, the relatively large degree of fusion gives the cat's pectoral muscles the appearance of a single muscle. This fusion makes it rather difficult to dissect, as they do not separate from each other easily.

The most superficial of the pectoral muscles, the **pectoantebrachialis**, has its origin on the manubrium. It inserts on fascia of the forearm and adducts the arm. It is *not found in humans.*

Posterior to the pectoantebrachialis is the much broader and triangle-shaped **pectoralis major**. Its origin is on the sternum and its insertion on the posterior humerus. It functions to adduct the arm. Posterior to the pectoralis major lies the **pectoralis minor**. This is the broadest and thickest muscle of the group. It extends posteriorly to the pectoralis major. It originates on the sternum, inserts near the proximal end of the humerus, and acts to adduct the arm.

The thin **xiphihumeralis** is posterior to the posterior edge of the pectoralis minor. It originates on the xiphoid process of the sternum and inserts by a narrow tendon on the humerus. It adducts and helps rotate the forelimb. It is *not found in humans.*

F. Deep Muscles of the Chest and Abdomen (Ventral View)

1. Serratus group

On the lateral thoracic wall, just ventral to the scapula, a large, fan-shaped muscle, the **serratus ventralis,** originates by separate bands from the ribs (see Figure 23.8). It

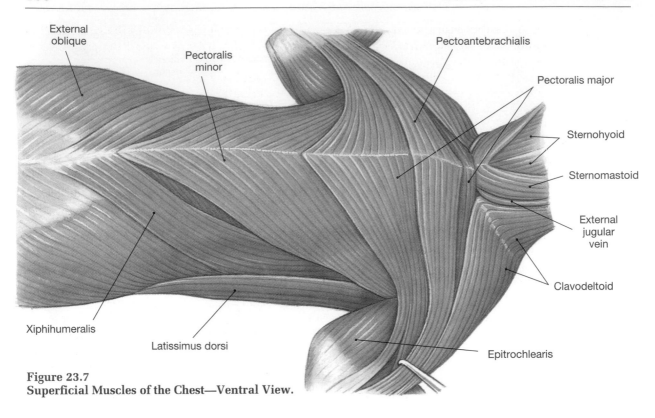

Figure 23.7
Superficial Muscles of the Chest—Ventral View.

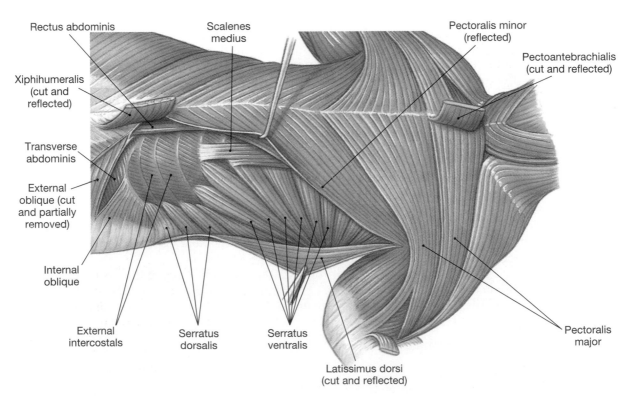

Figure 23.8
Deep Muscles of the Chest and Abdomen—Ventral View.

passes ventrally to the scapula and inserts on the vertebral border of the scapula. It is homologous to the serratus anterior of humans and functions to pull the scapula forward and ventrally.

The **serratus dorsalis** is a serrated muscle that lies medial to the serratus ventralis and overlies the anterior divisions of the sacrospinalis. Its origin is the middorsal cervical, thoracic, and lumbar raphe, and it inserts on the ribs. It acts to pull the ribs craniad and outward.

2. Scalenus

On the lateral surface of the trunk is a three-part muscle, the **scalenus,** which unites anteriorly. It originates on the ribs and inserts on the transverse processes of the cervical vertebrae. It acts to flex the neck and draw the ribs anteriorly.

3. Intercostal group

Deep to the external oblique, muscle fibers of the **external intercostals** run obliquely from the posterior border of one rib to the anterior border of the next rib. Their origin is on the caudal border of one rib, and their insertion is on the cranial border of the next rib. They lift the ribs during inspiration.

Toward the midline and deep to the external intercostals observe the **internal intercostals**. Bisect one external intercostal muscle to expose an internal intercostal. The fibers of the intercostals run at oblique angles: the internal from medial to lateral and the external from lateral to medial. Their origin is on the superior border of the rib below, and they insert on the inferior border of the rib above. They draw adjacent ribs together and depress ribs during active expiration.

G. Muscles of the Forelimb

1. Epitrochlearis

Observe the **epitrochlearis**, which is a broad, flat muscle covering the medial surface of the upper forelimb (see Figure 23.9). It appears to be an extension of the latissimus dorsi and originates from the fascia of the latissimus dorsi. It inserts on the olecranon process where it acts to extend the forelimb. It is *not found in humans.*

2. Biceps brachii

The **biceps brachii** is a convex muscle interior to the pectoralis major and minor on the ventromedial surface of the humerus. It originates on the scapula and inserts on the radial tuberosity near the proximal end of the radius. It functions as a flexor of the forelimb.

3. Brachioradialis

Observe a ribbonlike muscle on the lateral surface of the humerus (see Figure 23.10). This is the **brachioradialis**. Its origin is on the middorsal border of the humerus and its insertion on the distal end of the radius. It supinates the front paw.

4. Extensor carpi radialis

Deep to the brachioradialis, locate the **extensor carpi radialis**. There are two parts to this muscle: the shorter, triangular extensor carpi radialis brevis and, deep to it, the extensor carpi radialis longus. It originates from the lateral surface of the humerus above the lateral epicondyle and inserts on the bases of the second and third metacarpals. It causes extension at the carpal joints.

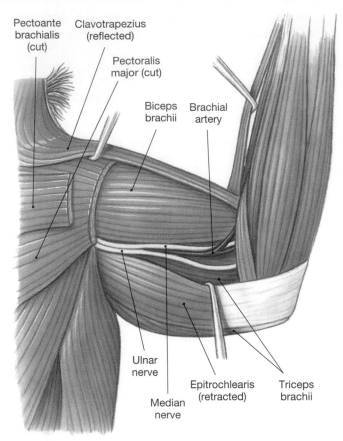

Figure 23.9
**Muscles of the Upper Forelimb—
Medial View.**

5. Pronator teres

The narrow muscle next to the extensor carpi radialis is the **pronator teres**. It runs from its point of origin on the medial epicondyle of the humerus and declines in size as it approaches insertion on the radius. It rotates the radius for pronation.

6. Flexor carpi radialis

The **flexor carpi radialis** is adjacent to the pronator teres. It originates on the distal end of the humerus and inserts on the second and third metacarpals. It acts to flex the wrist.

7. Palmaris longus

The large, flat muscle in the center of the medial surface of the forelimb is the **palmaris longus**. Its origin is on the medial epicondyle of the humerus, and it inserts on all digits, which it flexes.

8. Flexor carpi ulnaris

The flat muscle on the posterior edge of the forelimb is the **flexor carpi ulnaris**. It arises from a two-headed origin (the medial epicondyle of the humerus and the olecranon process) and inserts by a single tendon on the ulnar side of the carpals. It is a flexor of the wrist.

9. Brachialis

Locate the **brachialis** on the ventrolateral surface of the humerus (see Figure 23.10b). It originates on the lateral side of the humerus and inserts on the proximal end of the ulna. It functions to flex the forelimb.

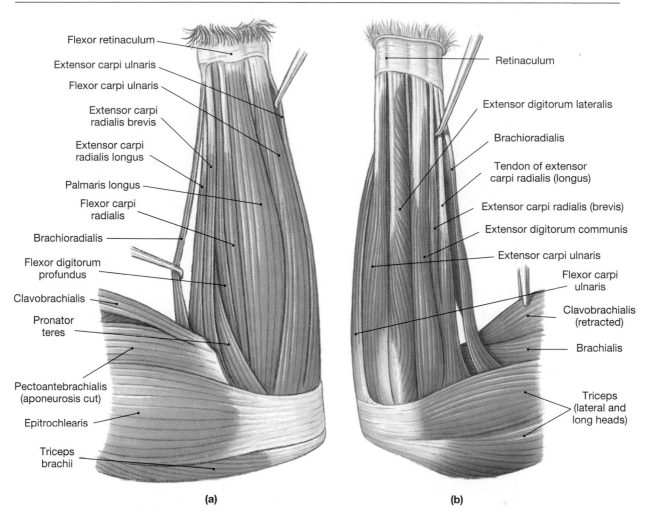

Figure 23.10
Muscles of the Forelimb.

(a) Medial View. **(b)** Lateral view.

10. Triceps brachii

On the lateral and posterior surfaces of the forelimb, observe the largest superficial muscle in the upper forelimb, the **triceps brachii**. It arises from three heads: The **long head** is the large muscle mass on the posterior surface and originates on the lateral border of the scapula. The **lateral head** lies next to the long head on the lateral surface, originating on the deltoid ridge of the humerus. The small **medial head** lies deep to the lateral head and originates on the shaft of the humerus. All the heads have a single insertion on the olecranon process of the ulna. Its function is to extend the forelimb.

11. Extensor carpi ulnaris

The blocklike muscle lateral to the extensor digitorum lateralis is the **extensor carpi ulnaris**. It originates from the lateral epicondyle of the humerus and inserts on the proximal end of the fifth metacarpal. It extends the ulnar side of the wrist and the fifth digit.

H. Muscles of the Thigh

1. Sartorius

The wide, superficial muscle covering the anterior half of the medial aspect of the thigh is the **sartorius** (see Figure 23.11). It originates on the ilium and inserts on the tibia. The sartorius adducts and rotates the femur and extends the tibia.

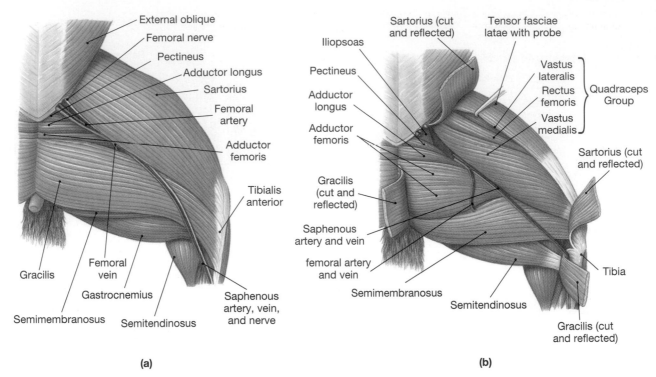

Figure 23.11
Muscles of the Thigh—Ventromedial View.
(a) Superficial muscles. **(b)** Deep muscles.

2. Gracilis

The broad muscle that covers the posterior portion of the medial aspect of the thigh is the **gracilis**. It originates from the ischium and pubic symphysis and inserts on the medial surface of the tibia. It adducts the thigh and draws it posteriorly.

3. Tensor fasciae latae

Posterior to the sartorius, locate a triangle-shaped muscle mass, the **tensor fasciae latae** (see Figure 23.12). This muscle originates from the crest of the ilium and inserts into the fascia lata. The tensor fasciae latae extends the thigh.

4. Gluteal group

Posterior to the tensor fasciae latae lies the **gluteus medius** (see Figure 23.12b). This is the largest of the gluteus muscles, and it originates on both the ilium and the transverse processes of the last sacral and first caudal vertebrae. It inserts on the femur and acts to abduct the thigh.

   Just posterior to the gluteus medius locate the **gluteus maximus**, a small, triangular hip muscle. It originates from the transverse processes of the last sacral and first caudal vertebrae and inserts on the proximal femur. It abducts the thigh.

5. Adductor group

Deep to the gracilis and anterior to the semimembranosus is the **adductor femoris** (see Figure 23.12b). It is a large muscle with an origin on the ischium and pubis. It inserts on the femur and acts to adduct the thigh.

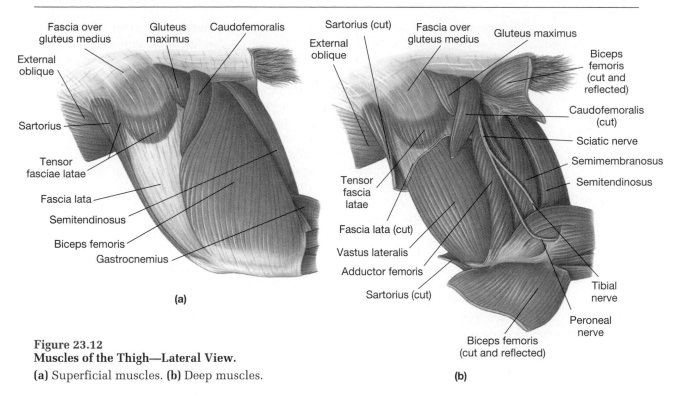

Figure 23.12
Muscles of the Thigh—Lateral View.
(a) Superficial muscles. **(b)** Deep muscles.

Anterior to the adductor femoris is the **adductor longus** muscle. It is a narrow muscle that originates from the ischium and pubis and inserts on the proximal surface of the femur. It adducts the thigh.

6. Pectineus

The **pectineus**, Figure 23.12, is anterior to the adductor longus. It is a deep, small muscle posterior to both the femoral artery and vein. It originates on the anterior border of the pubis and inserts on the proximal end of the femur. It functions to adduct the thigh.

7. Biceps femoris

Observe the large, broad muscle covering the lateral region of the thigh (see Figure 23.12). This is the **biceps femoris**. It originates on the ischial tuberosity and inserts on the tibia. The biceps femoris abducts the thigh and flexes the knee.

8. Semitendinosus

On the posterior aspect of the thigh, a straplike muscle, the **semitendinosus,** originates from the ischial tuberosity and inserts on the medial side of the tibia. It flexes the knee.

9. Semimembranosus

Observe a large, posterior muscle lying beneath the gracilis and medial to the semitendinosus (see Figure 23.12). This is the **semimembranosus**. It originates from the ischium and inserts on the medial epicondyle of the femur and medial surface of the tibia. It extends the thigh.

10. Quadriceps femoris group

Locate the group of four large muscles that comprise the **quadriceps femoris.** The quadriceps femoris covers about one-half the area of the thigh. Collectively these muscles insert into the patellar ligament and act as a powerful extender of the knee. Bisect the sartorius and free both borders of the tensor fasciae latae. Reflect these muscles and observe that the muscles of the quadriceps femoris converge and insert on the patella.

The large muscle on the anterolateral surface of the thigh is the **vastus lateralis**. It originates along the entire length of the lateral surface of the femur. The large muscle on the medial surface of the femur is the **vastus medialis** which originates on the shaft of the femur. The large, cylindrical **rectus femoris** lies between the medialis and lateralis muscles and originates on the acetabulum. The **vastus intermedius** lies deep to the rectus femoris. Bisect the rectus femoris and reflect it to expose the vastus intermedius. It originates on the shaft of the femur.

I.   Muscles of the Lower Hindlimb

1. Gastrocnemius

The **gastrocnemius** or calf muscle is on the posterior of the lower hindlimb (see Figure 23.13). It has two heads of origin, medial and lateral (the knee fascia and the distal end of the femur), which unite in the calcaneal tendon to insert on the calcaneus. It extends the foot.

2. Flexor digitorum longus

Between the gastrocnemius and the tibial bone is the **flexor digitorum longus**. This muscle has two heads of origin (the distal end of the tibia and the head and shaft of the fibula), and it inserts by four tendons onto the bases of the terminal phalanges. It acts as a flexor of the digits.

3. Tibialis anterior

On the anterior surface of the tibial bone, the **tibialis anterior** originates on the proximal ends of the tibia and fibula. It inserts on the first metatarsal and functions to flex the foot.

4. Soleus

Deep to the gastrocnemius but visible on the lateral surface of the calf is the **soleus** (see Figure 23.13). The soleus originates on the fibula and inserts on the calcaneus. It extends the foot.

5. Peroneus group

Deep to the soleus on the posterior and lateral surfaces locate the three **peroneus** muscles.

The **peroneus brevis** lies deep to the tendon of the soleus, originating from the distal portion of fibula and inserting on the base of the fifth metatarsal. The **peroneus longus** is a long, thin muscle that lies on the lateral surface, originating from the proximal portion of the fibula and inserting by a tendon which passes through a groove on the lateral malleolus and turns medially to attach to the bases of the metatarsals. The **peroneus tertius** lies along the tendon of the peroneus longus, originating on the fibula and inserting on the base of the fifth metatarsal. The peroneus longus and tertius flex the foot, while the peroneus brevis extends it.

6. Extensor digitorum longus

On the anterolateral border of the tibia, the **extensor digitorum longus** originates from the lateral epicondyle of the femur. It inserts by long tendons on each of five digits and functions to extends them.

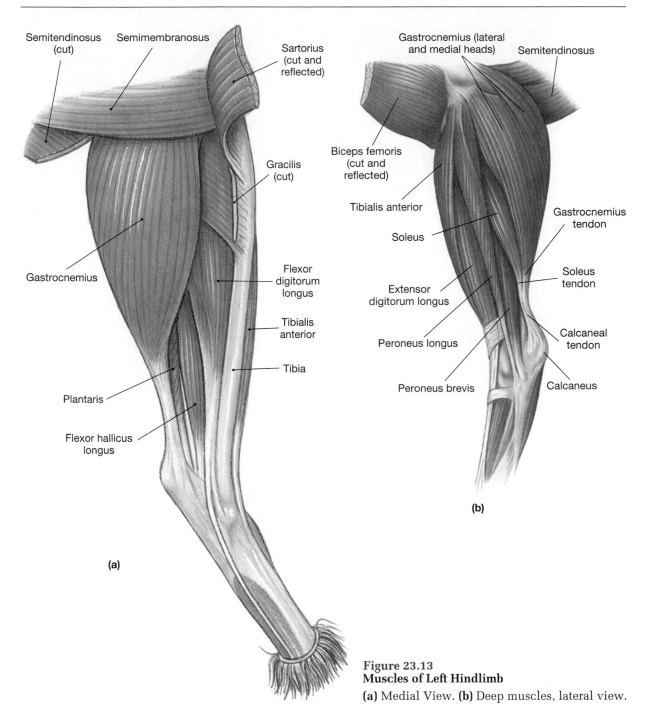

Semitendinosus (cut)

Semimembranosus

Sartorius (cut and reflected)

Gracilis (cut)

Flexor digitorum longus

Tibialis anterior

Tibia

Gastrocnemius

Plantaris

Flexor hallicus longus

(a)

Gastrocnemius (lateral and medial heads)

Semitendinosus

Biceps femoris (cut and reflected)

Tibialis anterior

Soleus

Gastrocnemius tendon

Soleus tendon

Extensor digitorum longus

Peroneus longus

Calcaneal tendon

Peroneus brevis

Calcaneus

(b)

Figure 23.13
Muscles of Left Hindlimb

(a) Medial View. **(b)** Deep muscles, lateral view.

MUSCLES OF THE CAT CHECKLIST

This is a list of bold terms presented in Exercise 23. Use this list for review purposes after you have completed the laboratory activities.

SUPERFICIAL MUSCLES OF THE BACK AND SHOULDER

• Figure 23.3

trapezius group
   spinotrapezius, acromiotrapezius, clavotrapezius
latissimus dorsi
levator scapulae ventralis
deltoid group
   spinodeltoid
   acromiodeltoid
   clavodeltoid

DEEP MUSCLES OF THE BACK AND SHOULDER

• Figure 23.4

supraspinatus
infraspinatus
teres major
rhomboideus group
   rhomboideus major, rhomboideus minor, rhomboideus capitis
splenius
subscapularis

SUPERFICIAL MUSCLES OF THE NECK

• Figure 23.5

sternomastoid, sternohyoid, digastric, mylohyoid, masseter

SUPERFICIAL MUSCLES OF THE ABDOMEN

• Figure 23.6

external oblique, internal oblique, transverse abdominis, rectus abdominis

SUPERFICIAL MUSCLES OF THE CHEST

• Figure 23.7

pectoralis group
   pectoantebrachialis, pectoralis major, pectoralis minor, xiphihumeralis

DEEP MUSCLES OF THE CHEST

• Figure 23.8

serratus group
   serratus ventralis, serratus dorsalis
scalenus
intercostal group
   external intercostals, internal intercostals

MUSCLES OF THE FORELIMB

• Figures 23.9 and 23.10

epitrochlearis, biceps brachii, brachioradialis, extensor carpi radialis, pronator teres, flexor carpi radialis, palmaris longus, flexor carpi ulnaris, brachialis, triceps brachii, extensor carpi ulnaris

MUSCLES OF THE THIGH

• Figures 23.11 and 23.12

sartorius
tensor fasciae latae
gluteal group
   gluteus medius, gluteus maximus
gracilis
adductor group
   adductor femoris, adductor longus
pectinus
biceps femoris
semitendinosus
semimembranosus
quadriceps femoris group
   quadriceps femoris, vastus lateralis, vastus medialis, rectus femoris, vastus intermedius

MUSCLES OF THE LOWER HINDLIMB

• Figure 23.13

gastrocnemius
flexor digitorum longus
tibialis anterior
soleus
peroneus group
   peroneus brevis, peroneus longus, peroneus tertius
extensor digitorum longus

MUSCLES OF THE CAT

Laboratory Report Exercise 23

A. Short-Answer Questions

1. Explain the differences between the cat and the human deltoid and trapezius muscles.

2. Describe the cat muscles of the neck.

3. How are the superficial chest muscles of the cat different from chest muscles in humans?

4. Describe the flexor and extensor muscles of the cat forelimb.

5. Compare cat and human muscles of the thigh and lower leg.

EXERCISE 24

Muscle Physiology

OBJECTIVES

On completion of this exercise, you should be able to:

- Relate the sliding filament theory to muscle contraction.
- Describe the individual molecular events of muscle contraction.
- Describe the anatomy at the neuromuscular junction.
- Explain the differences among a twitch, wave summation, and incomplete and complete tetanus.
- Demonstrate muscle fatigue.
- Demonstrate the differences between isometric and isotonic contractions.

WORD POWER

synaptic (synap—a union)
vesicle (vesic—a bladder)
tetanus (tetan—tense, rigid)

MATERIALS

weights (textbook)
stop watch (clock)

INTRODUCTION

Muscle and nerve tissues are excitable tissues which produce self-propagating electrical impulses called **action potentials**. These electrical impulses result from the movement of sodium and potassium ions through specific protein channels to cross the cell membrane. When a muscle fiber or neuron is at rest, there is a difference in the electrical charge inside and outside the cell membrane. This electrical difference is called the **resting membrane potential** and is between -70 and -90 mV at the inner membrane surface. Skeletal muscles have a resting potential of -85 mV.

When a neuron stimulates a muscle, the nerve action potential causes the neuron to release specific chemicals, collectively called **neurotransmitters**, which cause an action potential in the muscle fiber. Sodium channels open in the membrane, and sodium ions flood into the muscle fiber, causing the membrane to **depolarize** or become less negative. At the peak of depolarization the membrane is at $+30$ mV. At this millivoltage the sodium channels close and potassium channels open. Potassium ions exit the cell, and the fiber **repolarizes** to the resting potential. In summary, an electrical signal in the neuron causes release of a chemical signal, the neurotransmitter, which causes an electrical signal in the muscle fiber that results in contraction.

In this exercise you will investigate several types of contraction in your muscles. It is recommended that you review the microanatomy of skeletal muscles and the function of levers in Exercise 18 before performing these experiments.

Your laboratory may be equipped with an instrument called a physiograph which electrically stimulates muscles and records the characteristics of the contraction. Dissected frog muscles are often used for the muscle specimen. This exercise provides the background physiology necessary to perform such investigations.

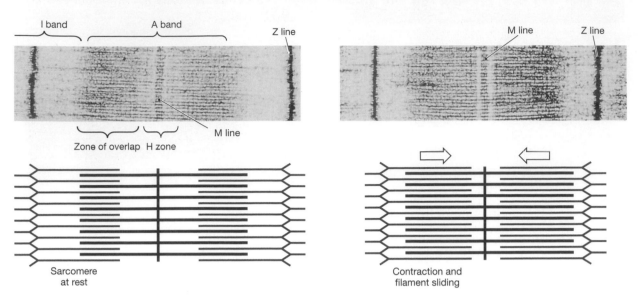

Figure 24.1
Changes in the Appearance of a Sarcomere during Contraction of a Skeletal Muscle Fiber.
During a contraction the A band stays the same width, but the Z lines move closer together and the I band gets smaller.

A.   Biochemical Nature of Muscle Contraction

The **sliding filament theory** is the current model of muscle contraction. The theory explains that when a muscle fiber contracts, filaments slide past one another, causing the muscle fiber to shorten. The filaments are arranged like links in a chain in repeating patterns called sarcomeres. Figure 24.1 illustrates the changes in a muscle fiber during contraction. Thin filaments, composed mostly of the protein actin, are pulled inward by the pivoting of the myosin heads of the thick filaments. As the thin filaments slide toward the middle of the sarcomere, the Z lines and I bands move closer together, the zone of overlap increases, and the H zone decreases. The A band remains constant in length during the contraction.

Motor neurons communicate with skeletal muscles at **neuromuscular junctions**. The end of the neuron's axon swells into a **synaptic knob** containing **synaptic vesicles** full of **acetylcholine (as-ē-til-KŌ-lēn)** (ACh), the **neurotransmitter** molecules used to stimulate the skeletal muscle to contract. When a nerve impulse, called a nerve **action potential**, reaches the synaptic knob, the knob's membrane becomes more permeable to interstitial calcium ions. Calcium diffuses into the synaptic knob and causes the synaptic vesicles to fuse with the knob's membrane and release the ACh into the **synaptic cleft**, the small gap between the neuron and the muscle fiber. The sarcolemma at the synaptic cleft is folded into a **motor end plate** and contains ACh receptors for binding of the neurotransmitter. When ACh binds to the receptor sites, the muscle fiber responds by producing a muscle action potential. This electrical current travels along the sarcolemma, down transverse tubules, and stimulates the release of stored calcium ions from the cisternae of the sarcoplasmic reticulum. Once calcium floods the sarcoplasm of the muscle fiber, contraction occurs. The muscle fiber produces an enzyme called **acetylcholinesterase** (AChE) which deactivates ACh and initiates muscle relaxation.

The contraction process involves five chemical steps: active site exposure, cross-bridge attachment, pivoting, cross-bridge detachment, and myosin return (activation). Figure 24.2 details the activities of each step.

Active site exposure occurs when calcium in the sarcoplasm binds to troponin molecules on the thin filaments. The active sites are where the myosin heads bind to actin molecules. While the muscle is at rest, troponin holds tropomyosin molecules in

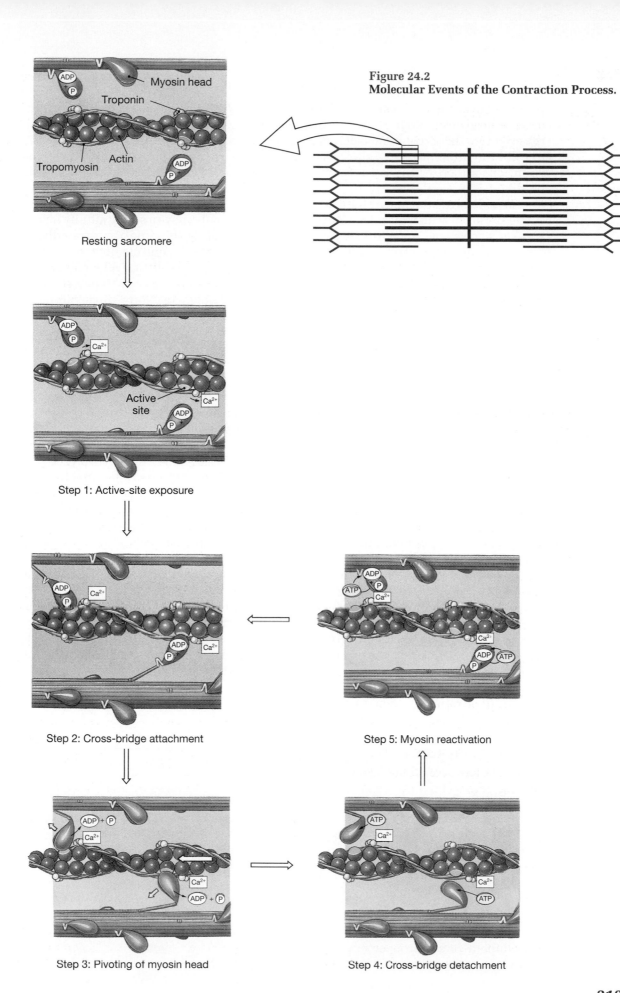

Figure 24.2
Molecular Events of the Contraction Process.

Resting sarcomere

Step 1: Active-site exposure

Step 2: Cross-bridge attachment

Step 3: Pivoting of myosin head

Step 4: Cross-bridge detachment

Step 5: Myosin reactivation

313

place to cover the active sites on actin molecules. If the active sites are covered, the myosin heads cannot attach to the thin filament and pull it inward. When calcium is released into the sarcoplasm, calcium ions bind to troponin and cause the tropomyosin to roll away from the active sites, exposing them to the myosin heads. As long as calcium ions are present in the sarcoplasm, the active sites remain exposed for myosin binding.

Cross-bridge attachment (attach) is the chemical binding of a myosin head to the active site on an actin molecule. Actin and myosin molecules have a strong affinity for one another, and so once the active sites are exposed, cross-bridge binding occurs rapidly.

The **pivoting** (pivot) step causes the sarcomeres, hence the muscle fiber, to shorten. Prior to contraction, the myosin head becomes energized by splitting ATP and storing the released energy. During pivoting, the head uses the stored energy to power stroke and pull the thin filaments toward the M line of the sarcomere. Notice in step three of Figure 24.2 that the angle of the myosin head has changed position as a result of the pivoting. After the pivot the head is depleted of stored energy and releases the ADP and phosphate into the sarcoplasm.

Cross-bridge detachment (detach) involves removing a myosin head from the active site so the head can be recocked for another attachment-pivot sequence. After the pivot and release of ADP an ATP molecule binds to the myosin head and causes the head to "back out" of the active site on the actin molecule.

The last step in muscle contraction is **myosin activation** (return). This step involves splitting of the new ATP on the myosin head to gain energy to recock the head and return it to the original prepivot position. The head is an ATPase and hydrolyzes ATP into ADP and released energy. As muscle contraction continues, the attach-pivot-detach-return sequence repeats until calcium is removed from the sarcoplasm and returned to the sarcoplasmic reticulum. Note that the first step, active site exposure, occurs only at the onset of contraction. The active sites remain exposed as long as calcium is bonded to troponin molecules and the tropomyosin molecules are away from the active sites.

The signal for **relaxation** of the muscle fiber is through the loss of neurotransmitter stimulation on the motor end plate. The muscle fiber secretes an enzyme called acetylcholinesterase which deactivates acetylcholine in the synaptic cleft and on the sarcolemma's ACh receptors. When ACh is inactivated, the sarcolemma no longer produces an action potential. Active transport mechanisms begin to pump calcium in the sarcoplasm back into the sarcoplasmic reticulum. As calcium leaves the troponin molecules, tropomyosin moves into place to block the active sites on actin to prevent myosin cross-bridge formation. The thin filaments slide out, and the sarcomeres and muscle fiber return to their initial relaxed length. Table 24.1 summarizes the steps involved in skeletal muscle contraction.

LABORATORY ACTIVITY   BIOCHEMICAL NATURE OF MUSCLE CONTRACTION

MATERIALS

muscle preparation (glycerinated muscle from a biological supply company)
glycerol (supplied with muscle prep)
ATP solution (supplied with muscle prep)
ion solution (supplied with muscle prep)
dissecting microscope
compound microscope
blank microscope slides
pipette or eye dropper
clean teasing needles

TABLE 24.1   **A Summary of the Steps Involved in Skeletal Muscle Contraction**

Key steps in the initiation of a contraction include the following:

1. At the neuromuscular junction (NMJ, ACh released by the synaptic terminal binds to receptors on the sarcolemma.

2. The resulting change in the transmembrane potential of the muscle fiber leads to the production of an action potential that spreads across its entire surface and along the T tubules.

3. The sarcoplasmic reticulum (SR) releases stored calcium ions, increasing the calcium concentration of the sarcoplasm in and around the sarcomeres.

4. Calcium ions bind to troponin, producing a change in the orientation of the troponin-tropomyosin complex that exposes active sites on the thin (actin) filaments. Myosin cross-bridges form when myosin heads bind to active sites.

5. The contraction begins as repeated cycles of cross-bridge binding, pivoting, and detachment occur, powered by the hydrolysi of ATP. These events produce filament sliding, and the muscle fiber shortens.

This process continues for a brief period, until:

6. Action potential generation ceases as ACh is broken down by acetylcholinesterase (AChE).

7. The SR reabsorbs calcium ions, and the concentration of calcium ions in the sarcoplasm declines.

8. When calcium ion concentrations approach normal resting levels, the troponin-tropomyosin complex returns to its normal position. This change re-covers the active sites and prevents further cross-bridge interaction.

9. Without cross-bridge interactions, further sliding cannot take place, and the contraction ends.

10. Muscle relaxation occurs, and the muscle returns passively to resting length.

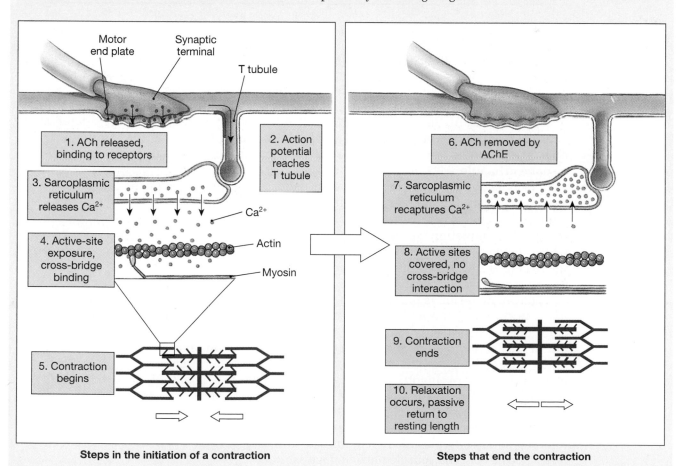

Steps in the initiation of a contraction   **Steps that end the contraction**

PROCEDURES

1. Place a sample of the muscle under the dissecting microscope and gently tease the fibers apart. Separate two or three individual fibers and transfer them to clean microscope slides. Add a drop of glycerol to each slide to prevent dehydration of the muscle fibers.

2. Add a coverslip to one of the slides and observe the muscle fibers with the compound microscope at low and high powers. Describe the appearance of these fibers. _____

   _____.

3. Choose another slide with teased muscle fibers. Without a coverslip, examine these fibers with the dissecting microscope. Place a millimeter ruler under the slide and measure the length of the relaxed muscle fibers. _____.

4. Add a drop of ATP solution to the fibers under the dissecting microscope and observe the fibers' response. After 30–45 s, measure the length of the fibers._____.

5. Repeat the observation in steps 3 and 4 with fresh muscle fibers and the ion solution instead of the ATP. After 30–45 s, measure the length of the fibers. _____.

6. Prepare another slide with teased muscle fibers. Repeat steps 3 and 4 using both the ATP and ion solutions. After 30–45 s, measure the length of the fibers. _____.

7. Dispose of the muscle preparations as indicated by your laboratory instructor.

B. **Types of Muscle Contraction**

In the preceding section, muscle contraction was presented relative to the events occurring inside a single muscle fiber. Muscle fibers do not act individually because a single motor neuron controls more than one muscle fiber. A motor neuron innervating large muscles of the thigh, for example, stimulates over 1000 muscle fibers. This group of fibers controlled by the same neuron is called a **motor unit** and can be considered a "muscle team" that contracts together when stimulated by the neuron. The muscle fibers are said to be "on" for contraction or "off" for relaxation, a concept called the **all-or-none principle**.

The type of contraction a muscle fiber performs is determined by the frequency of stimulation from the motor neuron. If a single action potential occurs in the neuron, only a small amount of ACh will be released and the muscle fiber will twitch. A **twitch** is a single stimulation-contraction-relaxation event in the fiber. Examine Figure 24.3 which displays **myograms** or recordings of a muscle contraction. The figure compares the twitching of different muscles. Each twitch has three events: the latent period and the contraction and relaxation phases.

The **latent period** is the time from the initial stimulation to the start of muscle contraction. During this brief period the muscle fiber is stimulated by ACh, releases calcium ions, exposes active sites, and attaches cross-bridges. No tension is produced during this period, and no pivoting occurs. The **contraction phase** involves shortening of the muscle fiber and the production of muscle tension or force. During this phase myosin heads are pivoting and cycling through the attach-pivot-detach-return sequence. As more calcium enters the sarcoplasm, more cross-bridges are formed, and tension increases as the thick filaments pull the thin filaments toward the center of the sarcomere. The **relaxation phase** occurs as AChE inactivates ACh and calcium ions are returned to the sarcoplasmic reticulum. The thin filaments passively slide back to their resting positions, and muscle tension decreases.

Figure 24.4 illustrates the effect of increasing the frequency or rate of stimulations on a skeletal muscle. If the muscle is stimulated again before it has completely relaxed from the first stimulation, the two contractions are summed. This phenomenon, called

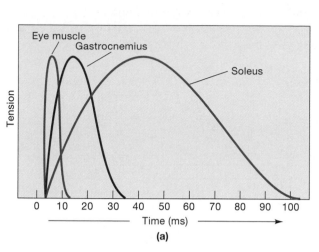 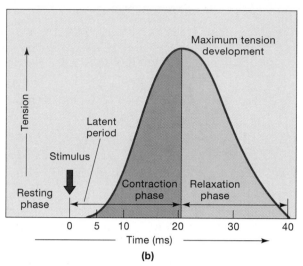

Figure 24.3
The Twitch and Development of Tension.

(a) Myogram showing differences in the time course of a twitch contraction in different skeletal muscles in the body. (b) Details of the time course of a single isometric twitch contraction in the gastrocnemius muscle. Note the presence of a latent period, which corresponds to the time needed for conduction of the action potential and the subsequent release of calcium ions by the sarcoplasmic reticulum.

wave summation, results in an increase in tension with each summation. Because the thin filaments have not returned to their resting length when the muscle is stimulated again, contractile force increases as more calcium is released into the sarcoplasm and more cross-bridges attach and pivot.

If the frequency of stimulation is increased further, the muscle produces peak tension with short cycles of relaxation. This type of contraction is called **incomplete tetanus** (see Figure 23.4b). If the rate of stimulation is such that the relaxation phase is completely eliminated, the contraction type is **complete tetanus**. During complete tetanus, peak tension is produced for a sustained period of time and results in a smooth, strong contraction. Most muscle work is accomplished by complete tetanus.

CLINICAL APPLICATION—TETANUS

Surely you have had a tetanus shot. Why is the injection called a tetanus shot when tetanus is a type of muscle contraction? Often an injury introduces a bacterium, *Clostridium tetani,* into the wound. The bacteria produces a toxin that binds to acetylcholine receptors in skeletal muscles and stimulates the muscle to contract. The enzyme acetylcholinesterase cannot inactivate the bacterial toxin, and the muscle remains in a painful tetanic contraction for an extended period of time. The term "lockjaw" indicates that the muscles used for mastication are often affected by the bacterial toxin. As a preventative measure, a tetanus shot contains human tetanus immune globulins that prevent the *Clostridium* organism from surviving in the body and producing toxins. Actual treatment for the infection and toxin is usually ineffective in preventing tetanus. Is your tetanus booster shot current?

Because all muscle work requires complete tetanus, the type of contraction does not determine the overall tension a muscle produces. Muscle strength is varied through the number of motor units activated. The process of **recruitment** stimulates more motor units to carry or move a load placed on a muscle. As recruitment occurs, more muscle

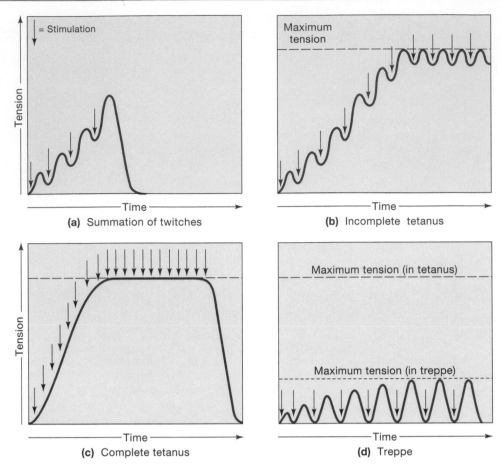

Figure 24.4
Effects of Repeated Stimulations.

(a) Summation of twitches occurs when successive stimuli arrive before relaxation has been completed. **(b)** Incomplete tetanus occurs if the rate of stimulation increases further. Tension production rises to a peak, and the periods of relaxation are very brief. **(c)** In complete tetanus the frequency of stimulation is so high that the relaxation phase has been entirely eliminated and tension plateaus at maximal levels. **(d)** Treppe is an increase in peak tension following repeated stimuli delivered shortly after completion of the relaxation phase of each twitch.

fibers are turned on and contract, thus tension increases. Imagine holding a book in your outstretched hand. If another book is added to the load, additional motor units are recruited to increase the muscle tension to support the added weight.

The myogram in Figure 24.4d is a recording of **treppe**, muscle contraction with complete relaxation between each stimulus. As the muscle is repeatedly stimulated, calcium ions accumulate in the cytosol, causing the first 30–50 contractions to increase in tension. Athletes always warm up before exercising and competing to ensure their muscular system is operating at peak performance.

Muscle fibers cannot contract indefinitely, and eventually they become **fatigued**. The force of contraction decreases as fibers lose the ability to maintain compete tetanic contractions. Fatigue is generally caused by declining energy and oxygen sources in the muscle and an accumulation of waste products. During intense muscle contraction, such as lifting a heavy object, the fiber becomes fatigued because of low ATP levels and a buildup of lactic acid, a by-product of anaerobic respiration. Joggers experience muscle fatigue as secondary energy reserves are depleted and damage accumulates in muscle fibers, especially in the sarcoplasmic reticulum and calcium-regulating mechanisms.

LABORATORY ACTIVITY   DEMONSTRATION OF MUSCLE FATIGUE

MATERIALS

heavy object (textbook)
stop watch (clock)

PROCEDURES

1. Extend your arm straight out in front of you, parallel to the floor.
2. Note the time and load your extended arm with the heavy book by placing it in your hand. Record the start time in Table 24.2 under Trial 1.
3. Hold the book with your arm straight for as long as possible. Once your arm starts to shake or your muscles ache, remove the book. Record the end time for Trial 1 in Table 24.2.

TABLE 24.2   Muscle Fatigue Demonstration

Trial	Start Time (s)	End Time (s)	Duration (s)
1	_____	_____	_____
2	_____	_____	_____
3	_____	_____	_____

4. Rest for 1 min and then repeat steps 1 through 3 as Trial 2. Record the start and end times in Table 24.2.

5. Rest for another minute and then repeat steps 1 through 3 as Trial 3. Record these times in Table 24.2.

6. Complete Table 24.2 by calculating the total time in seconds required for each trial. Enter the time for each trial in the Duration column in Table 24.2.

7. Plot on the graph the total time until fatigue for each trial. Label the *X* axis (horizontal axis) "Trial 1," "Trial 2," and "Trial 3," and the *Y* axis (vertical axis) "Time in seconds."

8. Interpret your experimental data. Why is there a difference in the time to fatigue for each trial?

_____

_____

C. Isometric and Isotonic Contractions

Two major types of complete tetanic contractions occur: isometric and isotonic, see Figure 24.5. **Isometric** (iso—same; metric—length) **contractions** occur when the muscle length is relatively constant while muscle tension changes. Because length is constant, no body movement occurs. Muscles for posture use isometric contractions to support the body weight. **Isotonic contractions** involve a constant tension while the length of the muscle changes. Consider picking up a book the flexing of your arm. Once you are holding the book, your arm muscles are "loaded" with the weight of the book. As you flex your arm, muscle tension changes minimally while muscle length varies greatly.

| LABORATORY ACTIVITY | DEMONSTRATION OF ISOMETRIC AND ISOTONIC CONTRACTIONS |

MATERIALS

heavy object (textbook)

PROCEDURES

1. Extend your arm straight out in front of you, parallel to the floor.

2. Palpate your extended biceps brachii muscle to feel the tension of the muscle. Also notice the length of the muscle. Record your observations in the Trial 1 column in Table 24.3.

3. Load your extended arm with a heavy book by placing it in your extended hand. Palpate your biceps brachii again and notice the degree of muscle tension and the length of the muscle. Record your observations in the Trial 2 column in Table 24.3.

4. Gently squeeze the loaded biceps brachii and repeatedly flex and extend your arm to move the book 4–6 in. each time. Record your muscle tension and length observations in the Trial 3 column in Table 24.3.

5. Describe each type of muscle contraction to complete Table 24.3.

TABLE 24.3   Isometric and Isotonic Contractions

	Trial 1	Trial 2	Trial 3
Tension	_____	_____	_____
Length	_____	_____	_____
Type of contraction	_____	_____	_____

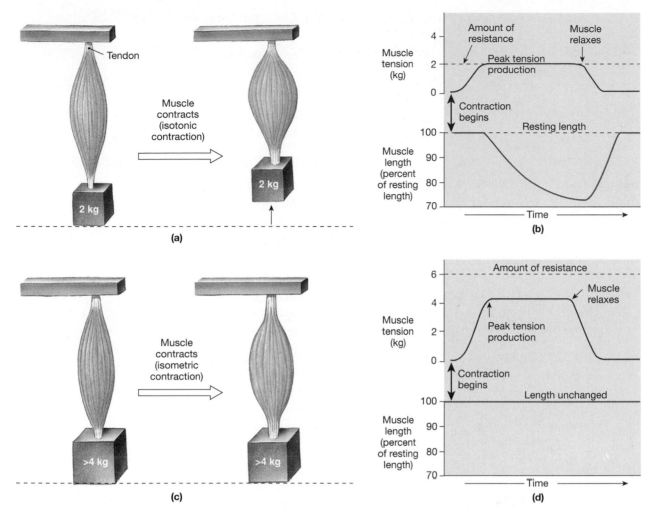

Figure 24.5
Isotonic and Isometric Contractions.

(a,b) This muscle is attached to a weight less than its peak tension capabilities. On stimulation, it develops enough tension to lift the weight. Tension remains constant for the duration of the contraction, although the length of the muscle changes. This is an example of isotonic contraction. **(c,d)** The same muscle is attached to a weight that exceeds its peak tension capabilities. On stimulation, tension will rise to a peak, but the muscle as a whole cannot shorten. This is an isometric contraction.

MUSCLE PHYSIOLOGY CHECKLIST

This is a list of bold terms presented in Exercise 24. Use this list for review purposes after you have completed the laboratory activities.

INTRODUCTION

neurotransmitter
depolarize
repolarize
action potential
transmembrane potential

BIOCHEMICAL NATURE OF CONTRACTION

• Figures 24.1 and 24.2, Table 24.1
sliding filament theory
neuromuscular junction
synaptic knob
synaptic vesicle
acetylcholine
synaptic cleft
motor end plate
active site exposure
cross-bridge attachment
pivoting
cross-bridge detachment
myosin activation
relaxation

TYPES OF MUSCLE CONTRACTION

• Figures 24.3 and 24.4
motor unit
all-or-none principle
myogram
twitch
latent period
contraction phase
relaxation phase
wave summation
incomplete tetanus
complete tetanus
recruitment
treppe
fatigue

ISOMETRIC AND ISOTONIC CONTRACTION

isometric contraction
isotonic contraction

MUSCLE PHYSIOLOGY

Laboratory Report Exercise 24

A. Matching

Match each term listed on the left with the correct description on the right.

1. _____ isometric contraction	A.	unresponsive time during contraction
2. _____ complete tetanus	B.	warming-up contraction
3. _____ repolarization	C.	time prior to tension
4. _____ all-or-none principle	D.	fusion of twitches
5. _____ fatigue	E.	contraction with rapid cycles of relaxation
6. _____ isotonic contraction	F.	contraction with no relaxation cycles
7. _____ refractory period	G.	response to a single stimulus
8. _____ depolarization	H.	length changes more than tension
9. _____ treppe	I.	muscle fibers are either "on" or "off"
10. _____ latent period	J.	shift in transmembrane potential toward 0 mV.
11. _____ wave summation		
12. _____ incomplete tetanus	K.	neurotransmitter
13. _____ twitch	L.	tension changes more than length
14. _____ recruitment	M.	reduction in contraction and performance
15. _____ acetylcholine	N.	return of membrane to resting potential
	O.	activating more motor units

B. Short-Answer Questions

1. Describe each phase of muscle contraction:
   a. exposure

   b. attach

   c. pivot

d. detach

e. return (activate)

2. Describe the events at the neuromuscular junction during muscle stimulation.

3. Distinguish between isometric and isotonic contractions.

4. Explain why muscles become fatigued.

5. Describe each of the following types of muscle contraction:
   a. wave summation

   b. treppe

   c. complete tetanus

   d. twitch

   e. incomplete tetanus

24A

Muscle Physiology Alternate

ELECTROMYOGRAPHY
Standard and Integrated EMG

OBJECTIVES

On completion of this exercise, you should be able to:

- Observe and record skeletal muscle tonus as reflected by a basal level of electrical activity associated with the muscle in a resting state.
- Record maximum clench strength for right and left hands.
- Observe, record, and correlate motor unit recruitment with increased power of skeletal muscle contraction.
- Listen to EMG "sounds" and correlate sound intensity with motor unit recruitment.

INTRODUCTION

When a muscle or nerve is stimulated, it responds by producing action potentials, electrical impulses that spread across the cell surface. An action potential in a muscle fiber activates the physiological events of contraction. The electrical stimulation is also passively conducted to the body surface by surrounding tissues. A pair of sensors placed on the skin can detect the electrical activity produced by the muscle. The impulse produced by individual muscle fibers is minimal; however, the combination of impulses from thousands of stimulated muscle fibers produces a measurable electrical change in the overlying skin. The electrodes are connected to an amplifier and the signals are graphed by a recorder to produce an **electromyogram (EMG),** a recording of the muscle's electrical activity. You are probably familiar with electrical tracings of the heart's electrical activity, the electrocardiogram or EKG, which will be studied in a latter exercise.

In this exercise, you will use the Biopac Student Lab system, shown in Figure 24A.1, to detect, record, and analyze series of muscle impulses. The system consist of three main components; sensors called **electrodes** and **transducers** that detect electrical impulses and other physiological phenomena, the MP30 Acquisition Unit that collects and amplifies data from the sensors, and a computer with software to record and interpret the EMG data. After applying electrodes to the skin over forearm muscles, you will clench your fist repeatedly with increasing force and the Biopac system will produce an EMG of the muscle impulses. By increasing the force of each fist clench the muscles are recruiting additional motor units to contract and produce more tension. For more background information it is recommended that you review the physiology of muscle stimulation, contraction, and recruitment in Exercise 24.

The EMG laboratory exercise is organized into four major procedures. The **"Set Up"** section describes where to plug in the electrode leads and how to apply the skin electrodes. The **"Calibration"** section adjusts the hardware so it can collect accurate physiological data. Once the hardware is calibrated the **"Data Recording"** section describes the steps to record the fist clench impulses. After saving the muscle data to computer disc, the **"Data Analysis"** section instructs you how to use the software tools to interpret and evaluate the EMG.

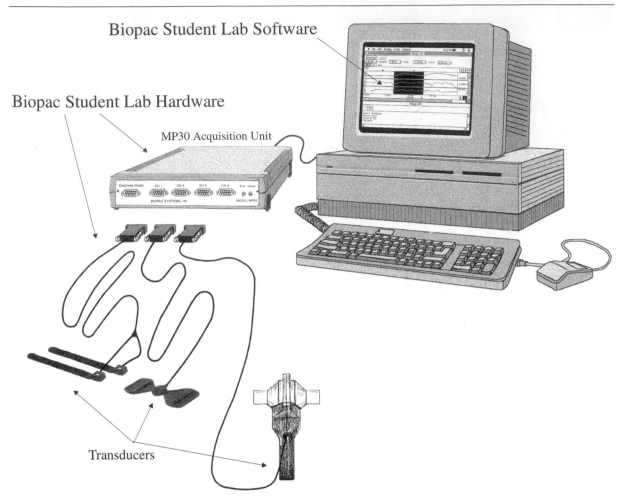

Biopac Student Lab Software

Biopac Student Lab Hardware

MP30 Acquisition Unit

Transducers

Figure 24A.1
Biopac Student Lab System Components.

      The Biopac Student Lab system is safe and easy to use but be sure to read the safety notices and to carefully follow the procedures as outlined in the laboratory activities. Under no circumstances should you deviate from the experimental procedures.

IMPORTANT SAFETY NOTICE

      BIOPAC Systems, Inc. instrumentation is designed for educational and research oriented life science investigations.

      BIOPAC Systems, Inc. does not condone the use of its instruments for clinical medical applications.

      Instruments, components, and accessories provided by BIOPAC Systems, Inc. are not intended for the cure, mitigation, treatment, or prevention of disease.

      The Biopac Student Lab is an electrically isolated data acquisition system, designed for biophysical measurements.

      Exercise extreme caution when applying electrodes and taking bioelectric measurements while using the Biopac Student Lab with other external equipment that also uses electrodes or transducers that may make electrical contact with the Subject.

      Always assume that currents can flow between any electrodes or electrical contact points.

      It is very important (in case of equipment failure) that significant currents are not allowed to pass through the heart. If electrocautery or defibrillation equipment is used, it is recommended that you disconnect the BIOPAC Systems, Inc. instrumentation from the Subject.

LABORATORY ACTIVITY   ELECTROMYOGRAPHY

MATERIALS

BIOPAC electrode lead set (SS2L)

BIOPAC disposable vinyl electrodes (EL503), 6 electrodes per subject

BIOPAC headphones (OUT1)

BIOPAC electrode gel (GEL1) and abrasive pad (ELPAD) or Skin cleanser or alcohol prep

Computer system: Macintosh®—minimum 68020 *or* PC running Windows 95®

Memory requirements: 4 MB RAM above and beyond what the operating system needs and any other programs that are running.

Biopac Student Lab software V3.0

BIOPAC acquisition unit (MP30)

BIOPAC wall transformer (AC100A)

BIOPAC serial cable (CBLSERA)

PROCEDURE—SET UP

1. Turn your computer **ON.**

   The desktop should appear on the monitor. If it does not appear, ask the laboratory instructor for assistance.

2. Make sure the BIOPAC MP30 unit is **OFF.**

3. **Plug the equipment in** as shown in Figure 24A.2.

   Electrode lead (SS2L)—CH 3.

   Headphones (OUT1)—back of unit

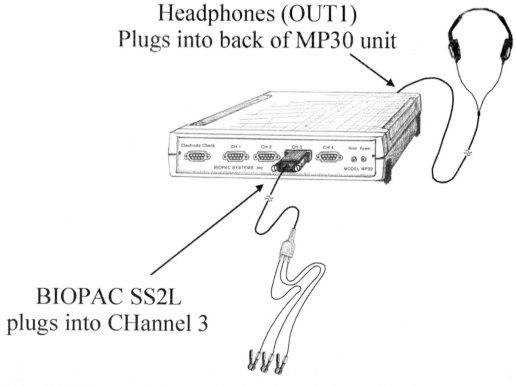

Headphones (OUT1)
Plugs into back of MP30 unit

BIOPAC SS2L
plugs into CHannel 3

Figure 24Λ.2
Biopac Cable Set Up.

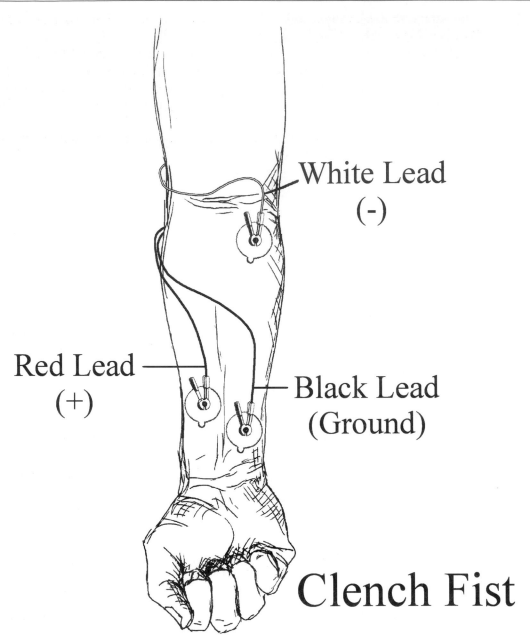

Figure 24A.3
Electrode Placement and Lead Attachment.
Fist is clenched for calibration procedure.

4. Turn **ON** the BIOPAC MP30 unit.
5. **Attach three electrodes** to the forearm, as shown in Figure 24A.3.

    For the first recording segment, select the **Subject's** dominant forearm (generally the right forearm if the subject is right-handed or the left forearm if the subject is left-handed) and attach the electrodes onto the forearm as shown; this will be **Forearm 1.**

    Use **Subject's** other arm for the second recording segment; this will be **Forearm 2.**

    Note: For optimal electrode adhesion, the electrodes should be placed on the skin at least 5 minutes before the start of the Calibration procedure.

6. Attach the electrode lead set (SS2L) to the electrodes, following the color code (see Figure 24A.3).

   Important: Make sure the electrode lead colors match Figure 24A.3.

   Each of the pinch connectors on the end of the electrode cable needs to be attached to a specific electrode. The electrode cables are each a different color, and you should follow Figure 24A.3 to ensure that you connect each cable to the proper electrode.

   The pinch connectors work like a small clothespin but will only latch onto the nipple of the electrode from one side of the connector.

7. **Start** the Biopac Student Lab Program.

8. Choose lesson **"L01-EMG-1"** and click **OK**.

9. Type in a unique **filename**.

   No two people have the same filename, so use a unique identifier, such as the subject's nickname or student ID#.

10. Click **OK**.

    This ends the Set Up procedure.

PROCEDURE—CALIBRATION

The Calibration procedure establishes the hardware's internal parameters (such as gain, offset, and scaling) and is critical for optimum performance. **Pay close attention to the Calibration procedure.**

1. Click on **Calibrate**.

   This will start the Calibration recording.

2. Read the dialog box and click **OK** when ready.

   A dialog box pops up when you click Calibrate, telling you how to prepare for the Calibration. The Calibration will not begin until you click **OK**.

3. **Wait** about two seconds, **clench** the fist as hard as possible (see Figure 24A.3), then **release**.

   The program needs a reading of your maximum clench to perform an auto-Calibration.

4. **Wait** for Calibration to stop.

   The Calibration procedure will last eight seconds and will stop automatically, so let it run its course.

5. **Check** the Calibration data.

   At the end of the eight-second Calibration recording, the screen should resemble Figure 24A.4.

   If similar, proceed to the Data Recording Section.

   If different, **Redo Calibration.**

   If your Calibration recording did not begin with a zero baseline (subject clenched before waiting two seconds), you need to repeat Calibration to obtain a reading similar to Figure 24A.4.

PROCEDURE—DATA RECORDING

1. Prepare for the recording.

   You will record two segments:

   a. Segment one records data from **Forearm 1**.

   b. Segment two records data from **Forearm 2**.

   In order to work efficiently, read this entire section so you will know what to do before recording.

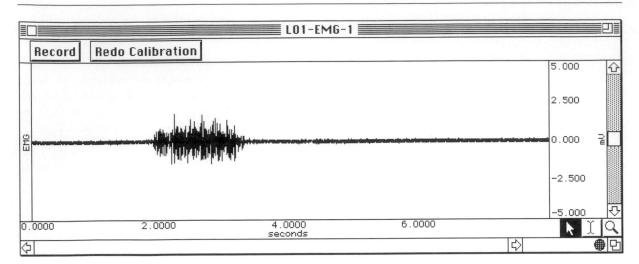

Figure 24A.4
Calibration EMG.

Check the last line of the journal and note the total amount of time available for the recording. Stop each recording segment as soon as possible so you don't waste recording time (time is memory).

Segment 1

2. Click on **Record.**

   You will begin to record data.

3. Clench-Release-Wait and repeat with increasing force to reach your maximum force on the fourth clench.

   Repeat a cycle of Clench-Release-Wait, holding for two seconds and waiting for two seconds after releasing before beginning the next cycle. Try to increase the strength in equal increments such that the fourth clench is the maximum force.

4. Click on **Suspend.**

   The recording should halt, giving you time to review the data and prepare for the next recording segment.

5. Review the data on the screen.

   If all went well, your data should look similar to Figure 24A.5.

   If *similar* and more recording segments are required, go to **Step 7.**

   If *different,* go to **Step 6.**

   The data would be *different* if the:

   a. Suspend button was pressed prematurely.

   b. Instructions were not followed.

6. Click on **Redo** if your data did not match Figure 24A.5, and **repeat Steps 2–5.**

   Click **Redo-** and repeat Steps 2–5. Note that once you press **Redo,** the data you have just recorded will be erased.

7. Remove the electrodes from the forearm.

   Remove the electrode cable pinch connectors, and peel off the electrodes. Throw out the electrodes (BIOPAC electrodes are not reusable). Wash the electrode gel residue from the skin, using soap and water. The electrodes may leave a slight ring on the skin for a few hours, which is quite normal.

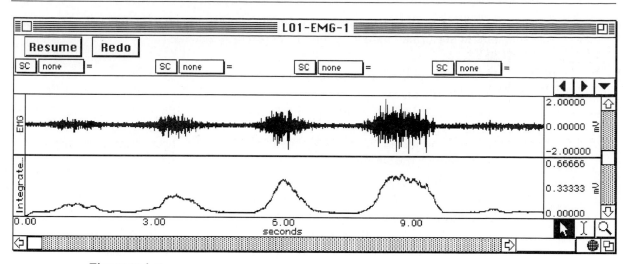

Figure 24A.5
Segment One.
EMG of four fist clenches with increasing force.

Segment 2

8. For **Forearm 2,** attach electrodes (per Set Up Step 5) to the subject's opposite arm. Refer to the Set Up section for proper electrode placement.

9. Click on **Resume.**

    A marker labeled "Forearm 2" will automatically be inserted when you click on Resume.

10. Clench-Release-Wait and repeat with increasing force to reach your maximum force on the fourth clench.

    Repeat a cycle of Clench-Release-Wait, holding for two seconds and waiting for two seconds after releasing before beginning the next cycle. Try to increase the strength in equal increments such that the fourth clench is the maximum force.

11. Click on **Suspend.**

    The recording should halt, giving you time to review the data for segment two.

12. Review the data on the screen.

    If all went well, your data should look similar to Figure 24A.5.

    If *similar,* go to **Step 14.**

    If *different,* go to **Step 13.**

    The data would be *incorrect* if the:

    a. Suspend button was pressed prematurely.

    b. Instructions were not followed.

13. If your data did not match Figure 24A.5, click on **Redo,** and **repeat Steps 9–12.**

    Click **Redo,"** and repeat Steps 9–12. Note that once you press **Redo,** the data you have just recorded will be erased.

14. Click on **Stop.**

    When you click on **Stop,** a dialog box comes up, asking if you are sure you want to stop the recording. Clicking "yes" will end the data recording segment and will automatically save the data. Clicking "no" will bring you back to the Resume or Stop options. This is simply one last chance to confirm you don't need to redo the last recording segment.

15. If you want to listen to the EMG signal, go to Step 16.

Listening to the EMG can be a valuable tool in detecting muscle abnormalities, and is performed here for general interest.

or

If you want to end the recording, go to Step 21.

16. **Subject** puts on the headphones.

    Note: When the Listen button is hit in the next step, it's possible that the volume through the headphones may be very loud due to system feedback. The volume cannot be adjusted, so you may have to position the headphones slightly off the ear to reduce the sound.

17. Click on **Listen.**

18. Experiment by changing the clench force as you watch the screen and listen.

    You will hear the EMG signal through the headphones as it is being displayed on the screen. The screen will display two channels: CH 3 EMG and CH 40 Integrated EMG.

    The data on the screen will not be saved.

    The signal will run until you press **Stop.**

    If others in your lab group would like to hear the EMG signal, pass the headphones around before clicking **Stop.**

19. Click on **Stop.**

    This will end the listening to EMG portion.

20. To listen again, click **Redo.**

    If another person wants to listen to the EMG, switch the headphones from the Subject to the new person and click Redo.

21. Click **Done.**

    A pop-up window with four options will appear. Make your choice, and continue as directed.

    If choosing the "Record from another subject" option:

    a. Attach electrodes per Set Up Step 5 and continue the entire lesson from Set Up Step 8.

    b. Each person will need to *use a unique filename.*

PROCEDURE—DATA ANALYSIS

1. Enter the **Review Saved Data** mode and choose the correct file.

   Note Channel Number (CH) settings in the small menu boxes as shown in Figure 24A.6.

Channel	Displays
CH 3	**Raw EMG**
CH 40	**Integrated EMG**

2. Set up your display window for optimal viewing of the first data segment.

   Figure 24A.7 shows a sample display of the first data segment recorded using the arm of the Subject's dominant hand.

   The following tools help you adjust the data window.

Autoscale horizontal	Horizontal (Time) Scroll Bar
Autoscale waveforms	Vertical (Amplitude) Scroll Bar
Zoom Tool	Overlap button
Zoom Previous	Split button

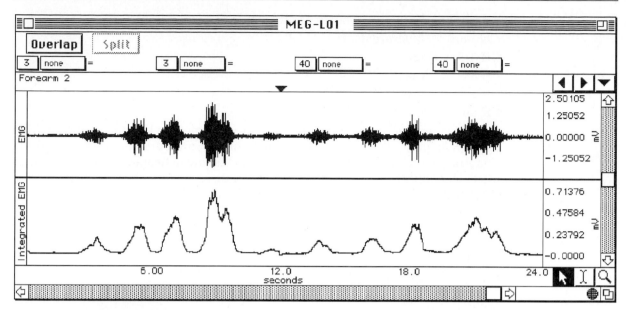

Figure 24A.6
Segment Two EMG.

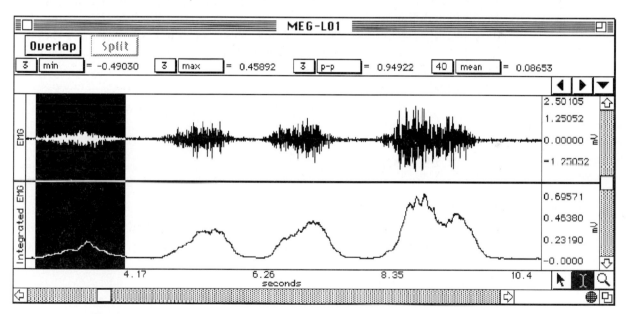

Figure 24A.7
Selection of EMG cluster for analysis.
Data for the highlighted EMG are displayed in the small menu boxes.

3. Set up the measurement boxes as follows:

   Channel   *Measurement*
   CH 3   **min**
   CH 3   **max**
   CH 3   **p-p**
   CH 40   **mean**

   The measurement boxes are above the marker region in the data window. Each measurement has three sections: channel number, measurement type, and result. The first two sections are pull-down menus that are activated when you click on them. The following is a brief description of these specific measurements:

   min: displays the minimum value in the selected area

   max: displays the maximum value in the selected area

   p-p: finds the maximum value in the selected area and subtracts the minimum value found in the selected area

   mean: displays the average value in the selected area

   The "selected area" is the area selected by the I-Beam tool (including the end-points).

4. Using the **I-Beam** cursor, select an area enclosing the first EMG cluster (Figure 24A.7). Record your data in section A of the EMG Lab Report.

5. Repeat Step 5 on each successive EMG cluster. Record your data in section A of the EMG Lab Report.

6. Scroll to the second recording segment. The second recording segment is Forearm 2, which begins after the first marker and represents the nondominant arm.

7. Repeat Steps 4 and 5 for the Forearm 2 data.

8. Save or print the data file.

   You may save the data to a floppy drive, save notes that are in the journal, or print the data file.

9. Exit the program.

   END OF DATA ANALYSIS

Name _____   Date _____

Section _____   Number _____

ELECTROMYOGRAPHY: STANDARD AND INTEGRATED EMG

Laboratory Report Exercise 24 Alternate

A. Data and Calculations

Subject Profile

Name _____   Height _____

Age _____   Weight _____

Gender: Male / Female

1. EMG Measurements

Cluster #	Forearm 1 (Dominant)				Forearm 2			
	Min	Max	P-P	Mean	Min	Max	P-P	Mean
	[3 min]	[3 max]	[3 p-p]	[40 mean]	[3 min]	[3 max]	[3 p-p]	[40 mean]
1								
2								
3								
4								

2. Use the mean measurement from the table above to compute the percentage increase in EMG activity recorded between the weakest clench and the strongest clench of Forearm 1.

Calculation: _____

Answer: _____%

B. Questions

1. Does there appear to be any difference in tonus between the two forearm clench muscles?

   _____ Yes   _____ No

   Would you expect to see a difference? Does the subject's sex influence your expectations? Explain._____

   _____

   _____

2. Compare the mean measurement for the right and left maximum clench EMG cluster. Are they the same or different? _____ Same   _____ Different

   Which one suggests the greater clench strength?

   _____ Right   _____ Left   _____ Neither

   Explain.

   _____

   _____

   _____

   _____

3. What factors in addition to sex contribute to observed differences in clench strength?

   _____

   _____

   _____

4. Explain the source of signals detected by the EMG electrodes.

   _____

   _____

   _____

   _____

5. What does the term *motor unit recruitment* mean?

   _____

   _____

   _____

   _____

   _____

6. Define *electromyography.*

   _____

   _____

   _____

Nervous System Overview

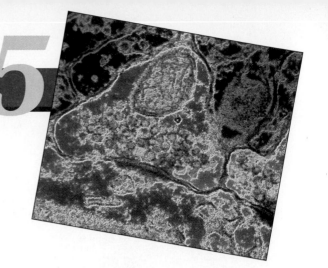

OBJECTIVES

On completion of this exercise, you should be able to:

- Outline the organization of the nervous system.
- Describe the general functions of the central and peripheral nervous systems.
- Compare sensory receptors and effectors.
- Compare the autonomic and somatic divisions of the efferent peripheral nervous system.
- List six types of neuroglia and a basic function for each type.
- Describe the cellular anatomy of a neuron.
- Generally discuss how a neuron communicates with other cells.

WORD POWER

afferent (afferen—carrying toward)
efferent (efferen—carrying away)
sensory (sens—a feeling)
somatic (somat—a body)
autonomic (auto—self)
synapse (synap—a union)

MATERIALS

compound microscope
prepared slides:
   multipolar neuron
   astrocyte
nervous system chart
neuron model

INTRODUCTION

The nervous system orchestrates body functions to maintain homeostasis. To accomplish this control, it must perform three vital tasks. First, it must detect changes in and around the body. Sensory receptors monitor specific environmental conditions and encode information about environmental changes as electrical impulses. Second, it must process the incoming sensory information and generate an appropriate motor response to adjust the activity of muscles and glands. Third, all sensory and motor activities must be orchestrated or integrated to achieve the balance of homeostasis.

A. Organization of the Nervous System

The nervous system is divided into two main components, the central nervous system (CNS) and the peripheral nervous system (PNS) (see Figure 25.1). The **central nervous system** consists of the **brain** and the **spinal cord**. The **peripheral nervous system** includes cranial and spinal nerves that communicate with the CNS. The PNS is responsible for providing the CNS with information concerning changes inside the body and in the surrounding environment. Sensory information is sent along peripheral nerves that join the CNS in the spinal cord or the brain. The CNS evaluates the sensory data and determines if muscle and gland activities should be modified in response to the environmental changes. Motor commands from the CNS are relayed to peripheral nerves that carry the commands to specific muscles and glands.

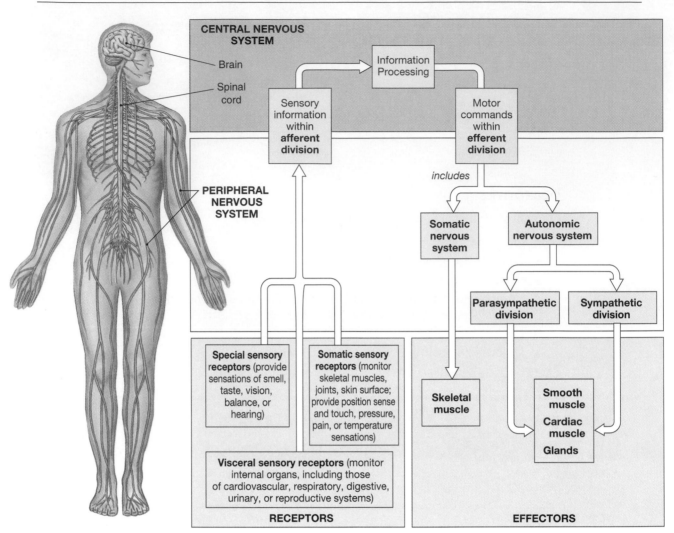

Figure 25.1
A Functional Overview of the Nervous System.

The PNS is subdivided into afferent and efferent divisions. The **afferent division** receives sensory information from **sensory receptors**, cells, and organs that detect changes in the body and the surrounding environment and sends the sensory informa- tion to the CNS for interpretation. The CNS decides the appropriate response to the sensory information and sends motor commands to the **efferent division**, which con- trols the activities of **effectors**, the muscles and glands of the body. The **somatic ner- vous system** conducts motor responses to skeletal muscles. The **autonomic nervous system**, which consists of the sympathetic and parasympathetic branches, sends com- mands to smooth and cardiac muscles and glands.

Peripheral nerves are subdivided into two groups according to the portion of the CNS with which they communicate. **Cranial nerves** are bundles of axons that extend off the brain. Foramina of the skull provide passageways out of the skull for the cranial nerves. **Spinal nerves** join the spinal cord at intervertebral foramina and pass into the extremities or the body wall. There are 12 pairs of cranial nerves and 31 pairs of spinal nerves, and each pair of nerves transmits specific information between the CNS and PNS. Functionally, all spinal nerves are "mixed" nerves and carry both sensory and motor signals. Cranial nerves are either entirely sensory nerves or mixed nerves. Although a nerve may transmit both sensory and motor impulses, a single neuron in the nerve produces only one type of signal.

What if an injury damaged portions of the afferent division of a peripheral nerve in the left leg? How would this affect the victim's sensory and motor functions?

LABORATORY ACTIVITY   ORGANIZATION OF THE NERVOUS SYSTEM

MATERIALS

Figure 25.1

PROCEDURE

1. Complete the following simulation of neural control by filling in the blanks:

    Imagine stepping on a tack. Sensory _____ in your foot send a pain signal to the CNS via the _____ division of the _____. Once information arrives in the CNS, the proper _____ response is sent to the _____ division of the PNS. These voluntary signals travel along the _____ nervous system to _____, the skeletal muscles receiving the motor commands.

B. Histology of the Nervous System

Two types of cells populate the nervous system: neurons and glial cells. Neurons are the communication cells of the nervous system and are capable of propagating and transmitting electrical impulses to respond to the ever-changing needs of the body. Glial cells have a supportive role in protecting and maintaining neural tissue.

Neuroglia cells

Neuroglia (noo-RŌ-glē-ah), or simply **glial** cells, are the most abundant cells in the nervous system. They protect, support, and anchor neurons in place. The CNS and PNS have different types of glial cells. The CNS has glial cells involved in production and circulation of cerebrospinal fluid. Both the CNS and PNS have glial cells that isolate and support neurons with white matter. Figure 25.2 highlights the various types of neuroglia.

The central nervous system has four distinct types of neuroglia: astrocytes, oligodendrocytes, microglia, and ependymal cells. **Astrocytes (AS-trō-sīts)** have many functions; they hold neurons in place and isolate them from other neurons. Astrocytes wrap feetlike extensions around blood vessels, creating a blood-brain barrier that prevents certain materials from passing out of the blood and into neural tissue (see Figure 25.3).

Oligodendrocytes (o-li-gō-DEN-drō-sīts) cover axons in the CNS with a fatty myelin sheath. Myelinated axons form white matter of the nervous system; nonmyelinated axons and cell bodies form gray matter. **Microglia (mī-KROG-lē-uh)** cells are phagocytes that remove microbes and cellular debris from neural tissue. **Ependymal (e-PEN-dī-mul)** cells line the ventricles of the brain and the central canal of the spinal cord. These cells contribute to the production of cerebrospinal fluid which circulates in the ventricles and in the central canal.

The peripheral nervous system has two types of neuroglia: satellite cells and Schwann cells. Where cell bodies cluster in groups called **ganglia**, **satellite cells** encase each cell body and isolate it from the interstitial fluid to regulate the neuron's chemical environment. **Schwann cells** wrap around neurons, and constitute the white matter of the PNS. The structure and importance of the myelin sheath is discussed in the next section.

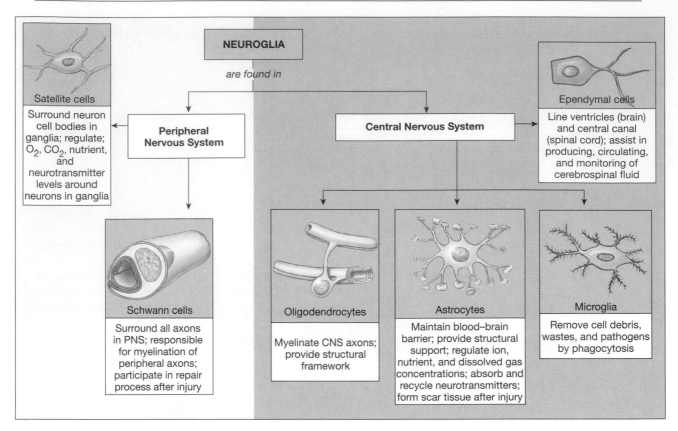

Figure 25.2
An Introduction to Neuroglia.

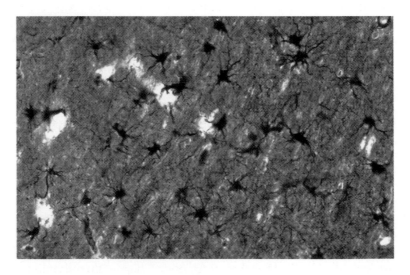

Figure 25.3
Astrocytes (400×).

LABORATORY ACTIVITY | MICROSCOPIC OBSERVATION OF NEUROGIA

MATERIALS

compound microscope
prepared slide of astrocytes

PROCEDURES

1. Examine a prepared microscope slide of astrocytes. Scan the slide at low magnification and locate a group of astrocytes.

2. Examine a single astrocyte at high magnification. Locate the nucleus and the numerous cellular extensions.

3. Draw an astrocyte in the space provided below.

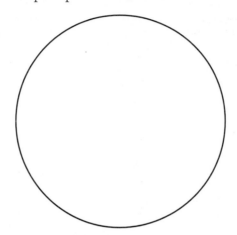

Neurons

Neurons are the longest cells of the body. Some neurons are 3 ft long and extend from the spinal cord to the toes. A **neuron** has three distinguishable features: dendrites, a cell body, and an axon. Examine the neuron illustrated in Figure 25.4. Many **dendrites** carry information into the large, rounded **cell body** or **soma** which contains the nucleus and organelles of the cell. The **perikaryon (per-i-KAR-ē-on)** surrounds the nucleus and contains **Nissl bodies**, groups of free and fixed ribosomes. One portion of the soma narrows into an **axon hillock** which extends at the **initial segment** into a single **axon**. The axon conducts impulses away from the soma toward other neurons or an effector. The axon may divide into several **collateral branches** which subdivide into smaller branches called **telodendria (te-lō-DEN-drē-uh)**. Each telodendrium ends at a **synaptic knob** that contains chemical messengers called **neurotransmitters** stored in **synaptic vesicles**. When an electrical impulse arrives at the synaptic knob, the neuron releases neurotransmitters which diffuse across a small space, called the **synaptic cleft**, and excite or inhibit the membrane of the neighboring neuron or effector.

Neurons may synapse with other neurons; skeletal, cardiac, and smooth muscle cells; and cells of glands. An individual neuron, however, may only synapse with one type of cell. For example, motor units in skeletal muscles are groups of muscle fibers that are all controlled by the same somatic motor neuron. The motor neuron branches and each muscle fiber receives a synaptic knob at the neuromuscular junction. Notice in Figure 25.4 that the neuron on the left is adjacent to a postsynaptic neuron on the right. In this cellular arrangement, the neuron on the let communicates through the synapse to the receiver, the postsynaptic neuron.

The axon of a neuron is usually covered with a fatty **myelin (MĪ-e-lin) sheath** which, in the peripheral nervous system, is produced by Schwann cells. As a Schwann cell wraps around a small section of axon, it squeezes the cytoplasm out of its extensions and encases the axon in multiple layers of membrane. Notice in Figure 25.5 that between the axon's **myelinated internodes** are gaps in the sheath called **nodes of Ranvier (RAHN-vē-ā)**. The membrane of the axon, the **axolemma**, is exposed at the nodes and permits a nerve impulse to rapidly arc from node to node. The **neurilemma (noo-ri-LEM-uh)** or outer layer of the Schwann cell covers the axolemma at the myelinated internodes. Myelin sheaths in the CNS are formed by oligodendrocytes.

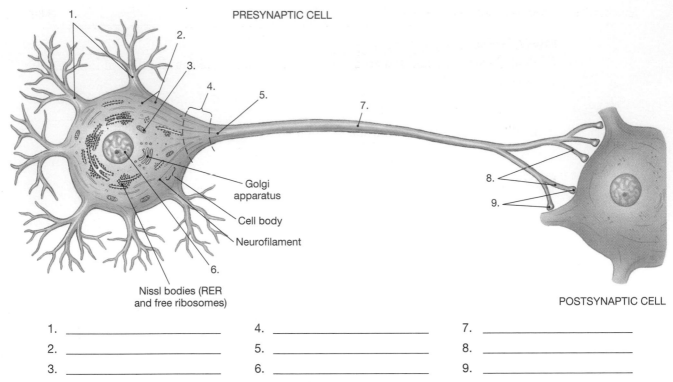

PRESYNAPTIC CELL

Golgi
apparatus

Cell body

Neurofilament

Nissl bodies (RER
and free ribosomes)

POSTSYNAPTIC CELL

1. _____ 4. _____ 7. _____
2. _____ 5. _____ 8. _____
3. _____ 6. _____ 9. _____

Figure 25.4
The Anatomy of a Multipolar Neuron.

A diagrammatic view of a neuron, showing major organelles.

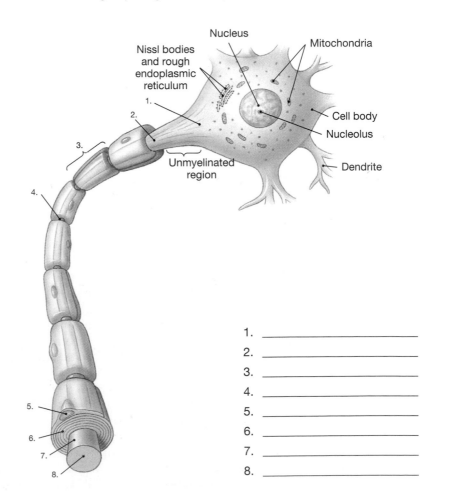

Nissl bodies
and rough
endoplasmic
reticulum

Nucleus

Mitochondria

Cell body

Nucleolus

Unmyelinated
region

Dendrite

1. _____
2. _____
3. _____
4. _____
5. _____
6. _____
7. _____
8. _____

Figure 25.5
Schwann Cells and Peripheral Axons.

Myelinated neuron showing
myelin sheath formed by
Schwann cells.

LABORATORY ACTIVITY   NEURONS

MATERIALS

compound microscope
prepared slides:
    giant multipolar neuron
    neuromuscular junction

PROCEDURES

1. Complete the labeling of Figures 25.4 and 25.5.
2. Examine a prepared microscope slide of a giant multipolar neuron, see Figure 25.6.
   a. Scan the slide at low magnification and locate several neurons.
   b. Select a single neuron and increase the magnification; identify the soma and nucleus.
   c. Can you distinguish between the dendrites and the axon?

   _____

   _____

3. Use Figure 25.7 as a guide and examine a prepared microscope slide of the neuromuscular junction.
   a. Scan the slide at low magnification and locate several telodendria and motor end plates.
   b. Increase the magnification and examine the end plate.

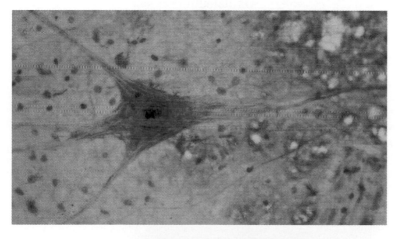

Figure 25.6
Multipolar Motor Neuron (400×).

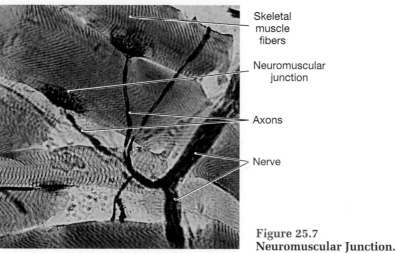

Skeletal
muscle
fibers

Neuromuscular
junction

Axons

Nerve

Figure 25.7
Neuromuscular Junction.

NERVOUS SYSTEM OVERVIEW CHECKLIST

This is a list of bold terms in Exercise 25. Use this list for review purposes after you have completed the laboratory activities.

ORGANIZATION OF THE NERVOUS SYSTEM

• Figure 25.1
central nervous system
  brain
  spinal cord
peripheral nervous system
afferent division
sensory receptors
efferent division
effectors
autonomic nervous system
  somatic nervous system
  cranial nerves
  spinal nerves

HISTOLOGY OF THE NERVOUS SYSTEM

• Figures 25.2–25.3
neuroglia
  glial cells
  astrocytes
  oligodendrocytes
  microglia
  ependyma
  ganglia
  satellite cell
  Schwann cell

neurons
• Figures 25.4–25.7
  dendrite
  cell body
  soma
  perikaryon
  Nissl bodies
  axon hillock
  initial segment
  axon
  myelin sheath
  axolemma
  neurilemma
  myelinated internode
  node of Ranvier
  collateral branch
  telodendria
  synaptic knob
  synaptic vesicle
  synaptic cleft
  neurotransmitter

Name _____   Date _____

Section _____   Number _____

NERVOUS SYSTEM OVERVIEW

Laboratory Report Exercise 25

A. Matching

Match the structure on the left with the correct description on the right.

1. _____ dendrite   A. free and fixed ribosomes
2. _____ axon   B. main branches of an axon
3. _____ perikaryon   C. outer Schwann cell membrane
4. _____ collateral branches   D. vacuole of neurotransmitters
5. _____ synaptic knob   E. brain and spinal cord
6. _____ axon hillock   F. fine branches of axon
7. _____ sensory receptor   G. axon region with Schwann cell
8. _____ central nervous system   H. forms blood-brain barrier
9. _____ Nissl bodies   I. directs impulses to cell body
10. _____ telodendria   J. surrounds nucleus
11. _____ astrocyte   K. swollen end of axon
12. _____ myelinated internode   L. connects soma to axon
13. _____ neurilemma   M. membrane of axon
14. _____ synaptic vesicles   N. detects changes in environment
15. _____ axolemma   O. single extension from cell body

B. Short-Answer Questions

1. Compare the central nervous system to the peripheral nervous system.

2. Describe the microscopic appearance of an astrocyte.

3. What type of molecules are stored in the synaptic knob?

4. List two functions of Schwann cells.

5. List the similarities and differences between spinal and cranial nerves.

6. List six types of neuroglial cells and indicate which cells are found in the CNS and PNS.

C. **Drawing**

Sketch a multipolar neuron, and label the dendrites, soma, axon hillock, initial segment, axon, collateral branches, telodendria, and synaptic knob.

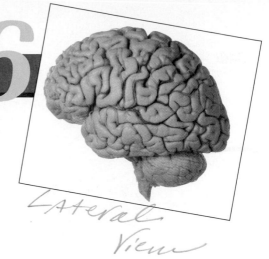

Lateral View

External Anatomy of the Brain

OBJECTIVES

On completion of this exercise, you should be able to:

- Name the three meninges that cover the brain.
- Describe the extensions of the dura mater.
- Identify the major regions of the brain and a basic function for each.
- Identify the surface features of each region of the brain.
- Identify the 12 pairs of cranial nerves.

WORD POWER

meninges (menin—membrane)
dura mater (dura—tough; mater—mother)
arachnoid (arachno—spider)
pia mater (pia—delicate)
cerebrum (cereb—the brain)
hemisphere (hemi—half)
folia (foli—leaf)

vermis (verm—worm)
medulla (medull—marrow)
pons (pons—bridge)
corpora (corp—body)
quadrigemina (quadrigemina—four twins)
thalamus (thala—chamber)

MATERIALS

brain models and charts
meningeal model
sheep brain
dissection equipment
gloves
safety glasses

INTRODUCTION

The brain is one of the largest organs in the body. It weighs approximately 3 lb and occupies the cranial cavity of the dorsal body cavity. Billions of synapses between neurons form a vast biological circuitry which no electronic computer will ever surpass. Every second the brain performs a huge number of calculations and interpretations and coordinates visceral activity to maintain homeostasis.

A. Protection and Support of the Brain

The brain and spinal cord of the central nervous system are protected by specialized membranes called **meninges (men-IN-jēz)**, a series of three membranes that anchor the CNS securely in place. Between certain meningeal layers is **cerebrospinal fluid** (CSF) which cushions the brain and prevents it from contacting the cranial bones during a head injury, much like a car's airbag prevents a passenger from hitting the dashboard. The **cranial meninges** surrounding the brain include the following layers: the **dura mater (DOO-ruh MÃ-ter)**, the **arachnoid (a-RAK-noyd)**, and the **pia (PI-uh) mater** (see Figure 26.1).

The **dura mater**, the outer meningeal covering, consists of two tissue layers. The **endosteal** portion is the outer layer and is fused with the periosteum of the flat cranial bones; the **meningeal** layer faces the arachnoid membrane. Between the inner and outer dural layers are large blood sinuses, collectively called **dural sinuses**, which drain blood from cranial veins into the jugular veins. The **superior** and **inferior sagittal sinuses** are large veins in the dura between the two hemispheres of the cerebrum.

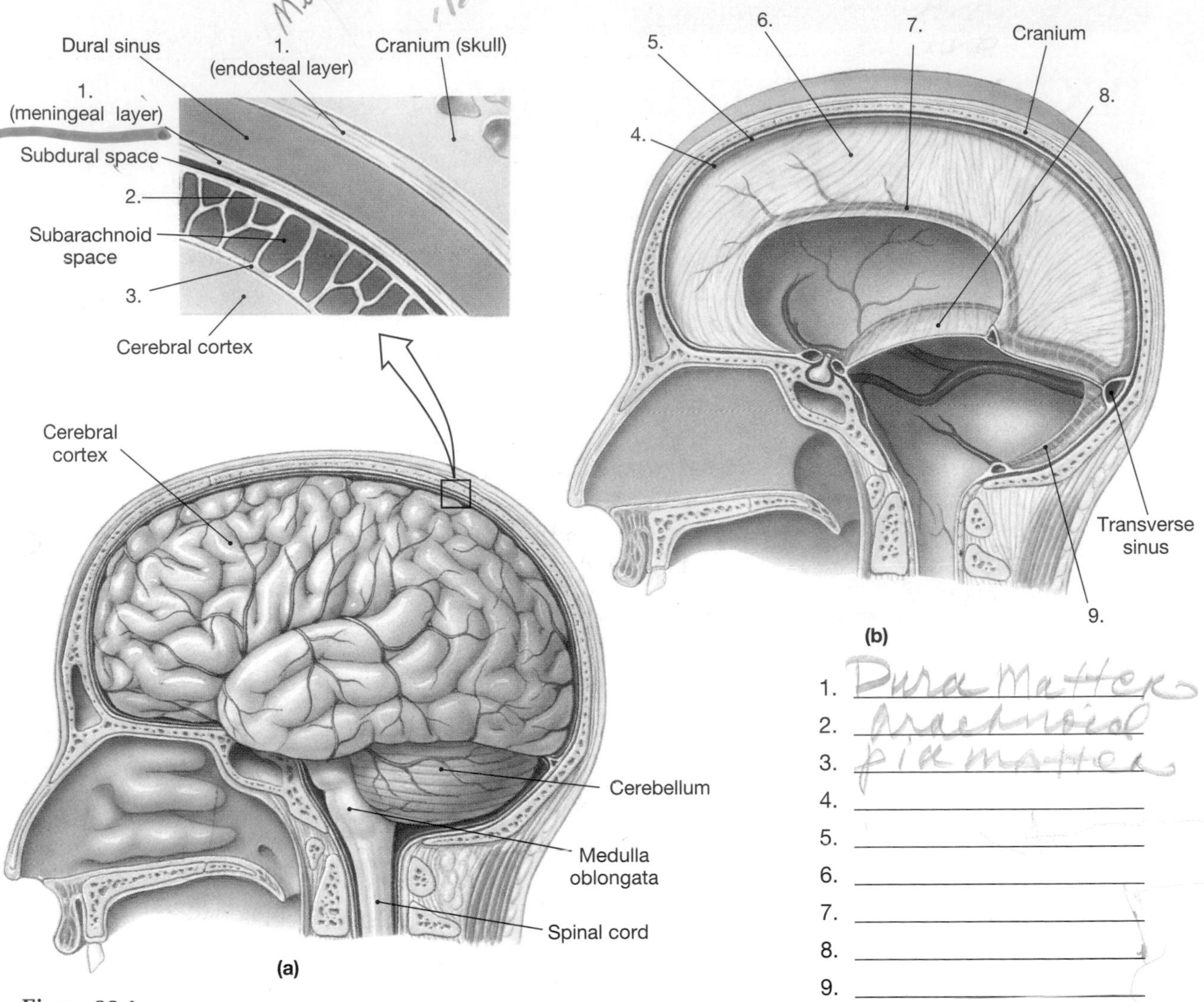

hard mother
Meninx — Dora madek — mostly collogen
Bone

(a)

(b)

Dural sinus
1. (endosteal layer)
Cranium (skull)
1. (meningeal layer)
Subdural space
2.
Subarachnoid space
3.
Cerebral cortex

Cerebral cortex
Cerebellum
Medulla oblongata
Spinal cord

5.
6.
7.
Cranium
4.
8.
Transverse sinus
9.

1. *Dura Matter*
2. *Arachnoid*
3. *pia matter*
4. _____
5. _____
6. _____
7. _____
8. _____
9. _____

Figure 26.1
The Relationship among the Brain, Cranium, and Meninges.
(a) A lateral view of the brain, showing its position in the cranium and the organization of the meninges. **(b)** A diagrammatic view, showing the orientation of the falx cerebri, tentorium cerebelli, and falx cerebelli.

The dura mater has extensions to further stabilize the brain. A midsagittal fold in the dura mater forms the **falx cerebri (FALKS ser-Ē-brē)** and separates the right and left hemispheres of the cerebrum. Posteriorly, the dura mater folds again as the **tentorium cerebelli (ten-TOR-ē-um ser-e-BEL-ē)** and separates the cerebellum from the cerebrum. The **falx cerebelli** is a dural fold between the hemispheres of the cerebellum. Locate these structures in Figure 26.1.

Deep to the dura mater is the **arachnoid,** named after the weblike connection this membrane has with the underlying pia mater. Between the dura and the arachnoid is the **subdural space**. The arachnoid forms a smooth covering over the brain.

Directly on the surface of the brain is the **pia mater** which contains many blood vessels for the brain. Between the arachnoid and pia mater is the **subarachnoid space** where **cerebrospinal fluid** circulates. The fluid cushions the brain through buoyancy and protects it from contacting the hard, bony wall of the cranium. Cerebrospinal fluid contains nutrients supplied by the blood to nourish the brain. The chambers in the brain, called ventricles, also contain CSF.

LABORATORY ACTIVITY   PROTECTIVE COVERINGS OF THE BRAIN

MATERIALS

brain models

charts

PROCEDURES

1. Label and review the meningeal anatomy presented in Figure 26.1.
2. Locate the dura mater, arachnoid, and pia mater on the available laboratory charts and models.

B. Major Regions and Surface Features of the Brain

The brain is divided into four major regions: (1) the **cerebrum (ser-ē-brum),** which is the large paired masses of the brain; (2) the **diencephalon (dī-en-SEF-a-lon),** which includes the thalamus and hypothalamus; (3) the **brain stem,** which is comprised of the midbrain, pons, and medulla oblongata; and (4) the **cerebellum.** Some anatomists divide the brain into six major regions by considering the components of the brain stem separately. Figure 26.2 highlights each region of the brain and summarizes its major functions.

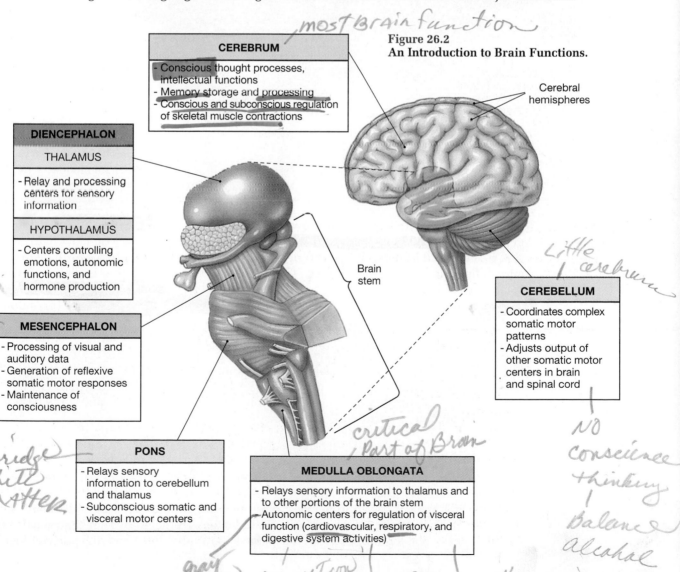

Figure 26.2
An Introduction to Brain Functions.

CEREBRUM
- Conscious thought processes, intellectual functions
- Memory storage and processing
- Conscious and subconscious regulation of skeletal muscle contractions

Cerebral hemispheres

DIENCEPHALON

THALAMUS
- Relay and processing centers for sensory information

HYPOTHALAMUS
- Centers controlling emotions, autonomic functions, and hormone production

MESENCEPHALON
- Processing of visual and auditory data
- Generation of reflexive somatic motor responses
- Maintenance of consciousness

Brain stem

CEREBELLUM
- Coordinates complex somatic motor patterns
- Adjusts output of other somatic motor centers in brain and spinal cord

PONS
- Relays sensory information to cerebellum and thalamus
- Subconscious somatic and visceral motor centers

MEDULLA OBLONGATA
- Relays sensory information to thalamus and to other portions of the brain stem
- Autonomic centers for regulation of visceral function (cardiovascular, respiratory, and digestive system activities)

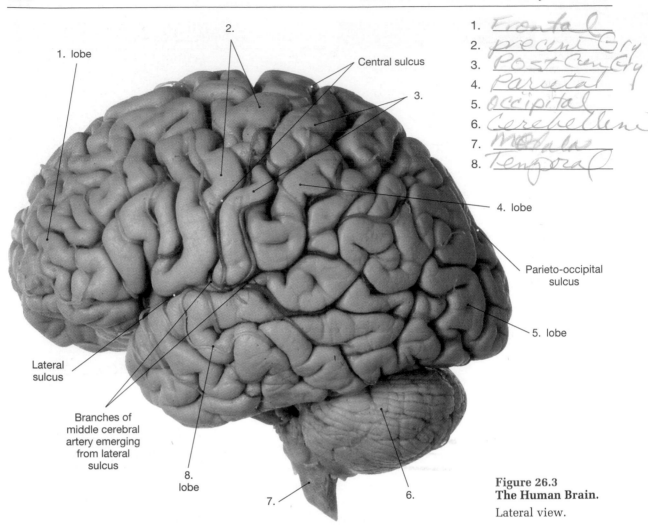

1. Frontal
2. precent Gry
3. Post Cen Gry
4. Parietal
5. occipital
6. Cerebellum
7. Medulla
8. Temporal

1. lobe
2.
Central sulcus
3.
4. lobe
Parieto-occipital
sulcus
5. lobe
Lateral
sulcus
Branches of
middle cerebral
artery emerging
from lateral
sulcus
8.
lobe
7.
6.

Figure 26.3
The Human Brain.
Lateral view.

Figure 26.3 shows a lateral view of the human brain. The **cerebrum**, or telencephalon, is the largest portion of the brain. In the superior view you can clearly see that it is divided into two lobes, the right and left **cerebral hemispheres**, by a deep fissure called the **longitudinal fissure**. The surface of each hemisphere is highly convoluted and folded as a result of its rapid embryonic growth. Each small fold is called a **gyrus (JĪ-rus)** (pl. gyri), and each small groove is called a **sulcus (SUL-cus)** (pl. sulci). Deeper grooves such as the one separating the two hemispheres are called **fissures**.

The **cerebellum** is under the posterior of the cerebrum. Directly anterior to the cerebellum are the components of the brain stem, the region where the brain and spinal cord meet. The slender **medulla oblongata**, or simply the **medulla**, connects the spinal cord to the brain. Superior to the medulla is the swollen **pons**. Above the pons is part of the **midbrain**, also called the mesencephalon. Anterior to the midbrain is the base of the **hypothalamus**, a portion of the diencephalon.

THE CEREBRUM

The cerebrum is the most complex part of the brain. Conscious thought, intellectual reasoning, memory processing and storage take place in the cerebrum.

The cerebrum consists of five lobes, most named for the overlying cranial bone (see Figure 26.3). The anterior cerebrum is the **frontal** lobe, and the prominent **central sulcus**, located approximately midposteriorly, separates the frontal lobe from the **parietal** lobe.

Figure 26.4
The Cerebral Hemispheres.
Major anatomical landmarks on the surface of the left cerebral hemisphere. Association areas are colored. To expose the insula, the lateral sulcus has been opened. The insert shows the multiple layers of neurons that are characteristic of the cerebral cortex.

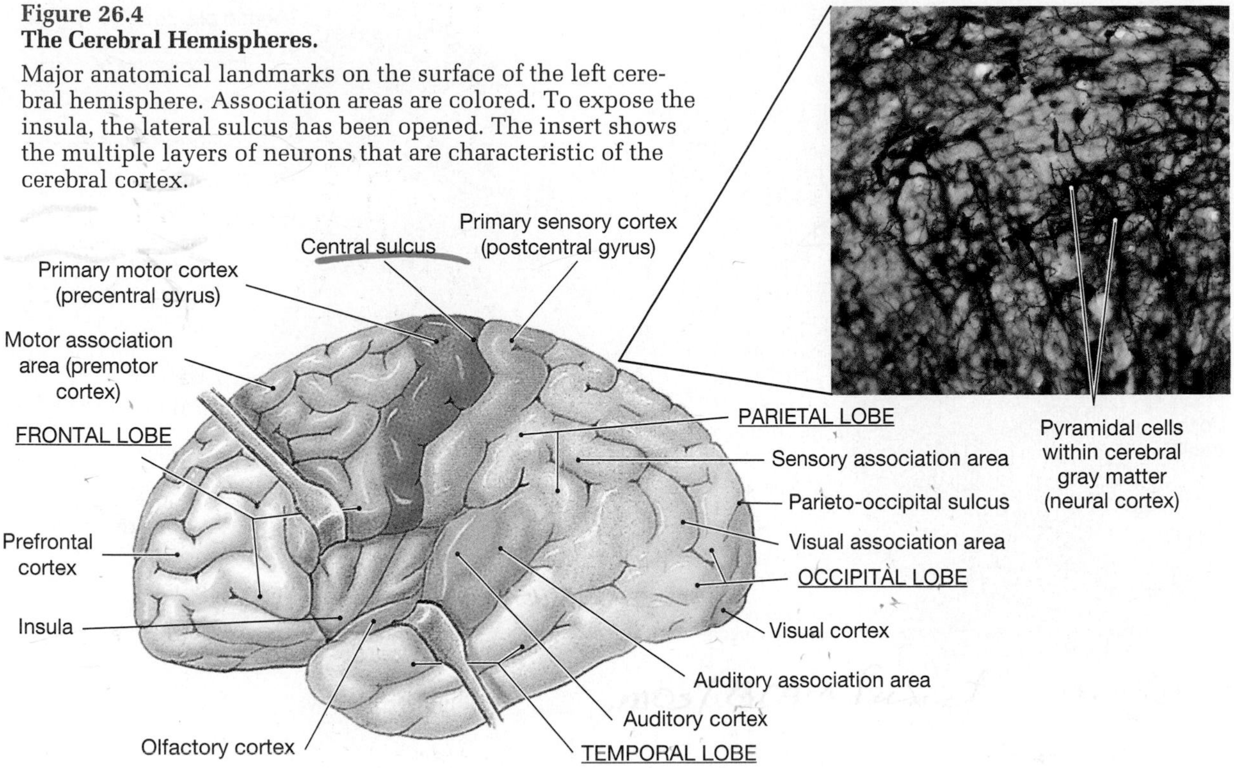

The **occipital** lobe corresponds to the position of the occipital bone of the posterior skull. The **lateral sulcus** defines the boundary between the large frontal lobe and the **temporal** lobe of the lower lateral cerebrum. Cutting into the lateral sulcus and peeling away the temporal lobe reveals a fifth lobe, the **insula.**

Regional specializations occur in the cerebrum (see Figure 26.4). The central sulcus divides motor and sensory regions of the cerebrum. Immediately anterior to the central sulcus is the **precentral gyrus.** This gyrus is the **primary motor cortex** where voluntary commands to skeletal muscles are generated. The **postcentral gyrus** on the parietal lobe is the **primary sensory cortex** where the general senses are perceived. Special senses such as sight and smell involve complex sensory information which requires more anatomical space in the brain. The occipital lobe contains the **visual cortex** where visual impulses from the eyes are interpreted. The temporal lobe houses the **auditory cortex** and the **olfactory cortex**.

Figure 26.4 also shows numerous **association areas**, regions that interpret sensory information from more than one sensory cortex or integrate motor commands into an appropriate response. The **premotor cortex** is a motor association area of the anterior frontal lobe. Auditory and visual association areas occur near the corresponding sensory cortex.

THE DIENCEPHALON

The diencephalon is embedded in the cerebrum and is only visible from the inferior aspect of the brain. Two major regions of the diencephalon are the **thalamus (THAL-a-mus)** and the **hypothalamus** (see Figure 26.5). The thalamus maintains a crude sense of awareness. All sensory impulses, except smell, pass into the thalamus and are relayed to the proper sensory cortex for interpretation. Nonessential sensory data are filtered out by the thalamus and do not reach the sensory cortex.

Only the exposed part of the diencephalon is the hypothalamus. Examine the inferior view of the brain as in Figure 26.7 and locate the pons and the midbrain. Just anterior to

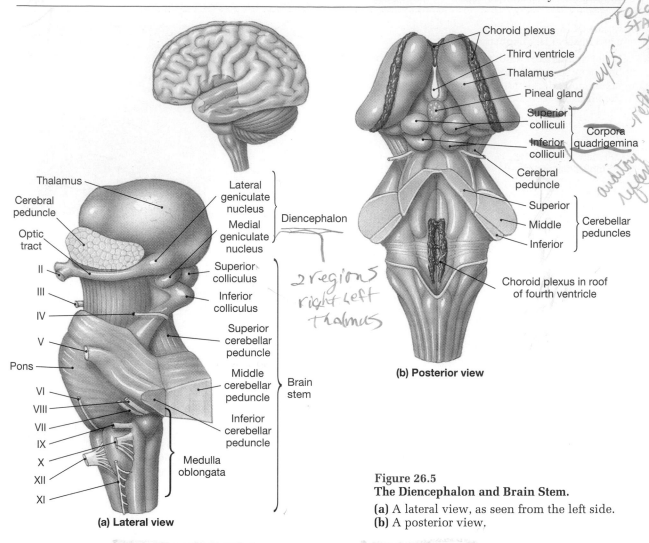

(handwritten margin notes)
Emotion
relay station sensory info
eyes
reflex centre
auditory reflex

Diencephalon

Thalamus
Cerebral peduncle
Optic tract
II
III
IV
V
Pons
VI
VIII
VII
IX
X
XII
XI

Lateral geniculate nucleus
Medial geniculate nucleus
Superior colliculus
Inferior colliculus
Superior cerebellar peduncle
Middle cerebellar peduncle
Inferior cerebellar peduncle
Medulla oblongata

Brain stem

(handwritten) 2 regions right left Thalmus

(a) Lateral view

Choroid plexus
Third ventricle
Thalamus
Pineal gland
Superior colliculi
Inferior colliculi
Corpora quadrigemina
Cerebral peduncle
Superior
Middle
Inferior
Cerebellar peduncles
Choroid plexus in roof of fourth ventricle

(b) Posterior view

Figure 26.5
The Diencephalon and Brain Stem.
(a) A lateral view, as seen from the left side.
(b) A posterior view.

the midbrain are a pair of rounded **mamillary (MAM-i-lar-ē) bodies** on the floor of the hypothalamus, which are hypothalamic nuclei that control eating reflexes for licking, chewing, sucking, and swallowing. The sheep brain has a single mammillary body, not a pair of bodies as in the human brain. Anterior to the mamillary bodies is the **infundibulum**, the stalk that attaches the **pituitary gland** to the hypothalamus. The **pineal gland** is a nipplelike structure superior to the midbrain.

THE BRAIN STEM

The midbrain

The midbrain is located inferior to the diencephalon. Although covered by the cerebrum, it can be observed by gently pushing down on the superior surface of the cerebellum and looking below the cerebrum. The **corpora quadrigemina (KOR-po-ra quad-ri-JEM-i-nuh)** is a series of four bulges next to the nipplelike pineal gland of the diencephalon (see Figure 26.5). The superior pair is the **superior colliculus (kol-IK-ū-lus)**, which functions as a visual reflex center which moves the eyeballs and the head to keep an object centered on the retina of the eye. The lower pair is the **inferior colliculus**, which functions as an auditory reflex center and moves the head to locate and follow sounds. The anterior midbrain between the pons and the hypothalamus consists of the **cerebral peduncles**, white fibers connecting the cerebral cortex with other parts of the brain.

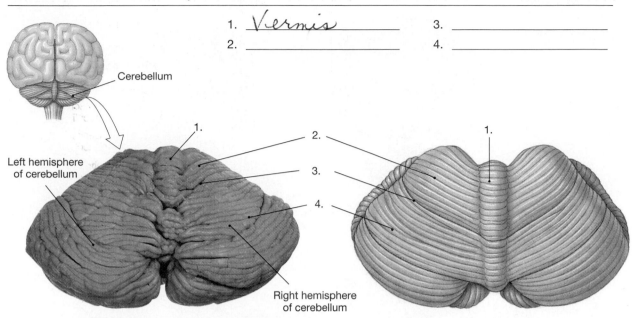

1. _Vermis_   3. _____

2. _____   4. _____

Cerebellum

Left hemisphere
of cerebellum

1.   2.   3.   4.   1.

Right hemisphere
of cerebellum

Figure 26.6
The Cerebellum.

The posterior, superior surface of the cerebellum, showing major
anatomical landmarks and regions.

The pons

The pons is superior to the medulla and functions as a relay station to direct sensory
information to the thalamus and cerebellum. It also contains certain sensory, somatic
motor, and autonomic cranial nerve nuclei. Locate the pons in Figure 26.5.

The medulla

The medulla connects the spinal cord with the brain (see Figures 26.5 and 26.7).
Descending communications between the brain and the spinal cord pass through the
medulla. The anterior surface of the medulla has two prominent folds called **pyramids**
where motor tracts cross over or decussate to the opposite side of the body. The right
and left sides of the brain have contralateral control of the opposite sides of the body
because sensory and motor tracts cross over or decussate from one side to the other in
the CWS. The medulla also functions as an autonomic center for visceral functions.
Nuclei in the medulla are vital *reflex centers* for the regulation of cardiovascular, respiratory, and digestive activities.

THE CEREBELLUM

The cerebellum (see Figure 26.6) is inferior to the occipital lobe of the cerebrum. Small
folds on the cerebellar cortex are called **folia (FŌ-lē-uh)**. Right and left **cerebellar hemispheres** are separated by a narrow **vermis**. Each cerebellar hemisphere consists of two
lobes: a smaller **anterior lobe** directly inferior to the cerebrum, and a **posterior lobe**.
The **transverse** or **primary fissure** separates the anterior and posterior cerebellar lobes.
In a sagittal section, a smaller **floculonodular (flok-ū-lō-NOD-ū-ler) lobe** is visible
internally where the fourth ventricle and the anterior wall of the cerebellum meet.

The cerebellum is primarily involved in the coordination of somatic motor functions, skeletal muscle contractions. Adjustments to postural muscles occur when
impulses from the vestibulocochlear cranial nerve (VIII) of the inner ear pass into the
floculonodular lobe where information concerning equilibrium is processed. Learned
muscle patterns, such as those involved in serving a tennis ball or playing the piano,
are stored and processed in the cerebellum.

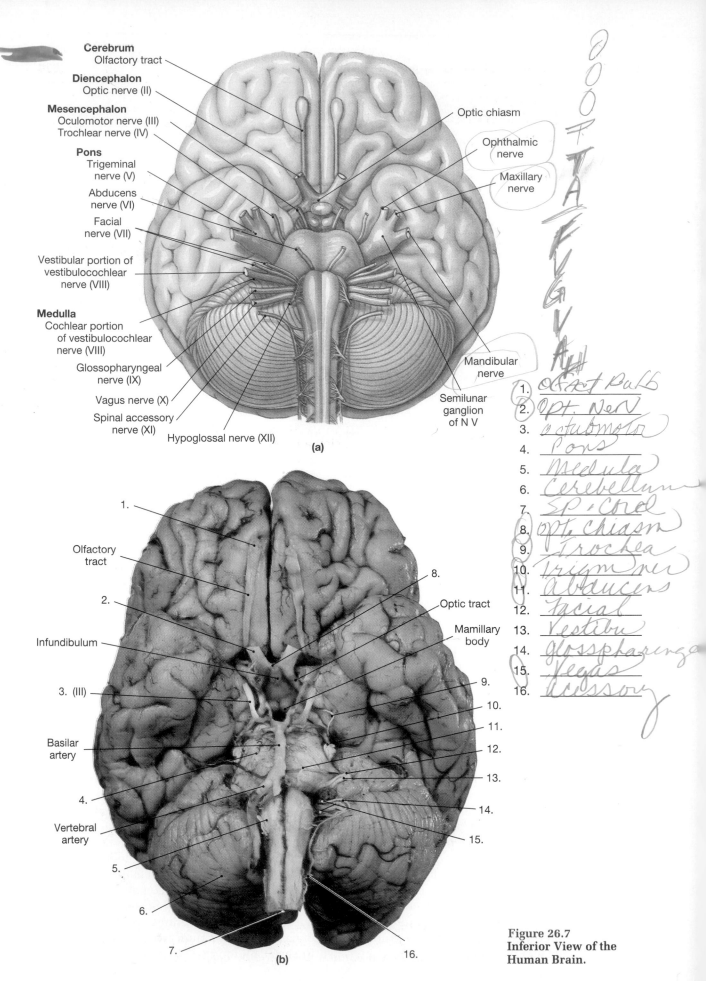

Cerebrum
Olfactory tract

Diencephalon
Optic nerve (II)

Mesencephalon
Oculomotor nerve (III)
Trochlear nerve (IV)

Pons
Trigeminal nerve (V)
Abducens nerve (VI)
Facial nerve (VII)
Vestibular portion of vestibulocochlear nerve (VIII)

Medulla
Cochlear portion of vestibulocochlear nerve (VIII)
Glossopharyngeal nerve (IX)
Vagus nerve (X)
Spinal accessory nerve (XI)
Hypoglossal nerve (XII)

Optic chiasm
Ophthalmic nerve
Maxillary nerve
Mandibular nerve
Semilunar ganglion of N V

(a)

1.
Olfactory tract
2.
Infundibulum
3. (III)
Basilar artery
4.
Vertebral artery
5.
6.
7.

8.
Optic tract
Mamillary body
9.
10.
11.
12.
13.
14.
15.
16.

(b)

Figure 26.7
Inferior View of the Human Brain.

354

1. Olfact bulb
2. Opt. Ner
3. octubmotor
4. Pons
5. Medula
6. Cerebellum
7. Sp. Cord
8. Opt. chiasm
9. Trochlea
10. Trigm ner
11. abducens
12. facial
13. vestibu
14. glosspharinga
15. Vegas
16. accssory

LABORATORY ACTIVITY EXTERNAL STRUCTURE OF THE BRAIN

1. Label Figures 26.3 and 26.6.
2. Locate each major region of the brain on lab models and charts.
3. Identify the surface features of each region on lab models.

C. **Cranial Nerves**

The brain and spinal cord of the central nervous system communicate with the rest of the body through cords of neurons called nerves. There are two major groups of nerves: spinal nerves and cranial nerves. As the name of each group implies, *spinal nerves* communicate with the CNS at the spinal cord, whereas cranial nerves interface with the brain. **Cranial nerves** emerge from the brain at specific anatomical locations and pass through various foramina of the skull to reach the peripheral structures they inner-vate. All spinal and cranial nerves occur in pairs. There are 31 pairs of spinal nerves and 12 pairs of cranial nerves.

Each cranial nerve pair is identified by name, Roman numeral, and type of infor-mation conducted. Some cranial nerves are entirely **sensory nerves**, but most are **mixed nerves** with both motor and sensory neurons. Cranial nerves that primarily conduct motor commands are considered **motor nerves** although they have a few sensory fibers to inform the brain about muscle tension and position.

The discussion of cranial nerves below describes both human and sheep brains. Figure 26.7 shows the position of each cranial nerve on the inferior surface of the brain. Table 26.1 summarizes the cranial nerves and includes the foramen each nerve passes through.

Olfactory nerve (I)

The olfactory nerve is composed of sensory fibers for the sense of smell and is located in the roof of the nasal cavity. The fibers terminate at the **olfactory bulb** which extends as the **olfactory tract**, part of the cerebrum, not the cranial nerve.

Optic nerve (II)

The optic nerve is a special sensory nerve that carries visual information. This nerve originates in the retina, the neural part of the eye that is sensitive to changes in light. The nerve is easy to identify on the human or sheep brain as the "X" at the **optic chiasm** inferior to the hypothalamus where some of the sensory fibers cross to the nerve on the opposite side of the brain. The optic nerve enters the thalamus which relays the visual signal to the occipital lobe. Some of the fibers enter the superior colliculus for visual reflexes.

Oculomotor nerve (III)

The oculomotor nerve innervates four extrinsic eyeball muscles—the superior, medial, and inferior rectus muscles and the inferior oblique muscle—and the levator palpebrae muscle of the eyelid. Autonomic motor fibers also control the intrinsic muscles of the iris and the ciliary body. The oculomotor nerve is located on the ventral midbrain just posterior to the optic nerve.

Trochlear nerve (IV)

The trochlear (TROK-lē-ar) motor nerve supplies motor fibers to the superior oblique muscle of the eye and originates where the midbrain joins the pons. The root of the nerve exits the midbrain on the lateral surface. Because it is easily cut or twisted off during

TABLE 26.1   The Cranial Nerves

Cranial Nerve (number)	Sensory Ganglion	Branch	Primary Function	Foramen	Innervation
Olfactory (I)			Special sensory	Cribriform plate of ethmoid bone	Olfactory epithelium
Optic (II)			Special sensory	Optic canal	Retina of eye
Oculomotor (III)			Motor	Superior orbital fissure	Inferior, medial, superior rectus, inferior oblique and levator palpebrae superioris muscles; intrinsic eye muscles
Trochlear (IV)			Motor	Superior orbital fissure	Superior oblique muscle
Trigeminal (V)	Semilunar		Mixed		Areas associated with the jaws
		Ophthalmic	Sensory	Superior orbital fissure	Orbital structures, nasal cavity, skin of forehead, upper eyelid, eyebrows, nose (part)
		Maxillary	Sensory	Foramen rotundum	Lower eyelid; superior lip, gums, and teeth; cheek, nose (part), palate and pharynx (part)
		Mandibular	Mixed	Foramen ovale	*Sensory:* inferior gums, teeth, lips; palate (part) and tongue (part). *Motor:* muscles of mastication
Abducens (VI)			Motor	Superior orbital fissure	Lateral rectus muscle
Facial (VII)	Geniculate		Mixed	Internal acoustic canal to facial canal; exits at stylomastoid foramen	*Sensory:* taste receptors on anterior 2/3 of tongue. *Motor:* muscles of facial expression, lacrimal gland, submandibular gland, sublingual salivary glands
Vestibulocochlear (Acoustic) (VIII)		Cochlear	Special sensory	Internal acoustic canal	Cochlea (receptors for hearing)
		Vestibular			Vestibule (receptors for motion and balance)
Glossopharyngeal (IX)	Superior (jugular) and inferior (petrosal)		Mixed	Jugular foramen	*Sensory:* posterior 1/3 of tongue; pharynx and palate (part); receptors for blood pressure, pH, oxygen, and carbon dioxide concentrations. *Motor:* pharyngeal muscles, and parotid salivary gland
Vagus (X)	Jugular and nodose		Mixed	Jugular foramen	*Sensory:* pharynx; auricle and external auditory canal; diaphragm; visceral organs in thoracic and abdominopelvic cavities. *Motor:* palatal and pharyngeal muscles, and visceral organs in thoracic and abdominopelvic cavities
Accessory (XI)		Medullary	Motor	Jugular foramen	Skeletal muscles of palate, pharynx, and larynx (with vagus nerve)
		Spinal	Motor	Jugular foramen	Sternocleidomastoid and trapezius muscles
Hypoglossal (XII)			Motor	Hypoglossal canal	Tongue musculature

removal of the dura, many sheep brains do not have this nerve intact. The superior oblique eye muscle passes through a trochlea or "pulley," hence the name of the nerve.

Trigeminal nerve (V)

The trigeminal (trī-JEM-i-nal) nerve is the largest of the cranial nerves. It is located on the lateral pons near the medulla and services much of the face. In life the nerve has three branches, the **opthalmic**, **maxillary**, and **mandibular branches**. The ophthalmic branch innervates sensory structures of the forehead, orbit, and nose. The maxillary branch contains sensory fibers for structures in the roof of the mouth including one-half of the maxillary teeth. The mandibular branch carries the motor portion of the nerve to the muscles of mastication. Sensory signals from the lower lip, gum, muscles of the tongue, and one-third of the mandibular teeth are also part of the mandibular branch. These branches are cut when the sheep brain is removed from the cranium and are not present on your study specimen.

Abducens nerve (VI)

The abducens (ab-DŪ-senz) nerve controls the lateral rectus extrinsic muscle of the eye. When this muscle contracts, the eyeball is abducted, hence the name "abducens." The nerve originates on the medulla and is positioned posterior and medial to the trigeminal nerve.

Facial nerve (VII)

It is located on the medulla, posterior and lateral to the abducens nerve. The facial nerve is a mixed nerve with sensory fibers for the anterior two-thirds of the tongue's taste buds and somatic and autonomic motor fibers. The somatic motor neurons innervate the muscles of facial expression such as the zygomaticus muscle. Visceral motor neurons control secretions of the salivary glands, lacrimal (tear) glands, and nasal mucous glands.

Vestibulocochlear nerve (VIII)

The vestibulocochlear nerve, also called the auditory nerve, is a special sensory nerve of the inner ear located on the medulla near the facial nerve. This nerve has two branches. The **vestibular branch** gathers information regarding the sense of balance from the vestibule and semicircular canals of the inner ear. The **cochlear branch** conducts auditory sensations from the cochlea, the organ of hearing in the inner ear.

Glossopharyngeal nerve (IX)

The glossopharyngeal (glos-ō-fah-RIN-jē-al) nerve is a mixed nerve of the tongue and throat. It supplies the medulla with sensory information from the posterior one-third of the tongue (remember the facial nerve innervates the anterior two-thirds of the tastebuds) and from the palate and pharynx. This nerve also conveys barosensory and chemosensory information from the carotid sinus and the carotid body where blood pressure and dissolved blood gases are monitored, respectively. Motor innervation by the glossopharyngeal nerve controls the pharyngeal muscles involved in swallowing and in the secretion of salivary glands. This nerve is difficult to identify on the sheep brain and may have been removed inadvertently during preparation of the specimen.

Vagus nerve (X)

The vagus (VĀ-gus) nerve is a complex nerve on the medulla which has mixed sensory and motor functions. Sensory information from the pharynx, diaphragm, and most of the internal organs of the thoracic and abdominal cavities ascends along the vagus nerve and synapses with autonomic nuclei in the medulla. The motor portion controls the involuntary muscles of the respiratory, digestive, and cardiovascular systems. It is the only cranial nerve to descend below the neck. The vagus nerve enters the ventral body cavity, however, it does not pass to the thorax via the spinal cord, rather it follows the musculature of the neck. Because this nerve regulates the activities of the internal organs of the thoracic and abdominal cavities, disorders of the nerve result in systemic disruption of homeostasis. Parasympathetic fibers in the vagus nerve control swallowing, digestion, heart rate, and respiratory patterns, and if this control is compromised, sympathetic stimulation goes unchecked and the organs respond as during exercise or stress. The cardiovascular and respiratory systems increase their activities, whereas the digestive system shuts down.

Spinal accessory nerve (XI)

The spinal accessory nerve, or just the accessory nerve, is a motor nerve controlling the skeletal muscles for swallowing and the sternocleidomastoid and trapezius muscles of the neck. It is the only cranial nerve with fibers originating from both the medulla and the spinal cord. Numerous threadlike branches from the medulla and the spinal cord unite in a single nerve.

Hypoglossal nerve (XII)

The hypoglossal (hī-pō-GLOS-al) nerve is positioned more to the midline than the other cranial nerves on the medulla. This motor nerve supplies motor fibers that control tongue movements for speech and swallowing.

LABORATORY ACTIVITY   CRANIAL NERVES AND NEURAL FUNCTION

MATERIALS

models
charts

PROCEDURE

1. Review the cranial nerves presented in Figure 26.7.
2. Locate each cranial nerve on lab models and charts.
3. Complete the following questions:
   a. Imagine watching a bird fly across your line of vision. What part of the brain is active in keeping an image of the moving object on the retina of the eyeball?

   _____

   b. You have just eaten a medium-sized pepperoni pizza and lain down to digest it. What cranial nerve stimulates the muscular activity of your digestive tract?

   _____

   c. As you throw a ball right-handed, in what part of the brain is the motor command initiated and where does it cross over to the right side of the body?

   _____

   d. You walk into an old building and immediately smell a dusty odor. After 20 min of exploring you realize that you no longer notice the smell. What part of the brain is responsible for this apparent loss of sensation?

   _____

   _____

    e. A child is preoccupied with a large cherry lollipop. What part of the child's brain is responsible for the licking and eating reflexes?

    _____

    f. Your favorite movie has made you cry yet again. What cranial nerve is responsible for the release of tears?

    _____

    g. A patient is brought into the emergency room with severe whiplash. He is not breathing and is losing cardiac function. What part of the brain has most likely been damaged?

    _____

    h. A woman is admitted to the hospital with *Bell's palsy* caused by an inflamed facial nerve. How does this affect the patient and how would you test her facial nerve?

    _____

    _____

D. Sheep Brain Dissection

The sheep brain, like all mammalian brains, is similar in structure and function to the human brain. One major difference between the human brain and that of other animals is the orientation of the brain stem. Humans are vertical animals and walk on two legs. The spinal cord is perpendicular to the ground, and so the brain stem must also be vertical. In four-legged animals the spinal cord and brain stem are parallel to the ground.

All vertebrate animals, sharks, fish, amphibians, reptiles, birds, and mammals, have a brain stem for basic body functions. These animals can learn through experience, a complex neurological process that requires higher-level processing and memory storage as in the human cerebrum. Imagine the complex motor activity necessary for locomotion in these animals.

Dissecting a sheep brain will enhance your study of models and charts of the human brain. Take your time during the dissection and follow the directions carefully. Refer to the manual and figures often during the procedures.

LABORATORY ACTIVITY   SHEEP BRAIN DISSECTION

MATERIALS

    gloves, safety glasses
    sheep brain (preferably with dura intact)
    dissection pan
    scissors
    scalpel
    forceps
    probe

PROCEDURES

I. Removal and Examination of the Meninges

Put on your gloves and safety glasses before you open a storage container of brains or handle a brain. A variety of preservatives are used on biological specimens, some of which may cause irritation to your skin and mucous membranes. Your lab instructor will inform you how to handle, dissect, and finally dispose of the sheep brain.

If your sheep brain does not have the dura mater, proceed to part II of the dissection and identify the external anatomy of the brain.

1. Examine the intact dura mater on the sheep brain (see Figure 26.8). Locate the **falx cerebri** and the **tentorium cerebelli** from the dorsal surface of the dura mater. How does the tissue of the falx cerebri compare to the dura covering the hemispheres?

2. Determine if your brain specimen still has the **ethmoid bone**, a mass of bone on the anterior frontal lobe. If the bone is present, slip your probe between the bone and the dura. Carefully pull the bone off the specimen, using your scissors to snip away any attached dura. Examine the removed ethmoid and identify the **crista galli**, the crest of bone where the meninges attach.

3. Insert a probe between the dura and the brain to gently separate the two. Use your scissors sparingly to cut the dura away from the base of the brain. Take care not to cut or remove a cranial nerve.

4. The dura is strongly anchored at the tentorium cerebelli. To detach the dura in one piece, first loosen the dura from the rest of the brain and gather it at the tentorium. Grasp the dura at the tentorium deep between the cerebellum and cerebrum and pull it straight out.

5. Open the detached dura and identify the falx cerebri and tentorium cerebelli. Proceed to part II when ready.

II. Identification of External Brain Anatomy

1. Examine the cerebrum and identify the frontal, parietal, occipital, and temporal lobes. The insula is a deep lobe and is not visible externally. How deep is the longitudinal fissure separating the right and left cerebral hemispheres? Observe the gyri and sulci on the cortical surface. Examine the surface between sulci for the arachnoid and pia mater.

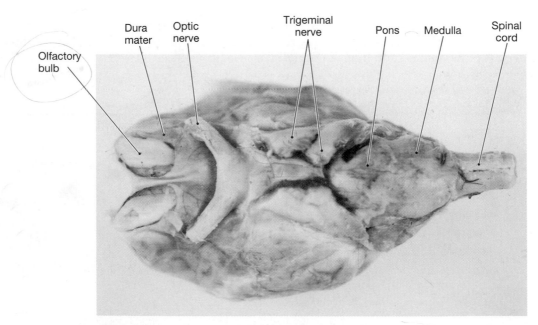

Figure 26.8
Ventral View of Sheep Brain Showing Dura Mater.

2. View the dorsal surface of the cerebellum. Compare the size of the folia with the gyri of the cerebrum. Unlike the human brain, the sheep cerebellum is not divided medially into two lateral hemispheres.

3. To examine the dorsal anatomy of the midbrain, hold the sheep brain as in Figure 26.9 and gently depress the cerebellum. The midbrain will then be visible between the cerebrum and cerebellum. Identify the medial pineal gland and the four elevated masses of the corpora quadrigemina. Distinguish between the superior colliculi and the inferior colliculi. What is the function of these masses?

4. Turn the brain to view the ventral surface of the brain as in Figure 26.10. Note how the spinal cord joins the medulla. Identify the pons and the cerebral peduncles of

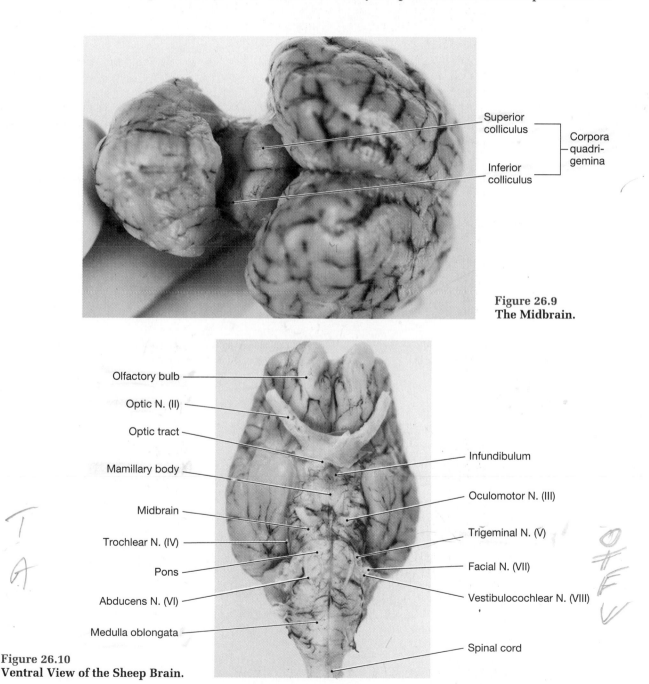

Figure 26.9
The Midbrain.

Figure 26.10
Ventral View of the Sheep Brain.

the midbrain. Locate the single mamillary body on the hypothalamus. The mamillary body of the human brain is a paired mass. The pituitary gland has most likely been removed from your specimen; however, you can still identify the stub of the infundibulum that attaches the pituitary to the hypothalamus.

5. Identify as many cranial nerves on your sheep brain as possible. Nerves I through III and V are usually intact and easy to identify. Your laboratory instructor may ask you to observe several sheep brains in order to study all the cranial nerves.

6. The internal anatomy of the sheep brain is presented in Exercise 27. Discard or store your sheep brain as indicated by your laboratory instructor. Proper disposal of all preserved material protects the local environment and is mandated by state and federal regulations.

EXTERNAL BRAIN ANATOMY CHECKLIST

This is a list of bold terms presented in Exercise 26. Use this list for review purposes after you have completed the laboratory activities.

PROTECTION AND SUPPORT

- Figure 26.1
cranial meninges
  dura mater
    endosteal layer
      dural sinus
      superior sagittal sinus
      inferior sagittal sinuses
    meningeal layer
  falx cerebri
  tentorium cerebelli
  falx cerebelli
  subdural space
  arachnoid
    subarachnoid space
    cerebrospinal fluid
  pia mater

CEREBRUM

- Figures 26.2–26.4
cerebrum
cerebral hemispheres
longitudinal fissure
gyrus
sulcus
fissures
frontal lobe
parietal lobe
occipital lobe
temporal lobe
insula
lateral sulcus
central sulcus
precentral gyrus
primary motor cortex
postcentral sulcus

primary sensory cortex
visual cortex
auditory cortex
olfactory cortex
association areas
premotor cortex

DIENCEPHALON

- Figures 26.2, 26.5, and 26.7
thalamus, hypothalamus
mamillary bodies
infundibulum
pituitary gland
pineal gland

BRAIN STEM

- Figures 26.2, 26.5, and 26.7
Midbrain
  corpora quadrigemina
  superior colliculus
  inferior colliculus
  cerebral peduncles
Pons and Medulla
  pons
  medulla
    pyramids

CEREBELLUM

- Figures 26.2, 26.3, 26.6, and 26.7
cerebellar hemispheres
  folia
vermis
anterior lobe, posterior lobe
transverse (primary) fissure
floculonodular lobe

CRANIAL NERVES

• Figures 26.5 and 26.7
sensory nerves
mixed nerves
motor nerves
olfactory nerve (I)
  olfactory bulb
  olfactory tract
optic nerve (II)
  optic chiasm
oculomotor nerve (III)
trochlear nerve (IV)
trigeminal nerve (V)
  opthalmic branch
  maxillary branch
  mandibular branch
abducens nerve (VI)
facial nerve (VII)
vestibulocochlear nerve (VIII)
  vestibular branch
  cochlear branch
glossopharyngeal nerve (IX)
vagus nerve (X)
spinal accessory nerve (XI)
hypoglossal nerve (XII)

SHEEP BRAIN

• Figures 26.8, 26.9, and 26.10
dura mater
falx cerebri
tentorium cerebelli
Cerebrum
longitudinal fissure
cerebral hemispheres
  frontal lobe
  parietal lobe
  temporal lobe
  occipital lobe
  insula
Diencephalon
thalamus
hypothalamus
mamillary body
pituitary gland
  infundibulum
pineal gland
Brain Stem
  Midbrain
    corpora quadrigemina
      superior colliculus
      inferior colliculus
  Pons
  Medulla
Cerebellum
  folia
Cranial Nerves I–XII

EXTERNAL ANATOMY OF THE BRAIN
Laboratory Report Exercise 26

A. Matching
Match each brain structure listed on the left with the correct description on the right.

1. _____ folia		A. ridge of cerebral cortex
2. _____ cerebrum		B. area where optic nerve crosses
3. _____ mamillary body		C. forms floor of the diencephalon
4. _____ longitudinal fissure		D. part of the midbrain
5. _____ gyrus		E. outer meningeal layer
6. _____ inferior colliculus		F. site of cerebrospinal fluid circulation
7. _____ optic chiasm		G. small folds on the cerebellum
8. _____ falx cerebri		H. cranial nerve III
9. _____ hypothalamus		I. narrow central region of cerebellum
10. _____ central sulcus		J. separates cerebellum and cerebrum
11. _____ cerebral peduncles		K. mass posterior to infundibulum
12. _____ dura mater		L. area of brain superior to medulla
13. _____ sulcus		M. lobe deep to the temporal lobe
14. _____ vagus nerve		N. divides motor and sensory cortex
15. _____ vermis		O. tissue between cerebral hemispheres
16. _____ subarachnoid space		P. contains five lobes
17. _____ pons		Q. part of the corpora quadrigemina
18. _____ tentorium cerebelli		R. cranial nerve X
19. _____ insula		S. cleft between cerebral hemispheres
20. _____ oculomotor nerve		T. furrow between gyrus

B. Short-Answer Questions
1. List the five major regions of the brain.

2. Which cranial nerves conduct the sensory and motor impulses of the eye?

3. List the location and specific anatomy of the corpora quadrigemina.

4. Describe the extensions of the dura mater.

5. What is the function of the precentral gyrus?

6. Which part of the brain is the visual cortex?

C. Labeling

Label Figure 26.11, which shows the inferior surface of the brain.

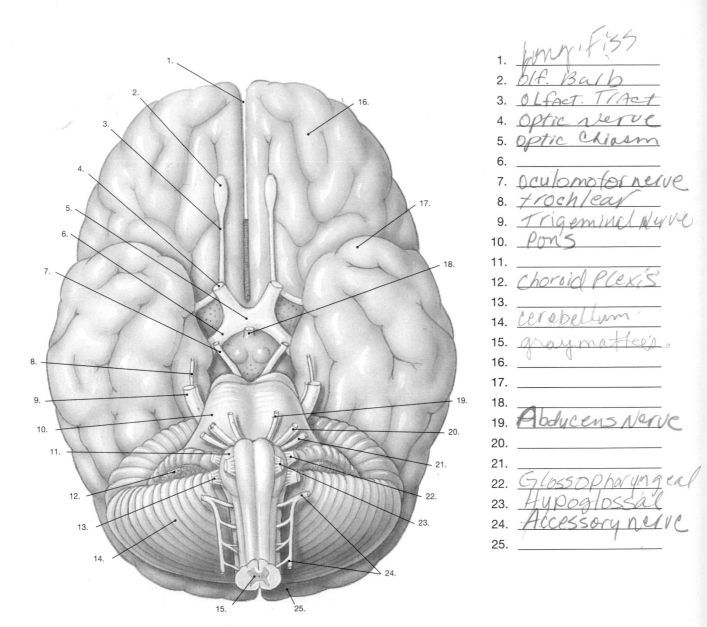

1. _long. fiss_
2. _olf. Balb_
3. _Olfact. Tract_
4. _optic nerve_
5. _Optic Chiasm_
6. _____
7. _oculomotor nerve_
8. _trochlear_
9. _Trigeminal Nerve_
10. _Pons_
11. _____
12. _choroid Plexis_
13. _____
14. _cerebellum_
15. _gray matters_
16. _____
17. _____
18. _____
19. _Abducens Nerve_
20. _____
21. _____
22. _Glossopharyngeal_
23. _Hypoglossal_
24. _Accessory nerve_
25. _____

Figure 26.11
The Inferior Aspect of the Brain.

External anatomy of the inferior surface, diagrammatic view.

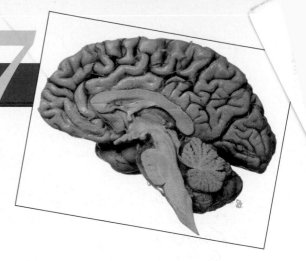

Internal Anatomy of the Brain

OBJECTIVES

On completion of this exercise, you should be able to:

- Identify the major internal features of the cerebrum, diencephalon, midbrain, brain stem, and cerebellum.
- Identify the anatomy of the human brain and the sheep brain in a frontal section.
- Trace a drop of cerebrospinal fluid through the ventricular system of the brain.

WORD POWER

septum pellucidum (septum—fence; pellu—transparent)

cerebral aqueduct (aquaduct—water canal)

choroid plexus (choroid—membrane)

corpus callosum (corp—body; callo—thick-skinned)

anterior commissure (commis—united)

fornix (forn—arch)

superior colliculus (collicul—small hill)

cerebral peduncles (peduncul—little foot)

arbor vitae (arbor—tree; vitae—life)

MATERIALS

human brain models
preserved sheep brain
dissection equipment
gloves
safety glasses

INTRODUCTION

To appreciate the internal architecture of the brain you will examine brain models and dissect a sheep brain along the midsagittal plane. Deeper structures of the cerebrum and diencephalon will be exposed by a frontal section. This exercise is organized for study of the human brain first, using laboratory models, followed by dissection of the sheep brain.

A. Sectional Anatomy of the Human Brain

Cerebrum

Figure 27.1 shows the midsagittal anatomy of the human brain. The cerebral hemispheres are connected by a tract of white matter called the **corpus callosum**. It is easily identified as the curved white structure at the base of the cerebrum. The inferior portion of the corpus callosum is the **fornix (FOR-niks)**, a white tract connecting deep structures of the limbic system, the "emotional" brain. The fornix narrows anteriorly and meets the **anterior commissure (kom-MIS-sur)**, another tract of white matter interconnecting the cerebral hemispheres.

Diencephalon

The diencephalon comprises the thalamus and hypothalamus, both visible in Figure 27.1. Inferior to the fornix is the most distinctive feature of the diencephalon, the round **intermediate mass**. The **thalamus** is shaped like a barbell with two oval masses on each end connected by the intermediate mass in the middle. The slight depression around

1. pre gyrus
2. corpus collosum
3. Anterior commisur
4. mam. Body
5. pons
6. med. ob.
7. postcent gyrus
8. Fornex
9. rt thalamus
10. pineal gand
11. thalamus
12. super colic
13. Inf colic
14. corp. quad.
15. aquiduct
16. cerebellum
17. 4th vent.

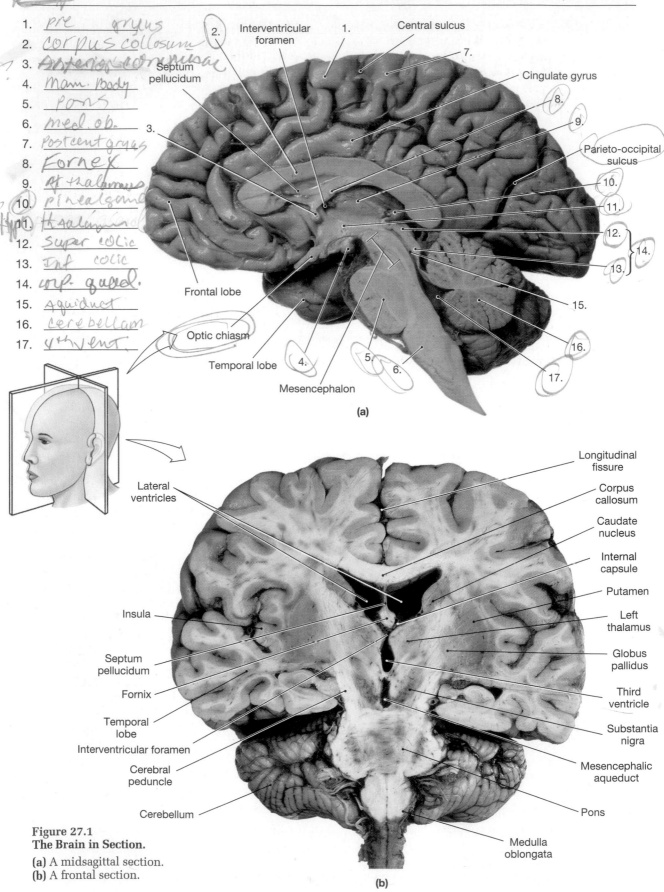

Figure 27.1
The Brain in Section.

(a) A midsagittal section.
(b) A frontal section.

the intermediate mass is the space of the **third ventricle** which passes through the diencephalon. Within the third ventricle lies a **choroid plexus,** a capillary network that produces cerebrospinal fluid. Below the thalamus and intermediate mass is the **hypothalamus.** Just posterior to the hypothalamus is the **mamillary body,** a small elevation anterior to the narrowing of the midbrain. Anterior to the mamillary body is the **infundibulum,** which supports the **pituitary gland.** Under the posterior margin of the corpus callosum is the conical **pineal gland.**

Midbrain

The midbrain is clearly visible in the midsagittal section in Figure 27.1. Posterior to the pineal gland is the **corpora quadrigemina** which includes the **superior colliculus** and the **inferior colliculus.** A long canal called the **cerebral aqueduct** passes anterior to these masses. **Cerebral peduncles** are located within the walls of the cerebral aqueduct. Their fibers extend to connect the pons with the cerebrum.

Cerebellum

The cerebellum is positioned posterior to the pons and medulla. In a sagittal section the white matter of the **cerebellum** is apparent. Because the tissue is highly branched, it is called the **arbor vitae,** the "tree of life" (see Figures 27.1 and 27.2). In the middle of the arbor vitae are the **cerebellar nuclei** which function in the involuntary regulation of skeletal muscle contraction. The fourth ventricle is anterior to the human cerebellum.

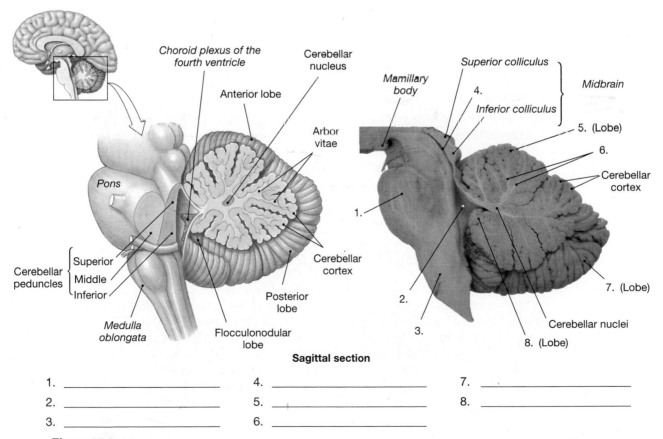

1. _____   4. _____   7. _____

2. _____   5. _____   8. _____

3. _____   6. _____

Figure 27.2
The Cerebellum.

A sectional view of the cerebellum, showing the arrangement of gray matter and white matter.

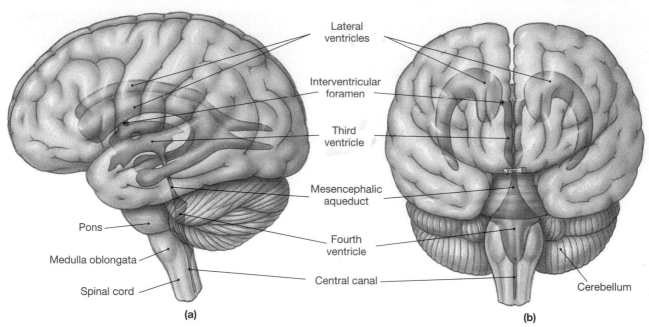

Figure 27.3
Ventricules of the Brain.

The orientation and extent of the ventricles as they would appear if the brain were transparent. **(a)** A lateral view. **(b)** An anterior view.

Each of the cerebellar hemispheres is divided into the **anterior lobe**, the **posterior lobe**, and the **flocculonodular lobe** which faces the fourth ventricle.

Brain stem

Pons and Medulla.   The enlarged area of the brain stem is the **pons**, located inferior to the cerebral peduncles. Below the pons is the **medulla,** which connects the spinal cord to the rest of the brain. Notice in Figure 27.1 how the brain stem tapers from the pons to the spinal cord. The **fourth ventricle** lies posterior to the brain stem between the medulla and the cerebellum.

Ventricular system of the brain

Deep in the brain are four chambers called **ventricles**, detailed in Figure 27.3. Each ventricle produces **cerebrospinal fluid** (CSF) which flows through the ventricles and enters the subarachnoid space surrounding the brain. Cerebrospinal fluid protects and nourishes the tissues of the brain. A pair of **lateral ventricles** extend deep into the cerebrum as horseshoe-shaped chambers. At the midline of the brain the lateral chambers are separated from one another by a thin membrane, the **septum pellucidum**. A brain sectioned at the midsagittal plane exposes this membrane. Cerebrospinal fluid circulates from the lateral ventricles through the **interventricular foramen** and enters the **third ventricle**. This small chamber surrounds the intermediate mass of the thalamus. CSF in the third ventricle passes through the **cerebral aqueduct (mesencephalic aquaduct)** of the midbrain and enters the **fourth ventricle** between the brain stem and the cerebellum. In the fourth ventricle two **lateral apertures** and a single **median aperture** direct CSF laterally to the exterior of the brain and into the subarachnoid space. CSF then circulates around the brain and spinal cord and then is reabsorbed from an **arachnoid granulation** which projects into the veins of the dural sinuses.

NOt

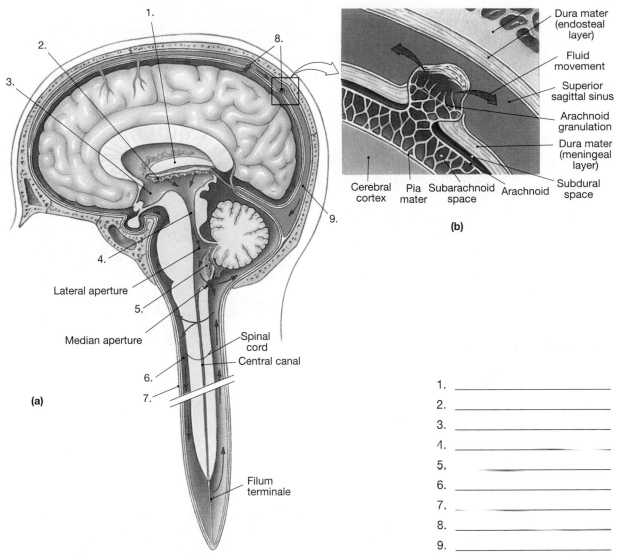

Dura mater
(endosteal
layer)

Fluid
movement

Superior
sagittal sinus

Arachnoid
granulation

Dura mater
(meningeal
layer)

Subdural
space

Cerebral   Pia   Subarachnoid   Arachnoid
cortex   mater   space

(b)

Lateral aperture

Median aperture

Spinal
cord

Central canal

Filum
terminale

(a)

1. _____
2. _____
3. _____
4. _____
5. _____
6. _____
7. _____
8. _____
9. _____

Figure 27.4
The Circulation of Cerebrospinal Fluid.

(a) A sagittal section indicating the sites of formation and routes of circulation of cerebrospinal fluid (red arrows). **(b)** The orientation of the arachnoid granulations.

STUDY TIP

Although all the ventricles produce CSF, it may help you remember the complete circulatory pathway if you start with a drop of CSF in a lateral ventricle and circulate it through the third and fourth ventricles.

Cerebrospinal fluid is produced in a specialized capillary network called the **choroid plexus** which is found in each chamber of the brain. The choroid plexus of the third ventricle passes through the interventricular foramen and expands to line the floor of the lateral ventricles, detailed in Figure 27.4. The choroid plexus of the fourth ventricle lies on the posterior wall of the ventricle. Each choroid plexus contributes to the formation of CSF. In the sheep brain, which is oriented on a horizontal axis, the fourth ventricle is inferior relative to the cerebellum.

No frontal section

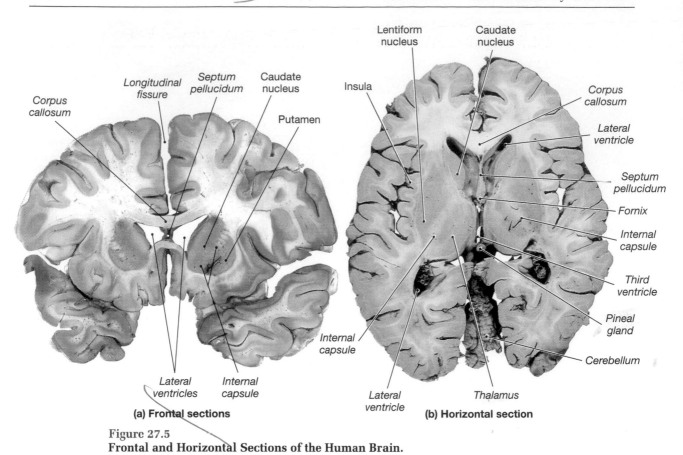

Figure 27.5
Frontal and Horizontal Sections of the Human Brain.

CLINICAL APPLICATION—HYDROCEPHALUS

The choroid plexes of an adult brain produces approximately 500 ml of cerebrospinal fluid daily. Since CSF is constantly being made, a volume equal to that produced must be removed from the CNS to prevent a buildup of fluid pressure in the ventricles. In an infant, if CSF production exceeds CSF reabsorption, the increase in cranial pressure expands the unfused skull, creating a condition called *hydrocephalus.* The result is an enlarged skull and possible brain damage due to high fluid pressure on the delicate neural tissues. Surgical installation of small tubes called shunts drains the excess CSF and reduces intracranial pressure.

Frontal and horizontal sections of the brain

Deep structures of the cerebrum and diencephalon are visible when the brain is sectioned on the frontal plane near the infundibulum, as in Figure 27.5. Notice in the figure how the corpus callosum transverses the brain to connect the two cerebral hemispheres. **Cerebral nuclei** and the **internal capsule** form the lateral walls of the lateral ventricles. These structures are involved in automating voluntary muscle contractions. The cerebral nuclei are masses of gray matter in each cerebral hemisphere. Each nucleus consists of a medial **caudate nucleus** and a lateral **putamen (pū-TĀ-men).** Between these gray structures lies the internal capsule, a band of white matter that connects the cerebrum to the diencephalon, brain stem, and the cerebellum. Locate these structures in the sectional views of the brain in Figure 27.5. Also, note the organization of **gray matter** and **white matter** in the brain. The covering of the cerebrum, the **cerebral cortex**, is gray matter. The corpus callosum and other bridging structures of the brain consist of white matter.

LABORATORY ACTIVITY   OBSERVATION OF THE HUMAN BRAIN

MATERIALS

human brain models and charts
ventricular system model
preserved and sectioned human brain (if available)

PROCEDURES

1. Review and label the anatomy presented in Figures 27.1–27.5.
2. Identify the structures of the cerebrum, diencephalon, midbrain, pons, medulla, and cerebellum. Which areas consist of gray matter?
3. Identify the components of the ventricular system and trace the circulation of cerebrospinal fluid on a brain model.

B. Dissection of the Sheep Brain

The external anatomy of the sheep brain and removal of the dura is presented in Exercise 26. In this exercise, you will section the sheep brain and study its internal organization.

LABORATORY ACTIVITY   DISSECTION OF THE SHEEP BRAIN

MATERIALS

gloves, safety glasses
sheep brain
dissection pan
scissors
large knife or scalpel
forceps, probe

PROCEDURES

Put on your gloves and safety glasses before you open a storage container of brains or handle a brain. A variety of preservatives are used on biological specimens, some of which may cause irritation to your skin and mucous membranes. Your lab instructor will inform you how to handle, dissect, and finally dispose of the sheep brain.

1. Lay the sheep brain in the dissecting pan. Place the blade of a large butcher knife in the anterior region of the longitudinal fissure, pointing the tip of the knife blade downward. With a single smooth motion, move the knife through the brain from anterior to posterior to separate the right and left halves along the midsagittal plane. If only a scalpel is available, attempt to dissect the brain with as few long, smooth cuts as possible.
2. Using Figure 27.6 as a guide, examine a section of the sheep brain and locate the following structures.

Cerebrum

Locate the **corpus callosum**, **anterior commissure**, and **fornix**, all white tracts interconnecting regions of the cerebrum. Gently slide a blunt probe inside the **lateral ventricle**. How deep does the ventricle extend into the cerebrum? Does the **septum pellucidum** cover the lateral ventricle between the corpus callosum and the fornix?

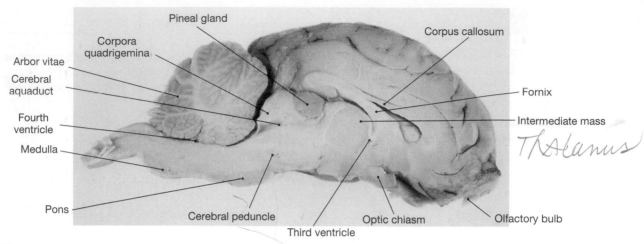

Figure 27.6
Midsagittal Section of the Sheep Brain.

Look inside the lateral ventricle and locate the **choroid plexus** which appears as a granular mass of tissue.

Diencephalon

Locate the **intermediate mass** surrounded by the **third ventricle**. The lateral walls of the third ventricle are the medial margins of the **thalamus**. Inferior to the intermediate mass is the **hypothalamus**. Identify the **infundibulum** and **pituitary gland** if they are still present on the sheep brain. The posterior mass of the hypothalamus is the **mamillary body**.

Cerebellum

The cerebellum of the sheep brain is located superior to the medulla and pons. In sagittal section the white matter of the **arbor vitae** is clearly visible. Between the cerebellum and brain stem is the **fourth ventricle**.

3. To view deep structures of the cerebrum and diencephalon make two frontal cuts on the sheep brain and examine the resulting slice of brain.

   Put the sheep brain halves together and lay them on the dissection pan with the ventral surface facing upward. Use a large knife to section the brain along a frontal plane through the infundibulum. Make another frontal cut to slice off a thin slab of brain. Lay the slab in the pan with the anterior surface upward, the area where you made your first cut. Using Figure 27.5 of the human brain, locate the following structures.

   Compare the distribution of **gray matter** and **white matter**. Notice how the **corpus callosum** joins each cerebral hemisphere. The **lateral ventricles** are separated by the thin **septum pellucidum**. Lateral to the ventricles are the gray nuclei of the **cerebral nuclei**. The **third ventricle** is inferior to the lateral ventricles.

4. Discard or store your sheep brain as indicated by your laboratory instructor. Proper disposal of all preserved material protects the local environment and is mandated by state and federal regulations.

BRAIN STEM

Midbrain

The **cerebral aqueduct** transverses the midbrain and separates the inferior **cerebral peduncles** from the upper midbrain. On the superior surface of the midbrain locate the **pineal gland** and the **corpora quadrigemina**. Distinguish between the **superior colliculus** and the **inferior collicullus** of the corpora quadrigemina.

Pons and Medulla

Posterior to the midbrain is the **pons** which can be identified as the swollen mass posterior to the cerebral peduncles. The pons narrows and joins the **medulla** which is connected to the **spinal cord**.

INTERNAL ANATOMY OF THE BRAIN CHECKLIST

This is a list of bold terms presented in Exercise 27. Use this list for review purposes after you have completed the laboratory activities.

CEREBRUM

• Figure 27.1
cerebrum
   corpus callosum
     fornix
anterior commissure

DIENCEPHALON

• Figure 27.1
diencephalon
thalamus
   intermediate mass
   third ventricle
hypothalamus
mamillary body
pineal gland

BRAIN STEM

Midbrain
• Figure 27.1
corpora quadrigemina
   superior colliculus
   inferior colliculus
cerebral aqueduct
cerebral peduncles

Pons and Medulla
• Figure 27.1
pons
medulla
   fourth ventricle

CEREBELLUM

• Figures 27.1 and 27.2
cerebellum
   arbor vitae
   anterior lobe
   posterior lobe
   floculonodular lobe

VENTRICULAR SYSTEM

• Figure 27.3
ventricles
   cerebrospinal fluid (CSF)
   choroid plexus
lateral ventricles
   septum pellucidum
interventricular foramen
third ventricle
cerebral aqueduct
fourth ventricle
   lateral apertures
   median aperture
arachnoid granulation

SECTIONAL ANATOMY OF THE BRAIN

• Figure 27.5
cerebral nuclei
   caudate nucleus
   putamen
internal capsule
white matter
gray matter
cerebral cortex

SHEEP BRAIN

• Figures 27.6

cerebrum
corpus callosum
anterior commissure
fornix
lateral ventricle
septum pellucidum
choroid plexus
diencephalon
intermediate mass
third ventricle
thalamus
hypothalamus
infundibulum
pituitary gland
mamillary body

cerebellum
arbor vitae
fourth ventricle
gray matter
white matter
brain stem
  midbrain
    cerebral aqueduct
    cerebral peduncles
    corpora quadrigemina
      superior colliculi
      inferior colliculi
  pons
  medulla

INTERNAL ANATOMY OF THE BRAIN

Laboratory Report Exercise 27

A. Matching

Match each brain structure listed on the left with the correct description on the right.

1. _____ third ventricle
2. _____ septum pellucidum
3. _____ thalamus
4. _____ corpus callosum
5. _____ pineal gland
6. _____ superior colliculus
7. _____ arachnoid granulation
8. _____ fornix
9. _____ hypothalamus
10. _____ cerebral aqueduct
11. _____ arbor vitae

A. part of the corpora quadrigemina
B. white tract between cerebral hemispheres
C. forms floor of the diencephalon
D. duct through the midbrain
E. white matter of cerebellum
F. site of cerebrospinal fluid reabsorption
G. white matter inferior to lateral ventricles
H. chamber of the diencephalon
I. nipplelike gland in diencephalon
J. forms the lateral walls of third ventricle
K. membrane that separates lateral ventricles

B. Short-Answer Questions

1. Describe the location of the four ventricles of the brain.

2. Describe the white tracts that interconnect the cerebral hemispheres.

3. Trace a drop of cerebrospinal fluid from a lateral ventricle to reabsorption at an arachoid granulation.

4. What is the function of the cerebral nuclei?

5. What is the function of the choroid plexus?

C. **Labeling**

Label Figure 27.7, a midsagittal close-up of the human brain.

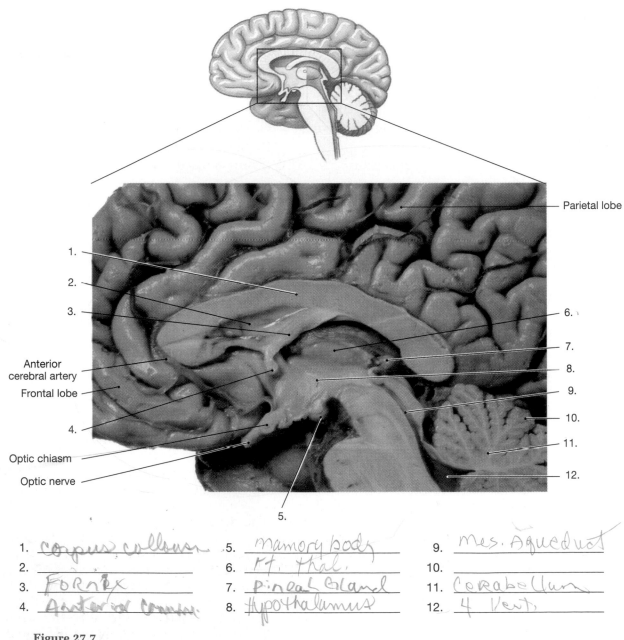

Figure 27.7
Detail of the Human Brain in Sagittal Section.
Midsagittal section through the diencephalon of human brain.

1. Corpus collosum   5. Mamory body   9. Mes. Aqueduct
2. _____   6. rt. thal.   10. _____
3. FORNIX   7. Pineal Gland   11. Cerebellum
4. Anterior comm.   8. Hypothalamus   12. 4 Vent.

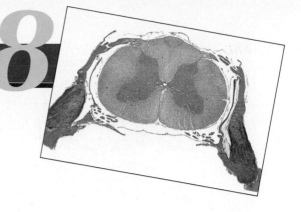

28

The Spinal Cord, Spinal Nerves, and Reflexes

OBJECTIVES

On completion of this exercise, you should be able to:

- Identify the major surface features of the spinal cord, including the spinal meninges.
- Identify the sectional anatomy of the spinal cord.
- Describe the organization and distribution of spinal nerves.
- List the events of a typical reflex arc.

WORD POWER

ganglion (gangia—knot on a string, swelling)
commissure (commis—united)
memnige (menin—membrane)
ramus (ram—branch)
sciatic (sciatic—of the hips)

MATERIALS

spinal cord models and charts
reflex hammer
microscope
microscope slides:
  spinal cord cross section
  peripheral nerve cross section

preserved cow spinal cord segment
dissection equipment
gloves
safety glasses

INTRODUCTION

The spinal cord connects the peripheral nervous system with the brain. In the PNS, peripheral nerves converge into spinal nerves and enter the spinal cord. Sensory neurons ascend the spinal cord and pass sensory information into the brain for interpretation. Motor neurons descend the spinal cord and carry motor commands from the brain to effectors, the muscles and glands of the body.

A. Gross Anatomy of the Spinal Cord

The spinal cord is continuous with the medulla of the brain. It descends approximately 45 cm (18 in.) in the spinal canal of the vertebral column and terminates between lumbar vertebrae L_1 and L_2. In young children, the spinal cord extends through most of the spine. After the age of 4, the spinal cord stops growing in length, yet the vertebral column continues to grow. By adulthood, the spinal cord is shorter than the spine and descends only to the level of the upper lumbar vertebrae.

Figure 28.1 presents the gross anatomy of the spinal cord. The diameter of the spinal cord is not constant along its length. Two swollen regions occur where spinal nerves of the limbs join the spinal cord. The **cervical enlargement** in the neck supplies nerves to the shoulders and arms. The **lumbar enlargement** occurs near the distal end of the cord where nerves supply the pelvis and lower limbs. Inferior to the lumbar enlargement, the spinal cord narrows and terminates at the **conus medullaris**. Spinal nerves fan out from the conus medullaris in a group called the **cauda equina (KAW-duh ek-WĪ-nuh)**, the "horse's tail." A thin thread of tissue, the **filum terminale**, extends past the conus medullaris to anchor the spinal cord in the sacrum.

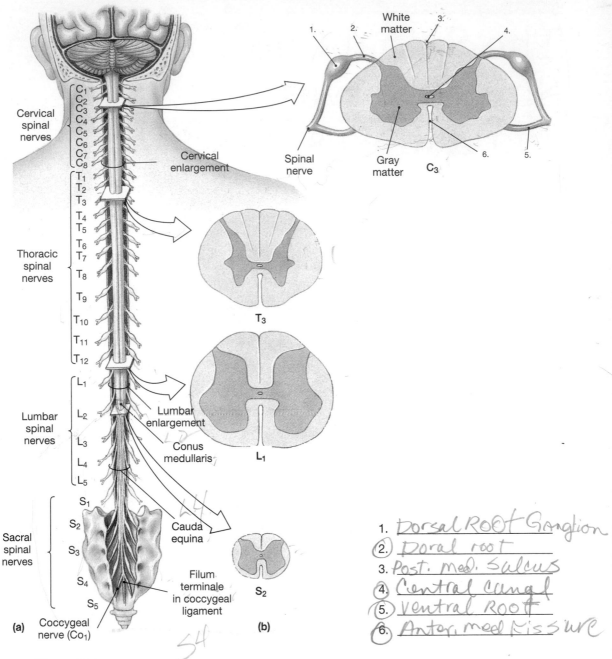

C1
C2
C3
C4
C5
C6
C7
C8

Cervical spinal nerves

Cervical enlargement

White matter

1.   2.   3.   4.

Spinal nerve   Gray matter   C3   6.   5.

T1
T2
T3
T4
T5
T6
T7
T8
T9
T10
T11
T12

Thoracic spinal nerves

T3

L1
L2
L3
L4
L5

Lumbar spinal nerves

Lumbar enlargement

Conus medullaris

L1

S1
S2
S3
S4
S5

Sacral spinal nerves

Cauda equina

Filum terminale in coccygeal ligament

S2

(a)   Coccygeal nerve (Co1)   (b)

1. Dorsal Root Ganglion
2. Dorsal root
3. Post. med. Sulcus
4. Central canal
5. Ventral Root
6. Anter. med Fissure

Figure 28.1
Anatomy of the Spinal Cord.

(a) Superficial anatomy and orientation of the adult spinal cord. The numbers to the left identify the spinal nerves and indicate where the nerve roots leave the vertebral canal. Note, however, that the adult spinal cord extends only to the level of vertebrae L_1 through L_2. (b) Inferior
views of cross sections through representative segments of the spinal cord, showing the
arrangement of gray and white matter.

The spinal cord is organized into 31 segments. Each segment has a
paired **spinal nerve** formed by the joining of two lateral extensions, the **dor-**
sal and **ventral roots**. The **dorsal root** contains sensory neurons ascending
into the spinal cord from sensory receptors. The **ventral root** consists of
motor neurons exiting the CNS and leading to effectors. The two roots join
to form the spinal nerve. Each spinal nerve is therefore a *mixed nerve* and

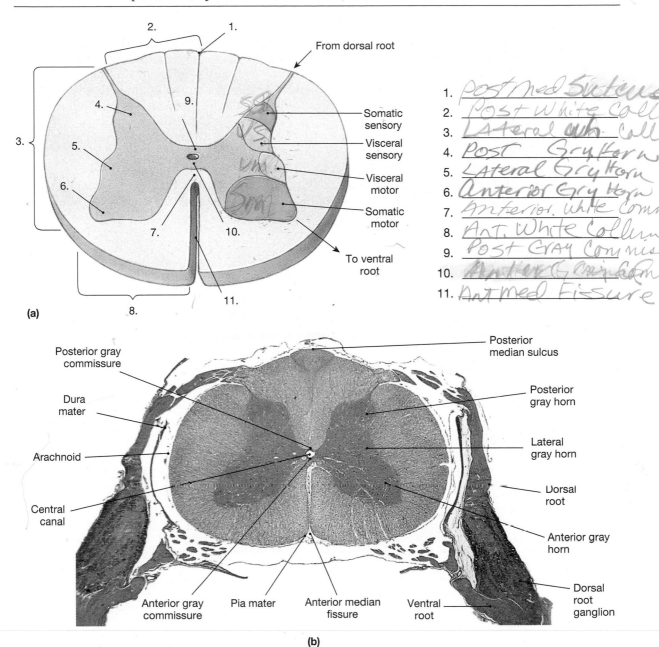

From dorsal root

Somatic
sensory

Visceral
sensory

Visceral
motor

Somatic
motor

To ventral
root

(a)

1. Post med Sulcus
2. Post White Collum
3. Lateral wh. Coll.
4. Post Gry Horn
5. Lateral Gry Horn
6. Anterior Gry Horn
7. Anterior White Commis
8. Ant. White Collum
9. Post Gray Commisur
10. Anterior Gray Commisur
11. Ant Med Fissure

Posterior
median sulcus

Posterior gray
commissure

Dura
mater

Posterior
gray horn

Arachnoid

Lateral
gray horn

Central
canal

Dorsal
root

Anterior gray
horn

Anterior gray   Pia mater   Anterior median   Ventral   Dorsal
commissure   fissure   root   root
                                                                        ganglion

(b)

Figure 28.2
The Sectional Organization of the Spinal Cord.

(a) The left half of this sectional view shows important anatomical landmarks and the major regions of white matter in the posterior white column. The right half indicates the functional organization of the gray matter in the anterior, lateral, and posterior gray horns. **(b)** A micrograph of a section through the spinal cord, showing major landmarks; compare with (a).

carries both sensory and motor information. The dorsal root swells at the **dorsal root ganglion** where cell bodies, or somae, of sensory neurons cluster. Locate the dorsal and ventral roots and the dorsal root ganglion in Figure 28.1.

Figure 28.2 illustrates the anatomy of the spinal cord in transverse section. The spinal cord is divided anteriorly by the deep **anterior median fissure** and posteriorly by the shallow **posterior median sulcus**. An H-shaped area called the **gray horns** contains many glial cells and neuron somae. Each horn of the gray matter has a specific type of neuron. The **posterior gray horns** carry sensory neurons into the spinal cord. The **lateral gray horns** contain visceral motor somae, and the **anterior gray horns** carry somatic motor neurons out of the cord. Axons may cross to the opposite side of the spinal cord at the cross bars of the horns, called the **anterior** and **posterior gray commissures**. Between the commissures is a hole, the **central canal**. This canal contains cerebrospinal fluid and is continuous with the fluid-filled ventricles of the brain.

Surrounding the gray horns are three masses of white matter: the **posterior, lateral,** and **anterior white columns**. The anterior white columns are connected by the **anterior white commissure.** Within each white column the myelinated axons form distinct bundles called **tracts**, or fasciculae, which are the equivalent of nerves inside the CNS. For example, in the posterior white column are ascending tracts of sensory neurons from receptors for touch, vibration, and pressure.

LABORATORY ACTIVITY   ANATOMY OF THE SPINAL CORD

MATERIALS

dissecting and compound microscopes
prepared slide: transverse (cross) section of spinal cord

PROCEDURES

1. Label and review the anatomy in Figures 28.1 and 28.2.
2. Locate each surface feature of the spinal cord on available lab models and charts.
3. Locate the internal anatomy of the spinal cord on a sectional spinal cord model.
4. Examine the microscopic features of the spinal cord in transverse section.
   a. For an orientation to the spinal cord slide, first view the slide at low magnification with a dissection microscope. Which features of the spinal cord enable you to distinguish anterior and posterior regions?

   _____

   b. Transfer your slide to a compound microscope. Move the slide around to survey the preparation at low magnification. What structures can you use as a landmark to distinguish between the posterior and anterior aspects of the spinal cord?

   _____

   c. Examine the central canal and the gray horns. Can you distinguish among the posterior, ventral, and anterior horns? Where are the gray commissures?

   _____

   d. Examine the white columns. What is the difference between gray and white matter in the central nervous system?

   _____

   e. Examine the outer layers surrounding the cross-sectioned spinal cord. Which of the meningeal layers can you locate?

   _____

   _____

DAP

B. Spinal Meninges

The spinal cord is protected within three layers of **spinal meninges (men-IN-jēz)**, highlighted in each part of Figure 28.3. The outer layer, the **dura mater (DOO-ruh MĀ-ter)**, is composed of a tough, fibrous connective tissue. The fibrous tissue attaches to the bony walls of the spinal canal and supports the spinal cord laterally. Superficial to

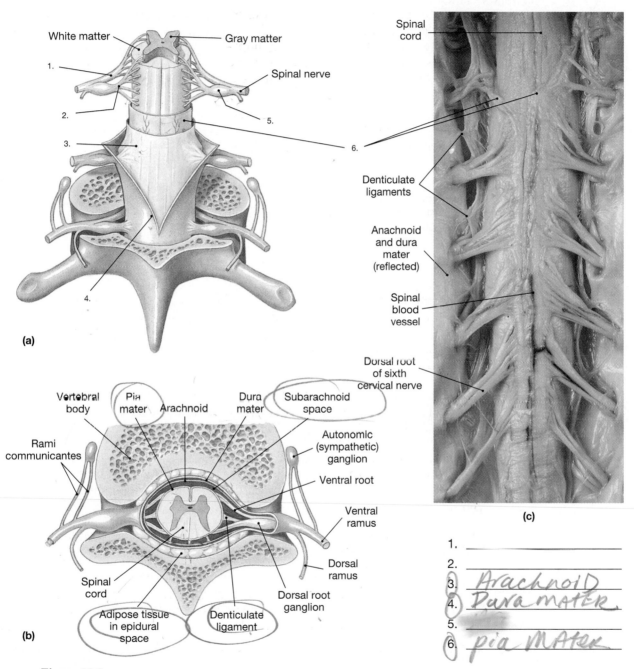

White matter
Gray matter
1.
Spinal nerve
2.
5.
3.
6.
4.
(a)

Spinal cord
Denticulate ligaments
Anachnoid and dura mater (reflected)
Spinal blood vessel
Dorsal root of sixth cervical nerve
(c)

Vertebral body
Pia mater   Arachnoid
Dura mater   Subarachnoid space
Rami communicantes
Autonomic (sympathetic) ganglion
Ventral root
Ventral ramus
Spinal cord
Dorsal ramus
Adipose tissue in epidural space
Denticulate ligament
Dorsal root ganglion
(b)

1. _____
2. _____
3. Arachnoid
4. Dura mater
5. _____
6. pia mater

Figure 28.3
The Spinal Cord and Spinal Meninges.

(a) Posterior view of the spinal cord, showing the meningeal layers, superficial landmarks, and the distribution of gray and white matter. **(b)** Sectional view through the spinal cord and meninges, showing the peripheral distribution of the spinal nerves. **(c)** Posterior view of the spinal cord and spinal nerve roots within the vertebral canal. The dura mater and arachnoid membrane have been removed; note the blood vessels in the delicate pia mater.

the dura mater is the **epidural space** which contains adipose tissue to pad the spinal cord. The **arachnoid (a-RAK-noyd)** is the second meningeal layer. A small cavity called the **subdural space** separates the dura mater and the arachnoid. The **subarachnoid space** contains cerebrospinal fluid which protects and cushions the spinal cord. The **pia mater** is the inner meningeal layer and lies directly over the neural tissue of the spinal cord. Blood vessels supplying the spinal cord are held in place by the thin pia mater. The pia mater extends laterally on each side of the spinal cord as the **denticulate ligament**. The tissue joins the dura mater to provide lateral support of the spinal cord. The filum terminale is also an extension of the pia mater and supports the spinal cord inferiorly.

CLINICAL APPLICATION

During childbirth, the expectant mother may request an **epidural block**, a procedure that introduces anesthesia in the epidural space. A thin needle is inserted between two lumbar vertebrae, and the drug is injected into the epidural space. The anesthetic numbs only the spinal nerves of the pelvis and lower limbs and reduces the discomfort the woman feels during the powerful labor contractions of her uterus.

A **spinal tap** is a procedure in which a needle is inserted into the subarachnoid space to withdraw a sample of cerebrospinal fluid. The fluid is then analyzed for the presence of microbes, wastes, and metabolites. To prevent injury to the spinal cord, the needle is inserted into the lower lumbar region inferior to the cord.

LABORATORY ACTIVITY   SPINAL MENINGES

MATERIALS

compound microscope

prepared slide: transverse (cross) section of spinal cord

PROCEDURES

1. Label and review the anatomy in Figure 28.3.
2. Locate the spinal meninges on the available lab models and charts.
3. Microscopic observation of the spinal meninges:
   a. Move the slide around to survey the preparation. Examine the outer layers surrounding the cross-sectioned spinal cord. Which of the meningeal layers can you locate?

   _____

   b. Add the spinal meninges in your previous drawing of the spinal cord cross section.

C. Spinal Nerves

Two types of nerves connect peripheral sensory receptors and effectors to the central nervous system: 12 pairs of **cranial nerves** and 31 pairs of **spinal nerves**. As their names indicate, cranial nerves connect with the brain and spinal nerves communicate with the spinal cord. Spinal nerves branch into **peripheral nerves** and make up axons of sensory and motor neurons.

Figure 28.3b illustrates the distribution of neurons in a spinal nerve. The nerve is formed by union of the dorsal and ventral roots. The spinal nerve exits the vertebral canal and branches into a **dorsal ramus**, which supplies the skin and muscles of the back, and a **ventral ramus**, which innervates the anterior and lateral skin and muscles. The ventral ramus has additional branches, called the **rami communicantes**, which carry autonomic neurons to the viscera. The rami communicantes lead to an **autonomic ganglion** where autonomic fibers synapse. A **white ramus** passes a preganglionic neuron into the autonomic ganglion where it synapses with a postganglionic neuron. A **gray ramus** carries the postganglionic fiber to the visceral organ the fiber innervates. The pathways of the autonomic nervous system will be discussed more fully in Exercise 29.

Examine Figure 28.4 which shows the spinal nerves emerging from the vertebral column and branching into **peripheral nerves.** Each spinal nerve passes through an intervertebral foramen between two adjacent vertebrae. Spinal nerves are numbered after the vertebra located inferior to each nerve. There are 8 **cervical nerves** (C_1 through C_8), 12 **thoracic nerves** (T_1 through T_{12}), 5 **lumbar nerves** (L_1 through L_5), 5 **sacral nerves** (S_1 through S_5), and a single **coccygeal nerve.** Although there are only 7 cervical vertebrae, there are 8 cervical spinal nerves. The eighth cervical nerve is located between vertebrae C_7 and T_1. Spinal nerves are mixed nerves and contain sensory, visceral motor, and somatic motor neurons.

In the cervical, lumbar, and sacral regions, groups of spinal nerves join in a network called a **plexus.** As muscles fuse during fetal development, the spinal nerves that supplied the individual muscles interconnect and create a plexus.

Cervical plexus

Eight cervical nerves supply structures of the neck, shoulder, arm, and diaphragm. The **cervical plexus** consists of branches of C_1 through C_4 and parts of C_5. This plexus innervates muscles of the larynx, the sternocleidomastoid, trapezius, and diaphragm muscles. The overlying skin of these muscles is also innervated by the cervical plexus.

Brachial plexus

The remaining cervical nerves are part of the **brachial plexus** which supplies the shoulder and arm. Major nerves of the brachial plexus include the axillary, radial, musculocutaneous, median, and ulnar nerves. The **axillary nerve** (C_5 and C_6) supplies the deltoid and teres minor muscles and the skin of the shoulder. The **radial nerve** (C_5 through T_1) controls the extensor muscles of the arm, forearm, and digits, as well as the skin over the posterior and lateral margins of the arm. The **musculocutaneous nerve** (C_5 through C_7) supplies the flexor muscles of the upper arm and the skin of the lateral forearm. The **median nerve** (C_6 through T_1) innervates the flexor muscles of the forearm and digits, the pronator muscles, and the lateral skin of the hand. The **ulnar nerve** (C_8 and T_1) controls the flexor carpi ulnaris muscle of the forearm, other muscles of the hand, and the medial skin of the hand. Notice how overlap occurs within the brachial plexus. For example, spinal nerve C_6 innervates both flexor and extensor muscles.

Lumbosacral plexus

The largest network is the **lumbosacral plexus** which includes nerves T_{12}, L_1 through L_4, and S_1 through S_4. Figure 28.4 presents the distribution of nerves in this plexus. The major nerves of the **lumbar** portion innervate the skin and muscles of the abdominal wall, genitalia, and thigh. The **genitofemoral nerve** supplies some of the external genitalia and the anterior and lateral skin of the thigh. The **lateral femoral cutaneous nerve** innervates the skin of the thigh from all aspects except the medial region. The **femoral nerve** controls the muscles of the anterior thigh and the adductor muscles of the thigh. The medial skin of the thigh is also supplied by the femoral nerve.

The **sacral** division of the lumbosacral plexus consists of two major nerves, the **sciatic nerve** and the **pudendal nerve.** The sciatic nerve descends the posterior leg and sends branches into the posterior thigh muscles and the musculature and skin of the

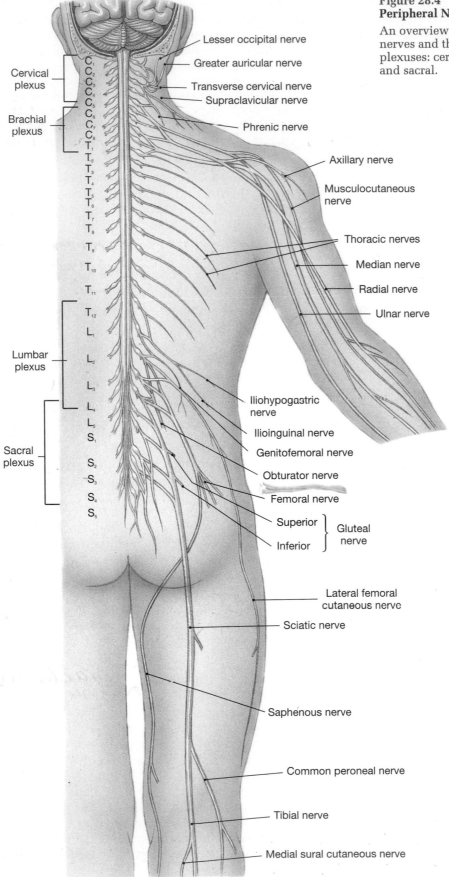

Cervical plexus

Brachial plexus

Lumbar plexus

Sacral plexus

C_1
C_2
C_3
C_4
C_5
C_6
C_7
C_8
T_1
T_2
T_3
T_4
T_5
T_6
T_7
T_8
T_9
T_{10}
T_{11}
T_{12}
L_1
L_2
L_3
L_4
L_5
S_1
S_2
S_3
S_4
S_5

Lesser occipital nerve

Greater auricular nerve

Transverse cervical nerve

Supraclavicular nerve

Phrenic nerve

Axillary nerve

Musculocutaneous nerve

Thoracic nerves

Median nerve

Radial nerve

Ulnar nerve

Iliohypogastric nerve

Ilioinguinal nerve

Genitofemoral nerve

Obturator nerve

Femoral nerve

Superior
Inferior
} Gluteal nerve

Lateral femoral cutaneous nerve

Sciatic nerve

Saphenous nerve

Common peroneal nerve

Tibial nerve

Medial sural cutaneous nerve

lower leg. The pudendal nerve supplies the muscular floor of the pelvis, the perineum, and parts of the skin of the external genitalia.

Peripheral nerves

Figure 28.5 illustrates the organization of a peripheral nerve, a branch of a spinal nerve. The nerve is compartmentalized by connective tissue much in the same way a skeletal muscle is organized. The peripheral nerve is wrapped in an outer covering called the **epineurium**. Beneath this layer is the **perineurium** which separates the axons into bundles called **fascicles**. Inside a fascicle the **endoneurium** surrounds each axon and isolates it from neighboring axons.

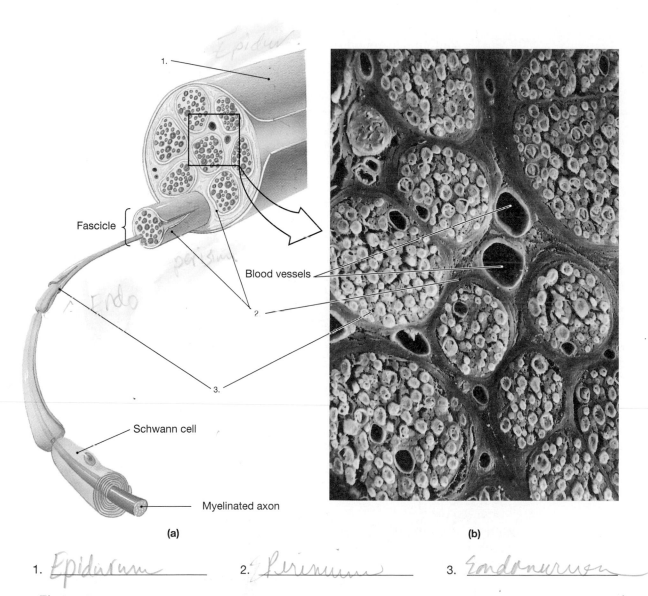

1.

Fascicle

Blood vessels

2

3.

Schwann cell

Myelinated axon

(a)   **(b)**

1. _Epidurum_   2. _Perineum_   3. _Endoneuron_

Figure 28.5
Organization of Peripheral Nerves.

(a) A typical peripheral nerve and its connective tissue wrappings. **(b)** A scanning electron micrograph showing the perineurium and endoneurium in great detail. (SEM × 340)

LABORATORY ACTIVITY   SPINAL NERVES

MATERIALS

compound microscope
prepared slide: peripheral nerve

PROCEDURES

1. Label and review the anatomy in Figures 28.4 and 28.5.
2. Locate each nerve plexus on available lab models and charts.
3. Locate the spinal nerves assigned by your laboratory instructor on the various lab models.
4. Microscopic observation of the **peripheral nerve**:
   a. Move the slide around to survey the preparation. Locate the epineurium. Is it continuous around the nerve?

   _____

   b. Examine a single fascicle. Can you distinguish the perineurium from the epineurium?

   _____

   c. Locate the individual axons inside a fascicle. Can you distinguish the myelin sheath of the axon from the endoneurium?

   _____

   d. Draw and label the peripheral nerve in the space provided.

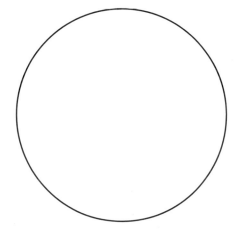

D. Spinal Reflexes

Reflexes are automatic neural responses to specific stimuli. Most reflexes have a protective function. Touch something hot and the withdrawal reflex removes your hand to prevent tissue damage. Shine a bright light into someone's eyes, and their pupils constrict to protect the retina from overstimulation. Reflexes cause rapid adjustments to maintain homeostasis. The central nervous system does minimal processing to respond to the stimulus. The sensory and motor components of a reflex are "prewired" and ini-

448 Text

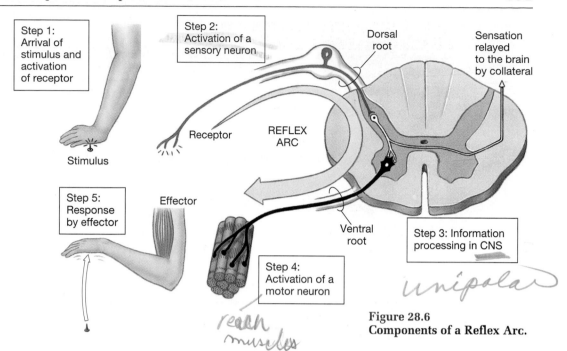

Step 1:
Arrival of
stimulus and
activation
of receptor

Step 2:
Activation of a
sensory neuron

Dorsal
root

Sensation
relayed
to the brain
by collateral

Receptor

REFLEX
ARC

Stimulus

Step 5:
Response
by effector

Effector

Step 3: Information
processing in CNS

Ventral
root

Step 4:
Activation of a
motor neuron

unipolar

reach
muscles

Figure 28.6
Components of a Reflex Arc.

tiate the reflex upon stimulation. Figure 28.6 depicts the five steps involved in a typi-cal reflex. First, a receptor is activated by a stimulus. The receptor in turn activates a sensory neuron which enters the central nervous system where the third step, informa-tion processing, occurs. The processing is performed at the synapse between the sen-sory and motor neurons. A conscious thought or recognition of the stimulus is not required to evaluate the sensory input of the reflex. The processing results in activation of a motor neuron which elicits the appropriate action, a response by the effectors. In this basic reflex arc only two neurons are involved, a sensory neuron and a motor neu-ron. Complex reflex arcs include interneurons between the sensory and motor neurons.

Many types of reflexes occur. **Innate** reflexes are the inborn responses of a newborn baby, such as grasping an object and suckling the breast for milk. **Cranial** reflexes have pathways in cranial nerves. **Visceral** reflexes pertain to the internal organs. **Spinal** reflexes process information in the spinal cord rather than the brain. **Somatic** reflexes involve skeletal muscles. Reflexes may be classified by the number of synapses involved in the reflex arc. A **monosynaptic** reflex is a reflex arc with only one synapse between the sensory and motor neurons. A **polysynaptic** reflex includes numerous interneurons between sensory and motor neurons. The response of a polysynaptic reflex is more complex and may include both stimulation and inhibition of muscles. Reflexes are used as a diagnostic tool to evaluate the function of specific regions of the brain and spinal cord.

You are probably familiar with the "knee jerk" or **stretch reflex** that occurs when the tendon over the patella is stimulated by hitting it with a rubber percussion hammer (see Figure 28.7). Tapping on the patellar tendon stretches receptors called **muscle spindles** in the quadriceps muscle group of the anterior thigh. This stimulus evokes a rapid motor reflex to contract the quadriceps and shorten the muscles.

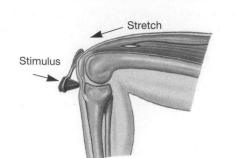

Figure 28.7
The Patellar Reflex.

(a) The patellar reflex is controlled by muscle spindles in the muscles that straighten the knee. A reflex hammer striking the muscle tendon provides the stimulus that stretches the spindle fibers. The response is an immediate increase in muscle tone and a reflexive kick. **(b)** The basic wiring of a monosynaptic reflex arc.

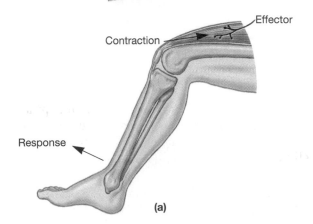

(a)

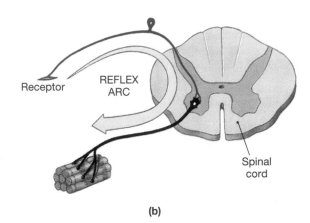

(b)

LABORATORY ACTIVITY   REFLEXES

MATERIALS

> reflex hammer (with rubber head)
> lab partner

PROCEDURES

1. **Stretch reflex** Have your lab partner sit down and cross their legs. Gently tap below the patella with the percussion hammer to stimulate the tendon of the rectus femoris muscle. What is the response? How might this reflex help maintain upright posture?

   _____

   _____

2. **Calcaneal reflex** This reflex tests the response of the calcaneal (Achilles) tendon. Have your lab partner hang their foot off a chair. Gently tap the calcaneal tendon with the percussion hammer. What is the response?

   _____

   _____

E. **Dissection of the Spinal Cord**

   Dissection of a sheep or cow spinal cord will provide you the opportunity to examine the meningeal layers and sectional anatomy. Only a small section of the spinal cord is necessary. If a spinal cord is unavailable in the laboratory, you may be able to acquire a portion of fresh cow spinal cord from your local butcher. Remember that the specimen in lab has most likely been fixed in preservatives that are irritating to your skin and mucous membranes. Disposal of the specimen requires certain safeguards and will be managed by your lab instructor.

LABORATORY ACTIVITY   SPINAL CORD DISSECTION

MATERIALS

>   gloves, safety glasses
>   segment of a sheep or cow spinal cord
>   dissection pan
>   scissors, scalpel
>   forceps, blunt probe

PROCEDURES

Put on your gloves and safety glasses before you open a container of preserved spinal cord segments or handle a spinal cord segment. A variety of preservatives are used on biological specimens, some of which will cause irritation to your skin and mucous membranes. Your lab instructor will tell you how to handle, dissect, and finally dispose of the spinal cord.

1. Lay the spinal cord on the dissection pan and cut a thin section about 2 cm (about $\frac{1}{2}$ in.) thick. Lay this cross section flat on the dissection pan to observe the internal anatomy. Use Figure 28.3 as a guide to help locate the anatomy of the spinal cord.

2. Identify the gray horns, central canal, and white columns. What type of tissue is found in the gray horns and the white columns? How can you determine the posterior margin of the cord?

   _____

   _____

3. Locate the spinal meninges by pulling the outer tissues away from the spinal cord with a forceps and blunt probe. Slip your probe between the meninges on the lateral spinal cord. Cut completely through the meninges and gently peel them back to expose the ventral and dorsal roots. How does the dorsal root differ in appearance from the ventral root?

   _____

   _____

4. Closely examine the meninges. With your probe, separate the arachnoid from the dura mater. With a dissection pin attempt to loosen a free edge of the pia mater. What function does each of these membranes serve?

   _____

   _____

5. Clean up your work area, wash the dissection pan, and discard your spinal cord specimen as indicated by your lab instructor.

SPINAL CORD, SPINAL NERVES, AND REFLEXES CHECKLIST

This is a list of bold terms presented in Exercise 28. Use this list for review purposes after you have completed the laboratory activities.

GROSS ANATOMY OF SPINAL CORD

• Figure 28.1
cervical enlargement
lumbar enlargement
conus medullaris
cauda equina
filum terminale
spinal nerve
dorsal root, ventral root, dorsal root ganglion

SECTIONAL ANATOMY OF SPINAL CORD

• Figure 28.2
anterior median fissure
posterior median sulcus
gray horns, posterior gray horn
lateral gray horn, anterior gray horn
anterior gray commissure, posterior gray commissure
posterior white column, lateral white column
anterior white column, anterior white commissure

SPINAL MENINGES

• Figure 28.3
spinal meninges
dura mater
epidural space, subdural space
arachnoid, subarachnoid space
pia mater
denticulate ligament

SPINAL NERVES

• Figures 28.3, 28.4, and 28.5
cranial nerve
spinal nerve
peripheral nerve
dorsal ramus
ventral ramus
rami communicantes
autonomic ganglion

white ramus
gray ramus
epineurium
fascicles
perineurium
endoneurium
cervical nerves (C_1–C_8)
thoracic nerves (T_1–T_{12})
lumbar nerves (L_1–L_5)
sacral nerves (S_1–S_5)
coccygeal nerve
cervical plexus
brachial plexus
axillary nerve
radial nerve
musculocutaneous nerve
median nerve
ulnar nerve
lumbosacral plexus
lumbar plexus
sacral plexus
genitofemoral nerve
lateral femoral cutaneous nerve
femoral nerve
sciatic nerve
pudendal nerve

SPINAL REFLEXES

• Figures 28.6 and 28.7
reflex
innated reflex
cranial reflex
visceral reflex
spinal reflex
somatic reflex
monosynaptic reflex
polysynaptic reflex
stretch reflex
spindles fibers
calcaneal reflex

Name _____   Date _____

Section _____   Number _____

SPINAL CORD, SPINAL NERVES, AND REFLEXES
Laboratory Report Exercise 28

A. Matching

Match each spinal cord structure listed on the left with the correct description on the right.

1. _____ lateral gray horn   A. site of cerebrospinal fluid circulation
2. _____ posterior median sulcus   B. sensory branch entering cord
3. _____ bundle of axons in nerve   C. surrounds axons of peripheral nerve
4. _____ rami communicantes   D. contains visceral motor somae
5. _____ subarachnoid space   E. tapered end of spinal cord
6. _____ ventral root   F. fascicle
7. _____ dorsal ramus   G. shallow groove of spinal cord
8. _____ posterior gray horn   H. posterior branch of a spinal nerve
9. _____ conus medullaris   I. motor branch exiting cord
10. _____ endoneurium   J. leads to an autonomic ganglion
11. _____ dorsal root   K. contains sensory somae

B. Short-Answer Questions

1. Describe the organization of white and gray matter in the spinal cord.

2. Describe the spinal meninges.

3. Discuss the major nerves of the brachial plexus.

4. List the five basic steps of a reflex.

5. Compare the types of neurons in the posterior, lateral, and anterior gray horns.

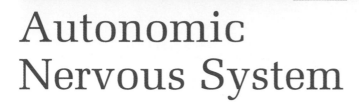

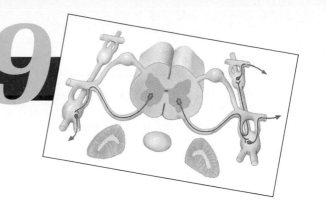

EXERCISE 29

Autonomic Nervous System

OBJECTIVES

On completion of this exercise, you should be able to:

- Outline a typical autonomic pathway.
- Compare the location of the preganglionic outflow from the CNS in the sympathetic and parasympathetic divisions.
- Compare the lengths of and the neurotransmitters released by each fiber in the sympathetic and parasympathetic divisions.
- Trace the sympathetic pathways into a chain ganglion, into a collateral ganglion, and into the adrenal medulla.
- Trace the parasympathetic pathways into cranial nerves III, VII, IX, and X and into the pelvic nerves.
- Compare the sympathetic and parasympathetic responses.

WORD POWER

autonomic (auto—self)
ganglion (gangli—knot on a string)
ramus (ram—branch)
intramural (intra—with; mura—wall)

MATERIALS

nervous system charts
spinal cord model

INTRODUCTION

All motor commands from the central nervous system are communicated to the efferent branch of the peripheral nervous system. This motor branch is divided into the somatic and autonomic motor systems. Voluntary motor signals to skeletal muscles are carried by the **somatic** division. Involuntary motor instructions to smooth muscles, heart muscle, and glands, collectively called **visceral effectors**, are relayed along the **autonomic** nervous system. If necessary, review the organization of the nervous system in Exercise 24.

The ANS is anatomically subdivided into the **sympathetic nervous system** and the **parasympathetic nervous system** (see Figure 29.1). The parasympathetic or **craniosacral (krā-nē-ō-SA-krul)** division is composed of efferent neurons exiting the cranium in certain cranial nerves and efferent neurons that descend the spinal cord and exit at the sacral level. The sympathetic or **thoracolumbar (tho-ra-kō-LUM-bar)** division includes efferent neurons that descend the spinal cord to the thoracic and upper lumbar levels.

All autonomic motor pathways consist of two groups of neurons: preganglionic and ganglionic. **Preganglionic neurons** descend from the brain into specific segments of the spinal cord and into certain cranial nerves. Preganglionic cell bodies, or somae, are located in the lateral gray horns of the spinal segments. Preganglionic axons, called **preganglionic fibers**, leave the CNS via the ventral roots. In the PNS the preganglionic fibers synapse with **ganglionic neurons** in bulblike structures called **autonomic**

Figure 29.1
Divisions of the ANS.

Anatomical divisions of the autonomic nervous system. At the cranial and sacral levels, the visceral efferent fibers from the CNS make up the parasympathetic, or craniosacral, division. At the thoracic and lumbar levels, the visceral efferent fibers that emerge form the sympathetic, or thoracolumbar, division.

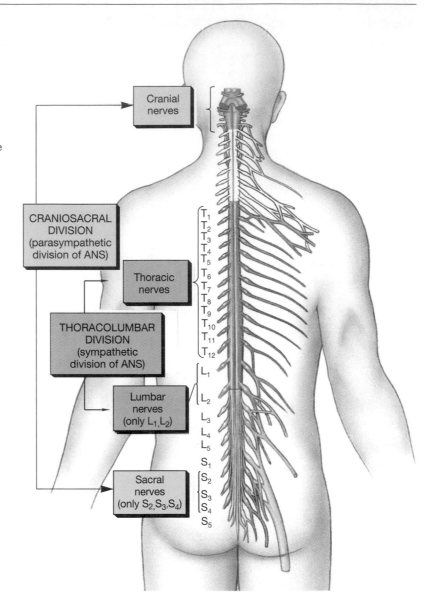

ganglia. Ganglionic axons, called **ganglionic fibers**, synapse with smooth muscles, the heart, and glands. Compare the lengths of the fibers and the locations of the autonomic ganglia for the sympathetic and parasympathetic pathways in Figure 29.2.

Sympathetic and parasympathetic subdivisions of the ANS differ in three major respects.

1. **Location of preganglionic neurons in the CNS**

   All autonomic preganglionic neurons originate in the brain stem. Where these neurons have their cell bodies and exit the CNS differs between the sympathetic and parasympathetic motor pathways. The sympathetic nervous system is also called the **thoracolumbar** division because the preganglionic neurons exit the spinal cord at segments T_1 through L_2. The parasympathetic nervous system or the **craniosacral** division and the preganglionic neurons exit the brain in cranial nerves or descend the spinal cord and exit the CNS in the sacral region.

CNS   PNS

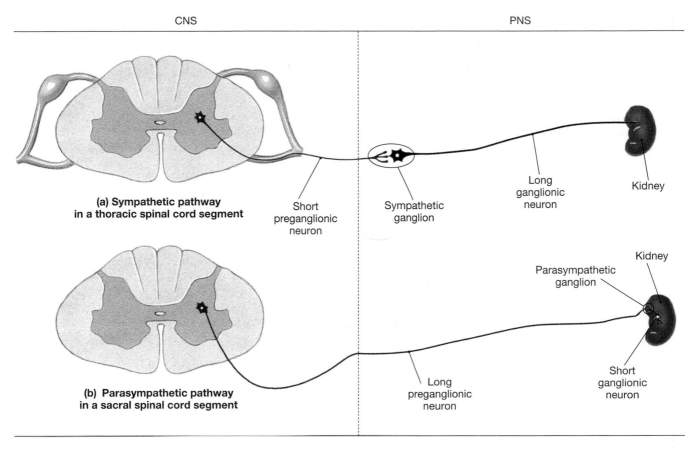

**(a) Sympathetic pathway
in a thoracic spinal cord segment**

Short preganglionic neuron

Sympathetic ganglion

Long ganglionic neuron

Kidney

Kidney

Parasympathetic ganglion

**(b) Parasympathetic pathway
in a sacral spinal cord segment**

Long preganglionic neuron

Short ganglionic neuron

**Figure 29.2
An Overview of ANS Pathways.**

(a) Sympathetic pathways consist of short preganglionic neurons that release ACh in sympathetic ganglia. They synapse with long ganglionic neurons that release NE at the visceral effector. The sympathetic response is generalized as a "fight on flight" response. **(b)** Parasympathetic pathways in the sacral spinal segments have long preganglionic neurons that exit the spinal cord in pelvic nerves instead of ventral roots. They release ACh in intramural ganglia in the walls of the viscera. Preganglionic neurons synapse with short ganglionic neurons that also release ACh. The general parasympathetic response is "rest and digest."

2. **Location of autonomic ganglia in the PNS**

   All autonomic ganglia are in the PNS. Their proximity to the CNS, however, differs in sympathetic ganglia and parasympathetic ganglia. Sympathetic ganglia, called sympathetic chain ganglia and collateral ganglia, are located close to the spinal cord. This anatomical location of the ganglia results in short preganglionic neurons and long ganglionic neurons. Parasympathetic ganglia are called intramural ganglia. Intramural means "within the walls," which perfectly describes the location of the ganglia, embedded in the walls of the visceral effectors. Parasympathetic preganglia are long and extend from the brain or the sacral spinal segments into the intramural ganglia. The ganglionic neuron is short because the intramural ganglia are already in the organ the ganglionic neuron innervates. Notice the difference in the location of the ganglia in Figure 29.2.

3. **Effect of the ganglionic neuron's neurotransmitter on the visceral effector**

   Although involuntary effectors have dual innervation, each division of the ANS releases a different neurotransmitter at the synapse with the effector. Sympathetic ganglionic neurons release norepinephrine (NE) at the effector synapse and cause a "fight-or-flight" response which increases the overall alertness of the individual. Increases in heart rate, blood pressure, and respiratory rate occur. Sweat glands begin to secrete, and digestive and urinary functions cease. Parasympathetic ganglionic neurons release acetylcholine (ACh) which slows the body for normal, energy-saving homeostasis. This "rest-and-digest" response decreases cardiovascular and respiratory activity and increases the processing of food and wastes.

Each division of the ANS shares the role of regulating autonomic function. Most visceral effectors have **dual innervation** with both sympathetic and parasympathetic nerves controlling their activity. Typically, one branch of the ANS stimulates the effector and the other branch acts antagonistically and inhibits the effector. However, because autonomic motor responses vary from organ to organ, it is difficult to categorize sympathetic or parasympathetic functions. Generally, sympathetic stimulation occurs during times of excitement, emotional stress, and emergencies and causes a fight-or-flight response. The sympathetic response is widespread in the body because each preganglionic fiber synapses with dozens of ganglionic neurons, thus fanning the sympathetic response. Sympathetic stimulation causes the adrenal medulla to releases epinephrine (E) which is transported by the blood to all body tissues where it increases the sympathetic response.

Parasympathetic stimulation, in general, conserves energy and causes a rest-and-digest response. It decreases the heart rate and respiratory rate and promotes muscle activities for peristalsis, defecation, and urination.

STUDY TIP

An easy way to remember the sympathetic fight-or-flight response is to consider how your organs must adjust their activities during an emergency situation, such as being chased by a wild animal. Sympathetic impulses increase your heart rate, blood pressure, and respiratory rate. Muscle tone increases in preparation for fighting off the animal or running for your life. Arterioles in skeletal muscles dilate, and more blood flows into the muscles to support their high activity level. Sympathetic stimulation decreases the activity of your digestive tract. An emergency situation is not the time to work at digesting your lunch! With digestive actions slowed, the body shunts blood from the abdominal organs and delivers it to the skeletal muscles. Once out of danger, sympathetic stimulation decreases and parasympathetic stimulation predominates to return your body functions to the routine "housekeeping" chores of digesting food, eliminating wastes, and conserving precious cell energy.

A. The Sympathetic (Thoracolumbar) Division

The organization of the sympathetic division of the ANS is diagrammed in Figure 29.3. Preganglionic neurons originate in the pons and the medulla of the brain stem. These autonomic motor neurons descend in the spinal cord to the thoracic and lumbar segments where their somae are located in the lateral gray horns. Preganglionic axons exit the spinal cord in ventral roots and pass into a spinal nerve which branches into an autonomic ganglion. In the ganglion, the preganglionic neuron synapses with a ganglionic neuron which innervates the visceral effector.

Because the sympathetic ganglia are located near the spinal cord, sympathetic preganglionic fibers are relatively short in length and sympathetic ganglionic fibers are relatively long. The short sympathetic preganglionic fibers release acetylcholine at the synapse in the ganglion with the ganglionic fiber. The effect of ACh at this synapse is always excitatory to the ganglionic fiber. The long ganglionic axons release norepinephrine onto the effector. The effect of the NE depends on the type of NE receptors present in the effector's cell membrane. Generally, the sympathetic response is to prepare the body for increased activity or a crisis situation, the fight-or-flight response which occurs during exercise, excitement, and emergencies.

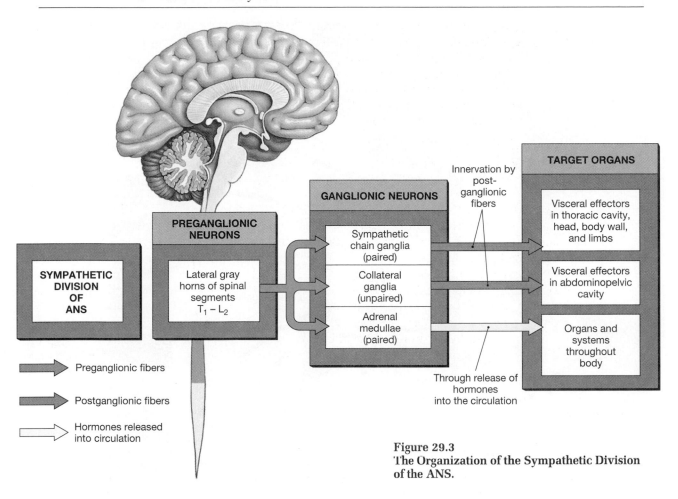

Figure 29.3
**The Organization of the Sympathetic Division
of the ANS.**

Sympathetic ganglia

Three types of sympathetic ganglia occur: sympathetic chain ganglia, collateral ganglia, and modified ganglia in the adrenal medulla (see Figure 29.4). **Sympathetic chain ganglia** are located lateral to the spinal cord and are also called **paravertebral ganglia**. Ganglionic neurons that exit the sympathetic chain innervate the visceral effectors of the thoracic cavity, head, body wall, and limbs.

Collateral ganglia are located anterior to the vertebral column and contain ganglionic neurons which lead to organs in the abdominopelvic cavity. Preganglionic fibers pass through the sympathetic chain ganglia without synapsing and join to form a network called the **splanchnic (SPLANK-nik) nerves**. This network divides and sends branches into the collateral ganglia where the preganglionic fibers synapse with ganglionic neurons. Axons of the ganglionic neurons synapse with the visceral effectors in the abdominopelvic cavity. The collateral ganglia are named after the adjacent blood vessels. The **celiac (SE-lē-ak) ganglion** supplies the liver, stomach, pancreas, and spleen. The **superior mesenteric ganglion** innervates the small intestine and parts of the large intestine. The **inferior mesenteric ganglion** controls most of the large intestine, the kidneys, the bladder, and the sex organs.

The adrenal glands are positioned on top of the kidneys. Each adrenal has an outer cortex layer which produces hormones and an inner **adrenal medulla** containing sympathetic ganglia and ganglionic neurons. During sympathetic stimulation, the ganglionic neurons in the medulla release epinephrine into the bloodstream and contribute to the overall sympathetic fight-or-flight response.

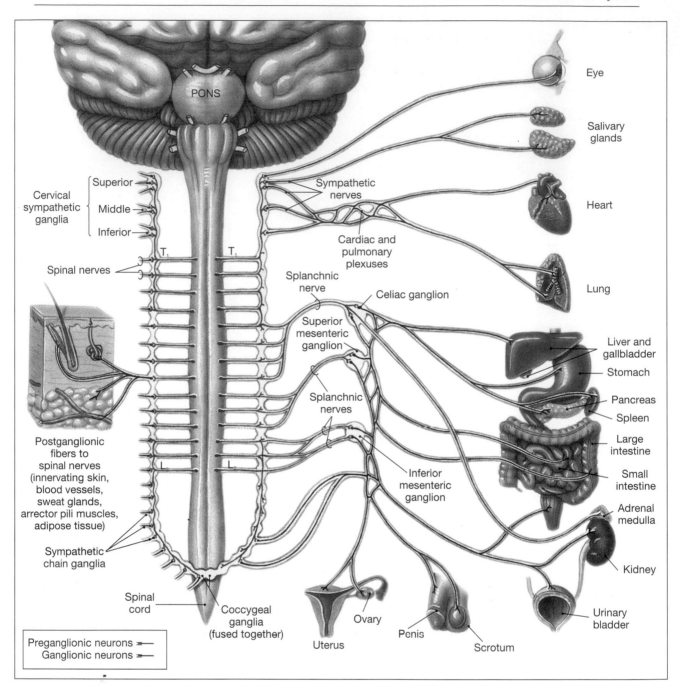

Figure 29.4
The Distribution of Sympathetic Innervation.

The distribution of sympathetic fibers is the same on both sides of the body. For clarity, the innervation of somatic structures is shown on the left, and the innervation of visceral structures on the right.

Overview of sympathetic pathways

Figure 29.4 outlines the general distribution of the sympathetic pathways. It illustrates the sympathetic chain ganglia positioned lateral to the spinal cord. For simplicity, only the somatic innervation is shown on the left and only visceral innervation is illustrated on the right of the spinal cord. In real life, sympathetic innervation is the same on both sides of the spinal cord. All sympathetic preganglionic neurons, represented in the figure by red lines, leave the thoracic and lumbar spinal segments, enter in a spinal nerve, and pass into the sympathetic chain ganglia. Ganglionic neurons, represented by the black lines, exit the ganglia and extend to the viscera.

Notice on the left of Figure 29.4 that all preganglionic fibers for somatic innervation synapse in a sympathetic chain ganglion. Visceral innervation, illustrated on the right of the figure, is more complex. The preganglionic neuron may not synapse with a ganglionic neuron in the sympathetic chain ganglion. Instead, it may lead into a collateral ganglion or enter another sympathetic chain ganglion to synapse with other ganglionic neurons. The axon of the ganglionic neuron exits the particular ganglion and travels to its target organ.

The sympathetic pathway to the adrenal medulla is generalized in Figure 29.4c. A long preganglionic neuron passes from the thoracic region of the spinal cord through several ganglia and enters the adrenal gland. In the adrenal medulla the preganglionic neuron synapses with a ganglionic neuron. The ganglionic neuron is not shown in the figure.

Sympathetic pathways in detail

Figure 29.5 details the sympathetic pathways involving each type of ganglion and the adrenal medulla. All sympathetic pathways start with a preganglionic neuron in the lateral gray horns which exits the spinal cord in the ventral root. The preganglionic axon leaves the ventral root and enters a **white ramus** to pass into the sympathetic chain ganglion.

Pathways involving the sympathetic chain ganglion

The pathway utilizing the sympathetic chain ganglia passes through the **rami communicantes**, the white and gray rami. Once the preganglionic axon enters the sympathetic chain ganglion via the white ramus, it usually synapses with a ganglionic neuron as shown in Figure 29.5a. The axon of the ganglionic neuron exits the sympathetic chain ganglion in either the gray ramus or an autonomic nerve. The gray ramus directs the ganglionic neuron into a spinal nerve which leads to general somatic structures such as blood vessels supplying skeletal muscles. A ganglionic axon in an autonomic nerve passes into the thoracic cavity to innervate the thoracic viscera.

Notice in the right side of Figure 29.5a that all the sympathetic chain ganglia are interconnected. A single preganglionic neuron may enter one sympathetic chain ganglion and branch into many different chain ganglia to synapse with up to 32 different ganglionic neurons. This fanning out of preganglionic neurons within the sympathetic chain ganglia contributes to the widespread effect that sympathetic stimulation has on the body.

Pathways involving the collateral ganglia

Figure 29.5b details the pathway involving collateral ganglia. Preganglionic neurons that synapse in the collateral ganglia do not synapse in the sympathetic chain ganglia. These axons pass through the chain ganglia and enter an autonomic nerve which leads into the collateral ganglia. In the collateral ganglia, the preganglionic axons synapse

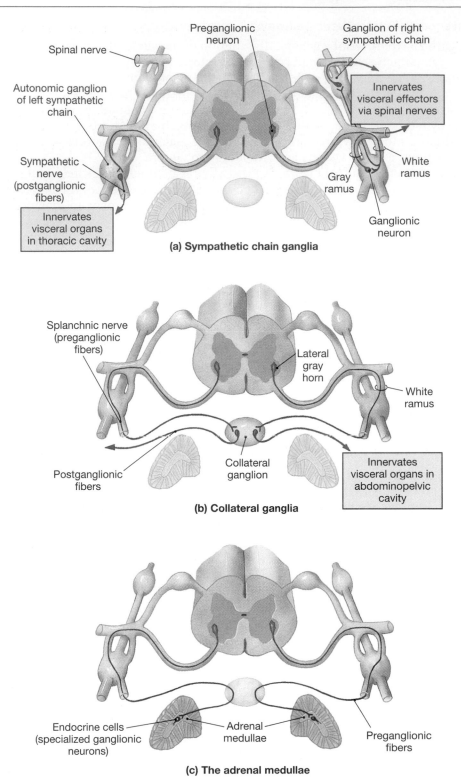

(a) Sympathetic chain ganglia

(b) Collateral ganglia

(c) The adrenal medullae

Figure 29.5
Sympathetic Pathways.

Superior views of sections through the thoracic spinal cord, showing the distribution of preganglionic and postganglionic fibers.

with ganglionic neurons. These neurons extend their axons into the visceral effectors in the abdominopelvic cavity.

Pathway to the adrenal medullae

The adrenal glands are a unique part of the sympathetic nervous system. Each adrenal gland is composed of an outer adrenal cortex and an inner **adrenal medulla**, the "emergency gland" of the body. Under sympathetic stimulation, the adrenal medulla secretes epinephrine and norepinephrine which contribute to the sympathetic response. Sympathetic neurons supplying the adrenal gland do not synapse in a sympathetic chain ganglion or a collateral ganglion. Instead, preganglionic axons penetrate deep into the adrenal gland and synapse with ganglionic neurons in the adrenal medulla. The ganglionic neurons secrete epinephrine and norepinephrine into the bloodstream for a widespread sympathetic response.

Table 29.1 at the end of the exercise compares the sympathetic and parasympathetic divisions of the ANS.

CLINICAL APPLICATION—STRESS AND THE ANS

Stress stimulates the body to increase sympathetic commands from the ANS. Appetite may decrease while blood pressure and general sensitivity to stimuli may increase. The individual may become irritable and have difficulty sleeping and coping with day-to-day responsibilities. Prolonged stress can lead to disease. Coronary diseases, for example, are more common in individuals who are employed in stressful occupations or live in stressful environments.

LABORATORY ACTIVITY   SYMPATHETIC PATHWAYS

1. Review the anatomy and pathways presented in Figures 29.1–29.5.

2. Use a chart of the nervous system and identify the lateral gray horns, ventral root, spinal nerve, white and gray rami, and sympathetic chain ganglion. What is the function of the white and gray rami?

   _____

   _____

3. Locate the collateral ganglia and the adrenal gland. Where are the collateral ganglia located relative to the spinal cord?

   _____

   _____

4. On the nervous system chart, trace the following sympathetic pathways:
   a. A preganglionic fiber synapsing in a collateral ganglion.
   b. A ganglionic fiber exiting a chain ganglion and passing into a spinal nerve.
   c. A preganglionic fiber synapsing in the adrenal medulla.

B. The Parasympathetic (Craniosacral) Division

The parasympathetic division of the autonomic nervous system contains preganglionic neurons in the brain stem and the sacral segments of the spinal cord (see Figures 29.6 and 29.7). The parasympathetic fibers are carried on cranial nerves III, VII, IX, and X or descend to the sacral level of the spinal cord.

In the brain, preganglionic neurons form autonomic nuclei which extend axons into the oculomotor, facial, glossopharyngeal, and vagus cranial nerves (III, VII, IX, X). The preganglionic axons synapse with ganglionic neurons in one of the **parasympathetic**

**Figure 29.6
The Organization of
the Parasympathetic
Division of the ANS.**

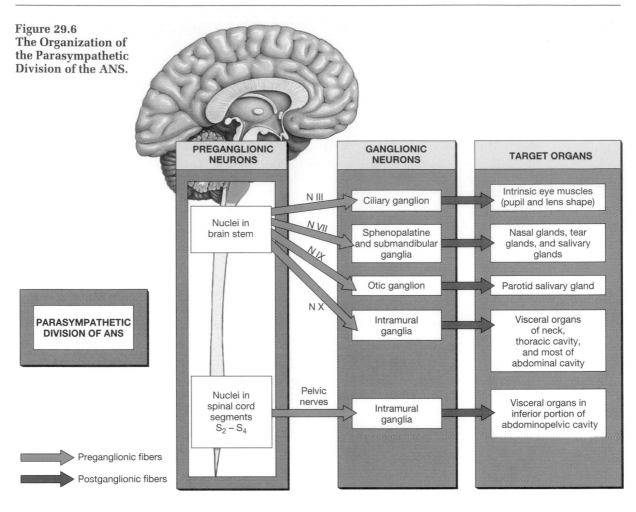

ganglia of the head: the **ciliary**, **sphenopalatine**, **submandibular**, or **otic ganglia**. The vagus nerve contains preganglionic fibers that synapse in **intramural ganglia**, ganglia located in the wall of the visceral organs. The target organ for each cranial nerve is included in Figures 29.6 and 29.7. If necessary, review the function of each of these cranial nerves in the previous exercise concerning the external anatomy of the brain.

The sacral portion of the parasympathetic division contains preganglionic neurons in the sacral segments of the spinal cord (S_2 through S_4). The cell bodies of the sacral preganglionic neurons cluster in an autonomic nucleus located in the lateral gray horns of the spinal cord. The preganglionic fibers remain separate from spinal nerves and exit from spinal segments S_2 through S_4 as **pelvic nerves**. Locate the pelvic nerves in Figure 29.7. The long preganglionic fibers extend into **intramural ganglia** located in the walls of the abdominal organs where they synapse with short ganglionic neurons that control the viscera. A network of preganglionic neurons, called the aortic plexus, occurs between the vagus nerve and the pelvic nerves. In the plexus, preganglionic sympathetic and parasympathetic neurons intermingle as they pass to their respective autonomic ganglia.

Because the intramural ganglia are located in the visceral walls, parasympathetic preganglionic neurons are long and ganglionic neurons are short. The preganglionic neurons release acetylcholine which is always excitatory to the ganglionic fiber. The short ganglionic fibers also release ACh. The effect of the ganglionic neuron's neurotransmitter depends on the type of ACh receptors present in the visceral effector's cell membrane. Generally, the parasympathetic response is a rest-and-digest response which slows body functions and promotes digestion and waste elimination. Table 29.1 and Figure 29.8 compare parasympathetic and sympathetic innervation.

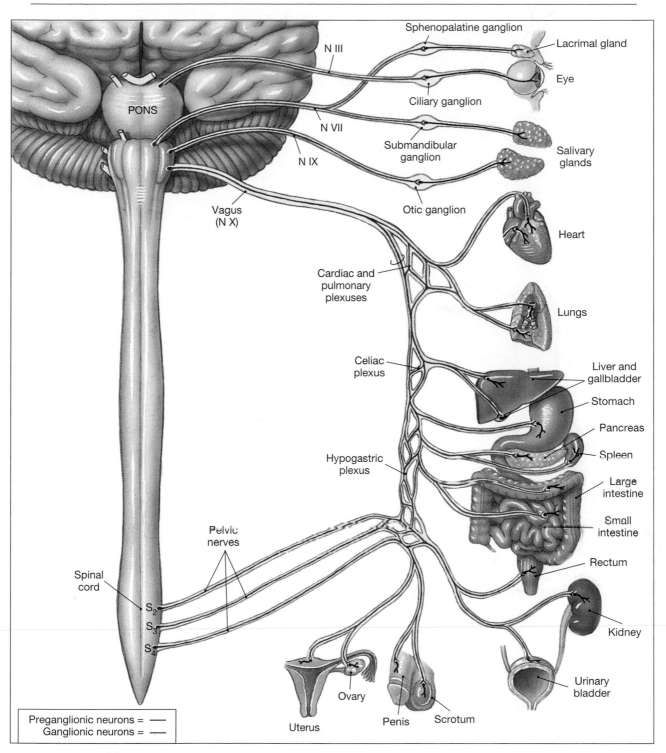

Figure 29.7
The Distribution of Parasympathetic Innervation.

TABLE 29.1	A Comparison of the Sympathetic and Parasympathetic Divisions of the ANS	
Characteristic	*Sympathetic Division*	*Parasympathetic Division*
Location of CNS visceral motor neuron	Lateral gray horns of spinal segments T_1–L_2	Brain stem and spinal segments S_2–S_4
Location of PNS ganglia	Near vertebral column	Typically intramural
Preganglionic fibers Length Neurotransmitter released	Relatively short Acetylcholine	Relatively long Acetylcholine
Postganglionic fibers Length Neurotransmitter released	Relatively long Normally NE; sometimes ACh	Relatively short Acetylcholine
Neuromuscular or neuroglandular junction	Varicosities and enlarged terminal knobs that release transmitter near target cells	Junctions that release transmitter to special receptor surface
Degree of divergence from CNS to ganglion cells	Approximately 1:32	Approximately 1:6
General function(s)	Stimulates metabolism; increases alertness; prepares for emergency ("fight or flight")	Promotes relaxation, nutrient uptake, energy storage, ("rest and digest")

Figure 29.8
The Organization of
the Parasympathetic
Division of the ANS.

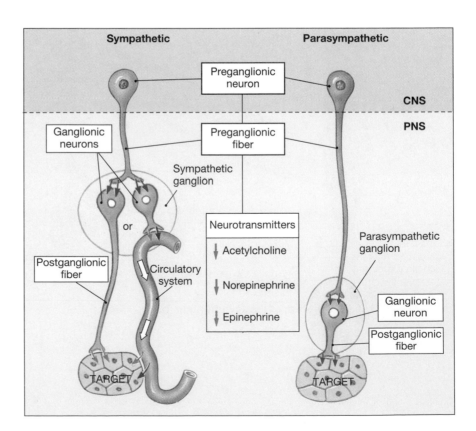

LABORATORY ACTIVITY PARASYMPATHETIC DIVISION

1. Review the anatomy and pathways presented in Figures 29.6 and 29.7.
2. Use a chart of the nervous system and identify the oculomotor, facial, glossopharyngeal, and vagus cranial nerves. On what part of the brain are these nerves located?

   _____

   _____

3. On the nervous system chart, trace the following parasympathetic pathways:
   a. A preganglionic fiber entering a pelvic nerve and traveling to the urinary bladder.
   b. The vagus nerve leading from the brain to the heart.
   c. A preganglionic fiber synapsing in the ciliary ganglion.

AUTONOMIC NERVOUS SYSTEM CHECKLIST

This is a list of bold terms presented in Exercise 29. Use this list for review purposes after you have completed the laboratory activities.

INTRODUCTION TO ANS

• Figures 29.1 and 29.2
somatic nervous system (SNS)
autonomic nervous system (ANS)
visceral effectors
sympathetic division of ANS
parasympathetic division of ANS
preganglionic neuron
preganglionic fiber
ganglionic neuron
ganglionic fiber
autonomic ganglion
dual innervation

SYMPATHETIC DIVISION

• Figures 29.1–29.5, 29.8, Table 29.1
thoracolumbar division of ANS
sympathetic chain ganglion
  paravertebral ganglion
  white ramus
  gray ramus
  rami communicantes
collateral ganglion
  splanchnic nerves
  celiac ganglion
  superior mesenteric ganglion
  inferior mesenteric ganglion
adrenal medulla

PARASYMPATHETIC DIVISION

• Figures 29.1, 29.2, 29.6–29.8, Table 29.1
craniosacral division of ANS
preganglionic neuron
preganglionic fiber
ganglionic neuron
ganglionic fiber
parasympathetic ganglia
ciliary ganglion
sphenopalatine ganglion
submandibular ganglion
otic ganglion
intramural ganglion
pelvic nerves

AUTONOMIC NERVOUS SYSTEM

Laboratory Report Exercise 29

A. Matching

Match each structure of the autonomic nervous system listed on the left with the correct description on the right.

1. _____ preganglionic neuron	A. parasympathetic division of ANS
2. _____ gray ramus	B. ganglia located in the wall of the viscera
3. _____ adrenal medulla	C. carries preganglionic fiber into a chain ganglion
4. _____ rami communicantes	D. has cell body located in autonomic ganglion
5. _____ thoracolumbar division of ANS	E. innervated by sympathetic preganglionic fibers
6. _____ collateral ganglia	F. parasympathetic nerves outflowing from sacral spinal segments
7. _____ intramural ganglia	G. white and gray rami
8. _____ white ramus	H. ganglia located anterior to spinal cord
9. _____ ganglionic neuron	I. cell body in lateral gray horn of spinal cord
10. _____ craniosacral division of ANS	J. carries postganglionic fiber into spinal nerve
11. _____ pelvic nerves	K. sympathetic division of ANS

B. Short-Answer Questions

1. Compare the outflow of preganglionic neurons from the CNS in the sympathetic and parasympathetic divisions of the ANS.

2. List four responses to sympathetic stimulation and four responses to parasympathetic stimulation.

3. Discuss the anatomy of the sympathetic chain ganglia. How do fibers enter and exit these important ganglia?

4. What cranial nerves are involved in the parasympathetic division of the ANS?

5. Compare the lengths of preganglionic and ganglionic neurons in the sympathetic and parasympathetic divisions of the ANS.

6. Compare the effect of the neurotransmitter from a sympathetic and a parasympathetic ganglionic fiber on the following autonomic effectors: smooth muscle in the digestive tract, cardiac muscle, and arterioles in skeletal muscles.

The Endocrine System

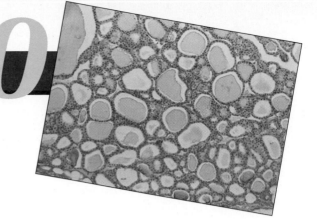

OBJECTIVES

On completion of this exercise, you should be able to:

- Compare the two regulatory systems of the body, the nervous and endocrine systems.
- Identify each endocrine gland on laboratory models.
- Describe the histological appearance of each endocrine gland.
- Identify each endocrine gland when viewed microscopically.

WORD POWER

adenohypophysis (adeno—a gland)
follicle (follicul—small bag)
parathyroid (para—near)
cortex (cort—bark or shell)
medulla (medull—pith)
acinar (acin—a berry)
corpus luteum (corpus—body; luteum—yellow)

MATERIALS

torso mannequins
endocrine charts
dissecting and compound
   microscopes
prepared slides:
   hypophysis
   thyroid

parathyroid
thymus
adrenal
pancreas
testes
ovary

INTRODUCTION

Two systems regulate homeostasis, the nervous and endocrine systems. These systems must coordinate their activities to maintain control of internal functions. The nervous system responds rapidly to environmental changes, sending electrical commands to the body that cause an immediate response. The duration of each electrical impulse is brief, measured in milliseconds. The endocrine system maintains long-term control. In response to stimuli, **endocrine glands** produce regulatory molecules called **hormones** which slowly cause a change in the metabolic activities of **target cells**, cells with membrane receptors for the hormone. The effect of the hormone is typically prolonged and lasts several hours.

The secretion of many hormones is regulated by negative feedback mechanisms. In **negative feedback**, a stimulus causes a response that reduces or removes the stimulus. An excellent example of a similar response is the operation of a central air conditioning (A/C) unit. When a room heats up, the warm temperature activates the thermostat on the wall, turning on the compressor of the A/C unit. Air flows in and cools the room and removes the stimulus, the warm temperature. Once the stimulus is completely removed, the A/C shuts off. Negative feedback is therefore a self-limiting mechanism.

An example of negative feedback control of hormonal secretion is the regulation of insulin, a hormone from the pancreas that lowers blood sugar concentration. When blood sugar levels are high, as they are after a meal, the pancreas secretes insulin.

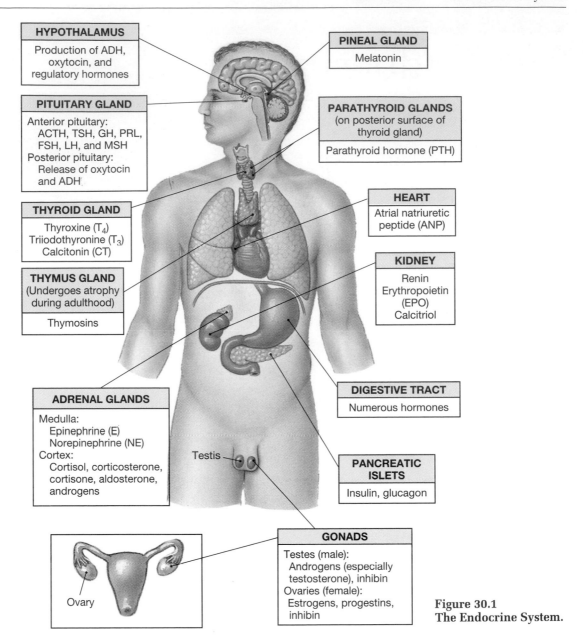

**Figure 30.1
The Endocrine System.**

Insulin stimulates cells to increase their sugar consumption and storage, thus lowering blood sugar concentration. As sugar levels return to normal, insulin secretion stops.

In this exercise we will study the pituitary gland, thyroid and parathyroid glands, thymus gland, adrenal glands, pancreas, and gonads. Figure 30.1 shows the location and the hormones secreted by each endocrine gland. Identify each one on all laboratory models and charts.

A. Pituitary Gland

Anatomy

The **pituitary gland** or **hypophysis (hī-POF-i-sis)** produces hormones that control the activity of other endocrine glands and is therefore called the "master gland." It is a

Figure 30.2
Anatomy and Orientation of the
Pituitary Gland. (LM ×62)

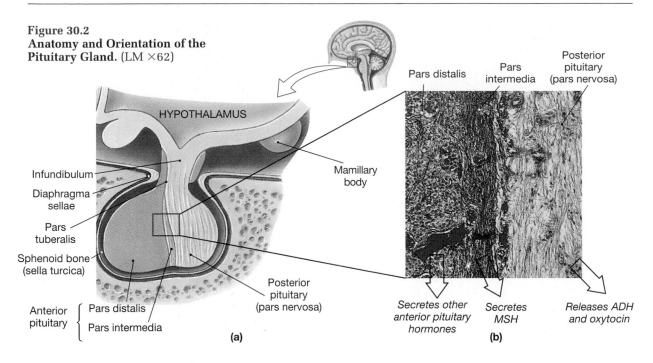

Pars distalis

Pars intermedia

Posterior pituitary (pars nervosa)

HYPOTHALAMUS

Infundibulum

Diaphragma sellae

Pars tuberalis

Sphenoid bone (sella turcica)

Mamillary body

Anterior pituitary { Pars distalis / Pars intermedia

Posterior pituitary (pars nervosa)

(a)

Secretes other anterior pituitary hormones

Secretes MSH

(b)

Releases ADH and oxytocin

small gland located in the sella turcica (Turk's saddle) of the sphenoid bone of the skull, immediately inferior to the hypothalamus of the brain. Note how in Figure 30.2 the sphenoid bone serves as a protective cradle around the gland. A stalk or **infundibulum (in-fun-DIB-ū-lum)** attaches the gland to the hypothalamus. Hormones and regulatory molecules from the hypothalamus travel down a plexus of blood vessels in the infundibulum to reach the hypophysis.

The pituitary gland contains two lobes, the anterior and posterior lobes. The anterior lobe, the **adenohypophysis (ad-ē-nō-hī-POF-i-sis)** or **pars distalis (dis-TAL-is)**, produces many hormones that target other endocrine glands, causing them to produce and secrete their hormones. A smaller part of the adenohypophysis, the **pars intermedia (in-ter-MĒ-dē-uh),** separates the anterior and posterior pituitary. It produces a single hormone, melanocyte-stimulating hormone (MSH). Figure 30.3 lists all the anterior pituitary hormones and their target organs. This figure also includes the regulatory hormones from the hypothalamus that control secretions of the adenohypophysis.

The posterior lobe, the **neurohypophysis (noo-rō-hī-POF-i-sis)** or **pars nervosa**, does not produce hormones; the hypothalamus produces antidiuretic hormone (ADH) and oxytocin (OT) which pass down the infundibulum to enter the neurohypophysis for storage and release.

Histology

The adenohypophysis is populated by a variety of cell types, each secreting a different hormone. Compared to the neurohypophysis, it contains more cells and looks denser. Most of the cells stain a reddish color. The neurohypophysis consists of axons from hypothalamic neurons.

**Figure 30.3
Pituitary Hormones
and Their Targets.**

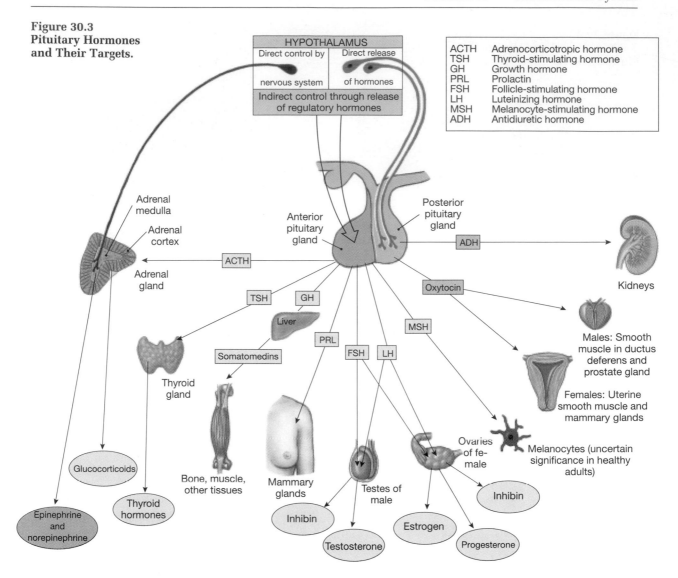

HYPOTHALAMUS

Direct control by nervous system | Direct release of hormones

Indirect control through release of regulatory hormones

ACTH	Adrenocorticotropic hormone
TSH	Thyroid-stimulating hormone
GH	Growth hormone
PRL	Prolactin
FSH	Follicle-stimulating hormone
LH	Luteinizing hormone
MSH	Melanocyte-stimulating hormone
ADH	Antidiuretic hormone

Adrenal medulla

Adrenal cortex

Anterior pituitary gland

Posterior pituitary gland

ADH

Kidneys

Adrenal gland

ACTH

Oxytocin

TSH   GH

Liver

MSH

Males: Smooth muscle in ductus deferens and prostate gland

Somatomedins

Thyroid gland

PRL

FSH   LH

Females: Uterine smooth muscle and mammary glands

Ovaries of fe-male

Melanocytes (uncertain significance in healthy adults)

Glucocorticoids

Bone, muscle, other tissues

Mammary glands

Testes of male

Inhibin

Thyroid hormones

Inhibin

Estrogen

Epinephrine and norepinephrine

Testosterone

Progesterone

LABORATORY ACTIVITY   MICROSCOPIC OBSERVATIONS
OF THE PITUITARY GLAND

MATERIALS

dissecting and compound microscope
prepared slide: pituitary gland

PROCEDURES

1. Obtain a slide of the pituitary gland. The slide usually has both lobes of the gland.
2. Use the dissecting microscope to survey the pituitary gland slide at low magnification. Can you distinguish between the two lobes of the gland?
3. Examine the slide at low and medium powers with the compound microscope. Identify the adenohypophysis and neurohypophysis, noting the different cell arrangements in each.

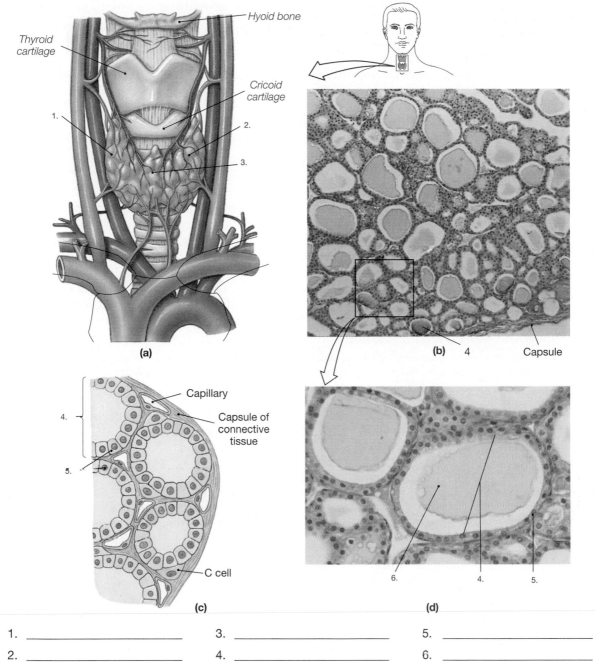

Hyoid bone

Thyroid
cartilage

Cricoid
cartilage

1.

2.

3.

(a)

(b) 4   Capsule

Capillary

Capsule of
connective
tissue

4.

5.

C cell

(c)

6.   4.   5.

(d)

1. _____   3. _____   5. _____

2. _____   4. _____   6. _____

Figure 30.4
The Thyroid Gland.

(a) Location and gross anatomy of the thyroid. **(b)** Histological organization of the thyroid. (LM ×99)
(c) Diagrammatic view of a section through the wall of the thyroid gland. **(d)** Histological details of
the thyroid gland, showing thyroid follicles. (LM ×211)

B. Thyroid Gland

Anatomy

The **thyroid gland** is located in the anterior aspect of the neck, directly inferior to the
thyroid cartilage (Adam's apple) of the larynx and just superior to the trachea. This
gland consists of two lateral lobes connected by a central mass, the **isthmus** (IS-mus).
Locate the thyroid gland in Figures 30.1 and 30.4a. This gland produces two groups

of hormones associated with the regulation of cellular metabolism and calcium homeostasis.

Histology

The thyroid gland is very distinctive. It is composed of oval **follicles** embedded in connective tissue. Locate the follicles in the micrograph in Figure 30.4b. Each one is composed of a single layer of cuboidal epithelium called **follicular cells**. The lumen of each follicle is filled with a **colloid**, a glycoprotein substance called **thyroglobulin (thī-rō-GLOB-ū-lin)** that stores thyroid hormones. Adjacent to the follicles are larger **C cells**, also called parafollicular cells.

Hormones

Follicular cells produce the hormones **thyroxine** (T_4) and **triiodothyronine** (T_3), hormones that regulate metabolic rate. These hormones are secreted into the lumen of the follicles as the glycoprotein thyroglobulin, which stores them. Thyroglobulin is later reabsorbed by the follicular cells and released into the bloodstream.

The C cells of the thyroid gland produce the hormone **calcitonin** (**CT**) which decreases blood calcium levels. Calcitonin stimulates osteoblasts in bone tissue to store calcium in bone matrix and lower fluid calcium levels. It also inhibits osteoclasts in bone from dissolving bone matrix and releasing calcium.

What if you were diagnosed as having hyperthyroidism? Your thyroid gland is overly active and secretes too much thyroid hormone. How would your body respond to this imbalance?

LABORATORY ACTIVITY   MICROSCOPIC OBSERVATIONS OF THYROID

MATERIALS

dissecting and compound microscope
prepared slide: thyroid gland

PROCEDURES

1. Label the features of the thyroid gland in Figure 30.4.
2. Scan the thyroid slide with the dissecting microscope. What structures are visible at this magnification?
3. Use a compound microscope to view the thyroid slide at low and medium powers. Locate the following structures: thyroid follicle, follicular cells, thyroglobulin, and C cells.
4. In the space provided sketch several thyroid follicles as observed at medium magnification.

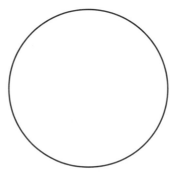

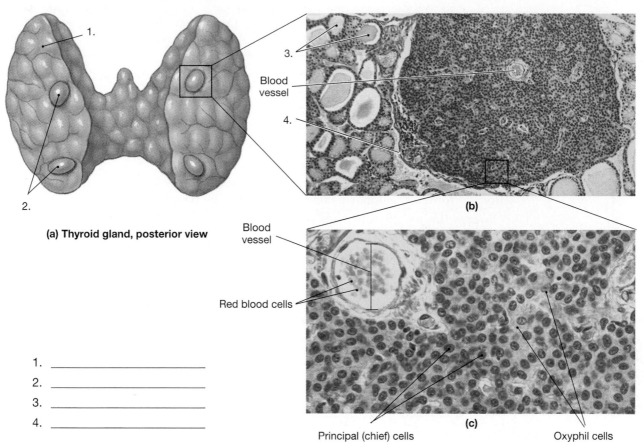

(a) Thyroid gland, posterior view

Blood vessel

Blood vessel

Red blood cells

Principal (chief) cells   (c)   Oxyphil cells

1. _____
2. _____
3. _____
4. _____

Figure 30.5
The Parathyroid Glands

(a) The location of the parathyroid glands on the posterior surfaces of the thyroid lobes. **(b)** Both parathyroid and thyroid tissues. (LM × 94) **(c)** Parathyroid cells. (LM × 685)

C. Parathyroid Glands

Anatomy

The **parathyroid glands** consist of two pairs of oval masses on the posterior surface of the thyroid gland. Each one is isolated from the underlying thyroid tissue by the parathyroid **capsule**. Locate the parathyroid glands on the laboratory models and label them in Figures 30.1 and 30.5.

Histology

The parathyroid glands are composed of **chief cells** and **oxyphil cells**. The chief cells stain dark, are more numerous, and cluster together in strings or cords. The cells with less stain in the cytoplasm are oxyphil cells. See Figure 30.5 for details on the histology of the parathyroid glands.

Hormones

Parathyroid glands produce **parathormone (PTH)**, also called **parathyroid hormone**, which is antagonistic to calcitonin from the thyroid gland. Parathormone increases the blood calcium level by stimulating osteoclasts in bone to dissolve small areas of bone matrix and release calcium ions into the blood. PTH also stimulates calcium uptake in the digestive system and reabsorption of calcium from the filtrate in the kidneys.

LABORATORY ACTIVITY   MICROSCOPIC OBSERVATIONS
                       OF THE PARATHYROID GLANDS

MATERIALS

dissecting and compound microscope
prepared slide: parathyroid gland

PROCEDURES

1. Label the parathyroid tissue in Figure 30.5.

2. Examine the parathyroid slide with the dissecting microscope. Scan the gland for thyroid follicles that may be on the slide near the parathyroid tissue.

3. Observe the parathyroid slide at low and medium powers with the compound microscope. Locate the dark stained chief cells and the light stained oxyphil cells.

4. In the space provided, sketch the parathyroid gland as observed at medium magnification.

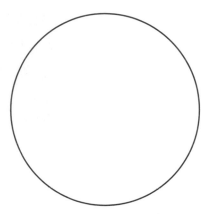

D. Thymus Gland

Anatomy

The **thymus gland** is located inferior to the thyroid gland, in the thoracic cavity posterior to the sternum. Because hormones secreted by the thymus gland facilitate development of the immune system, the gland is larger and more active in youngsters than in adults.

Histology

The thymus gland is organized into many **lobules**, separated by **septae.** Each lobule consists of a dense outer **cortex** and a light-staining central **medulla**. The cortex is populated by reticular epithelial cells that secrete the thymic hormones. The medulla is characterized by the presence of oval **thymic corpuscles** (Hassall's corpuscles) which are surrounded by cells called **lymphocytes**. Adipose and other connective tissues are abundant in an adult thymus since the function and size of the gland decrease after puberty. Label the thymus gland in Figure 30.6.

Hormone

The epithelial cells of the thymus gland produce several hormones, however, the function of only one, **tymosin,** is understood. Tymosin is essential in the development and maturation of the immune system. Removal of the gland during early childhood usually results in a greater susceptibility to acute infections.

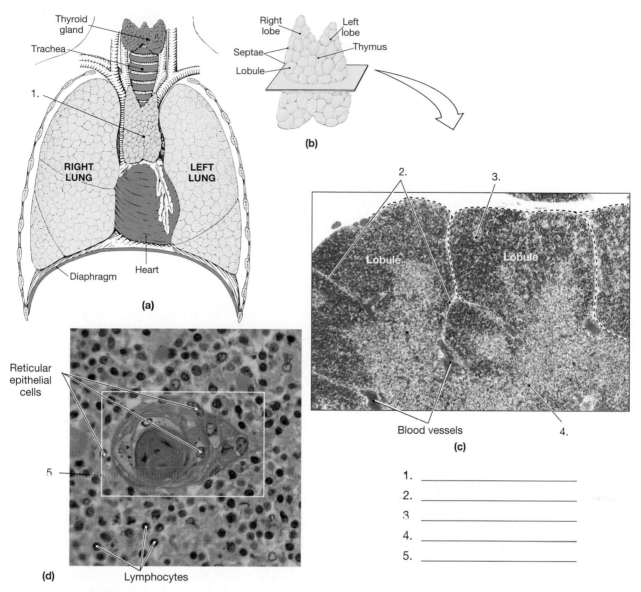

Figure 30.6
The Thymus.

(a) Appearance and position of the thymus on gross dissection; note its relationship to other organs in the chest. (b) Anatomical landmarks on the thymus. (c) A low-power light micrograph of the thymus. Note the fibrous septae that divide the thymic tissue into lobes resembling interconnected lymphatic nodules. (LM ×40) (d) At higher magnification the unusual structure of Hassall's corpuscles can be examined. The small cells in view are lymphocytes in various stages of development. (LM ×532)

LABORATORY ACTIVITY	MICROSCOPIC OBSERVATION OF THE THYMUS GLAND

MATERIALS

dissecting and compound microscopes
prepared slide: thymus gland

PROCEDURES

1. Label the thymus gland features in Figure 30.6.

2. Scan a slide of the thymus gland with the dissecting microscope and distinguish between the cortex and the medulla.

3. Examine the thymus slide with the compound microscope at low magnification to locate a stained thymic corpuscle. Increase the magnification and examine the corpuscle. The cells surrounding the corpuscles are lymphocytes.

4. In the space provided, sketch the thymus gland as observed at medium magnification.

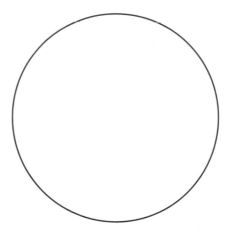

E. Adrenal Glands

Anatomy

Above each kidney is an **adrenal gland**, also called a **suprarenal gland (soo-pra-RĒ-nal)** because of its location superior to the kidney (see Figure 30.1). An **adrenal capsule** protects and attaches the adrenal gland. The adrenal gland is organized into two major regions: the outer cortex and the inner medulla. The cortex is differentiated into three distinct regions, each producing specific hormones. Label the cortex and medulla in Figure 30.7.

Histology and hormones

The adrenal **cortex** comprises three distinct layers or zones, the zona glomerulosa, the zona fasciculata, and the zona reticularis. The **zona glomerulosa (glō-mer-ū-LŌ-suh)** is the outermost cortical region. Cells in this area are stained dark and arranged in oval clusters. The zona glomerulosa secretes a group of hormones called **mineralocorticoids** which regulate, as their name implies, mineral or electrolyte concentrations of body fluids. A good example is **aldosterone,** which stimulates the kidney to reabsorb more sodium from the fluid being processed into urine.

The next layer, the **zona fasciculata (fa-sik-ū-LA-tuh)**, is made up of larger cells organized in tight columns. These cells contain large amounts of lipid, making them appear lighter than the surrounding cortical layers (see Figure 30.7). The zona fasciculatus produces a group of hormones called **glucocorticoids** which are involved in fighting stress, increasing glucose metabolism, and preventing inflammation. **Cortisol** and **cortisone** are commonly used in creams to control irritating rashes and allergic responses of the skin.

The deepest layer of the cortex, next to the medulla, is the **zona reticularis (re-tik-ū-LAR-is)**. Cells in this area are small and loosely linked together in chainlike structures. Interestingly, the zona reticularis produces **androgens**, male sex hormones. Both males and females produce small quantities of androgens in the zona reticularis.

A woman is diagnosed with a hyperactive zona reticularis, resulting in andrenogenital syndrome. How will the increased levels of androgen affect her body?

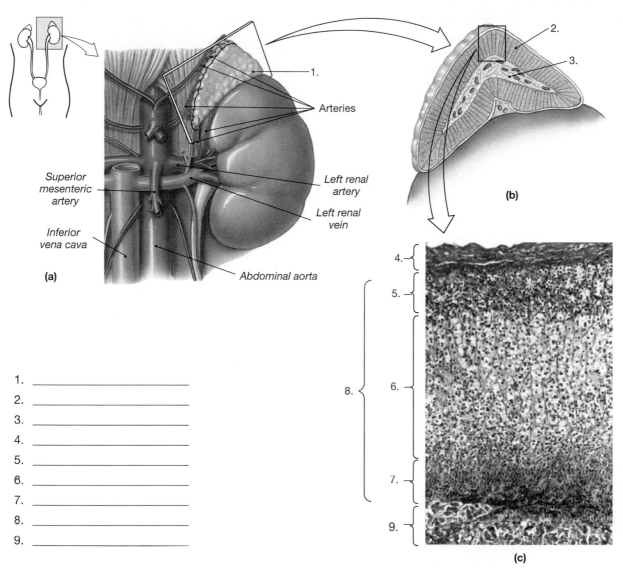

1. _____
2. _____
3. _____
4. _____
5. _____
6. _____
7. _____
8. _____
9. _____

Figure 30.7
The Adrenal Gland.

(a) Superficial view of the kidney and adrenal gland. (b) An adrenal gland, showing the sectional plane for part (c). (c) Light micrograph, with the major regions identified.

The adrenal **medulla** is the innermost part of the gland. The cells of the medulla are modified sympathetic neurons that secrete **epinephrine** and **norepinephrine**, hormones involved in the sympathetic fight-or-flight response. The many blood vessels in the medulla give the tissue a dark-red coloration.

LABORATORY ACTIVITY   MICROSCOPIC OBSERVATIONS OF THE ADRENAL GLANDS

MATERIALS

   dissecting and compound microscopes
   prepared slide: adrenal gland

PROCEDURES

1. Label the components of the adrenal gland in Figure 30.7.
2. Examine a slide of the adrenal gland with the dissecting microscope and distinguish among the capsule, adrenal cortex, and medulla.
3. Observe the adrenal gland with the compound microscope and differentiate among the three layers of the adrenal cortex.
4. In the space provided, sketch the adrenal gland showing the details of the three cortical layers and the medulla.

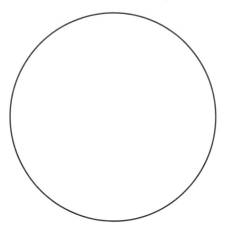

F. Pancreas

Anatomy

The **pancreas** lies between the stomach and the duodenum of the small intestine. The gland is a "double gland," performing both exocrine and endocrine functions. Exocrine glands have ducts or tubules that transport secretions to the body surface or to an exposed internal surface. Exocrine cells of the pancreas secrete digestive enzymes, buffers, and other molecules into pancreatic ducts for transport to the duodenum. Endocrine portions of the pancreas produce hormones that regulate blood sugar metabolism.

Histology

Figure 30.8 details the histology of the pancreas which is composed mainly of exocrine **acinar cells** which secrete pancreatic juice for digestive processes. These cells are dark-stained and grouped together in small ovals or columns. Connective tissues and pancreatic ducts are dispersed in the tissue. Embedded in the acinar cells are isolated clusters of **pancreatic islets** or islets of Langerhans (LAN-ger-hanz). Each islet houses four types of endocrine cells, **alpha cells**, **beta cells**, **delta cells**, and **F cells**. These cells are difficult to distinguish with routine staining techniques and will not be individually examined.

Hormones

Pancreatic hormones regulate carbohydrate metabolism. Alpha cells in the pancreatic islets secrete the hormone **glucagon (GLOO-ka-gon)** which raises blood sugar concentration by catabolizing glycogen, the stored form of carbohydrate, into glucose for cellular respiration. This process is called glycogenolysis. Beta cells secrete **insulin (IN-suh-lin)** which accelerates glucose uptake by cells and glycogenesis, the formation of glycogen. Insulin lowers blood sugar concentration by promoting the removal of sugar from the bloodstream.

Although advertisements on television encourage us to eat a candy bar for quick energy, some individuals actually feel depressed after eating chocolate and other high-sugar snacks. What is the cause of their "sugar depression"?

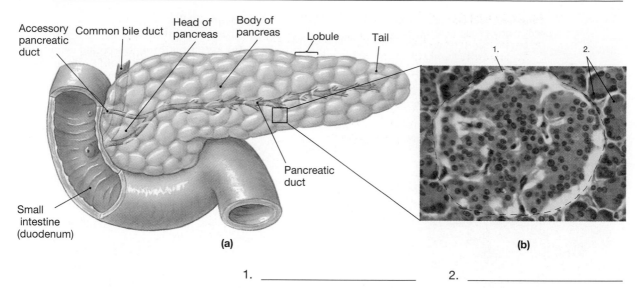

1. _____   2. _____

Figure 30.8
The Endocrine Pancreas.

(a) Gross anatomy of the pancreas. **(b)** Pancreatic islet surrounded by exocrine-secreting cells.
(LM ×276)

LABORATORY ACTIVITY   MICROSCOPIC OBSERVATIONS OF THE PANCREAS

MATERIALS

   compound microscope
   prepared slide: pancreas

PROCEDURES

1. Label the histology of the pancreas in Figure 30.8.

2. Use a compound slide and locate the dark-stained acinar cells and the oval pancreatic ducts of the pancreas. Identify the clusters of pancreatic islets, the endocrine portion of the gland.

3. In the space provided, sketch the pancreas, labeling the pancreatic islets and the acinar cells.

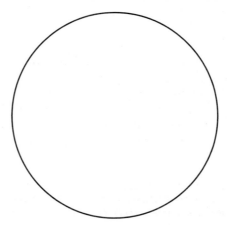

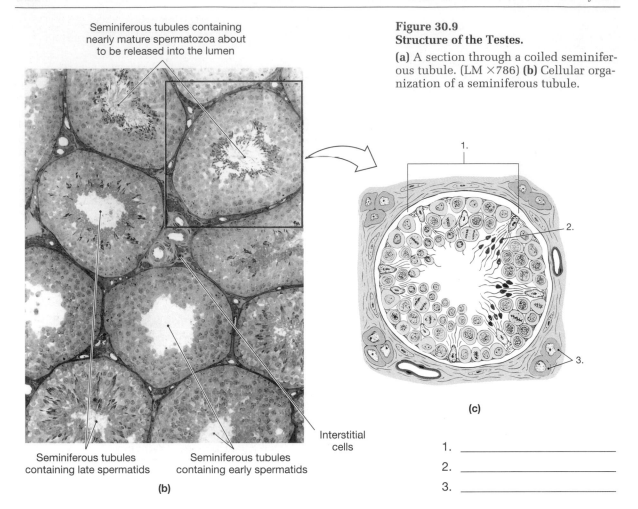

Seminiferous tubules containing
nearly mature spermatozoa about
to be released into the lumen

Figure 30.9
Structure of the Testes.

(a) A section through a coiled seminiferous tubule. (LM ×786) **(b)** Cellular organization of a seminiferous tubule.

Interstitial
cells

(c)

Seminiferous tubules
containing late spermatids

Seminiferous tubules
containing early spermatids

(b)

1. _____

2. _____

3. _____

G. Gonads

Gonads are the male and female reproductive organs that produce gametes, the spermatozoa and ova that fuse and start a new life. The **testes** and **ovaries** secrete hormones
to regulate development of the reproductive system and maintenance of the sexually
mature adult. Locate the gonads in Figure 30.1.

Anatomy and histology

Testes are the male gonads, located outside the body in the pouchlike scrotum. A testis or
testicle is made up of coiled **seminiferous (se-mi-NIF-e-rus) tubules** which produce spermatozoa. Figure 30.9 illustrates seminiferous tubules in cross section with spermatozoa
in the tublar lumen. **Interstitial cells**, located between the seminiferous tubules, are
endocrine cells and secrete the male sex hormone **testosterone (tes-TOS-ter-ōn)**, the hormone that produces and maintains secondary male sex characteristics such as facial hair.

The **ovaries** are the female gonads, located in the pelvic cavity. During each ovarian
cycle, a small group of immature eggs or oocytes begins to develop an outer capsule of
follicular cells. Eventually, one follicle becomes a **Graafian (GRAF-ē-an) follicle**, a
fluid-filled bag containing an oocyte for release at ovulation (see Figure 30.10). The
Graafian follicle is a temporary endocrine structure and secretes the hormone **estrogen**
to prepare the uterus for implantation of a fertilized egg. After ovulation, the ruptured

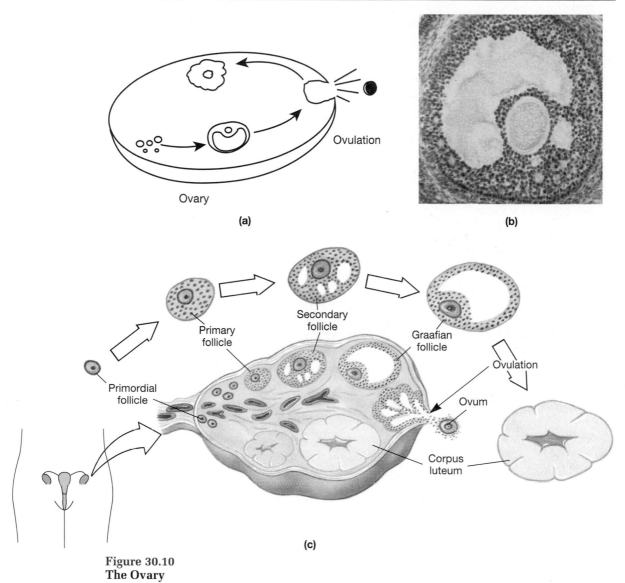

Figure 30.10
The Ovary

(a) Overview of ovarian cycle highlighting the Graafian follicle, ovulation, and corpus luteum. **(b)** Micrograph of Graafian follicle. **(c)** Follicular development during the ovarian cycle.

Graafian follicle becomes the **corpus luteum (LOO-tē-um)**, another temporary endocrine structure which secretes **progesterone (pro-JES-ter-ōn)** that promotes further thickening of the uterine wall.

LABORATORY ACTIVITY   MICROSCOPIC OBSERVATIONS OF THE GONADS

MATERIALS

compound microscope
prepared slides:
    testis
    ovary

PROCEDURES
Testis
1. Label the structures of the testis in Figure 30.9.
2. Scan a slide of the testis at low magnification and identify the many seminiferous tubules. Increase the magnification to locate interstitial cells between the seminiferous tubules.
3. Sketch a cross section of a testis detailing the seminiferous tubules and interstitial cells.

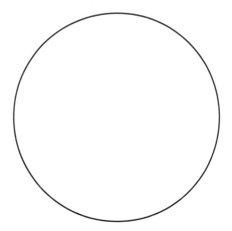

Ovary
1. Label the ovarian structures in Figure 30.10.
2. Scan the ovary slide at low power to locate the large Graafian follicle. Identify the developing ovum inside the follicle.
3. In the space provided, sketch the Graafian follicle.

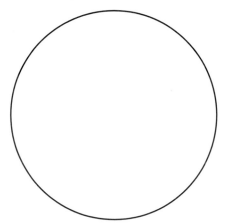

ENDOCRINE SYSTEM CHECKLIST

This is a list of bold terms presented in Exercise 30. Use this list for review purposes after you have completed the laboratory activities.

INTRODUCTION

- Figure 30.1
endocrine glands
hormone
target cell
negative feedback

PITUITARY GLAND

- Figures 30.1–30.3
pituitary gland
infundibulum
adenohypophysis
pars intermedia
neurohypophysis

THYROID AND PARATHYROID GLANDS

- Figures 30.4 and 30.5
thyroid gland
isthmus
   thyroid follicles
   follicular cells
   colloid
   C cells (parafollicular cells)
Hormones
   thyroxine (T_4), triidothyronine (T_3),
   calcitonin (CT)
parathyroid gland
   parathyroid capsule
   chief cells
   oxyphil cells
Hormone
   parathyroid hormone (PTH)

THYMUS GLAND

- Figures 30.6
thymus gland
   lobules
   septae
   cortex
   medulla
   thymic corpuscles
Hormone
   thymosin

ADRENAL GLAND

- Figure 30.7
adrenal (suprarenal) gland
adrenal capsule
cortex
   zona glomerulosa
   zona fasciculata
   zona reticularis
medulla
Hormones:
cortex:
   mineralocorticoids, aldosterone, glucocorticoids, cortisol, cortisone, androgens
medulla:
   epinephrine, norepinephrine

PANCREAS

- Figure 30.8
pancreas
   acinar cells
   pancreatic islets
   alpha cells
   beta cells
   delta cells
   F-cells
Hormones:
   glucogon, insulin

GONADS

• Figures 30.9 and 30.10

testes
  seminiferous tubules
  interstitial cells
Hormone:
  testosterone

ovary
  oocyte
  follicular cells
  Graafian follicle
  corpus luteum
Hormones:
  estrogen
  progesterone

ENDOCRINE SYSTEM

Laboratory Report Exercise 30

A. Matching

Match each endocrine structure listed on the left with the correct description on the right.

1. _____ thyroid follicle	A.	contains an ovum
2. _____ adrenal medulla	B.	four oval masses on posterior thyroid gland
3. _____ thymic corpuscle		
4. _____ seminiferous tubules	C.	neurohypophysis
5. _____ zona glomerulosa	D.	produces insulin
6. _____ parathyroid gland	E.	cells between the thyroid follicles
7. _____ acinar cells	F.	inner cortical layer of adrenal gland
8. _____ Graafian follicle	G.	contains hormones in a colloid
9. _____ adenohypophysis	H.	pituitary gland
10. _____ master gland	I.	develops from ruptured Graafian follicle
11. _____ target cell		
12. _____ pancreatic islets	J.	innermost part of gland above kidney
13. _____ interstitial cells	K.	stalk of pituitary gland
14. _____ zona reticularis	L.	outer cortical layer of adrenal gland
15. _____ pars nervous	M.	produce spermatozoa
16. _____ C cells	N.	exocrine cells of pancreas
17. _____ infundibulum	O.	anterior pituitary gland
18. _____ corpus luteum	P.	found in thymus gland
	Q.	produces testosterone
	R.	cell that responds to a specific hormone

B. Short-Answer Questions

1. Describe how negative feedback regulates the secretion of most hormones.

2. Why is the pituitary gland called the master gland of the body?

3. Describe the hormones produced by the three layers of the adrenal cortex.

4. Describe which structures are involved in the ovulation of an egg.

5. What are the endocrine functions of the pancreas?

6. Compare the histology of an adult thymus with that of an infant.

7. How is blood calcium regulated by the endocrine system?

C. Drawing

Draw a section of the thyroid gland and detail several thyroid follicles.

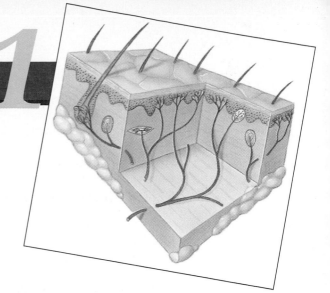

General Senses, Gustation, and Olfaction

OBJECTIVES

On completion of this exercise, you should be able to:

- List the receptors for the general senses and the special senses.
- Discuss the distribution of cutaneous receptors.
- Describe the two-point discrimination test.
- Describe the location and structure of taste buds and papillae.
- Describe the location and structure of the olfactory receptors.
- Identify the microscopic features of taste buds and olfactory epithelium.
- Explain why olfaction accentuates gustation.

WORD POWER

proprioceptor (propri—one's own)
nociceptor (noci—to injure)
gustatory cells (gusto—taste)
papillae (papilla—nipple)
circumvallate papillae (vallate—to surround)
fungiform (fungi—fungus; form—shaped)
filiform (filiform—thread)

MATERIALS

compass with millimeter scale
compound microscope
prepared slides:
    olfactory epithelium
    taste buds
phenylthiocarbamide papers
diced apple and onion
wintergreen oil
isopropyl alcohol

INTRODUCTION

Changes in the internal and external environments are detected by special cells called sensory receptors. Each type of receptor is sensitive to a specific stimulus. The taste buds of the tongue, for example, are stimulated by dissolved chemicals and not by sound waves or light rays. Receptors may be grouped into two broad sensory categories, general senses and special senses. The **general senses** have simple neural pathways and include the following sensations: touch, temperature, pain, chemical and pressure detection, and body position or proprioception. The **special senses** have complex pathways, and the receptors are housed in specialized organs. The special senses include gustation (taste), olfaction (smell), vision, audition (hearing), and equilibrium. In this exercise you will study the receptors of the general senses and the receptors of two special senses, taste and smell.

A variety of senses are included as general senses because of the simple structure of their receptors. A sensory neuron, which may be wrapped in a connective tissue sheath, monitors a specific region called a **receptive field**. Overlap in these fields enables the brain to detect where a stimulus was applied to the body. The neuron of the receptive field is connected to a specific area of the sensory cortex. This neural connection is called a **labeled line**, and the CNS interprets sensory information entirely on the basis of the line over which it arrives. Labeled lines identify the location of the stimulus. The cerebral cortex cannot tell the difference between true and false sensations. For example, when you rub your eyes, flashes of light are sometimes visible. The eye rubbing activates the optic nerve, and the sensory cortex interprets this false impulse as a visual signal.

Nerve impulses are similar to bursts of messages over a telegraph wire. The pattern of action potentials is called **sensory coding** and provides the CNS with information such as intensity, duration, variation, and movement of the stimulus.

Receptors monitor their receptive field by two methods. **Tonic** receptors are always active. Pain receptors are tonic receptors and are always "on." **Phasic** receptors are usually inactive and are turned "on" with stimulation. These receptors provide information on the rate of change of a stimulus. Root hair plexuses and Meissner's and Pacinian corpuscles are all phasic receptors for touch.

Additionally, receptors display **adaptation**, a reduction in sensitivity to repeated stimulus. When a receptor is stimulated, it first responds strongly, and then the response declines. Peripheral adaptation is the decline in response to stimuli at receptors. This adaptation reduces the amount of sensory information the CNS must process. Central adaptation occurs in the CNS. Inhibition along a sensory pathway reduces sensory infomation. Phasic receptors are fast adapting, tonic receptors are slow adapting.

A. The General Senses

Receptors for body position are called **proprioceptors (prō-prē-ō-SEP-tors)**. **Muscle spindle** receptors in muscles and **Golgi tendon organs** in tendons near joints inform the brain about muscle tension and joint position. **Thermoreceptors** have a wide distribution in the body, including the dermis, skeletal muscles, and the hypothalamus of the brain, our internal thermostat. Pressure receptors or **baroreceptors** convey signals about fluid and gas pressures. These receptors are usually simple nerve endings in blood vessels and the lungs that monitor pressure changes. **Chemoreceptors** monitor changes in chemical concentrations in body fluids. The **carotid bodies** within the base of the internal carotid arteries consist of baroreceptive and chemoreceptive neurons.

The skin has a variety of touch receptors as illustrated in Figure 31.1. **Meissner's (MĪS-nerz) corpuscles** are nerve endings sensitive to touch and are located in the dermal papillae of the skin. **Merkel's (MER-kelz) disks** are embedded in the epidermis and are very responsive to touch. **Free nerve endings** are dendrites of **nociceptors (nō-sē-SEP-tors)** or pain receptors in the epidermis. **Root hair plexuses** are dendrites of sensory neurons wrapped around hair roots. These receptors are stimulated when an insect, for example, touches and moves a hair on the skin. Pressure-sensitive receptors include **Pacinian (pa-SIN-ē-an) corpuscles** and **Ruffini (roo-FĒ-nē) corpuscles**. These receptors are located deep in the dermis.

Cutaneous receptors are not evenly distributed in the body. Some areas are densely populated with a particular receptor, while others have a few or none of that receptor. This explains why your fingertips, for example, are more sensitive to touch than your scalp. The *two-point discrimination test* is used to map the distribution of touch receptors on the skin. A compass with two points is used to determine the distance between cutaneous receptors. The compass points are gently pressed into the skin, and the subject decides if one or two points are felt. If the sensation is that of a single compass point, then only one receptor has been stimulated by the compass. By gradually increasing the distance between the points until two distinct sensations are felt, the density of the receptor population in that region can be measured.

Pacinian corpuscles
are large receptors
most sensitive to
pulsing or vibrating
stimuli; they are
common in the skin of
the fingers, breasts,
and external genitalia
and in joint capsules,
mesenteries, and the
wall of the urinary
bladder.

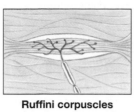

Ruffini corpuscles
detect pressure and
distortion of the dermis.

Merkel Cells

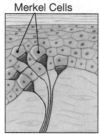

Merkel's discs
are nerve endings that
contact specialized
epithelial cells *(Merkel cells)*
sensitive to fine touch.

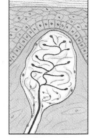

**Meissner's
corpuscles** provide
touch sensations from
the eyelids, lips,
fingertips, nipples,
and external genitalia.

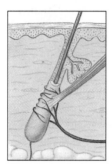

Free nerve endings
of the **root hair
plexus** are distorted
by movement of
the hairs.

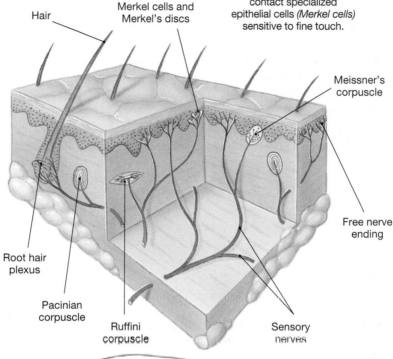

Hair

Merkel cells and
Merkel's discs

Meissner's
corpuscle

Free nerve
ending

Root hair
plexus

Pacinian
corpuscle

Ruffini
corpuscle

Sensory
nerves

Free nerve endings
are found between
epidermal cells and
are sensitive to a
variety of stimuli.

Figure 31.1
Tactile Receptors in the Skin.

LABORATORY ACTIVITY   TESTS FOR GENERAL SENSATIONS

Two-Point Discrimination Test
MATERIALS

compass with millimeter scale

PROCEDURES

1. Gently place the compass on the tip of an index finger of your laboratory partner. Slightly spread the compass points and replace them on the same area of the fingertip. Repeat this procedure until the subject feels two distinct points. Record the results in the area provided in Table 31.1.

2. Reset the compass and test the back of the hand, the back of the neck, and the lateral sides of the nose. Record the data in Table 31.1.

TABLE 31.1 Two-Point Discrimination Test	
Area of the Skin	*Minimum Distance for Two-Point Detection (mm)*
Back of neck	
Back of hand	
Tip of index finger	
Side of nose	

B. The Special Senses: Gustation and Olfaction

Gustation

The receptors for **gustation (GŪS-tă-shun)** or taste are located in **taste buds** covering the surface of the tongue, the pharynx, and the soft palate. Taste buds are inside elevations called **papillae (pa-PIL-lē)**, detailed in Figure 31.2. The base of the tongue has several circular papillae forming an inverted "V" called **circumvallate (sir-kum-VAL-āt)** papillae. These papillae contain taste buds that are sensitive to bitter substances. The tip and sides of the tongue contain buttonlike papillae called **fungiform** papillae which are stimulated by salty, sweet, and sour substances. Approximately two-thirds of the anterior portion of the tongue is covered with **filiform** papillae which provide a rough surface for the movement of food. Filiform papillae do not contain taste buds.

A taste bud contains up to 20 gustatory cells. Each cell has a small hair or microvillus that projects through a small **taste pore**. Contact with dissolved food stimulates the microvilli to produce gustatory impulses. Sensory information from the taste buds is carried to the brain by parts of three cranial nerves: the vagus nerve (X) which serves the pharynx, the facial nerve (VII) which serves the anterior two-thirds of the tongue, and the glossopharyngeal nerve (IX) which serves the posterior one-third of the tongue.

LABORATORY ACTIVITY   GUSTATORY RECEPTORS

MATERIALS

compound microscope
prepared slide: taste buds

PROCEDURES

1. Obtain a slide of the tongue showing a taste bud in cross section. Using Figure 31.2, identify the papillae and taste buds.
2. The sense of taste is a genetically inherited trait. Not all individuals can perceive the same substance. The chemical phenylthiocarbamide (PTC) tastes bitter to some individuals and sweet to others. Others, however, cannot taste the chemical. Approximately 30% of the population are nontasters of PTC.
   a. Take a strip of PTC paper, place on your tongue, and chew it several times. Are you a taster or a nontaster?
   b. Students will record their results on the chalkboard. Calculate the percentage of tasters verses nontasters and complete Table 31.2.

Olfaction

Olfactory receptors are located in the roof of the nasal cavity. Most of the air we inhale passes straight through the nasal cavity and into the pharynx. Sniffing increases our sense of smell by pulling air higher into the nasal cavity where the olfactory receptors are found. Locate the olfactory receptors in Figure 31.3.

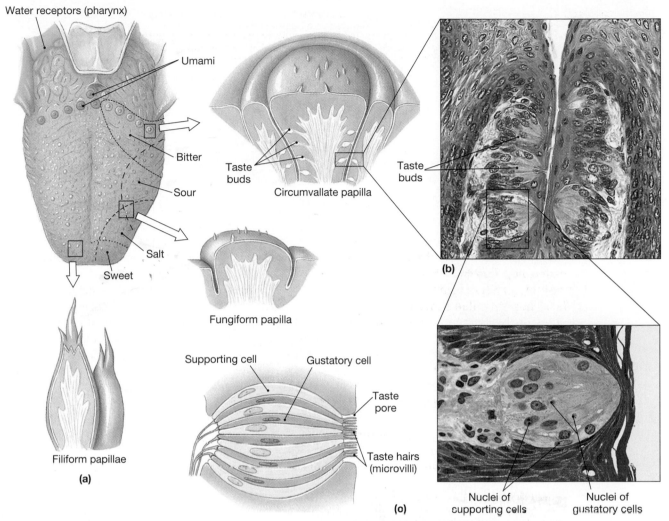

Figure 31.2
Gustatory Reception.

(a) Taste receptors are located in taste buds, which form pockets in the epithelium of fungiform or circumvallate papillae. **(b)** Taste buds in a circumvallate papilla. (LM × 280) **(c)** A taste bud, showing receptor (gustatory) cells and supporting cells. The diagrammatic view shows details of the taste pore that are not visible in the micrograph. (LM × 650)

TABLE 31.2	PTC Taste Experiment
Tasters (%)	*Nontasters (%)*

Olfactory cells have many cilia that are sensitive to airborne molecules. **Goblet cells** interspersed among the olfactory cells secrete a protective coating of mucus over the cilia. For a substance to be smelled it must diffuse through the mucus before it can stimulate the cilia of the olfactory receptors. Our sense of smell is drastically reduced by colds and allergies because mucus production increases in the nasal cavity and blocks the diffusion of molecules to the olfactory receptors. Olfactory receptors are quick to adapt to a repeated stimulus. When a new odor is present, the nose is immediately capable of sensing the new molecule. If the receptors were fatigued instead of

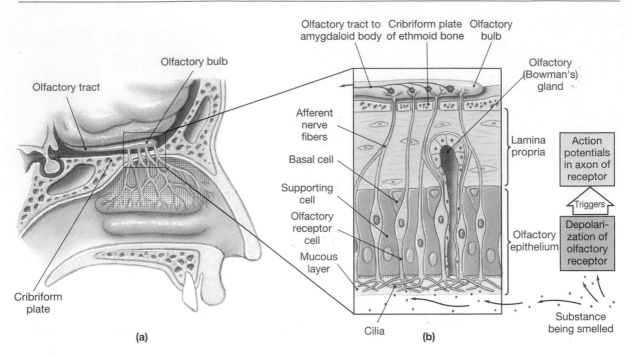

Figure 31.3
The Olfactory Organs.

(a) The structure of the olfactory organ on the left side of the nasal septum. **(b)** An olfactory receptor is a modified neuron with multiple cilia extending from its free surface.

adapted, you would not be receptive to new stimuli once the receptors reached exhaustion. We experience olfactory adaptation when we visit a friend's house. At first, we notice that the house smells different than ours. Soon, however, we adapt to the different smell and are receptive to new smells.

LABORATORY ACTIVITY OLFACTORY RECEPTORS

I. Microscopic Observation of Olfactory Epithelium
MATERIALS

> compound microscope
> prepared slide: olfactory epithelium

PROCEDURES

1. Obtain a slide of olfactory receptors. Identify the supporting and olfactory cells as seen in Figure 31.3.

II. Adaptation in Olfactory Receptors

In this experiment you will determine the length of time it takes for your olfactory epithelium to adapt to a particular odor. Use care when smelling the vials. Waft your hand over the open vial and wave the odor toward your nose.

MATERIALS

> oil of wintergreen
> isopropyl alcohol
> laboratory partner

TABLE 31.3	Olfactory Adaptation		
Vial	*Start Time*	*Stop Time*	*Time to Adaptation*
Wintergreen oil	_____	_____	_____
Isopropyl alcohol	_____	_____	_____

PROCEDURES

1. Hold the vial of wintergreen oil near your face and waft the fumes toward your nose. Ask your lab partner to record the time in Table 31.3.
2. Breathe through your nose to smell the wintergreen oil. Continue wafting and smelling until you no longer sense the odor. Have your partner record the time it took for adaptation to occur.
3. Immediately following loss of sensitivity to the wintergreen oil, smell the vial of alcohol. Explain how you can smell the second vial but could no longer smell the first.

_____

4. Repeat procedures a–c using the alcohol as the first vial. Is there a difference in adaptation to the two substances?

_____

III. Experiment with Olfactory and Gustatory Receptors

The sense of taste is thousands of times more sensitive when gustatory and olfactory receptors are stimulated simultaneously. The following experiment demonstrates the effect of smell on the sense of taste. You will place small cubes of apple or onion on your laboratory partner's dry tongue and ask if he or she can detect which type of cube is present.

MATERIALS

diced onion
diced apple
paper towels
laboratory partner

PROCEDURES

1. Dry the surface of your partner's tongue with a clean paper towel.
2. Have the partner hold his or her nose and close his or her eyes.
3. Place a cube on your partner's dry tongue.
4. Your partner is to identify the cube as apple or onion when it is placed on the tongue, immediately after chewing, and upon opening the nostrils. Responses to the different food cubes should be recorded in Table 31.4 as "yes" or "no."

TABLE 31.4	Gustatory/Olfactory Sensations		
Food	*Dry Tongue*	*After Chewing*	*Open Nostrils*
Apple	_____	_____	_____
Onion	_____	_____	_____

SENSORY ANATOMY
AND PHYSIOLOGY CHECKLIST

This is a list of the bold terms presented in Exercise 31. Use this list for review purposes after you have completed the laboratory activities.

RECEPTORS OF THE GENERAL SENSES

• Figure 31.1
receptive field
labeled line
sensory coding
   phasic receptors
   tonic receptors
adaptation

proprioceptors
   muscle spindles
   Golgi tendon organs
thermoreceptors
baroreceptors
chemoreceptors
   carotid bodies
mechanoreceptors
   Meissner's corpuscle
   Merkel's disc
   free nerve endings
   nociceptors
   root hair plexus
   Pacinian corpuscle
   Ruffini corpuscles

GUSTATION

• Figure 31.2
papillae
   taste bud
      circumvallate papillae
      fungiform papillae
      filiform papillae
   taste pore

OLFACTION

• Figure 31.3
olfactory receptors
   goblet cells

Name _____   Date _____

Section _____   Number _____

SENSORY ANATOMY AND PHYSIOLOGY
Laboratory Report Exercise 31

A. Matching
Match each sense on the left with the correct receptor on the right.

1. _____ fine touch of epidermis   **A.** Pacinian corpuscle
2. _____ deep pressure   **B.** Meissner's corpuscle
3. _____ fine touch of dermis   **C.** root hair plexus
4. _____ pain   **D.** muscle spindles
5. _____ epidermal contact   **E.** merkel's discs
6. _____ body position   **F.** free nerve endings

B. Short-Answer Questions

1. Describe the two-point discrimination test and explain how it demonstrates receptor density in the skin.

2. Explain the concept of a receptive field.

3. Why does a cold affect your sense of smell?

4. Where are the receptors for taste located?

5. Describe an experiment that demonstrates how the senses of taste and smell are linked.

6. Draw the tongue and identify where the taste buds for sweet, bitter, sour, and salty compounds are concentrated.

7. Several minutes after applying cologne, you notice that you cannot smell the fragrance. Explain this response by your olfactory receptors.

Anatomy of the Eye

OBJECTIVES

On completion of this exercise, you should be able to:

- Identify and describe the accessory structures of the eyeball.
- Identify and explain the actions of the six extrinsic eye muscles.
- Describe and identify the external and internal components of the eye.
- Describe the three tunics of the eyeball.
- Identify the microscopic organization of the retina.
- Identify the structures of a dissected cow eye.

WORD POWER

canthus (kanthos—corner of the eye)
lacrimal (lacrima—tear)
palpebral (palpebra—eyelid)
caruncle (caruncula—small fleshy mass)
sclera (scler—hard)
fovea (fovea—cuplike depression or pit)

MATERIALS

eyeball in orbit model
compound microscope
prepared slide of the retina
sectioned eye model
preserved cow eyes
dissection equipment
safety glasses, gloves

INTRODUCTION

The eyes are complex and highly specialized sensory organs. We can see the pale light of the stars and the intense, bright blue of the sky. To function in such a wide range of light conditions, the retina of the eye has two types of receptors, rods for night vision and cones for bright light and color vision. Because the level and intensity of light are always changing, the eye must regulate the size of the pupil, which allows light to enter the eye. To move the eye, six oculomotor muscles surround the eye. Four out of the twelve cranial nerves control the muscular activity of the eye and transmit sensory signals to the brain. A sophisticated system of tear production and drainage keeps the surface of the eye clean and moist. In this exercise you will examine the anatomy of the eye and dissect a cow eye.

A. External Eye Anatomy

The human eye is a spherical organ measuring about 2.5 cm (1 in.) in diameter. Only about one-sixth of the eyeball is visible between the eyelids, and the rest is recessed in the bony orbit of the skull. The corners of the eye are called the **lateral canthus (KAN-thus)** and **medial canthus**. A red, fleshy structure in the medial canthus of the eye, the **lacrimal caruncle (KAR-ung-kul)**, contains modified sebaceous and sweat glands. Secretions from the caruncle accumulate in the medial canthus during long periods of sleep. Locate these structures in Figure 32.1.

443

Figure 32.1
External Features and Accessory Structures of the Eye.

(a) Gross and superficial anatomy of the accessory
structures. **(b)** Details of the organization of the lacrimal
apparatus.

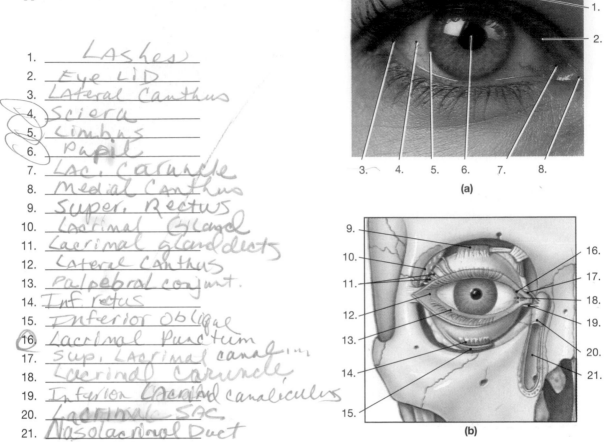

1. _Lashes_
2. _Eye LiD_
3. _Lateral Canthus_
4. _Sclera_
5. _Limbus_
6. _Pupil_
7. _Lac. Caruncle_
8. _Medial Canthus_
9. _Super. Rectus_
10. _Lacrimal Gland_
11. _Lacrimal gland ducts_
12. _Lateral Canthus_
13. _Palpebral conjunct._
14. _Inf. rectus_
15. _Inferior oblique_
16. _Lacrimal Punctum_
17. _Sup. Lacrimal canaliculi_
18. _Lacrimal Caruncle_
19. _Inferior Lacrimal canaliculus_
20. _Lacrimal Sac_
21. _Nasolacrimal Duct_

(a)

(b)

   The **accessory structures** of the eye are the eyelids, eyebrows, eyelashes, lacrimal
apparatus, and six extrinsic eye muscles. The eyelids or **palpebrae (pal-PĒ-brē)** are
skin-covered muscles that protect the surface of the eye. Blinking the eyelid keeps the
surface of the eye lubricated and clean. A thin mucous membrane called the **conjunctiva (kon-junk-TĪ-vuh)** covers the underside of the eyelids (**palpebral conjunctiva**) and
reflects over the anterior surface of the eyeball to the perimeter of the corneal epithelium (**ocular conjunctiva**). It secretes a mucus that prevents friction and moistens the
surface of the eyeball. Conjunctivitis is an inflammation of the conjunctival membrane
caused by bacteria, dust, smoke, air pollutants, or other such irritants. The **eyelashes**
are short hairs that project from the border of each eyelid. Eyelashes and **eyebrows** protect the eyeball from foreign objects such as perspiration and dust and partially shade
the eyeball from the sun. **Sebaceous ciliary glands**, located at the base of the hair follicles, help lubricate the eyeball. An infection of these glands is called a sty.
   The **lacrimal apparatus** (see Figure 32.1) consists of the lacrimal glands, lacrimal
canals, the lacrimal sac, and the nasolacrimal duct. The **lacrimal glands** are superior
and lateral to each eyeball. Each contains 6 to 12 excretory **lacrimal ducts** which
deliver a slightly alkaline solution, called lacrimal fluid or **tears**, onto the anterior surface of the eye. This fluid contains an antibacterial enzyme called lysozyme which
attacks any bacteria that may be on the surface of the eye. The tears move medially
across the surface of the eye and enter two small openings of the medial canthus called
the **lacrimal puncta** and pass into two ducts, the **lacrimal canals** which lead to an
expanded portion of the **nasolacrimal duct** called the **lacrimal sac**. The nasolacrimal

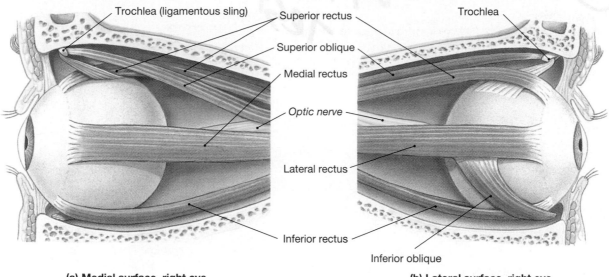

Trochlea (ligamentous sling)
Superior rectus
Superior oblique
Medial rectus
Optic nerve
Lateral rectus
Inferior rectus
Trochlea

(a) Medial surface, right eye

Inferior oblique

(b) Lateral surface, right eye

1. _Sup. rect_
2. ____ rect_
3. ____ Oblig._
4. ____ Oblique_
5. _Lat. Rect._
6. _Inf. Ret._

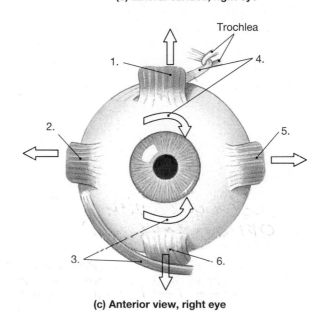

Trochlea

1.
4.
2.
5.
3.
6.

(c) Anterior view, right eye

Figure 32.2
Extrinsic Eye Muscles.

duct drains the tears into the nasal cavity. Tears clean, moisten, and lubricate the surface of the eye.

Six extrinsic (external) eye muscles control the movements of the eye (see Figure 32.2). The **superior rectus**, **inferior rectus**, **medial rectus**, and **lateral rectus** are straight muscles that move the eyeball up and down and side to side. The **superior** and **inferior oblique** muscles attach diagonally on the eyeball. The superior oblique has a tendon passing through a trochlea ("pulley") located on the upper orbit. The superior oblique muscle rolls the eye downward and the inferior oblique rolls it upward.

B. Internal Eye Anatomy

The eyeball is divided anatomically into three tunics or layers, each detailed in Figure 32.3. The outermost layer is called the **fibrous tunic** because of the abundance of dense connective tissue. The **sclera (SKLER-uh)** is the white part of the fibrous tunic covering all of the eyeball except the transparent **cornea (KOR-nē-uh)** where light enters the eye. The sclera resists punctures and maintains the shape of the eye. The cornea consists primarily of many layers of collagen fibers which are densely packed. Corneal transplants

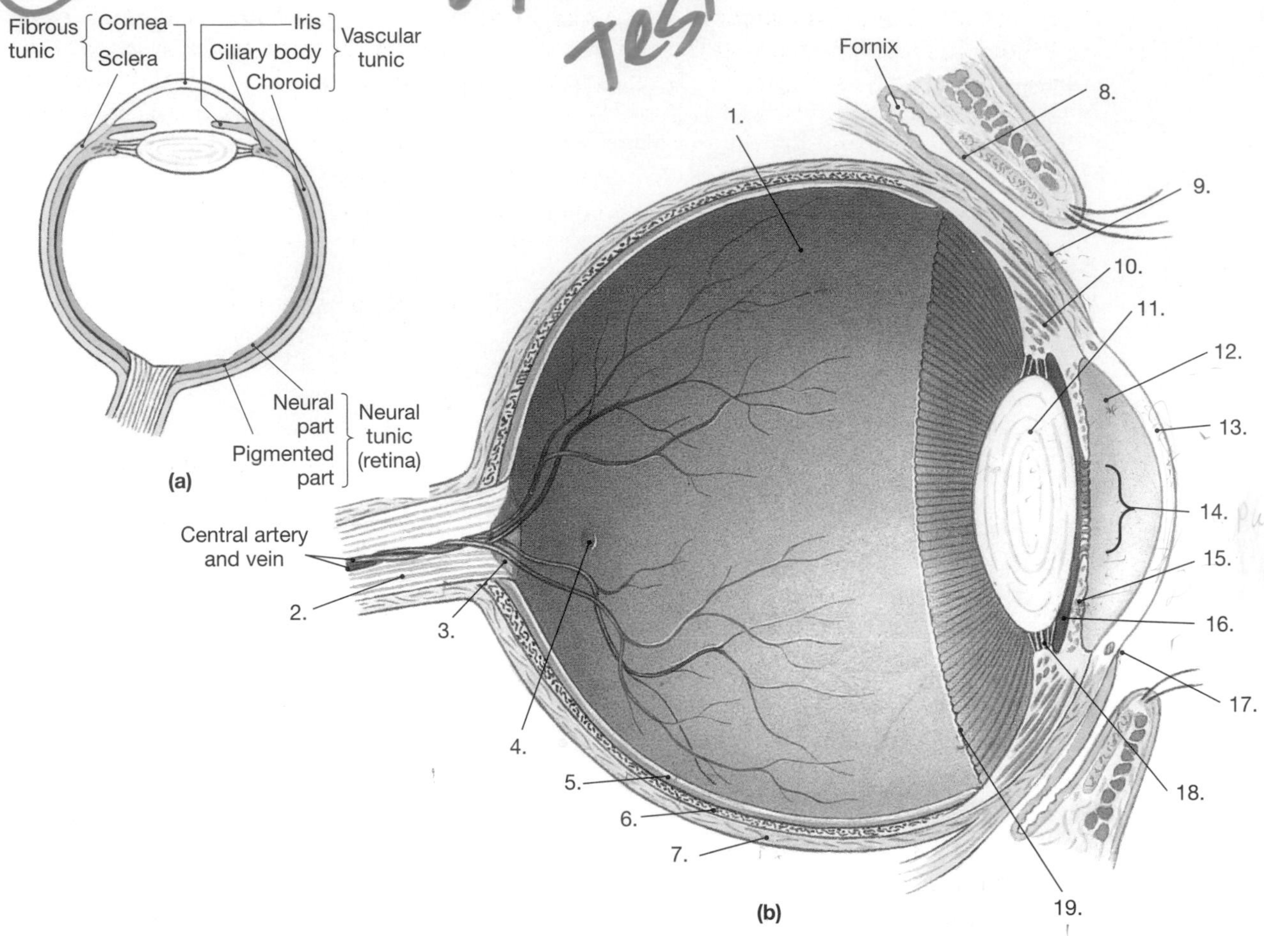

LAb
TEST

Figure 32.3
Sectional Anatomy of the Eye.

(a) The three layers, or tunics, of the eye. **(b)** Major anatomical landmarks and features viewed in a sagittal section through the left eye.

1. posterior cavity
2. optic nerve
3. optic Disc
4. Fovea
5. Retna Neural Part (retna)
6. Choroid
7. sclera

8. Eye LiD
9. LimBus
10. Ciliary Body
11. Lens
12. anterior Cavity
13. cornia
14. pupile

15. Iris
16. Posterior Chamber
17. Ocular Conjunctive
18. Suspensory Ligamn
19. Ora serrata of Len

are possible because the cornea is avascular and antibodies that would cause rejection of a foreign tissue do not circulate in this area. The **limbus** is the border between the sclera and the cornea. The **scleral venous sinus** (canal of Schlemm), a small passageway that communicates with veins of the eye, is located in this border.

The **vascular tunic** or **uvea** is the middle layer of the eye. The most-posterior portion, the **choroid**, is highly vascularized and contains a dark pigment (melanin) which absorbs light to prevent reflection. Anteriorly, the uvea is modified into the **ciliary body** and the pigmented **iris**. The ciliary body contains two structures, the **ciliary**

process and the **ciliary muscle**. The ciliary process is a series of folds with thin **suspensory ligaments** extending to the lens. The ciliary muscle adjusts the shape of the **lens** for near and far vision. The front of the iris is pigmented and has a central aperture called the **pupil**. Circular and radial muscles of the iris change the diameter of the pupil to regulate the amount of light striking the retina. The autonomic nervous system controls both sets of muscle fibers. In bright lights and for close vision, parasympathetic activation causes the circular fibers to constrict the pupils. In low light and for distant vision, sympathetic activation causes the radial fibers to contract, which dilates the pupils and allows more light into the eye.

The **retina** (see Figure 32.4) is also called the **neural tunic** because it contains receptors and two layers of sensory neurons. The anterior margin of the retina is the **ora serrata (Ō-ra ser-RA-tuh)** and appears as a jagged edge, much like a serrated knife.

The **lens** divides the eyeball into an **anterior cavity**, an area between the lens and the cornea, and a **vitreous cavity**, an area between the lens and the retina. The anterior cavity is further subdivided into an **anterior chamber** between the iris and the cornea and a **posterior chamber** between the iris and the lens. Capillaries of the ciliary processes form a watery fluid called **aqueous humor**, which is secreted into the posterior chamber. This fluid circulates through the pupil and into the anterior chamber where it is reabsorbed into the blood. The aqueous humor helps to maintain the intraocular pressure of the eyeball and supply nutrients to the avascular lens and cornea. The second, larger cavity of the eyeball is the **vitreous cavity**. This cavity contains the **vitreous body**, a clear jellylike substance that holds the retina against the choroid layer and prevents the eyeball from collapsing.

Receptors of the retina are called **photoreceptors** and are stimulated by rays of light composed of photons. Two types of receptors occur, rods and cones. **Cones** are stimulated by moderate or bright illumination and respond to different wavelengths or colors of light. Our sharpest vision or visual acuity is attributed to cones. **Rods** are sensitive to low illumination and to motion. These receptors are not stimulated by different wavelengths (colors), and our night vision is in black and white. The rods and cones face a **pigmented epithelium** on the posterior of the eyeball. Light passes through the retina, reflects off the pigmented epithelium, and strikes the photoreceptors. The photoreceptors communicate with a layer of sensory neurons called **bipolar cells**. These cells pass the visual signal on to a layer of **ganglion cells** whose axons converge to an area called the **optic disk** or blind spot where the **optic nerve** exits the eyeball. **Horizontal cells** form a network that inhibits or facilitates communication between the visual receptors and the bipolar cells. A equivalent cell called an **amacrine (AM-a-krin) cell** accentuates communication between bipolar and ganglion cells.

The optic disk lacks photoreceptors and is a "blind spot" in your field of vision. Because the visual fields of the eyes overlap, the blind spot is "filled in" and is not noticeable. Lateral to the optic disk is an area of high cone density called the **macula lutea (LOO-tē-uh)**, which means "yellow spot." In the center of the macula lutea is a small depression called the **fovea (FŌ-vē-uh)**. The fovea is the area of sharpest vision because of the high density of cones. There are no rods in the macula or the fovea, but outside these areas their numbers increase toward the periphery of the retina.

LABORATORY ACTIVITY   THE EYE

MATERIALS

eye models
eye charts
compound microscope
prepared slide: retina

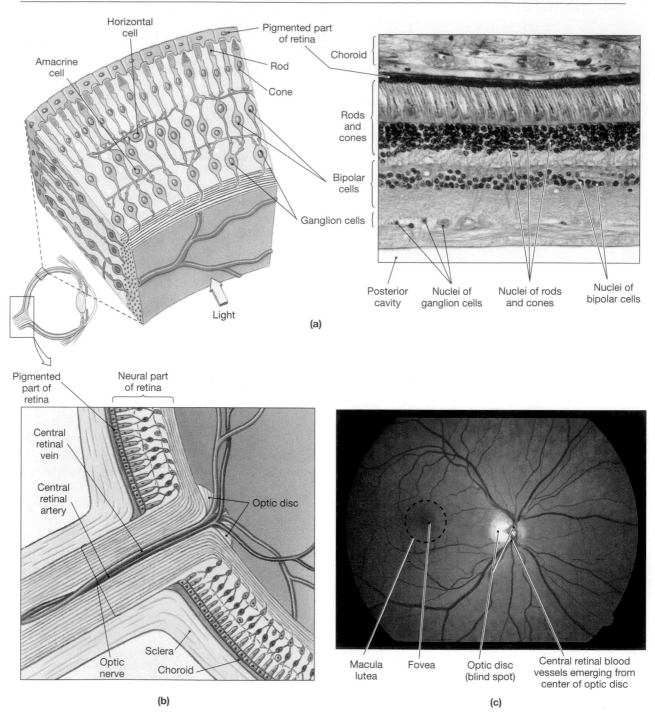

Figure 32.4
The Organization of the Retina.

(a) The cellular organization of the retina. The photoreceptors are closest to the choroid rather than near the posterior cavity (vitreous chamber). (LM × 290) **(b)** The optic disc in diagrammatic horizontal section. **(c)** A photograph of the retina as seen through the pupil.

PROCEDURES

1. Examine the eye models and charts available in the laboratory and locate each structure discussed in the manual. Label the structures of the eye in Figures 32.1–32.3.
2. Study the prepared histology slide of the retina. Use Figure 32.4 as a reference and identify the following:
   a. pigment epithelium
   b. photoreceptor (rods and cones) layer
   c. bipolar cells
   d. ganglion cells

LABORATORY ACTIVITY   DISSECTION OF THE COW EYE

The anatomy of the cow eye is similar to the human eye. The fibrous tunic, the sclera, is difficult to cut and care must be used while dissecting.

MATERIALS

fresh or preserved cow eyes

dissecting pan, dissecting scissors,

blunt probe, scalpel, gloves, safety glasses

PROCEDURES

1. Examine the external features of the eyeball. Note the amount of adipose tissue that serves to cushion the eye. Identify the optic nerve (II) as it leaves the eyeball, the remnants of the extrinsic muscles of the eye, the conjunctiva, the sclera, and the cornea. The cornea, which is normally transparent, will be opaque if the eye has been preserved.
2. Securely hold the eyeball and with the scissors remove the adipose and the extrinsic muscles from the surface of the eyeball, taking care not to remove the optic nerve (II).
3. Lay the eyeball on the dissecting pan and with a sharp scalpel make an incision about 0.6 cm ($\frac{1}{4}$ in.) back from the cornea. Continue the cut around the circumference of the eyeball with scissors, being sure to maintain the $\frac{1}{4}$-in. distance back from the cornea.
4. Carefully separate the anterior and posterior portions of the eyeball. The vitreous humor should stay with the posterior portion of the eyeball.
5. Examine the anterior portion of the eyeball. Place a blunt probe between the lens and the ciliary processes and carefully lift the lens up a little. The halo of delicate transparent filaments between the lens and the ciliary processes is formed by the suspensory ligaments. Notice the ciliary body where the suspensory ligaments originate and the heavily pigmented iris with the pupil in its center.

   Locate the vitreous body in the posterior portion of the eyeball. The retina is the tan-colored membrane that is easily separated from the heavily pigmented choroid. Examine the optic disk where the retina attaches to the posterior of the eyeball. The choroid has a greenish-blue membrane, the **tapetum lucidum**, which improves night vision in animals such as sheep and cows; humans do not have this structure.

ANATOMY OF THE EYE CHECKLIST

This is a list of the bold terms presented in Exercise 32. Use this list for review purposes after you have completed the laboratory activities.

EXTERNAL ANATOMY OF EYE

- Figures 32.1 and 32.2

lateral and medial canthus

lacrimal caruncle

accessory structures:

    palpebrae

    conjunctiva(palpebral and ocular)

    eyelashes

    eyebrows

    sebaceous ciliary glands

    lacrimal apparatus

    lacrimal glands

    lacrimal ducts

    tears

    lacrimal puncta

    lacrimal canal

    lacrimal sac

    nasolacrimal duct

    superior rectus muscle

    inferior rectus muscle

    medial rectus muscle

    lateral rectus muscle

    superior oblique muscle

    inferior oblique muscle

INTERNAL ANATOMY OF EYE

- Figure 32.3

fibrous tunic

    sclera

    cornea

    limbus

    scleral venous sinus (canal of Schlemm)

vascular tunic

uvea

choroid

    ciliary body

    iris

    ciliary process

    ciliary muscle

    suspensory ligaments

nervous tunic retina

lens, pupil

anterior cavity

    anterior and posterior chambers

    aqueous humor

vitreous cavity

vitreous body

RETINA (DETAILED)

- Figure 32.4

pigment layer

photoreceptors (rods and cones)

bipolar neurons

ganglion neurons

horizontal cells

amacrine cells

optic nerve

optic disk

macula lutea

fovea

ora serrata

tapetum lucidum (cow eye)

ANATOMY OF THE EYE

Laboratory Report Exercise 32

A. **Matching**

Match the description on the left with the correct eye structure on the right.

1. _____ transparent part of fibrous tunic	**A.** limbus
2. _____ thin filaments attached to lens	**B.** iris
3. _____ corners of eyes	**C.** vitreous body
4. _____ produces tears	**D.** optic disk
5. _____ drains tears into nasal cavity	**E.** ciliary muscle
6. _____ adjusts shape of lens	**F.** caruncle
7. _____ border of sclera and cornea	**G.** canthus
8. _____ jellylike substance of eye	**H.** cornea
9. _____ depression in retina	**I.** sclera
10. _____ area lacking photoreceptors	**J.** aqueous humor
11. _____ white of eyeball	**K.** retina
12. _____ watery fluid of anterior eye	**L.** lacrimal gland
13. _____ regulates light entering eye	**M.** suspensory ligaments
14. _____ contains photoreceptors	**N.** fovea
15. _____ red structure in medial eye	**O.** nasolacrimal duct

B. **Short-Answer Questions**

1. Describe the three layers of the eye.

2. List the flow of tears from the lacrimal gland to the nasal cavity.

3. Describe the cavities of the eye.

4. Describe how the muscles of the pupil regulate the amount of light entering the eye.

5. List the six extrinsic muscles of the eye.

C. Labeling

Label Figure 32.5, a horizontal section of the eye.

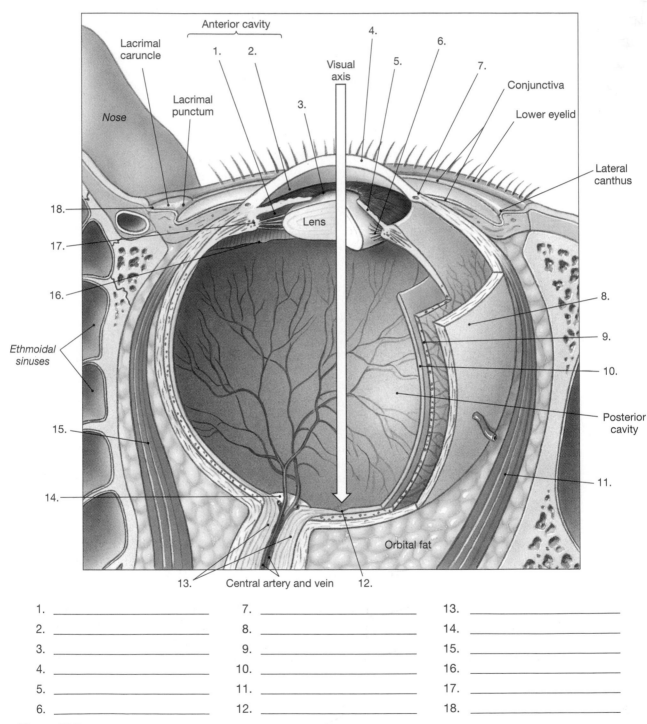

Figure 32.5
The Sectional Anatomy of the Eye.

A horizontal section of the right eye.

1. _____
2. _____
3. _____
4. _____
5. _____
6. _____
7. _____
8. _____
9. _____
10. _____
11. _____
12. _____
13. _____
14. _____
15. _____
16. _____
17. _____
18. _____

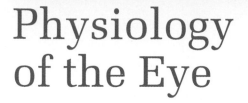

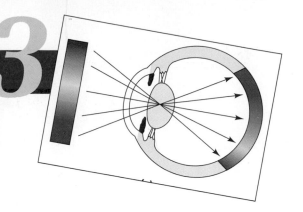

Physiology of the Eye

OBJECTIVES

On completion of this exercise, you should be able to:

- Explain why a blind spot exists in the eye and describe how it is mapped.
- Demonstrate the use of a Snellen eye chart and an astigmatism chart.
- Explain the terms myopia, hyperopia, presbyopia, and astigmatism.
- Describe how to measure accommodation.
- Discuss the role of convergence in near vision.

WORD POWER

emmetropic (emmetros—according to measure)
astigmatism (stigma—point)
presbyopia (presby—old man)
myopic (myo—to shut)
hyperopia (hyper—above, over)

MATERIALS NEEDED

Snellen eye chart
astigmatism eye chart
pen or pencil

A. Visual Acuity

Sharpness of vision or **visual acuity** is tested with a Snellen eye chart. This chart contains black letters of various sizes printed on white cardboard. Letters of a certain size are seen clearly by a person with normal vision at a given distance. A person with a visual acuity of 20/20 is considered to have normal or **emmetropic (em-e-TRŌ-pik)** vision. The lines of letters are marked with numerical values such as 20/20 and 20/30. A reading of 20/30, for example, indicates that you can see at 20 ft what an emmetropic eye can see at 30 feet; 20/30 vision is not as sharp as 20/20 vision. An eye that focuses an image in front of the retina is **myopic (mi-Ō-pik)** or nearsighted and can see close objects clearly. If an image is focused behind the retina, the eye is **hyperopic (HĪ-per-ō-pik)** or farsighted and can see only distant objects clearly. Corrective lenses are used to adjust for both conditions.

   Astigmatism (Ah-STIG-mah-tizm) is a reduction in sharpness of vision due to an irregular cornea or lens. These uneven surfaces bend or **refract** light rays incorrectly, resulting in blurred vision. The astigmatism chart has 12 sets of three lines laid out in a circular arrangement resembling a clock. Remove your glasses or contacts before performing this test.

LABORATORY ACTIVITY   VISUAL TESTS

I. Visual Acuity Test

The **Snellen eye chart**, see Figure 33.1, should be mounted on a wall at eye level, and a strip of masking tape should be placed on the floor to mark a distance of 20 ft from the

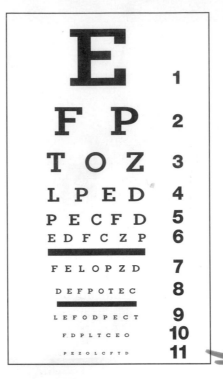

Figure 33.1
Snellen Eye Chart.

chart. You will perform this test in groups of two. One person stands next to the eye chart and indicates which line is to be read. The other stands at the 20-ft mark and tests his or her vision.

MATERIALS

Snellen eye chart
20-ft mark from chart

PROCEDURES

1. Stand at the 20-ft mark and cover one of your eyes with either a cupped hand or an index card. Read a line where you can easily focus on all the letters. Continue to view lines of progressively smaller letters until you cannot accurately read them. Record in Table 33.1 the value of the smallest line that you can read without errors. This line is your visual acuity for that eye, 20/20 for example.
2. Repeat this process with your other eye. Record your data in Table 33.1.
3. Now, using both eyes, read the lowest line that you can see clearly. Record your data in Table 33.1.
4. If you wear glasses or contacts, repeat the test while wearing your corrective lenses. Record your data in Table 33.1.

TABLE 33.1   Visual Acuity

	Acuity (without glasses)	Acuity (with glasses)
Left eye		
Right eye		
Both eyes		

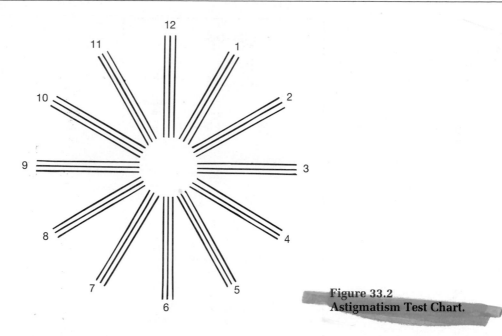

Figure 33.2
Astigmatism Test Chart.

II. Astigmatism Test

The astigmatism eye chart, shown in Figure 33.2, should be mounted on a wall at eye level. You will perform this test in groups of two. If you wear glasses, take the test with your glasses both on and off.

MATERIALS

astigmatism chart

PROCEDURES

1. Look at the white circle in the center of the chart.
2. If all the radiating black lines appear equally sharp and equally black, you have no astigmatism.
3. If some lines appear blurred or are not consistently dark, you have astigmatism.

III. Blind Spot Mapping

The optic disk or blind spot is an area of the retina lacking photoreceptors. Normally the blind spot is not seen because the visual fields of the eyes overlap and fill in the missing information from the blind spot.

MATERIALS

Figure 33.3

PROCEDURES

1. Hold Figure 33.3 2 in. from your face. Close your left eye and stare at the cross.
2. Slowly move the page away from your face. The dot will disappear when its image falls on your blind spot.

B. Accommodation

For objects closer than 20 ft from the eye the lens must bend light rays together so that they focus a sharp image on the retina. This process is called **accommodation** and is detailed in Figure 33.4.

Figure 33.3
The Optic Disc.

Close your left eye and stare at the cross with your right eye, keeping the cross in the center of your field of vision. Begin with the page a few inches away from your eye, and gradually increase the distance. The dot will disappear when its image falls on the blind spot, at your optic disc. To check the blind spot in your left eye, close your right eye and repeat this sequence while you stare at the dot.

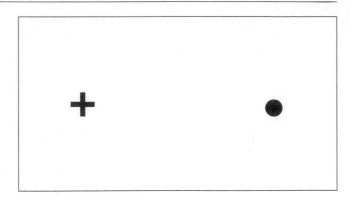

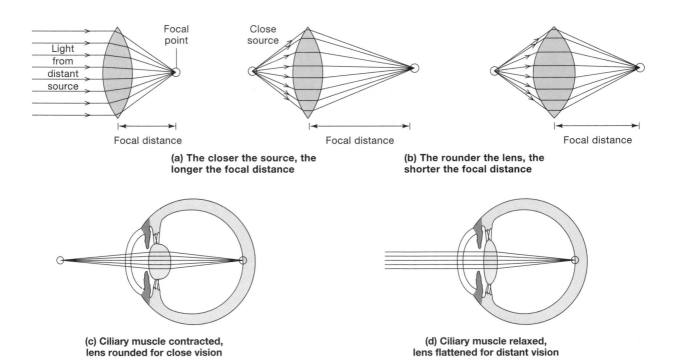

(a) The closer the source, the longer the focal distance

(b) The rounder the lens, the shorter the focal distance

(c) Ciliary muscle contracted, lens rounded for close vision

(d) Ciliary muscle relaxed, lens flattened for distant vision

Figure 33.4
Image Formation and Visual Accommodations.

(a) A lens refracts a light toward a specific point. The distance from the center of the lens to that point is the focal distance of the lens. Light from a distant source arrives with all the light waves traveling parallel to one another. Light from a nearby source, however, is still diverging or spreading out from its source when it strikes the lens. Note the difference in focal distance after refraction. **(b)** The rounder the lens, the shorter the focal distance. **(c)** When the ciliary muscle contracts, the suspensory ligaments allow the lens to round up. **(d)** For the eye to form a sharp image, the focal distance must equal the distance between the center of the lens and the retina. The lens compensates for variations in the distance between the eye and the object in view by changing its shape. When the ciliary muscle relaxes, the ligaments pull against the margins of the lens and flatten it.

Objects 20 ft or further from the eye have parallel light rays that enter the eye. The ciliary muscle is relaxed, and the ciliary body is behind the lens. This causes the suspensory ligaments to pull the lens flat for proper refraction of the parallel light rays. Objects closer than 20 ft to the eye have divergent or spreading light rays that require more refraction to focus them on the retina. The ciliary muscle contracts, and the ciliary body shifts forward, releasing the tension on the suspensory ligaments which causes the lens to bulge and become rounded. The spherical lens increases refraction, and the divergent rays are bent into focus on the retina. Reading and other activities requiring near vision cause eyestrain and fatigue because of the contraction of the ciliary muscle for accommodation. The lens gradually loses its elasticity as we age and causes a form of far-sightedness called **presbyopia (prez-bē-Ō-pē-uh)**. Many individuals have difficulty reading small type by the age of 40 and may require reading glasses to correct for the reduction in accommodation.

LABORATORY ACTIVITY   ACCOMMODATION DETERMINATION

Accommodation is determined by measuring the closest distance from which one can see an object in sharp focus, the **near point** of vision. A simple test for near-point vision involves moving an object toward the eye until it becomes blurred.

MATERIALS

pencil
ruler

PROCEDURES

1. Hold a pencil with the eraser up approximately 2 ft from your eyes and look at the ribs on the metal eraser casing.
2. Close one eye and slowly bring the pencil toward your open eye.
3. Measure the distance from the eye just before the metal casing blurs. Record your measurements in Table 33.2.
4. Repeat the procedure with the other eye.
5. To focus on near objects the eyes must rotate medially, a process called convergence. To observe convergence, stare at the pencil from arm's length and slowly move it closer to your eyes. Which way do your eyes move?
6. Compare near-point distances with classmates of various ages. Record this comparative data in Table 33.2.
7. Why does accommodation change as we age?

TABLE 33.2	Near-Point Determination		
Name and Age	**Right Eye**	**Left Eye**	**Both Eyes**
_____	_____	_____	_____
_____	_____	_____	_____
_____	_____	_____	_____

PHYSIOLOGY OF THE EYE CHECKLIST

This is a list of the bold terms presented in Exercise 33. Use this list for review purposes after you have completed the laboratory activities.

PHYSIOLOGY OF THE EYE

• Figures 33.1–33.4
visual acuity
emmetropic
myopic
hyperopic

astigmatism
refract
Snellen eye chart
accommodation
presbyopia
near point

PHYSIOLOGY OF THE EYE

Laboratory Report Exercise 33

A. Matching

Match the term on the left with the correct description on the right.

1. __G__ emmetropic   **A.** farsighted
2. __D__ astigmatism   **B.** age-related farsightedness
3. __A__ hyperopic   **C.** adjustment of lens shape
4. __F__ myopic   **D.** irregular cornea
5. __B__ presbyopia   **E.** bending of light rays
6. __H__ visual acuity   **F.** nearsighted
7. __E__ refraction   **G.** normal eye
8. __C__ accommodation   **H.** sharpness of vision

B. Short-Answer Questions

1. Explain the difference between viewing a distant and a close object.

2. Why does accommodation decrease with age?

3. Describe an experiment that demonstrates the presence of a blind spot on the retina.

4. Explain how the Snellen eye chart is used to determine visual acuity.

5. Describe each of the following visual acuities:
   a. 20/20

   emm

   b. 20/25

   c. 20/15

   d. 20/200

6. Give an example of the refraction of light.

Anatomy of the Ear

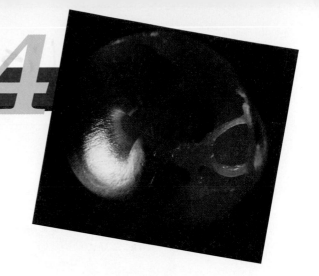

OBJECTIVES

On completion of this exercise, you should be able to:

- Identify and describe the components of the external, middle, and inner ear.
- Describe the anatomy of the cochlea.
- Describe the components of the semicircular canals and the vestibule and their roles in static and dynamic equilibrium.

WORD POWER

otolith (oto—ear; lith—stone)
pinna (pinna—wing)
scala (scala—staircase)
concha (concha—shell)
cupula (cupa—tub)
ossicle (ossicle—bone)

MATERIALS

ear models
auditory ossicles model
inner ear model
cochlea cross-section model
compound microscope
prepared slide of cochlea

INTRODUCTION

The ear is divided into three regions: the external or outer ear, the middle ear, and the inner ear. The external and middle ear direct sound waves to the inner ear for hearing. The inner ear has three compartments, each containing a specialized population of receptors. The cristae of the semicircular canals and the maculae of the vestibule inform the CNS of our position in the world compared with gravity and our movements. The spiral organ of the cochlea converts vibrations of sound waves into nerve impulses for auditory sensations.

A. The External Ear

The flap or **pinna** of the outer ear funnels sound waves into the **external auditory canal**, a tubular chamber that delivers sounds to the **tympanic membrane** (eardrum). The tympanic membrane is a thin sheet of fibrous connective tissue stretched across the canal. The canal contains wax-secreting cells called **ceruminous glands** and many hairs that prevent dust and debris from entering the ear. Locate the structures of the external ear in Figure 34.1.

B. The Middle Ear

The middle ear (Figure 34.1) is a small space inside the temporal bone. It is connected to the back of the throat (nasopharynx) by the **auditory tube,** also called the pharyngo-tympanic tube or Eustachian tube. This tube equalizes pressure between the external air and the cavity of the middle ear. Three small bones called **auditory ossicles** transfer vibrations from the external ear to the internal ear. The **malleus** is connected to the tympanic membrane, and the **incus** joins the malleus to the third bone, the **stapes,**

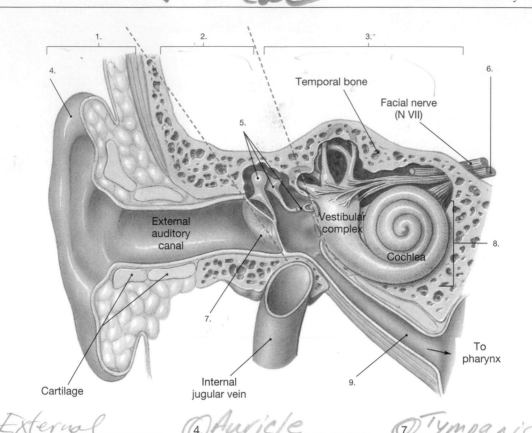

Temporal bone

Facial nerve
(N VII)

External
auditory
canal

Vestibular
complex

Cochlea

Cartilage

Internal
jugular vein

To
pharynx

1. *External*
2. *Middle*
3. *Inner*
4. *Auricle*
5. *Auditory ossicles*
6. *Vest. Nerve*
7. *Tympanic Membrane*
8. *Boney Lab. of In Ear*
9. *Auditory Tube*

Figure 34.1
Anatomy of the Ear.
The orientation of the external, middle, and inner ear.

which connects to the inner ear. Vibrations of the tympanic membrane are transferred to the malleus which conducts them to the incus and stapes. The stapes in turn pushes on the oval window of the cochlea to stimulate the auditory receptors.

C. The Inner Ear

The inner ear consists of three major regions: a helical cochlea, an elongated vestibule, and three semicircular canals. Each region is organized into two fluid-filled chambers: an outer **bony labyrinth** lined by an inner **membranous labyrinth**. The osseous labyrinth is embedded in the temporal bone and contains **perilymph**. The membranous labyrinth is filled with **endolymph**.

The **anterior**, **lateral**, and **posterior semicircular canals**, shown in Figure 34.2a, are positioned perpendicular to one another. Together they function as an organ of dynamic balance and equilibrium maintenance during motion. Inside the membranous labyrinth, the base of each semicircular canal has a swollen **ampulla** containing the **crista**. Each crista is composed of hair cells (receptors) and supporting cells covered by a gelatinous cap called the **cupula (KŪ-pū-la)**. Movement of the head causes the endolymph to push or pull on the cupula and bend or stretch the embedded hair cells.

The **vestibule (VES-ti-būl)** is an area between the semicircular canals and the cochlea. The membranous labyrinth in this area contains two sacs, the **utricle (Ū-tri-KUL)** and the **saccule (SAK-ūl)**, which contain **maculae (MAK-ū-lē)**, receptors for stationary equilibrium. Like the cristae, the maculae have hair cells and a gelatinous

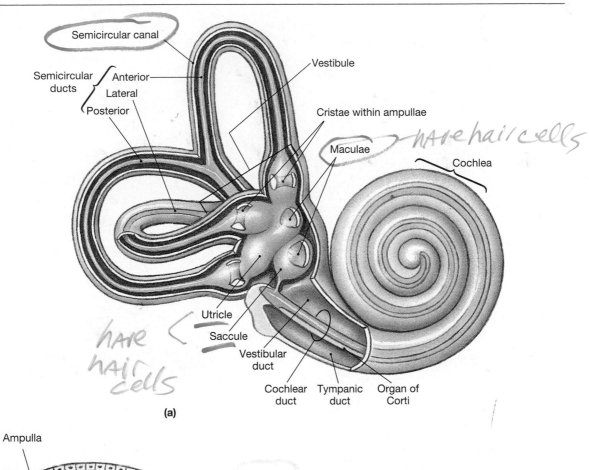

have hair cells (handwritten)

hare hair cells (handwritten)

(a)

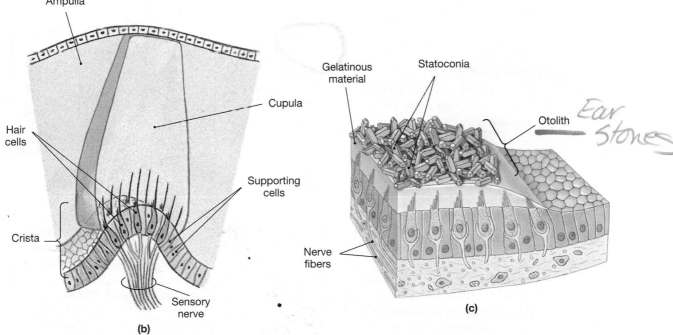

Ear Stones (handwritten)

(b)

(c)

Figure 34.2
The Inner Ear.

(a) The bony and membranous labyrinths. Areas of the membranous labyrinth containing sensory receptors (cristae, maculae, and the organ of Corti) are shown in purple. **(b)** A cross section through the ampulla of a semicircular duct. **(c)** The structure of a macula from the saccule.

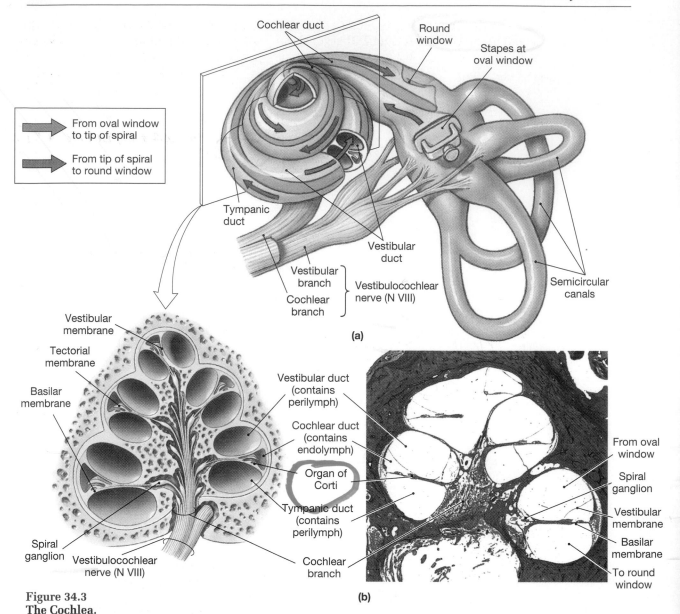

Figure 34.3
The Cochlea.

(a) The structure of the cochlea. (b) Diagrammatic and sectional views of the cochlear spiral.

covering. Embedded in the gelatinous membrane are calcium carbonate crystals called **statoconia.** The gelatinous membrane and the statoconia collectively are called an **otolith,** which means "ear stone." When the head is tilted, the statoconia and membrane shift their position and the hair cells in the utricle and saccule are stimulated. Impulses from the maculae are passed to sensory neurons in the **vestibular branch** of the vestibulocochlear nerve (cranial nerve VIII).

The **cochlea (KOK-lē-uh)** is the coiled region of the inner ear. It consists of three tubes rolled up like a shell. Inside the cochlea, the **cochlear duct,** or scala media, contains hair cells that are sensitive to vibrations caused by sound. The cochlear duct is part of the **membranous labyrinth** and is filled with endolymph. Surrounding the cochlear duct are the **vestibular duct** (scala vestibule) and the **tympanic duct** (scala tympani) of the **bony labyrinth.** These ducts are filled with perilymph. The floor of the cochlear duct is the **basilar membrane** where the hair cells occur. The **vestibular membrane** separates the cochlear and vestibular ducts. The ducts follow the helix of the cochlea, and the vestibular and tympanic ducts interconnect at the tip of the spiral.

To pass vibrations of sound to the inner ear, the stapes of the middle ear is connected to the vestibular duct at the membrane called the **oval window.** When the stapes vibrates against the oval window, pressure waves are produced in the ducts of the cochlea. The waves stimulate the hair cells in the cochlear duct then pass into the tympanic duct where the **round window** stretches to dissipate the wave energy.

The cochlear duct contains the sensory receptor for hearing called the **spiral organ** or the **organ of Corti** (see Figure 34.4). This receptor consists of **hair cells,** supporting cells, and a gelatinous membrane called the **tectorial (tek-TOR-ē-al) membrane.** The hair cells rest on the basilar membrane, and their long stereocilia extend into the endolymph and contact the tectorial membrane. Sound waves cause fluid movement in

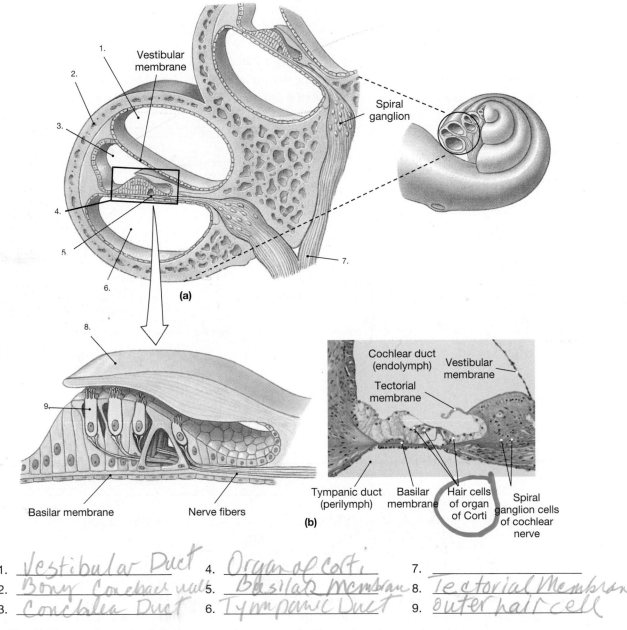

(a)

(b)

Basilar membrane   Nerve fibers

Cochlear duct (endolymph)   Vestibular membrane
Tectorial membrane

Tympanic duct (perilymph)   Basilar membrane   Hair cells of organ of Corti   Spiral ganglion cells of cochlear nerve

Vestibular membrane
Spiral ganglion

1. _Vestibular Duct_   4. _Organ of Corti_   7. _____
2. _Bony Cochlear wall_ 5. _Basilar Membran_ 8. _Tectorial Membrane_
3. _Cochlea Duct_   6. _Tympanic Duct_   9. _outer hair cell_

Figure 34.4
The Cochlea and Organ of Corti.

(a) Structure of the cochlea as seen in section. **(b)** The three-dimensional structure of the tectorial membrane and hair cell complex of the organ of Corti.

thc cochlea, and the hair cells are stimulated as they are pushed against the tectorial membrane. The hair cells synapse with sensory neurons in the **cochlear branch** of the vestibulocochlear nerve (VIII), which transmits the impulses to the auditory cortex of the brain. The **spiral ganglia** contains cell bodies of sensory nuerons in the cochlear branch.

LABORATORY ACTIVITY   THE EAR

MATERIALS

ear models and charts
compound microscope
prepared slide of cochlea

PROCEDURES

1. Label the structures of the ears in Figures 34.1 and 34.4.
2. Identify the structures of the ear on laboratory charts and models.
3. Examine a prepared slide of a cross section of the cochlea and identify the structures shown in Figure 34.3 and 34.4.
4. In the space below, sketch a cross section of the cochlea.

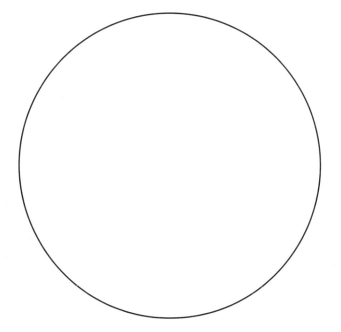

ANATOMY OF THE EAR CHECKLIST

This is a list of the bold terms presented in Exercise 34. Use this list for review purposes after you have completed the laboratory activities.

EXTERNAL EAR

• Figure 34.1
pinna
external auditory canal
  tympanic membrane
ceruminous glands

MIDDLE EAR

• Figure 34.1
auditory ossicles
  malleus
  incus
  stapes
auditory tube (pharyngotympanic tube)

INNER EAR

• Figures 34.1–34.4
bony labyrinth
  perilymph
membranous labyrinth
  endolymph

anterior semicircular canal
lateral semicircular canal
posterior semicircular canal
  ampulla
    crista
    cupula

vestibule
  utricle
  saccule
    macula
      otoliths
        gelatinous membrane
        statoconia
vestibulocochlear nerve (VIII)
  vestibular branch
  cochlear branch
cochlea

COCHLEA (DETAILED)

• Figure 34.3 and 34.4
cochlear duct
  spiral organ (of Corti)
    hair cells
      stereocilia
  tectorial membrane
  basilar membrane

vestibular duct
  vestibular membrane
  oval window
tympanic duct
round window

Name _____   Date _____

Section _____   Number _____

ANATOMY OF THE EAR

Laboratory Report Exercise 34

A. Matching

Match the description on the left with the correct structure on the right.

1. _____ eardrum	**A.** statoconia	
2. _____ receptors in semicircular canals	**B.** basilar membrane	
3. _____ coiled region of inner ear	**C.** tectorial membrane	
4. _____ outer layer of inner ear	**D.** membranous labyrinth	
5. _____ receptor for hearing	**E.** cupula	
6. _____ receptors in vestibule	**F.** bony labyrinth	
7. _____ attachment site for stapes	**G.** tympanic membrane	
8. _____ jellylike substance of crista	**H.** cochlea	
9. _____ contains endolymph	**I.** semicircular canals	
10. _____ membrane over auditory receptors	**J.** round window	
11. _____ chamber inferior to organ of Corti	**K.** crista ampullaris	
12. _____ perpendicular loops of inner ear	**L.** spiral organ	
13. _____ membrane of tympanic duct	**M.** oval window	
14. _____ membrane supporting spiral organ	**N.** maculae	
15. _____ crystals of maculae	**O.** tympanic duct	

B. Short-Answer Questions

1. Describe the three major regions of the ear.

2. How are sound waves passed on to the tympanic membrane?

3. Describe the receptors for equilibrium.

4. Describe the receptors for hearing.

C. **Drawing**

Sketch a cross section of the cochlea. Label the ducts, membranes, and the spiral organ.

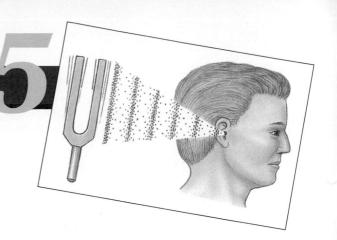

Physiology of the Ear

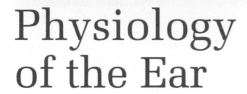

OBJECTIVES

On completion of this exercise, you should be able to:

- Explain the difference between static and dynamic equilibrium by performing various comparative tests.
- Test your range of hearing by using tuning forks.
- Compare the conduction of sound through air versus bone (Rinne test).

WORD POWER

macula (macula—spot)
utricle (utriculus—leather bag)
saccule (sacculus—small sac)

proprioceptors (proprius—one's own)
cristae (crista—crest)

MATERIALS

Weber's test: tuning forks (50–100 cps, 1,000 cps)
Rinne test: tuning forks (1,000 cps, 5,000 cps)

INTRODUCTION

The inner ear serves two unique functions: balance and hearing. Imagine life without a sense of balance. We would not be able to stand, let alone play sports or enjoy a brisk walk. Balance, or **equilibrium,** feeds the brain a constant stream of information detailing the body's position compared to the ground. Although we live in a three-dimensional world, our bodies function in two dimensions: front to back and side to side. Since our sense of the third dimension, top to bottom, is "ground based," we can get disoriented in deep water or while piloting a plane. **Vertigo,** or motion sickness, may occur when the CNS receives conflicting sensory information from the inner ear, the eyes, and other receptors. For example, while reading in a moving car, the eyes are concentrating on the steady book and the inner ear is responding to the motion of the car. The CNS receives the opposing sensory signals and may respond with vomiting, dizziness, sweating, and other symptoms of motion sickness.

Have you ever stood at your microwave oven and listened to your popcorn pop, waiting to push the stop button for a perfect batch? The ear is a dynamic sense organ capable of hearing multiple sound waves simultaneously. We can talk with a friend while listening to music and still hear the phone ring. The receptors for hearing, the **auditory** receptors, are located in the cochlea of the inner ear. The cochlea is coiled like a snail shell and contains delicate hair cells that are sensitive to sound.

A. Equilibrium

The receptors for equilibrium are in the membranous labyrinth of the vestibule of the inner ear. Each semicircular canal has a group of hair cells called the **crista** located at a base of one side of the canal's loop. The crista is the receptor for **dynamic equilibrium**

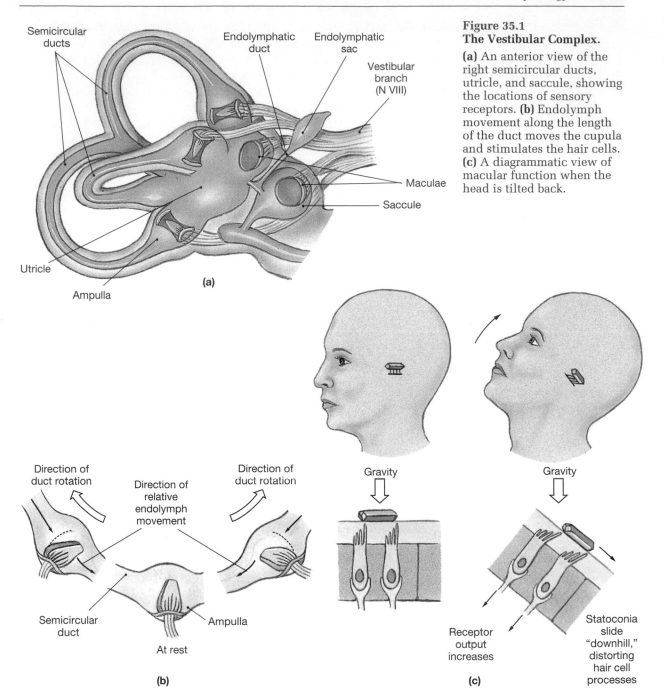

Figure 35.1
The Vestibular Complex.

(a) An anterior view of the right semicircular ducts, utricle, and saccule, showing the locations of sensory receptors. (b) Endolymph movement along the length of the duct moves the cupula and stimulates the hair cells. (c) A diagrammatic view of macular function when the head is tilted back.

and responds to rotational movements such as spinning in a chair (see Figure 35.1). The utricle and saccule have hair cells organized into maculae. Each macula has an **otolith** comprised of hair cells and a **gelatinous membrane** with embedded **statoconia** crystals. The maculae are receptors for **static equilibrium** and sense changes in body position compared to gravity, such as while in an upside-down postion. When the head and body suddenly move, the otoliths, which are heavier than the hair cells, lag behind and distort hair cells, causing them to produce sensory impulses.

Visual awareness of the surroundings enhances the sense of equilibrium by comparing your body position relative to surrounding stationary objects. Loss of visual references to the ground and other stationary objects usually results in a loss of balance. **Nystagmus (nis-TAG-mus)** is the trailing of the eyes slowly in one direction followed by their rapid movement in the opposite direction. You will look for this event when the subject is spinning in a swivel chair in the next experiment. As body position changes, nystagmus occurs to produce a brief, crisp image on the retina. Initially, the eyes move slowly in the direction opposite the rotation and then quickly jump in the other direction. If the body is rotating to the right, the eyes will first slowly track to the left and then rapidly veer to the right. This cycle of slow and fast eye movements produces a brief, stationary image on the retina instead of an image that is blurred because of the movement.

LABORATORY ACTIVITY   Equilibrium Tests

I. Equilibrium and Vision
MATERIALS

laboratory partner

PROCEDURES

1. Perform the following with your lab partner. Be prepared to help your partner if he or she losses balance and starts to fall.
2. Stand on both feet in a clear area of the room. Raise one foot and then close your eyes. Record your observations in Table 35.1.

TABLE 35.1   **Equilibrium Tests**

Standing Position	Observation
On one foot, eyes open	
On one foot, one eye closed	
On one foot, both eyes closed	

II. Nystagmus

This experiment involves spinning a volunteer, the subject, in a chair to observe his or her eye movements. Do not use anyone who is prone to motion sickness. The observer will spin the subject and record eye reflexes. To prevent accidents, four other individuals will be needed to hold the base of the chair steady and to serve as spotters to help the subject stay in the chair during the experiment. The subject should remain in the chair for several minutes after the experiment to regain normal equilibrium.

MATERIALS

swivel chair
laboratory partner

PROCEDURES

1. Seat the subject in a swivel chair with the spotters supporting the base with their feet.
2. Instruct the subject to stare straight ahead with both eyes.
3. Spin the chair clockwise for 10 revolutions (one rotation per second).
4. Quickly and carefully stop the chair and observe the subject's eye movements. Record your observations in Table 35.2.

TABLE 35.2   Nystagmus Test

Rotation and Eyes	Observation
Clockwise, eyes open	
Clockwise, eyes closed	
Counterclockwise, eyes open	
Counterclockwise, eyes closed	

5. Keep the subject seated for approximately 2 min to regain balance.
6. Using other subjects, repeat the experiment with a counterclockwise rotation. Additionally, ask one subject to close his or her eyes during the rotation and then open them when the chair stops. Record your observations in Table 35.2.

B. Hearing

We rely on our sense of hearing for communication and for an awareness of events in our immediate surroundings. Sounds are produced by vibrating objects that cause air molecules to compress and decompress (see Figure 35.2). The number of compressed air waves that occur in one second is the **frequency,** or pitch, of the sound. Faster vibrations produce a higher pitch than an object vibrating more slowly. The unit **Hertz** (Hz) is used for the frequency, or number of cycles per second, of the compression waves. Humans can hear sounds from about 20 to 20,000 Hz. The volume, or **amplitude,** of a sound is measured in **decibels** (Db). The higher or "taller" a sound wave, the higher the decibel of the sound.

Hearing occurs when sound waves enter the external auditory canal and strike the tympanic membrane, causing the membrane to vibrate at the same frequency of the sound waves (see Figure 35.3 and Table 35.3). The tympanic membrane converts the sound wave energy into mechanical energy and causes the auditory ossicles to vibrate.

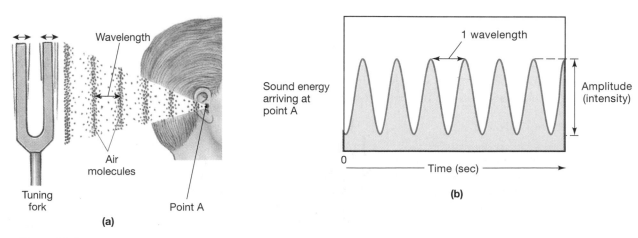

Figure 35.2
Sound Production.

(a) Sound waves generated by a tuning fork travel through the air as pressure waves. The frequency of a sound wave is the number of waves that pass a fixed reference point in a given time. **(b)** Frequencies are reported in cycles per second (cps), or hertz (Hz).

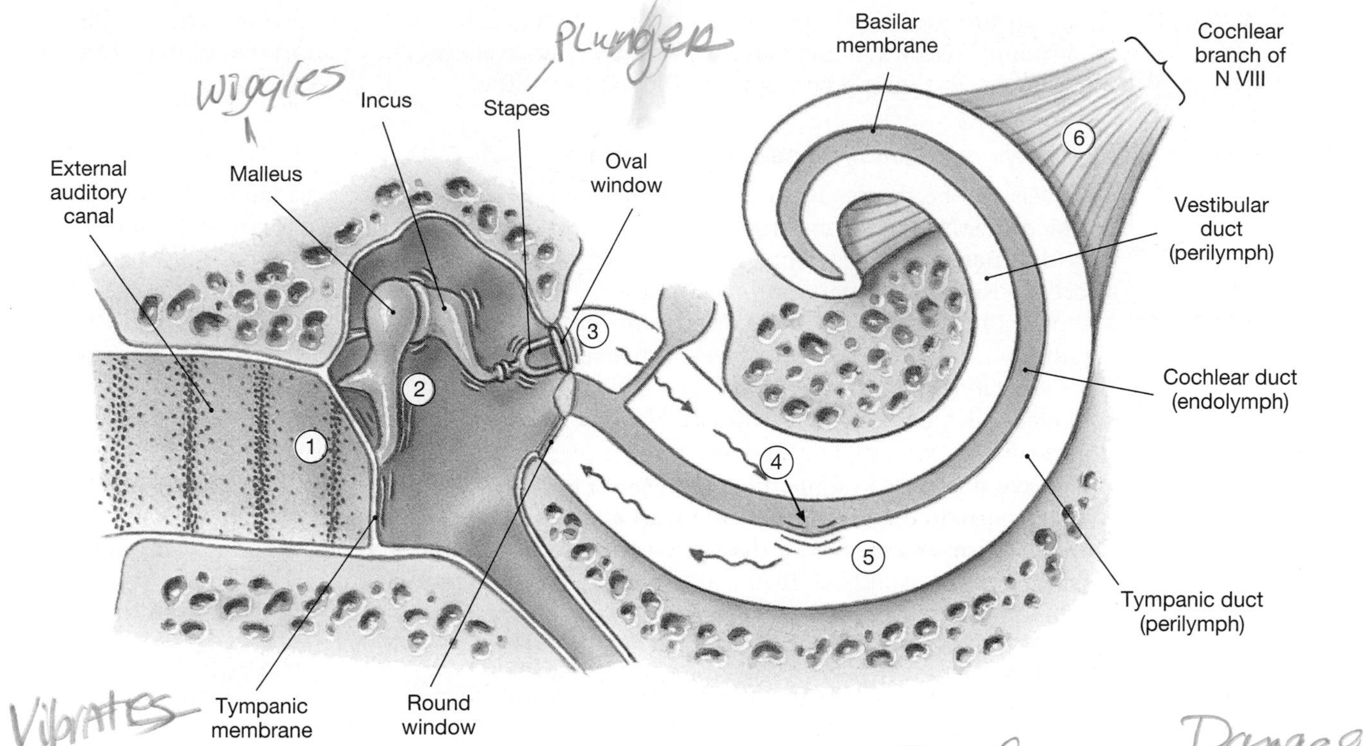

Figure 35.3
The Physiology of Hearing.
Steps involved in the reception of sound and the process of hearing. Table 35.3 summarizes each step.

TABLE 35.3 Steps in the Production of an Auditory Sensation (See Figure 35.3)

1. Sound waves arrive at the tympanic membrane.
2. Movement of the tympanic membrane causes displacement of the auditory ossicles.
3. Movement of the stapes at the oval window establishes pressure waves in the perilymph of the vestibular duct.
4. The pressure waves distort the basilar membrane on their way to the round window of the tympanic duct.
5. Vibration of the basilar membrane causes vibration of hair cells against the tectorial membrane.
6. Information concerning the region and intensity of stimulation is relayed to the CNS over the cochlear branch of N VIII.

The malleus, which is connected to the tympanic membrane, vibrates and moves the incus, which moves the stapes. The stapes is attached to the oval window of the cochlea. Vibrations of the stapes move the oval window, creating pressure waves in the perilymph of the vestibular duct. The pressure waves correspond to the sound waves that initially hit the tympanic membrane. The waves pass through the cochlear duct and into the tympanic duct, causing the basilar membrane to vibrate. The pressure waves are dissipated by the stretching of the round window.

The vibrations in the basilar membrane push specific hair cells into the tectorial membrane. The cells fire sensory impulses when their stereocilia touch the tectorial membrane and bend. The auditory ingormation is passed over the cochlear branch of the vestibulocochlear nerve.

Deep sounds have long sound waves that stimulate the distal portion of the basilar membrane. High frequency sounds, such as glass breaking, have short waves that stimulate the basilar membrane close to the oval window.

Middle ear conduction verses nerve deafness

Deafness can be the result of many factors, not all of which are permanent. Two categories of deafness are discernable, conduction deafness and nerve deafness. **Conduction deafness** involves damage to the tympanic membrane or one or more of the auditory ossicles. Proper conduction produces vibrations heard equally in both ears. If a conduction problem exists, sounds are normally heard better in the unaffected ear. Conduction tests with tuning forks, however, cause *the ear with the deafness to hear the sound louder than the normal ear.* This is due to an increased sensitivity to sounds in the ear with conduction deafness. Hearing aids are often used to correct for conduction deafness.

Nerve deafness is a result of damage to the cochlea or the cochlear nerve. Repetitive exposure to excessively loud noises such as music and machinery can damage the delicate spiral organ. Nerve deafness cannot be corrected and results in a permanent loss of hearing, usually within a specific range of frequencies (tones).

In the following tests, sound vibrations from tuning forks will be conducted through the bones of the skull, bypassing normal conduction by the external and middle ear. Since the inner ear is surrounded by the temporal bone, vibrations will be transmitted directly from the bone into the cochlea.

LABORATORY ACTIVITY   HEARING TESTS

Note: A quiet environment is necessary for conducting hearing tests. The subject should use his or her index finger to close the ear opposite the ear being tested. Strike the tuning fork on the heel of your hand and hold it next to the subject's ear.

I. Weber's Test
MATERIALS

> tuning forks (100 cps and 1,000 cps)
> laboratory partner

PROCEDURES

1. Have your partner correctly strike the fork and place its base on top of your head above your ears. Do you hear the vibrations from the tuning fork? Is it louder in one ear than in the other? _____

2. Repeat the procedures using the 1,000-cps tuning fork. Do you hear one tuning fork better than the other? _____

II. Rinne Test
MATERIALS

> tuning forks (1,000 cps and 5,000 cps)
> laboratory partner

PROCEDURES

1. Have your partner sit down and find the mastoid process behind his or her right ear.
2. Strike the tuning fork and place the base of its handle against the mastoid process. This bony process will conduct the sound vibrations of the tuning fork to the inner ear.
3. When your partner indicates he or she can no longer hear the tuning fork, quickly hold the fork close to the external ear. In normal hearing the tuning fork should still be audible. If middle ear deafness exists, no sound will be heard. Repeat this test on the left ear.
4. Repeat steps 1 through 3 with the 5,000-cps tuning fork. Record your observations in Table 35.4.

TABLE 35.4	**Hearing Tests Results**	
	Left Ear	*Right Ear*
1,000 cps	_____	_____
5,000 cps	_____	_____

...QUICKCHECK

PHYSIOLOGY OF THE EAR CHECKLIST

This is a list of the bold terms presented in Exercise 35. Use this list for review purposes after you have completed the laboratory activities.

EQUILIBRIUM

• Figure 35.1

dynamic equilibrium
- semicircular canals
  - ampullae
    - cristae

static equilibrium
- utricle and saccule
  - maculae
    - otoliths

nystagmus

HEARING

• Figures 35.2 and 35.3

hearing

cochlea
- organ of Corti

conduction deafness

nerve deafness

Weber's test

Rinne test

Name _____   Date _____

Section _____   Number _____

PHYSIOLOGY OF THE EAR

Laboratory Report Exercise 35

A. Matching

Match the description on the left with the correct structure on the right.

1. _____ swollen area of semicircular canal	A.	vestibule
2. _____ site of auditory receptors	B.	spiral organ
3. _____ vibrates oval window	C.	cochlea
4. _____ amplifies sound	D.	nerve deafness
5. _____ contains the macula	E.	eye movements during rotation
6. _____ dissipates sound energy	F.	ampulla
7. _____ organ of Corti	G.	stapes
8. _____ loss of nerve function	H.	round window
9. _____ damage to tympanic membrane	I.	tympanic membrane
10. _____ nystagmus	J.	conduction deafness

B. Short-Answer Questions

1. Describe the process of hearing.

2. If conduction deafness exists, why is the sound of a tuning fork heard best in the deaf ear?

3. List the three bones of the middle ear in order from the tympanic membrane to the oval window of the inner ear.

4. Describe the receptors for dynamic and static equilibrium.

5. Explain the phenomenon of nystagmus.

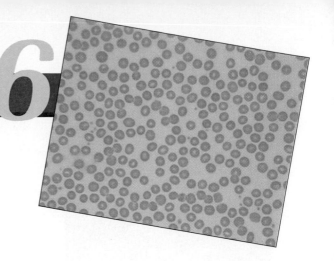

The Blood

OBJECTIVES

On completion of this exercise, you should be able to:

- List the function of the blood.
- Describe each component of the blood.
- Distinguish each type of blood cell on a blood smear slide.
- Describe the antigen-antibody reactions of the ABO and Rh blood groups.
- Safely collect a blood sample using the blood lancet puncture technique.
- Safely type a sample of blood to determine the ABO and Rh blood types.
- Correctly perform a hematocrit test on a sample of blood.
- Correctly perform a coagulation test on a sample of blood.
- Discuss how to properly discard of blood contaminated wastes.

WORD POWER

erythrocyte (erythro—red)
leukocyte (leuko—white)
neutrophil (...phil—loving)
lymphocyte (lympho—watery)
agglutination (agglutinate—to stick together)
hemolytic (hemo—blood)
coagulation (coagul—to curdle)

MATERIALS

compound microscope
prepared slide of blood smear
gloves and safety glasses
bleach solution in spray bottle
paper towels
sterile blood lancets
sterile alcohol prep pads
disposable blood typing plates or microscope slides

biohazardous waste container
wax marker
clean toothpicks
anti-A and anti-B typing sera
heparinized capillary tubes
centrifuge
seal ease clay
hematocrit tube reader

INTRODUCTION

Blood is a composite fluid connective tissue that flows through the vessels of the vascular system. In response to injury, blood has the intrinsic ability to change from a liquid to a gel so as to clot and stop bleeding. Blood comprises cells and cell pieces that are collectively called the formed elements. These cells are carried in an extracellular fluid called blood plasma. Blood has many diverse functions, all relating to supplying your cells with essential materials and maintaining the internal environment.

Red blood cells transport oxygen to trillions of cells in your body. The blood controls the chemical composition of all interstitial fluid by regulating pH and electrolyte levels. White blood cells are part of the immune system and protect you from microbes by producing antibody molecules and phagocytizing foreign cells.

A. Blood Plasma

Figure 36.1 details the composition of whole blood. A sample of blood is approximately 55% plasma and 45% formed elements. **Plasma (PLAZ-muh)** is approximately

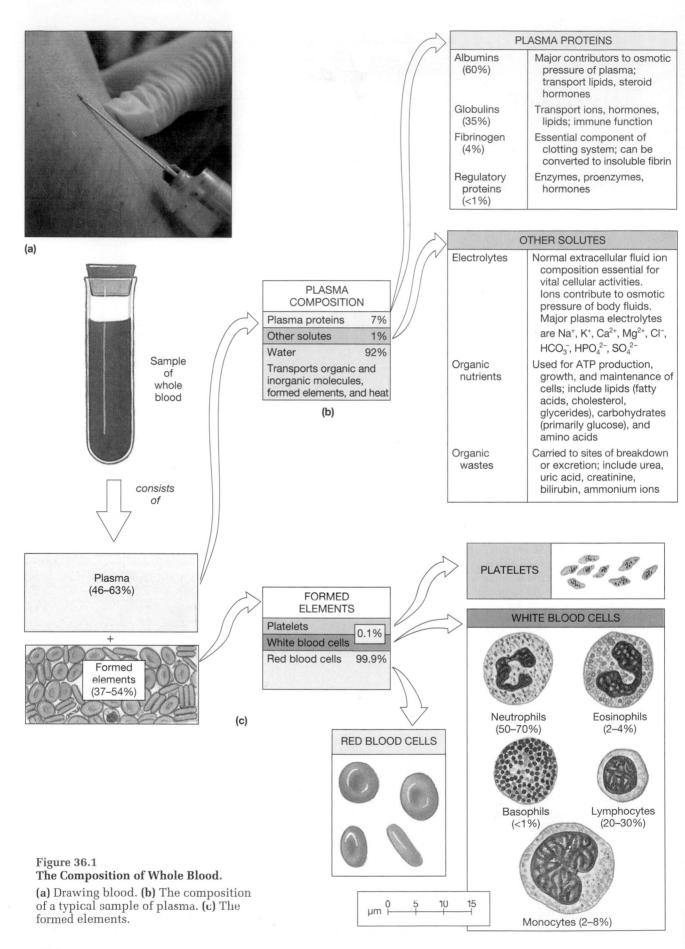

PLASMA PROTEINS

Albumins (60%)	Major contributors to osmotic pressure of plasma; transport lipids, steroid hormones
Globulins (35%)	Transport ions, hormones, lipids; immune function
Fibrinogen (4%)	Essential component of clotting system; can be converted to insoluble fibrin
Regulatory proteins (<1%)	Enzymes, proenzymes, hormones

PLASMA COMPOSITION

Plasma proteins	7%
Other solutes	1%
Water	92%
Transports organic and inorganic molecules, formed elements, and heat	

(b)

OTHER SOLUTES

Electrolytes	Normal extracellular fluid ion composition essential for vital cellular activities. Ions contribute to osmotic pressure of body fluids. Major plasma electrolytes are Na^+, K^+, Ca^{2+}, Mg^{2+}, Cl^-, HCO_3^-, HPO_4^{2-}, SO_4^{2-}
Organic nutrients	Used for ATP production, growth, and maintenance of cells; include lipids (fatty acids, cholesterol, glycerides), carbohydrates (primarily glucose), and amino acids
Organic wastes	Carried to sites of breakdown or excretion; include urea, uric acid, creatinine, bilirubin, ammonium ions

Sample of whole blood

consists of

Plasma (46–63%)

+

Formed elements (37–54%)

FORMED ELEMENTS

Platelets	0.1%
White blood cells	
Red blood cells	99.9%

PLATELETS

WHITE BLOOD CELLS

Neutrophils (50–70%)　Eosinophils (2–4%)

Basophils (<1%)　Lymphocytes (20–30%)

Monocytes (2–8%)

RED BLOOD CELLS

μm 0 5 10 15

(c)

Figure 36.1
The Composition of Whole Blood.

(a) Drawing blood. **(b)** The composition of a typical sample of plasma. **(c)** The formed elements.

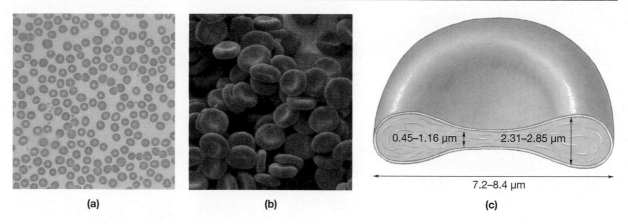

Figure 36.2
Anatomy of Red Blood Cells.

(a) When viewed in a standard blood smear, red blood cells appear as two-dimensional objects because they are flattened against the surface of the slide. (LM ×320) **(b)** A scanning electron micrograph of red blood cells reveals their three-dimensional structure quite clearly. (SEM ×1195) **(c)** A sectional view of a mature red blood cell, showing average dimensions.

92% water and contains proteins to regulate the osmotic pressure of blood, proteins for the clotting process and antibody proteins are part of the immune system to protect your body from invading pathogens. Electrolytes, hormones, nutrients, and some blood gases are transported in the blood plasma. Later in this exercise, you will do a hematocrit test that separates the formed elements from the plasma.

B. The Formed Elements

The formed elements may be organized into three groups of cells: the red blood cells or erythrocytes, the white blood cells or leukocytes, and the platelets (see Figure 36.1). When stained, each group is easy to identify with a microscope. The reddish colored cells are erythrocytes, the stained cells are leukocytes, and the small cell fragments between the erythrocytes and leukocytes are platelets.

Erythrocytes

Erythrocytes (e-RITH-rō-sīts) are commonly called red blood cells or RBCs. They are the most abundant of all blood cells. Erythrocytes are biconcave discs and microscopically appear thinner and lighter staining in the middle (see Figure 36.2). The biconcave shape most importantly increases the surface area on the red blood cell for rapid gas exchange between the blood and the tissues of the body. Their shape also allows them to flex and squeeze through narrow capillaries. Erythrocytes are red and lack a nucleus. White blood cells, however, have a nucleus.

The major function of red blood cells is to transport blood gases. Oxygen is picked up in the lungs and carried to the tissue cells of the body. While supplying the cells with oxygen, the blood acquires carbon dioxide from the cells. The plasma and red blood cells convey the carbon dioxide to the lungs for removal during exhalation. To accomplish the task of gas transport, each RBC contains millions of hemoglobin (Hb) molecules. **Hemoglobin (HĒ-mō-glō-bin)** is a complex protein molecule with four iron atoms that allow oxygen and carbon dioxide molecules to loosely bind to the Hb.

Leukocytes

Leukocytes (LOO-kō-sīts) are white blood cells (WBCs) that contain a darkly stained nucleus. The nucleus often is branched into several lobes, as shown in Figure 36.3. Leukocytes lack hemoglobin and therefore do not transport blood gases. Leukocytes can

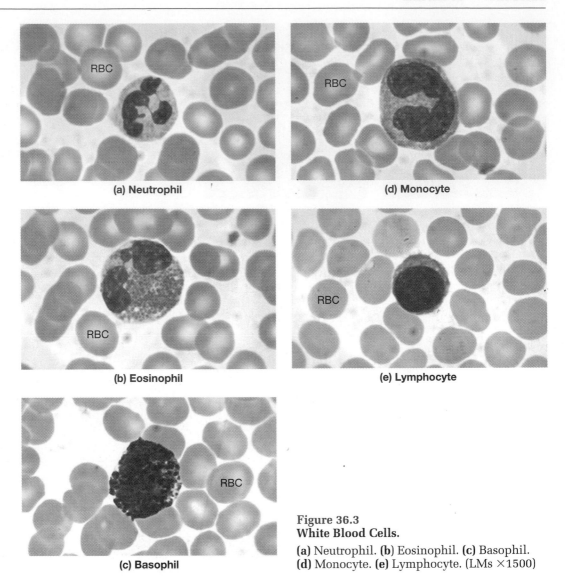

Figure 36.3
White Blood Cells.

(a) Neutrophil. (b) Eosinophil. (c) Basophil.
(d) Monocyte. (e) Lymphocyte. (LMs ×1500)

pass between the endothelial cells of capillaries and enter the interstitial spaces of tissues. Most leukocytes are phagocytes and are part of the immune system.

Two broad classes of leukocytes occur: **granular leukocytes** and **agranular leukocytes**. The granular leukocytes, collectively called **granulocytes**, have granules in their cytoplasm and include the neutrophils, eosinophils, and basophils. Agranular leukocytes, which include the monocytes and lymphocytes, have few cytoplasmic granules. Figure 36.3 presents a micrograph of each leukocyte.

Neutrophils   The most common leukocytes are the **neutrophils (NOO-trō-filz)** that comprise up to 70% of the entire white blood cell population. These granulocytes are also called **polymorphonuclear (pol-ē-mōr-fō-NOO-klē-ar)** leukocytes because their nucleus is complex and branches into two to five lobes. Neutrophils have many small cytoplasmic granules that stain pale purple, visible in Figure 36.3a.

Neutrophils are the first leukocytes to arrive at a wound site to begin the process of infection control. They phagocytize bacteria and release hormones called **leukotrienes** that attract other phagocytes, such as eosinophils and monocytes, to the site of injury.

Neutrophils are short-lived and survive in the blood for up to 10 hours. Active neutrophils in a wound may live only 30 minutes until they succumb to the toxins released by bacteria they have ingested.

Eosinophils   **Eosinophils (ē-ō-SIN-ō-filz)** are identified by the presence of medium-sized granules that stain an orange-red color, as shown in Figure 36.3b. Eosinophils are similar in size to neutrophils. The nucleus is conspicuously segmented into two lobes. Eosinophils are phagocytes that engulf bacteria and other microbes that the immune system has coated with antibodies. They also contribute to decrease the inflammatory response at a wound or site of infection. Approximately 3% of the circulating leukocytes are eosinophils.

Basophils   Less than 1% of circulating leukocytes are **basophils (BĀ-sō-filz)**. These scarce cells are distinguished by large cytoplasmic granules that stain a very dark blue. The granules are so large and numerous that the nucleus is obscured as in the basophil illustrated in Figure 36.3c. Basophils are smaller than neutrophils and eosinophils. They are sometimes difficult to locate on a blood smear because of their low frequency.

Basophils migrate to an injured tissue and release histamines which cause vasodilation and heparin to prevent blood clotting. Mast cells in the tissue respond to these molecules and induce local inflammation.

Monocytes   Approximately 2% to 8% of circulating white blood cells are monocytes. On a blood smear slide, monocytes are roundish and may have small extensions, much like an amoeba. These large cells are distinguished by a darkly stained kidney-bean-shaped nucleus surrounded by pale blue cytoplasm (see Figure 36.3d). **Monocytes (MON-ō-sīts)** are agranular leukocytes; however, ingested material such as phagocytized bacteria and debris will stain and appear similar to granules.

Monocytes are wanderers. They leave the bloodstream between the capillary endothelium and patrol the body tissues in search of microbes and worn-out tissue cells. They are second to neutrophils in arriving at a wound site. As neutrophils die from phagocytizing bacteria, the monocytes phagocytize the neutrophils.

Lymphocytes   **Lymphocytes (LIM-fō-sīts)** are the smallest of the leukocytes and are approximately the size of a red blood cell (see Figure 36.3e). The distinguishing feature of lymphocytes is a large nucleus that occupies almost the entire cell, leaving room for only a small halo of pale blue cytoplasm around the edge of the cell. Lymphocytes are abundant in the bloodstream and compose 20% to 30% of all circulating leukocytes. As their name suggests, lymphocytes are the main cells populating the lymphatic system. Many lymphocytes occur in lymph nodes, glands, and other lymphatic structures.

Several types of lymphocytes exist; however, they cannot be individually distinguished with a light microscope. Generally, lymphocytes provide immunity from microbes and defective cells by two methods. **T-cell lymphocytes** attach to and destroy a foreign cell in a cell-mediated response by releasing cytotoxic chemicals to kill the invaders. **B-cell lymphocytes** become sensitized to a specific antigen, then manufacture and pour antibodies into the bloodstream. The antibodies attach to and help destroy foreign antigens.

Platelets

Platelets (PLĀT-lets) are small cellular pieces produced in the red bone marrow when huge megakaryocyte cells fragment. Platelets lack a nucleus and other organelles. They survive in the bloodstream for a brief time and are involved in the coagulation or clotting of blood. Locate the platelet in Figure 36.3b.

| LABORATORY ACTIVITY | OBSERVATIONS OF ERYTHROCYTES, LEUKOCYTES, AND PLATELETS |

MATERIALS

compound microscope

human blood smear slide (Wright's or Giesma stained)

PROCEDURES

1. Obtain a human blood smear slide, and place it on the stage of the microscope. Move the slide so that the blood smear is over the aperture in the stage.

2. Adjust the magnification of the microscope to low power. Slowly turn the coarse focus knob until you can clearly see blood cells. Now use the fine focus adjustment as you examine individual cells on the slide.

3. Scan the slide at low magnification. Notice the abundance of red blood cells. The dark stained cells are the various leukocytes.

4. Increase magnification, and examine each type of formed element. Which cells have cytoplasmic granules?

_____

5. Sketch each blood cell and complete Table 36.1 as you observe each blood cell.

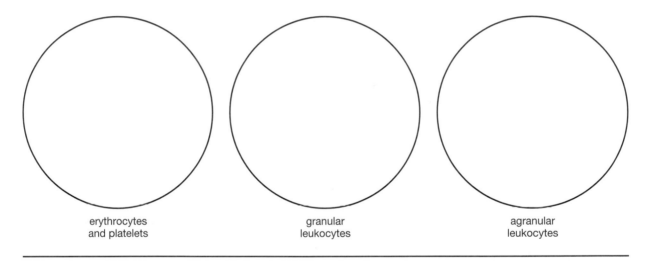

| erythrocytes and platelets | granular leukocytes | agranular leukocytes |

TABLE 36.1   **A Summary of the Formed Elements of the Blood**

Cell	Functions	Appearance
Erythrocyte		
Neutrophil		
Eosinophil		
Basophil		
Monocyte		
Lymphocyte		
Platelet		

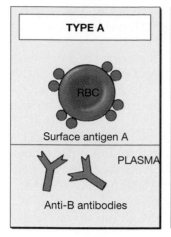
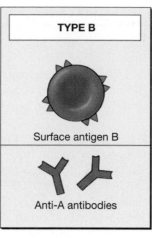
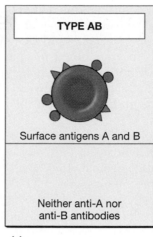
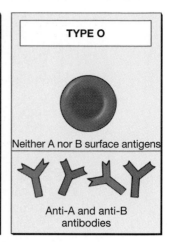

(a)

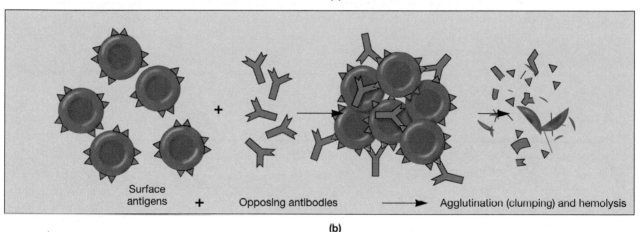

Surface antigens **+** Opposing antibodies ⟶ Agglutination (clumping) and hemolysis

(b)

Figure 36.4
Blood-Typing and Cross-Reactions.
The blood type depends on the presence of surface antigens (agglutinogens) on RBC surfaces. **(a)** The plasma contains antibodies (agglutinins) that will react with foreign surface antigens. The relative frequencies of each blood type in the U.S. population are indicated in Table 36.5. **(b)** In a cross-reaction, antibodies that encounter their target antigens lead to agglutination and hemolysis of the affected RBCs.

C. ABO and RH Blood Groups

Your blood type is determined by genes you inherited from your parents, and it does not change during your lifetime. Each blood type is due to the presence or absence of specific molecules on the surface of the erythrocytes. These molecules, called **antigens**, are like cellular name tags that inform your immune system that your cells belong to "self" and are not "foreign." The surface antigens on erythrocytes are called **agglutinogens (a-gloo-TIN-ō-jenz).**

Each blood group also has **antibody** molecules, called **agglutinins (a-GLOO-ti-ninz),** that are present in the blood plasma. The antibodies (agglutinins) will attach to antigens (agglutinogens) of foreign blood cells and will cause hemolysis of the cells.

More than 50 different blood systems occur in the human population. In this section, you will study the two most common blood systems, the ABO and Rh blood groups. Since each blood group is controlled by a unique gene, your ABO blood type does not influence your Rh blood type.

The ABO blood group

Figure 36.4 illustrates the characteristics of the ABO blood groups. There are four blood types in the ABO system: types A, B, AB, and O. Two antigens, A and B, occur in different combinations that determine the ABO blood type. Type A blood has the A antigen on the cell surface, blood type B has the B antigen, type AB has both A and B antigens, and type O blood has neither A nor B antigens.

Present in your blood plasma are antibodies, which will not react to your own blood. If a different ABO blood type is transfused into your blood stream, your antibodies will attach to the antigens of the foreign blood cells and will cause the cells to clump, or **agglutinate**. Hemolysis, or bursting, then occurs to destroy the introduced blood cells.

Type A blood has A antigens on the membrane surface and type anti-B antibodies in the plasma. The anti-B antibody will not react with the type A antigen but it does provide immunity against other blood types. Type B blood cells are covered with the B antigen, whereas the plasma contains the anti-A antibodies. AB blood is unique in that each red blood cell contains both the A and B cellular antigens on the membrane surface. The AB blood lacks the anti-A and the anti-B plasma antibodies. Although the surface antigens are absent in type O blood it has both anti-A and anti-B plasma antibodies.

To determine an ABO blood type, the presence of antigens is detected by adding drops of **anti-sera** that contain either the anti-A or the anti-B antibodies. The serum antibodies will react with the corresponding antigens on the RBC surface. For example, the anti-A antibodies will react with the A antigens found in type A blood. The blood will agglutinate as the antibodies react with the antigens. Complete Table 36.2 before you type a blood sample.

TABLE 36.2	A Summary of ABO Blood Characteristics	
Blood Type	*Antigen*	*Antibody*
Type A	_____	_____
Type B	_____	_____
Type AB	_____	_____
Type O	_____	_____

Rh blood groups

The Rh blood group is named after the Rhesus monkey, the research animal in which the blood group was first identified. Rh blood groups include Rh positive (+) and Rh negative (−) blood. Although this blood system is separate from the ABO system, the two are usually used together to identify a blood type. For example, a blood sample may be A+ or A−.

Unlike the ABO system with two cellular antigens, the Rh system has only one cellular antigen, called the D antigen. A single Rh antibody designated anti-D, also occurs. The D antigen is only present on RBCs that are Rh positive; Rh negative blood cells lack the D antigen. Since Rh positive blood has the D antigen, it must lack the anti-D antibody. Interestingly, the Rh negative individual also lacks the anti-D antibody. However, if the Rh negative blood is exposed to the Rh D antigen in Rh positive blood, the immune system of the Rh negative person will produce the anti-D antibody. This becomes clinically significant in cases of pregnancy with Rh incompatibilities between the mother and her fetus. This incompatibility is covered in more detail in the Clinical Application box. Complete Table 36.3 before doing the Rh blood typing test.

TABLE 36.3   A Summary of Rh Blood Characteristics

Blood Type	Cellular Antigen	Plasma Antibody (before sensitization)	Plasma Antibody (after sensitization)
Rh positive	_____	_____	_____
Rh negative	_____	_____	_____

CLINICAL APPLICATION

If an expectant mother is Rh negative and her baby is Rh positive, a potentially life-threatening Rh incompatibility exists for the baby. Normally, the blood from the fetus does not mix with the mother's bloodstream. The umbilical cord of the baby connects to the placenta where fetal blood capillaries exchange gases, wastes, and nutrients with the mother's blood. If internal bleeding occurs and the mother is exposed to the D antigens in her baby's Rh positive blood, she will produce the anti-D antibody. These anti-D antibodies will cross the membranes of the placenta and enter the fetal bloodstream where they will hemolyze the fetal blood cells of this fetus and any other future Rh+ fetuses. This Rh factor is called the **hemolytic disease of the newborn**, or **erythroblastosis fetalis (e-rith-rō-blas-TŌ-sis fē-TAL-is)**. A dosage of anti-Rh antibodies, called RhoGam, may be given to the mother during pregnancy and after delivery to destroy any Rh positive fetal cells in her bloodstream. This will prevent her from developing anti-D antibodies in her blood.

LABORATORY ACTIVITY   ABO AND RH BLOOD TYPING

MATERIALS

gloves and safety glasses

biohazardous waste container

sterile blood lancet (disposable)

sterile alcohol prep pads (disposable)

disposable blood typing plates or sterile microscope slides

wax pencil (if using microscope slides instead of typing plates)

clean toothpicks

anti-A, anti-B, and anti-D blood typing sera

warming box for Rh reaction

paper towels

bleach solution in spray bottle

SAFETY TIPS

1. Refer to Exercise 1, "Laboratory Safety," and review the handling and disposal of blood.
2. Some infectious diseases are spread by contact with blood. Follow all instructions carefully, and protect yourself by wearing gloves and working only with your own blood.
3. Materials contaminated with blood must be disposed of properly. Your instructor will inform you of methods to dispose of blood lancets, slides, alcohol prep, and toothpicks.

Combined ABO and Rh Blood Typing

Your laboratory instructor may ask for a volunteer to "donate" blood to demonstrate the process of blood typing. Alternatively, many biological supply companies sell simulated blood typing kits that contain a bloodlike solution and anti-sera. These kits contain no human or animal blood products and safely show the principles of typing human blood.

PROCEDURES: Blood Sample Collection

1. Wash both hands thoroughly with soap, then dry hands with a clean paper towel. Obtain an additional paper towel to place blood-contaminated instruments on while collecting a blood sample. Wear gloves while collecting and examining blood. If collecting a sample from yourself, wear a glove on the hand used to hold the lancet.

2. Open a sterile alcohol prep pad, and clean the tip of your index finger. Be sure to thoroughly disinfect the entire fingertip, including the sides of the finger. Place used prep pad on the extra paper towel.

3. Open a sterile blood lancet to expose only the sharp tip. Do not use an old blood lancet, even if it was used on one of your own fingers. Use the sterile tip immediately; do not allow time for the sterile tip to inadvertently become contaminated.

4. With a swift motion, jab the point of the lancet into the lateral surface of the finger tip. Place the used blood lancet on the paper towel until it can be disposed of in a biohazard container.

5. Gently squeeze a drop of blood to each individual depression on the blood typing plate or to the circles on the slide. If necessary, slowly "milk" the finger to work more blood out of the puncture site.

PROCEDURES: ABO and Rh Typings

1. Add a drop of anti-A serum to the A blood sample. Do not allow blood to touch and contaminate the dropper bottle. Repeat the process by adding a drop of anti-B serum to the B sample and a drop of anti-D to the D sample of blood.

2. Immediately and gently mix each drop of anti-serum into the blood with a clean toothpick. To prevent cross contamination of the blood samples, use a separate clean toothpick to stir each sample. Place all used toothpicks on the paper towel until they can be disposed of in a biohazard container.

3. Place the slide or plate on the warming box, and agitate the blood samples by rocking the box carefully back and forth for two minutes.

4. Examine the blood drops for clumping or agglutination that is visible with the unaided eye, and compare your blood sample to Figure 36.5, which details the antigen-antibody reaction for each blood type. Agglutination results when the antibodies in the anti-serum react with the matching antigen on the red blood cells. Type A blood agglutinates with the anti-A serum, and type B blood agglutinates with the anti-B serum. Type AB blood agglutinates with both anti-A and anti-B sera, and type O blood does not agglutinate with either sera. If the sample agglutinates with the anti-D serum, the blood is Rh positive, whereas no agglutination shows the blood is Rh negative. The anti-D agglutination reaction is often weaker and less easily observed than the A and B agglutination reactions. A microscope may help you observe the anti-D reaction.

5. Record your blood typing results in Table 36.4 by indicating yes or no for the presence of agglutination. Collect blood typing data from several classmates to

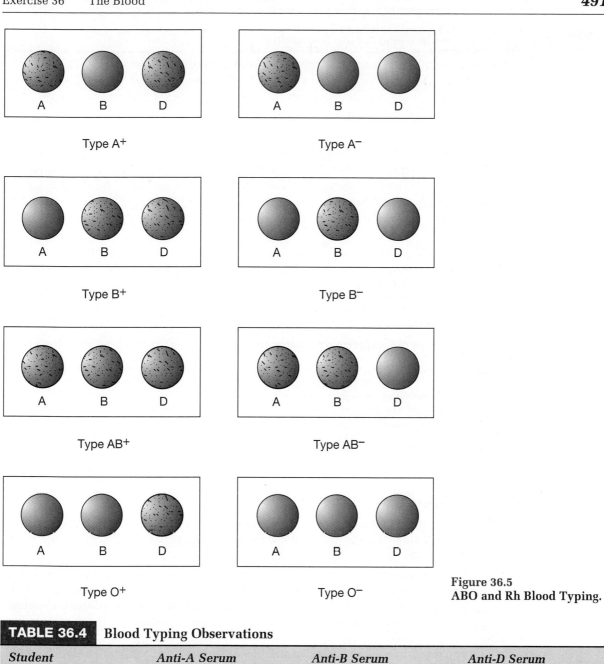

Type A⁺   Type A⁻

Type B⁺   Type B⁻

Type AB⁺   Type AB⁻

Type O⁺   Type O⁻

Figure 36.5
ABO and Rh Blood Typing.

TABLE 36.4	Blood Typing Observations		
Student	*Anti-A Serum*	*Anti-B Serum*	*Anti-D Serum*

compare agglutination responses among blood types. How does your blood type compare to the distribution of human blood types as presented in Table 36.5?

_____

6. Dispose of blood lancets, slides, toothpicks, towels, and gloves in biohazard containers as described by your instructor.

TABLE 36.5	Differences in Blood Group Distribution				
	Percentage with Each Blood Type				
Population	*O*	*A*	*B*	*AB*	*Rh⁺*
U.S. (average)	46	40	10	4	85
Caucasian	45	40	11	4	85
African-American	49	27	20	4	95
Chinese	42	27	25	6	100
Japanese	31	39	21	10	100
Korean	32	28	30	10	100
Filipino	44	22	29	6	100
Hawaiian	46	46	5	3	100
Native North American	79	16	4	<1	100
Native South American	100	0	0	0	100
Australian Aborigines	44	56	0	0	100

DISPOSAL OF MATERIALS AND DISINFECTION OF WORK SPACE

1. On completion of the blood typing exercise, dispose of all blood-contaminated materials in the appropriate biohazard box. A box for sharp objects may be available to dispose of blood lancets, toothpicks, and glass microscope slides.
2. Your lab instructor may ask you to disinfect your workstation with a bleach solution. Wear your gloves and safety glasses while wiping the surfaces clean.
3. Lastly, remove your gloves and dispose of them in the biohazard box. Remember to wash your hands after disposing of all materials.

D. Packed Red Cell Volume (Hematocrit)

The **hematocrit (he-MA-tō-krit)** test, or packed cell volume (PCV), measures the volume of packed formed elements, the cells and platelets in the blood. Since red blood cells far outnumber all the other formed elements, the test mainly measures the volume of red blood cells. To perform the test, a drop of blood is collected in a heparinized capillary tube. Heparin is an anticoagulant and prevents blood from clotting inside the capillary tube so that the red blood cells can be separated from the plasma. A microcentrifuge is used to spin the blood sample at a very high speed for approximately four minutes. The gravitational forces produced by the rotation of the centrifuge pack the cells into one end of the capillary tube, therefore the term *packed cell volume.* On completion of the centrifugation, the sample is placed on a tube reader, and the percentage of packed blood cells is measured.

The hematocrit results provide information regarding the oxygen-carrying capacity of the blood. A low hematocrit value indicates that the blood has fewer RBCs to transport oxygen. Average hematocrit values for males range from 40% to 54%; for women the range is lower, from 37% to 47%.

LABORATORY ACTIVITY   HEMATOCRIT DETERMINATION

MATERIALS

> gloves and safety glasses
> paper towels
> sterile blood lancet (disposable)

heparinized capillary tubes
microcentrifuge
biohazardous waste disposal container
bleach solution in spray bottle
sterile alcohol prep pads (disposable)
seal-ease clay
tube reader

PROCEDURES: Hematocrit

Your laboratory instructor may ask for a single volunteer to "donate" blood for demonstration of the hematocrit test.

1. Review the safety tips in the previous section. Obtain a blood sample as previously instructed.

2. Gently squeeze a drop of blood out of your finger. Excess pressure forces interstitial fluid into the blood that may alter your hematocrit reading. If you are having difficulty obtaining a blood sample, lance your finger again with a clean, sterile lancet.

3. Place a sterile heparinized capillary tube on the drop of blood. Orient the open end of the tube downward, as shown in Figure 36.6, to allow the blood to flow into the tube. Fill the tube at least two-thirds full with blood.

4. Carefully seal *one* end of the tube with the seal-ease clay, as shown in Figure 36.7. Do not force the delicate capillary tube into the clay, for it may break and cause you to jam glass into your hand. Clean any blood off the clay container with the bleach solution and a paper towel.

5. Place the tube in the microcentrifuge with the clay end toward the outer margin of the chamber. Because the centrifuge spins at high speeds, the chamber must be balanced by placing tubes evenly in the chamber. Counterbalance your capillary tube by placing another sample directly across from yours. An empty tube with clay may be used if another student blood sample is not available for centrifugation.

6. Screw the inner cover on with the centrifuge wrench. Do not overtighten the lid. Close the outer lid and push the latch in.

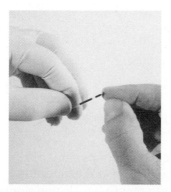

Figure 36.6
Filling Capillary Tube with Blood.

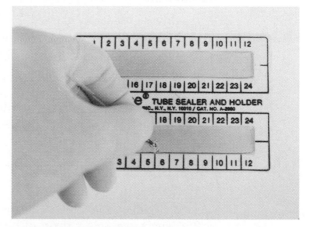

Figure 36.7
Plugging a Capillary Tube with Clay.

7. Set the timer to four to five minutes, and allow the centrifuge to spin. Do not attempt to open or stop the centrifuge while it is turning. Always keep loose hair and clothing away from the centrifuge.

8. After the centrifuge turns off and stops spinning, open the lid and the inner safety cover to remove the capillary tube. Your blood sample should have clear plasma at one end of the tube and packed red cells at the other end.

9. Place the capillary tube in the tube reader. Because there is a variety of tube readers, your instructor will show how to use the reader that is available in your laboratory.

10. What is the hematocrit of your blood sample? Is it within the normal hematocrit range?

_____

_____

11. Describe the appearance of your blood plasma.

_____

_____

12. Dispose of blood lancets, slides, toothpicks, towels, and gloves in biohazard containers as described by your instructor.

E. The Coagulation of Blood

If a sample of blood is removed from the body and allowed to sit for three to four minutes, the blood will change from a liquid to a gel state. This process is called **coagulation (cō-ag-ū-LĀ-shun)** or clotting and prevents excessive blood loss. The coagulation process is a complex chemical chain-reaction beyond the scope of this laboratory exercise. Generally, when you cut yourself, proteins circulating in the blood are activated by enzymes. Through a series of reactions, the protein fibrinogen is converted into an insoluble form called **fibrin**. The fibrin molecules occur in long threads that form a net to trap platelets and plug the wound. In the coagulation test you will determine how fast these reactions occur in your blood. A non-heparinized capillary tube is used to permit coagulation inside the tube. At 30 sec time intervals you will carefully break the tube and look for the presence of fibrin threads.

LABORATORY ACTIVITY   COAGULATION TIME

MATERIALS

gloves and safety glasses
biohazardous waste container
sterile blood lancet (disposable)
sterile alcohol prep pads (disposable)
paper towels
bleach solution in spray bottle
non-heparinized capillary tubes
small metal file

PROCEDURES: Coagulation Time

1. Review the safety tips in the previous section. Obtain a blood sample as previously instructed. Wear gloves while collecting and examining blood.

2. Gently squeeze a drop of blood out of your finger. If you are having difficulty obtaining a blood sample, lance your finger again with a clean, sterile lancet.

3. Place a sterile non-heparinized capillary tube on the drop of blood. Orient the open end of the tube downward to allow the blood to flow into the tube. Fill the tube at least 2/3 full with blood. Once the capillary tube is prepared note the time on a watch or clock.

4. Lay the blood sample on a paper towel. Every 30 sec break the capillary tube as follows:

    a. Hold the tube flat on the paper towel and gently scratch the glass of the tube with an edge of the metal file.

    b. Place your thumbs and index fingers on each side of the scratch and break the capillary tube by slowly bending the tube away from you.

    c. Slowly separate the two broken ends of the tube and look for the presence of a thin fibrin thread. Repeat steps a–c until fibrin is observed. Note the time when the coagulation occurred.

5. How much time was required for the fibrin thread to form? How does fibrin contribute to the coagulation process?

DISPOSAL OF MATERIALS AND DISINFECTION OF WORK SPACE:

1. Upon completion of the exercise, dispose of all blood contaminated materials in the appropriate biohazard box. A box for sharp objects may be available to dispose of blood lancets and capillary tubes.

2. Your lab instructor may ask you to disinfect your work station with a bleach solution. Wear your gloves and safety glasses while wiping the surfaces clean.

3. Lastly, remove your gloves and dispose of them in the biohazard box. Remember to wash your hands after disposing of all materials.

BLOOD CHECKLIST

FORMED ELEMENTS OF BLOOD

- Figures 36.1–36.3

erythrocytes (RBC)
   biconcave discs
   hemoglobin
leukocytes (WBC)
granular leukocytes
   neutrophil
      polymorphonuclear
   eosinophil
   basophil
agranular leukocytes
   monocyte
   lymphocyte
      T-cell lymphocyte
      B-cell lymphocyte
platelet

BLOOD TESTS

- Figures 36.4–36.7

antigen (agglutinogen)
antibody (agglutinin)
agglutination
anti-serum

ABO blood group
   type A (A antigen, anti-B antibody)
   type B (B antigen, anti-A antibody)
   type AB (AB antigen, no plasma antibody)
   type O (no surface antigen, both anti-A and anti-B antibodies)

Rh blood group
   Rh positive (D antigen, no plasma antibody)
   Rh negative (no surface antigen, produces anti-D antibody if exposed to Rh D antigen

hematocrit (PVC)

Coagulation
   fibrin

THE BLOOD

Laboratory Report Exercise 36

A. Matching
Match each term or structure listed on the left with the correct description on the right.

1.	_____ erythrocyte	A.	eosinophil
2.	_____ polymorphonuclear cell	B.	molecules on membrane surface
3.	_____ granular leukocytes	C.	has A antigens and anti-B antibodies
4.	_____ leukocyte	D.	has the Rh antigen
5.	_____ antibodies	E.	neutrophil
6.	_____ type A blood	F.	lacks the Rh antigen
7.	_____ Rh positive	G.	red blood cell
8.	_____ red-orange stained blood cell	H.	contain cytoplasmic granules
		I.	reacts with a membrane molecule
9.	_____ type B blood	J.	has B antigens and anti-A antibodies
10.	_____ Rh negative	K.	white blood cell
11.	_____ antigens		

B. Short-Answer Questions

1. Describe what would happen if type A blood were transfused into the bloodstream of someone with type B blood.

2. What happens in the blood of an Rh negative individual who has been exposed to Rh positive blood?

3. What is the main function of red blood cells?

4. List the five types of leukocytes, and describe the function of each.

5. Describe how to do a hematocrit test. What are the average hematocrit values for males and females?

6. Describe how to type blood to detect the ABO and Rh blood groups.

7. Describe how to test the coagulation time of a blood sample.

C. **Drawing**

   Complete each typing slide by indicating where agglutination occurs.

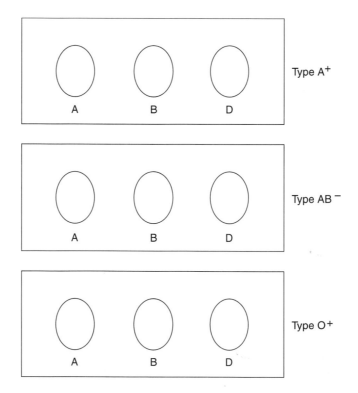

Anatomy of the Heart

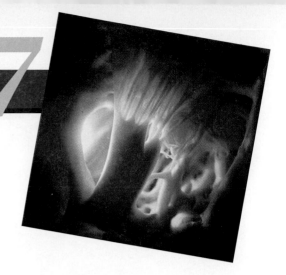

OBJECTIVES

On completion of this exercise, you should be able to:

- Identify the gross external and internal anatomy of the heart.
- Identify and discuss the function of the valves of the heart.
- Identify the major blood vessels of the heart.
- Trace a drop of blood through the heart, identifying both circuits.
- Identify the vessels of coronary circulation.
- Identify the components of the conduction system of the heart.
- Discuss the differences between a prenatal and a postnatal heart.
- Identify the anatomy of a dissected sheep heart.

WORD POWER

pericardium (peri—around)
myocardium (myo—muscle)
endocardium (endo—inner)
atrium (atri—vestibule)
ventricle (ventricul—belly)
septum (septi—fence)
pulmonary (pulmo—lung)
systemic (systema—system)

MATERIALS

torso model
heart models
cardiac muscle tissue slide
compound microscope
sheep heart
dissection scissors
blunt probe
gloves
safety glasses

INTRODUCTION

The heart beats approximately 100,000 times daily to send blood flowing into miles of blood vessels, providing nutrients and regulating substances, gases, and removal of wastes for all body cells. All organ systems of the body depend on the cardiovascular system, and so damage to the heart often results in widespread disruption of homeostasis.

For a drop of blood to complete one circuit through the body it must be pumped twice by the heart. Each passage through the heart directs blood into a vascular system that loops back to the opposite side of the heart. Deoxygenated blood enters the right side of the heart and is pumped to the lungs to become saturated with oxygen and have its carbon dioxide removed. This oxygenated blood then flows to the left side of the heart which must produce enough pressure to force blood to flow to all your body tissues.

Your anatomical studies in this unit include the histology of cardiac muscle tissue, external and internal heart structures, blood flow through the heart, and fetal heart structures. Since all mammals have a four-chambered heart, you will dissect a sheep heart to reinforce your observations of the human heart.

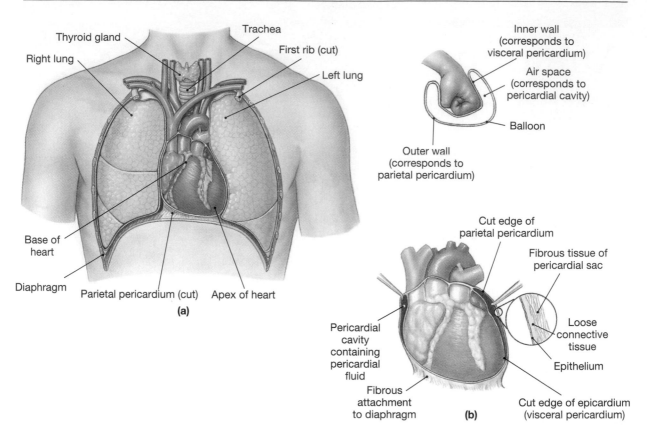

Figure 37.1
Location of the Heart in the Thoracic Cavity.
(a) Position of heart relative to other thoracic organs. **(b)** Detail of the pericardium.

A. Location of Heart and Anatomy of Heart Wall

Figure 37.1 illustrates the location of the heart within the mediastinum of your thoracic cavity. Blood vessels join the heart at the **base** positioned medially in the mediastinum. Because the left side of the heart has more muscle mass, the **apex** at the inferior tip of the heart is more on the left side of the thoracic cavity. A line traced from your right shoulder to your left hip would pass through the axis of your heart, separating the right and left chambers.

Within the mediastinum the heart lies inside a space, the **pericardial (per-i-KAR-dē-al) cavity**. This cavity is formed by the **pericardium**, a serous membrane. Recall from your earlier studies that all serous membranes are double membranes with parietal and visceral layers. The pericardial cavity contains **serous fluid** to reduce friction during muscular contraction.

Examine Figure 37.1b and locate the pericardial cavity. The outer **parietal pericardium** attaches the heart in the mediastinum, and the inner **visceral pericardium** or **epicardium** lines the surface of the heart. Imagine pushing your fist into an inflated balloon. The balloon around your fist represents the visceral pericardium, and the rest of the balloon represents the parietal pericardium. The space inside the balloon is the pericardial cavity.

Both layers of the pericardium contain a **mesothelial (mes-ō-thē-lē-al) layer**, a simple epithelium that lines the pericardial cavity and secretes serous fluid. Locate this layer in Figure 37.2. **Loose connective tissue** attaches the mesothelium to the underly-

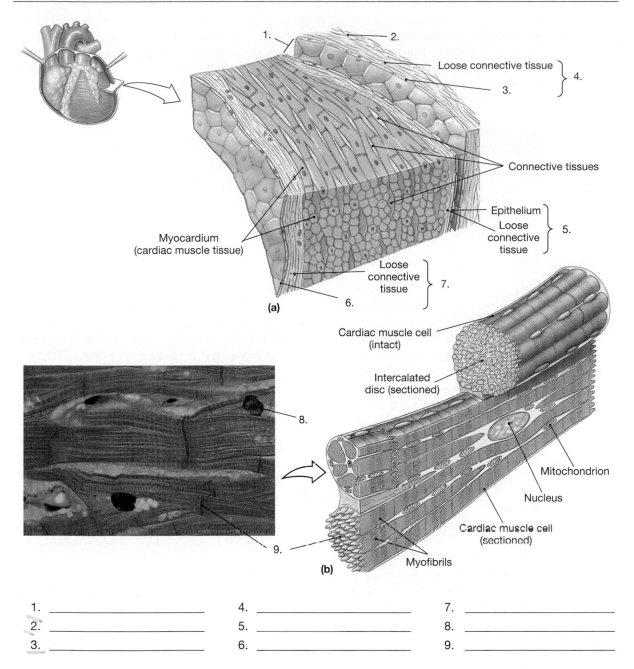

Loose connective tissue

Connective tissues

Epithelium
Loose connective tissue

Myocardium
(cardiac muscle tissue)

Loose connective tissue

(a)

Cardiac muscle cell
(intact)

Intercalated disc (sectioned)

Mitochondrion

Nucleus

Cardiac muscle cell
(sectioned)

Myofibrils

(b)

1. _____   4. _____   7. _____
2. _____   5. _____   8. _____
3. _____   6. _____   9. _____

Figure 37.2
The Heart Wall.

(a) A diagrammatic section through the heart wall showing the relative positions of the epicardium, myocardium, and endocardium. **(b)** Sectional and diagrammatic view of cardiac muscle tissue. Cardiac muscle cells are smaller than skeletal muscle fibers and also have a single, centrally placed nucleus, branching interconnections between cells, and intercalated discs. (LM ×575)

ing structures of the pericardium. The parietal pericardium is also composed of an outer **fibrous** layer that holds the heart in the thoracic cavity and prevents it from over-expanding. Notice the fibrous attachment at the diaphragm in Figure 37.1.

The heart wall is organized into three layers: the epicardium, myocardium, and endocardium. Locate each layer of the heart wall in Figure 37.2a. The **epicardium** is the same structure as the visceral pericardium. The **myocardium** is composed of cardiac muscle tissue and constitutes most of the heart wall. Histologically, the myocardium is composed of **striated** muscle cells called **cardiac muscle cells** or **cardiocytes**. Examine the micrograph in Figure 37.2b. Each cardiocyte is **uninucleated**, containing a single nucleus, and is branched. Cardiocytes interconnect at these branches with **intercalated discs**. Deep to the myocardium is the **endocardium**, a thin layer of endothelial tissue lining the chambers of the heart.

LABORATORY ACTIVITY   SUPERFICIAL ANATOMY OF THE HEART

MATERIALS

heart models and specimens
compound microscope
prepared slide of cardiac muscle

PROCEDURES

1. Label and review the anatomy presented in Figures 37.1 and 37.2.
2. Identify the layers of the heart wall on laboratory models and specimens.
3. Examine the microscopic structure of cardiac muscle.
   a. Obtain a microscope and a slide of cardiac muscle tissue. Focus the microscope on the heart tissue, starting at low magnification. Use Figure 37.2b for reference while observing the slide.
   b. Increase the magnification and locate several cardiocytes. Note the single nucleus in each cell. Intercalated discs are dark-stained lines where cardiocytes connect together. Do you see any branched cardiocytes?
   c. Sketch several cardiocytes and intercalated discs in the space provided below. Use low and high magnifications for your sketches.

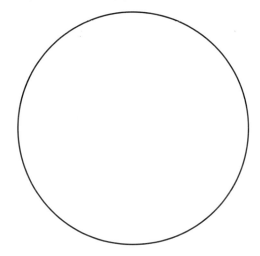

B. Chambers and Vessels of the Heart

All mammalian hearts have four chambers and are anatomically divided into right and left sides with each side having an upper and a lower chamber. The upper chambers are the **right atrium (A-trē-um)** and the **left atrium**, and the lower chambers are the **right ventricle (VEN-tri-kul)** and the **left ventricle**. Lining the inside of the right atrium are muscular ridges, the **pectinate muscle**. The wall between the atria is called the **interatrial septum**, and the **ventricles** are separated by the **interventricular septum**. Folds of muscle tissue called **trabeculae carneae (tra-BEK-ū-lē CAR-nē-ē)** occur on the inner surface of each ventricle. Locate these structures in Figure 37.3.

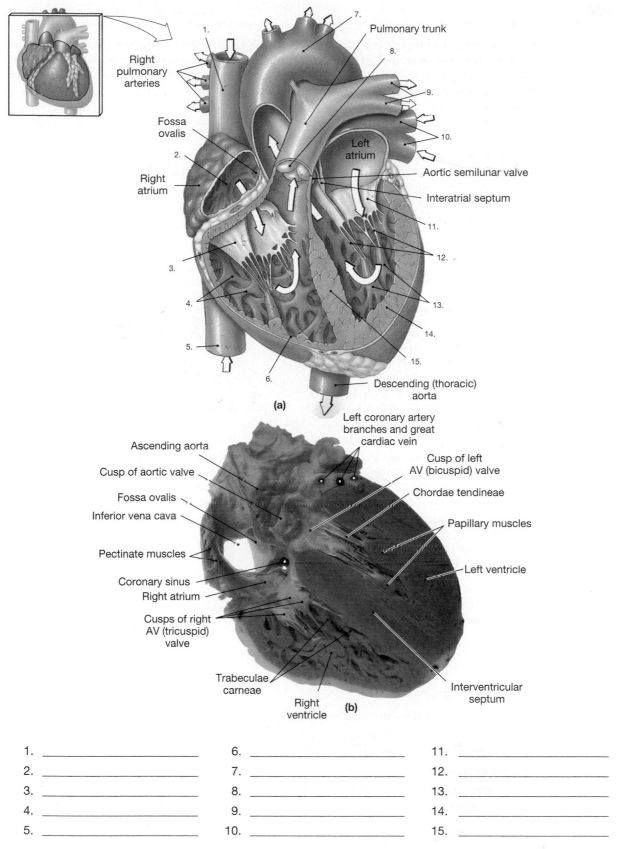

Right
pulmonary
arteries

Pulmonary trunk

1.

7.

8.

9.

10.

Fossa
ovalis

Left
atrium

Right
atrium

Aortic semilunar valve

2.

Interatrial septum

11.

12.

3.

13.

4.

14.

15.

5.

6.

Descending (thoracic)
aorta

(a)

Left coronary artery
branches and great
cardiac vein

Ascending aorta

Cusp of left
AV (bicuspid) valve

Cusp of aortic valve

Chordae tendineae

Fossa ovalis

Papillary muscles

Inferior vena cava

Pectinate muscles

Left ventricle

Coronary sinus

Right atrium

Cusps of right
AV (tricuspid)
valve

Interventricular
septum

Trabeculae
carneae

Right
ventricle

(b)

1. _____
2. _____
3. _____
4. _____
5. _____

6. _____
7. _____
8. _____
9. _____
10. _____

11. _____
12. _____
13. _____
14. _____
15. _____

Figure 37.3
Sectional Anatomy of the Heart.

(a) Diagrammatic frontal view of the internal heart. **(b)** Photograph of a human heart in frontal section.

The atria are "receiving chambers" and fill with blood returning to the heart in veins. Blood in the atria flows into the ventricles which are "pumping chambers." The ventricles fill with blood and then squeeze their walls together to pressurize the blood and squirt it into two large arteries for distribution to the lungs and body tissues.

Figure 37.4 highlights the external anatomy of the heart. Each atrium is covered externally by a flap, the **auricle (AW-ri-kul)**. Fat tissue and blood vessels occur along grooves in the heart wall. The **coronary sulcus** is a deep groove between the right atrium and ventricle and extends to the posterior surface. The boundary of the right and left ventricles is marked anteriorly and posteriorly by the **anterior interventricular sulcus** and the **posterior interventricular sulcus**. Coronary blood vessels are also shown in Figure 37.4 which is discussed in another section of this exercise.

STUDY TIP

Remember anatomical position when observing right and left structures on figures and models of the heart. For example, in Figure 37.3 the right atrium is on the left side of the illustration. The figure depicts a heart in anatomical position in front of you. Also, notice that most heart models and textbook figures are color-coded. Red designates a vessel carrying oxygenated blood, and blue is used for vessels with deoxygenated blood. Remember that arteries transport blood away from the heart and veins transport blood toward the heart.

The right atrium receives deoxygenated blood from the **superior vena cava (VĒ-nā KĀ-va)** and the **inferior vena cava.** Identify the venae cavae in Figure 37.3. From the right atrium blood flows into the right ventricle which then pumps the deoxygenated blood to the lungs by way of the **pulmonary trunk**. In the lungs the deoxygenated blood is converted to oxygenated blood as gases diffuse between the blood and the lungs. Four **pulmonary veins** drain the lungs and return the oxygenated blood from the lungs to the left atrium. Blood flows from the left atrium into the left ventricle which then pumps the oxygenated blood into the **aorta**. The aorta branches into the major systemic arteries which transport oxygenated blood to all body tissues. In the body tissues blood is converted to deoxygenated blood and then returns to the right atrium in one of the venae cavae to repeat the double circuit.

LABORATORY ACTIVITY   CHAMBERS AND VESSELS OF THE HEART

MATERIALS

heart models and specimens

PROCEDURES

1. Label and review the anatomy presented in Figures 37.3 and 37.4.
2. Locate each structure on lab materials.

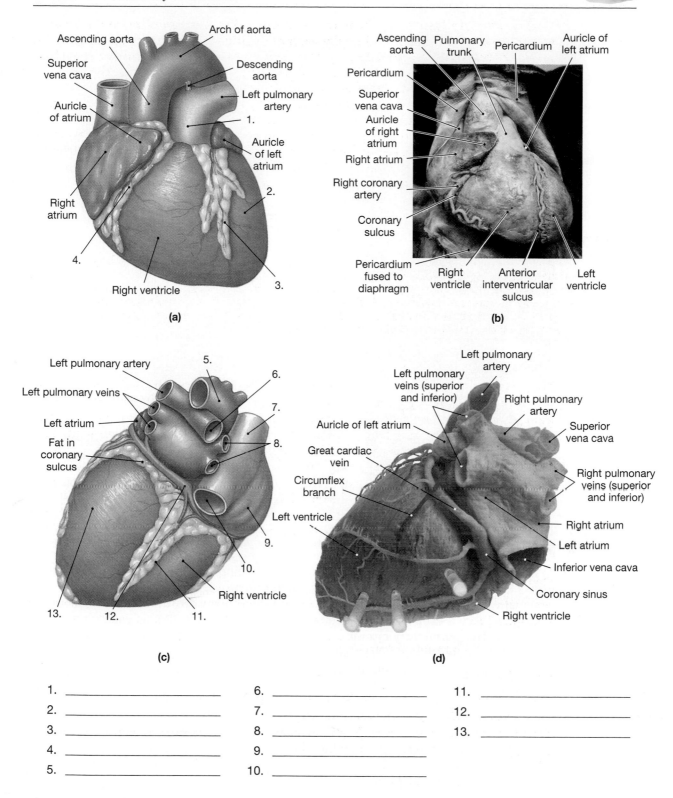

(a) Anterior view.

Arch of aorta

Ascending aorta

Superior vena cava

Descending aorta

Left pulmonary artery

Auricle of atrium

1.

Auricle of left atrium

2.

Right atrium

4.

3.

Right ventricle

(a)

Ascending aorta   Pulmonary trunk   Pericardium   Auricle of left atrium

Pericardium

Superior vena cava

Auricle of right atrium

Right atrium

Right coronary artery

Coronary sulcus

Pericardium fused to diaphragm   Right ventricle   Anterior interventricular sulcus   Left ventricle

(b)

Left pulmonary artery

5.

6.

Left pulmonary veins

7.

Left atrium

8.

Fat in coronary sulcus

Left ventricle

9.

10.

Right ventricle

13.   12.   11.

(c)

Left pulmonary artery

Left pulmonary veins (superior and inferior)

Right pulmonary artery

Auricle of left atrium

Superior vena cava

Great cardiac vein

Right pulmonary veins (superior and inferior)

Circumflex branch

Right atrium

Left atrium

Inferior vena cava

Coronary sinus

Right ventricle

(d)

1. _____	6. _____	11. _____
2. _____	7. _____	12. _____
3. _____	8. _____	13. _____
4. _____	9. _____	
5. _____	10. _____	

Figure 37.4
Surface Anatomy of the Heart.

(a) Anterior view. **(b)** Human heart in anterior view. **(c)** Posterior view. **(d)** Human heart, posterior view.

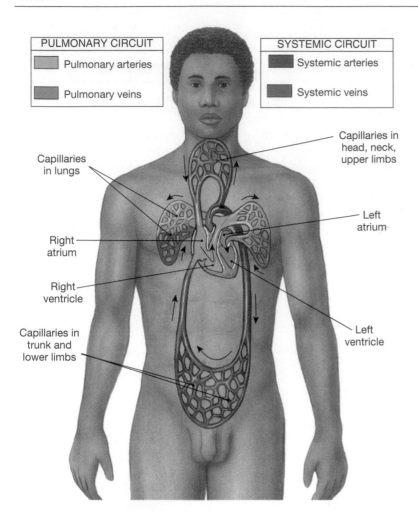

PULMONARY CIRCUIT
Pulmonary arteries
Pulmonary veins

SYSTEMIC CIRCUIT
Systemic arteries
Systemic veins

Capillaries in head, neck, upper limbs

Capillaries in lungs

Right atrium

Right ventricle

Capillaries in trunk and lower limbs

Left atrium

Left ventricle

Figure 37.5
An Overview of the Cardiovascular System.

Driven by the pumping of the heart, blood flows through separate pulmonary and systemic circuits. Each circuit begins and ends at the heart and contains arteries, capillaries, and veins.

C. Pulmonary and Systemic Circulation

As a double pump the heart delivers blood into two circulatory pathways. Each pathway begins in a ventricle which delivers blood into a series of arteries, capillaries, and veins and then concludes with the return of blood to the atrium opposite the initial ventricle. Trace a drop of blood through the heart in Figure 37.5 as each pathway is discussed. The **pulmonary circuit** directs deoxygenated blood to the lungs for conversion to oxygenated blood. The right ventricle receives the deoxygenated blood and pumps it into an artery, the pulmonary trunk, which conducts it to the capillaries of the lungs for oxygenation. Four pulmonary veins drain the lungs and return the oxygenated blood to the left atrium. Blood from the left atrium enters the left ventricle for the **systemic circuit**. The left ventricle pumps oxygenated blood to the aorta which conducts blood to all major systemic arteries of the organ systems. After flowing through the capillaries of the organs, deoxygenated blood is returned to the right atrium via the venae cavae. Blood passes through the heart twice: the right ventricle pumps it to the lungs and then the left ventricle pumps it to the body tissues.

LABORATORY ACTIVITY   PULMONARY AND SYSTEMIC CIRCULATION

MATERIALS

heart models and specimens

PROCEDURES

1. Trace a drop of blood through the pulmonary and systemic circuits in Figure 37.5.
2. Locate the vessels of the heart on all lab material.

D. Valves of the Heart

To control and direct blood flow, the heart has two pairs of atrioventricular and semilunar valves. The **atrioventricular (AV) valves** are located between the atria and ventricles on each side of the heart. These valves prevent blood from reentering the atria when the ventricles contract. The right atrioventricular valve has three flaps or cusps and is also called the **tricuspid (trī-kus-pid) valve**. The left atrioventricular valve has two cusps and is called the **bicuspid valve** or the **mitral (MĪ-tral) valve**. Observe in Figure 37.3 that the cusps of each AV valve have small cords, the **chordae tendineae (KOR-dē TEN-di-nē-ē)**, which are attached to **papillary (PAP-i-ler-ē) muscles** on the floor of the ventricles. When the ventricles contract, the AV valves are held closed by the papillary muscles pulling on the chordae tendineae.

The other pair of valves are the **pulmonary** and **aortic semilunar valves** located at the base of each vessel. Semilunar valves are composed of three small cusps and prevent backflow of blood into the ventricles when the ventricles are relaxed. Identify each pair of valves in Figure 37.3.

Figure 37.6a illustrates the function of the AV and the semilunar valves. When a ventricle contracts, in a phase called ventricular systole, pressure forces the atrioventricular valve closed. To keep this valve from reversing like an umbrella in a strong wind, the papillary muscles contract and pull on the chordae tendineae to secure the cusps. Increased pressure then forces the semilunar valve open and blood flows into the artery. Figure 37.6b represents valve function during ventricular diastole, the relaxation phase of the ventricle. Blood flows from the atrium into the relaxed ventricle through the open atrioventricular valve. Note in the figure that the papillary muscles and the chordae tendineae are relaxed. The semilunar valve is closed to prevent backflow of blood from the artery into the ventricle.

STUDY TIP

The atrioventricular and semilunar valves generally work in opposition. When one pair of valves is open, the other pair is closed or is preparing to close. If the atrioventricular valves are open, for example, the semilunar valves are closed.

LABORATORY ACTIVITY   HEART VALVES

MATERIALS

heart models and specimens

PROCEDURES

1. Review valve anatomy in Figures 37.3. and 37.6.
2. Identify each valve on all laboratory material.

E. Coronary Circulation

To produce the pressure required to push blood through the vascular system the heart can never completely rest. The heart has a vascular system called the **coronary circulation** which supplies the myocardium with the oxygen necessary for muscle contraction.

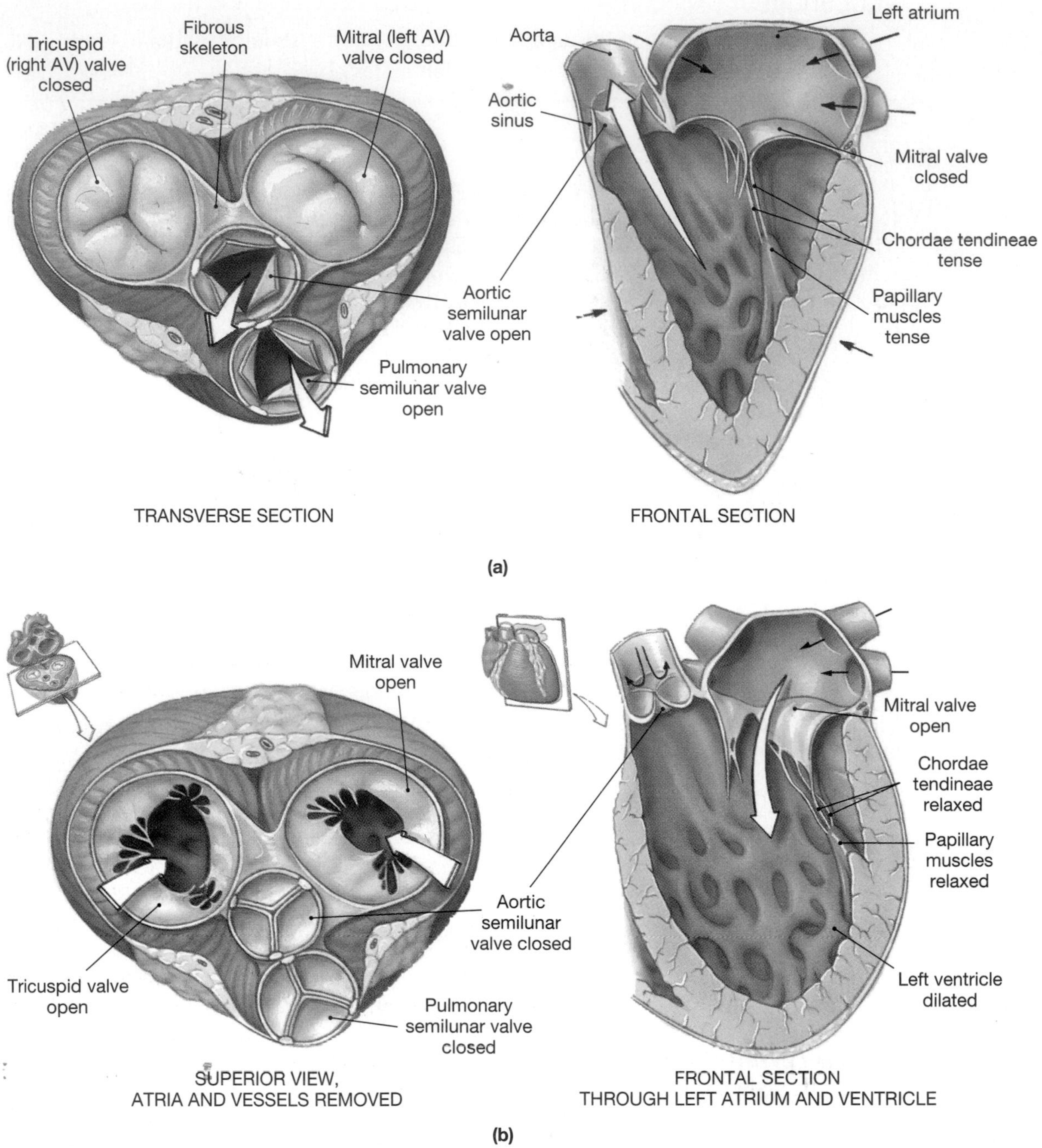

TRANSVERSE SECTION

FRONTAL SECTION

(a)

SUPERIOR VIEW,
ATRIA AND VESSELS REMOVED

FRONTAL SECTION
THROUGH LEFT ATRIUM AND VENTRICLE

(b)

Figure 37.6
Valves of the Heart.

(a) The appearance of the cardiac valves during ventricular systole (contraction), when the AV valves are closed and the semilunar valves are open. In the frontal section, note the attachment of the bicuspid valve to the chordae tendineae and papillary muscles. **(b)** Valve position during ventricular diastole (relaxation), when the AV valves are open and the semilunar valves are closed. Note that the chordae tendineae are slack and the papillary muscles are relaxed.

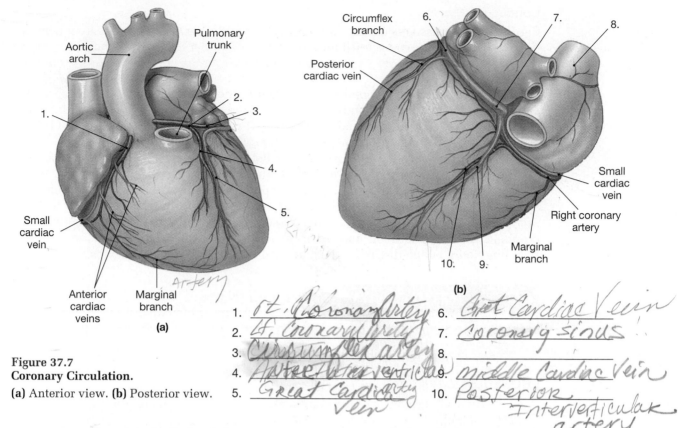

Figure 37.7
Coronary Circulation.
(a) Anterior view. (b) Posterior view.

1. rt. Coronary Artery
2. Lt. Coronary Artery
3. Circumflex Artery
4. Anterior interventricular
5. Great Cardiac Vein
6. Great Cardiac Vein
7. Coronary Sinus
8. _____
9. middle Cardiac Vein
10. Posterior Interventricular artery

Examine the base of the aorta in Figure 37.7a and identify the right and left **coronary arteries** which branch off and penetrate the myocardium to the outer heart wall. The right coronary artery follows the coronary sulcus, a groove along the right atrium. The **posterior interventricular artery** branches off the right coronary artery and supplies posterior regions of the myocardium (see Figure 37.7b). The left coronary artery branches into a **circumflex artery** and an **anterior interventricular artery** (left anterior descending artery). The circumflex artery follows the left side of the heart and enters the posterior coronary sulcus with the right coronary artery. The anterior interventricular artery lies within the anterior interventricular sulcus.

Cardiac veins collect deoxygenated blood from the myocardium. The **great cardiac vein** follows along the anterior interventricular sulcus and curves around the left side of the heart. The **middle cardiac vein** drains most of the posterior myocardium, and the **small cardiac vein** drains the upper right area of the heart. The cardiac veins merge to form the **coronary sinus** situated in the posterior region of the coronary sulcus. The coronary sinus is a large vein that empties into the right atrium. Recall that the right atrium also receives deoxygenated blood from the venae cavae. Locate the coronary veins in Figure 37.7.

LABORATORY ACTIVITY   CORONARY CIRCULATION

MATERIALS

heart models and specimens

PROCEDURES

1. Identify the coronary blood vessels on lab materials.
2. Label and review the coronary blood vessels in Figure 37.7.

F. Conduction System

Cardiac muscle tissue is unique in that it is *autorhythmic*, producing its own contraction and relaxation phases without stimulation from nerves. Nerves may increase or decrease the heart rate, yet a living heart removed from the body continues to contract on its own.

Figure 37.8 details the **conduction system** of the heart. Special cells called **nodal cells** produce and conduct electrical currents to the myocardium which coordinate the heart's contraction. The pacemaker of the heart is the **sinoatrial (sī-nō-Ā-trē-al) (SA) node**, located where the superior vena cava empties into the upper right atrium. Nodal cells in the SA node self-excite faster than other areas of the heart and thus set the pace for the heart's contraction. The **atrioventricular (AV) node** is located on the lower medial floor of the right atrium. The SA node stimulates the AV node which then spreads the impulse toward the ventricles through the **atrioventricular bundle** (bundle of His). The atrioventricular bundle passes into the interventricular septum and branches into right and left **bundle branches**. The bundle branches divide into fine **Purkinje fibers** which distribute the electrical impulses to the cardiocytes.

LABORATORY ACTIVITY   CONDUCTION SYSTEM

MATERIALS

heart models and specimens

PROCEDURES

1. Identify all structures of the conduction system on lab materials.
2. Label and review the conduction system in Figure 37.8.

**Figure 37.8
The Conduction System of the Heart.**

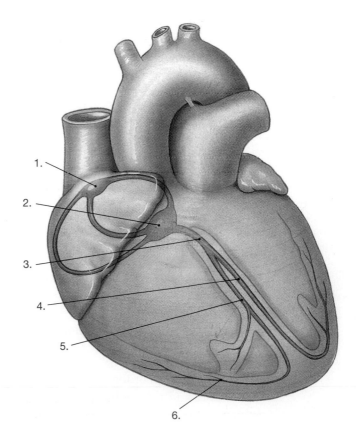

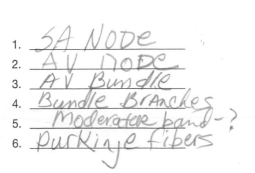

1. _SA NODE_
2. _AV NODE_
3. _AV Bundle_
4. _Bundle Branches_
5. _Moderator band - ?_
6. _Purkinje fibers_

G. The Fetal Heart

A fetus receives oxygen from the mother through the placenta, a vascular organ that connects the fetus to the wall of the mother's uterus. During development, the fetal lungs are filled with amniotic fluid, and for efficiency some of the blood is shunted away from the fetal pulmonary circuit by several structures as shown in Figure 37.9. The **foramen ovale** is a hole in the interatrial wall. Much of the blood entering the right atrium from the inferior vena cava passes through the foramen ovale to the left atrium and avoids the right ventricle and the pulmonary circuit. Some of the blood that does enter the pulmonary trunk may bypass the lungs through a connection with the aorta, the **ductus arteriosus**. At birth, the foramen ovale closes and becomes a depression on the interatrial wall, the **fossa ovalis**. The ductus arteriosus closes and becomes the **ligamentum arteriosum**.

LABORATORY ACTIVITY | THE FETAL HEART

MATERIALS

heart models and specimens

PROCEDURES

1. Label and review the fetal structures in Figure 37.9.
2. Identify the fossa ovalis and ligamentum arteriosum on lab models.

**Figure 37.9
Fetal Circulation.**

(a) Blood flow to and from the placenta. **(b)** Blood flow through the neonatal (newborn) heart.

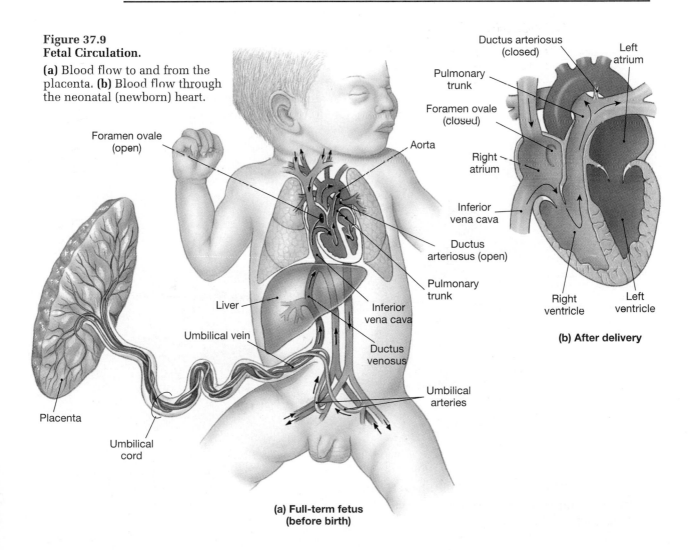

**(a) Full-term fetus
(before birth)**

(b) After delivery

H. **Sheep Heart Dissection**

The sheep heart, like all mammalian hearts, is similar in structure and function to the human heart. One major difference is the position of the great vessels joining the heart. In four-legged animals, the inferior vena cava has a posterior connection to the heart instead of an inferior attachment as in humans.

Dissecting a sheep heart will enhance your studies of models and charts of the human heart. Take your time while dissecting and follow the directions carefully.

LABORATORY ACTIVITY | SHEEP HEART DISSECTION

MATERIALS

gloves
safety glasses
fresh or preserved sheep heart
dissecting pan
dissecting scissors
blunt probe

PROCEDURES

1. Put on gloves and safety glasses. Read the safety information in Exercise 1.
2. Wash the sheep heart with cold water to flush out preservatives and blood clots. Minimize your skin and mucous membrane exposure to the preservatives.
3. Carefully follow the instructions below. Cut into the heart only as instructed.

I. **External Anatomy**

1. Figure 37.10 details the external anatomy of a sheep heart. Examine the surface of the heart for the **pericardium**. Often this serous membrane has been removed in

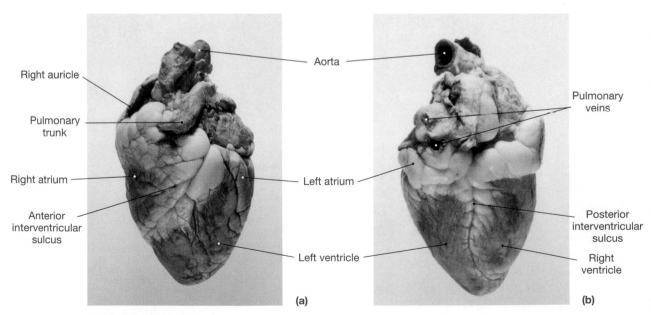

Figure 37.10
External Anatomy of the Sheep Heart.
(a) Anterior view (b) posterior view.

preserved specimens. Carefully use your scalpel to scrape the outer heart muscle to loosen the **epicardium**. Next, locate the anterior surface by orienting the heart so the **auricles** face you. Under the auricles are the **right** and **left atria**. Note the **base** of the heart above the atria where the large blood vessels occur. The **apex** is the inferior tip of the heart. Squeeze gently above the apex to locate the **right** and **left ventricles**. Locate the **anterior interventricular sulcus**, the fat-laden groove between the ventricles. Carefully remove some of the fat tissue to locate coronary blood vessels. Identify the other grooves, the **coronary sulcus** between the right atrium and ventricle and the **posterior interventricular sulcus** between the ventricles on the posterior surface.

2. Identify the **pulmonary trunk** and the **aorta**. The pulmonary trunk is anterior to the aorta. If the pulmonary trunk was cut long, you may be able to identify the right and left **pulmonary arteries** branching off the trunk. Posterior to the pulmonary trunk is the aorta. The **brachiocephalic artery** is the first major branch of the aorta and is often intact in preserved material.

3. Follow along the inferior margin of the right auricle to the posterior surface. The prominent vessel at the termination of the auricle is the **superior vena cavae**. At the base of this vessel is the **inferior vena cava**. Next, examine the posterior aspect of the left atrium and find the four **pulmonary veins**. You may need to carefully remove some of the fatty tissue around the superior region of the left atrium to locate the veins.

II. Internal Anatomy

Follow the instructions carefully to improve your dissection technique and identification of structures. Use Figure 37.11 as a reference to the internal anatomy of the sheep heart.

1. Place a blade of your dissection scissors into the **superior vena cava** and cut down into the upper portion of the **right atrium**. Cut approximately $\frac{1}{2}$ in. or until you have exposed the **tricuspid valve**. Do not cut through the tricuspid valve.

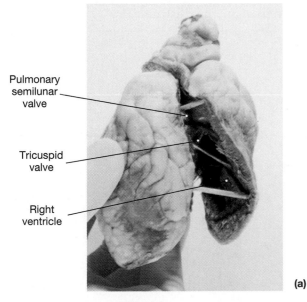

Pulmonary semilunar valve

Tricuspid valve

Right ventricle

(a)

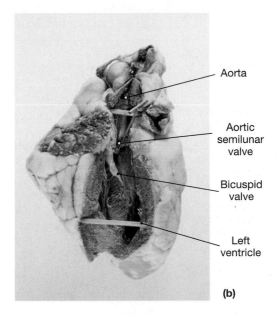

Aorta

Aortic semilunar valve

Bicuspid valve

Left ventricle

(b)

Figure 37.11
Internal Anatomy of the Sheep Heart.
(a) Right ventricle **(b)** left ventricle.

2. Examine the right atrium for the comblike **pectinate muscles** lining the inner wall. Identify the opening of the **coronary sinus**, located between the superior and inferior venae cavae.

3. To observe the function of the **tricuspid valve** slowly fill the right atrium with water. Watch the water drain into the lower right ventricle. Gently squeeze the outer wall of the right ventricle and observe the cusps of the tricuspid valve closing. Drain the water from the heart to continue the dissection.

4. Use your dissection scissors and continue cutting the right atrium. Cut through the tricuspid valve to the inferior part of the right ventricle. Open the ventricle with your fingers and identify the **chordae tendineae** and **papillary muscles** of the tricuspid valve. Note the folds of **trabeculae carneae** along the inner ventricular walls. Squeeze the outer wall of the left ventricle to help locate the **interventricular septum**.

5. In the right ventricle, carefully insert a blunt probe into the opening of the **pulmonary trunk**. Slide the probe into the vessel until it exits the pulmonary trunk outside the heart. Remove the probe.

6. To expose the **pulmonary semilunar valve** use scissors to cut through the right ventricular wall to the pulmonary trunk. Completely open the pulmonary trunk by cutting the entire length of the vessel. Spread the trunk open and observe the small cusps of the pulmonary semilunar valve at the base of the vessel.

7. Next, dissect the left side of the heart. Place the pointed blade of your dissecting scissors into a **pulmonary vein** and cut into the left atrium to expose the **bicuspid valve**. Fill the left atrium with water and gently squeeze the outer wall of the left ventricle to observe the cusps of the **bicuspid valve**.

8. Continue cutting through the left atrium past the bicuspid valve and into the **left ventricle**. Review the anatomy of the left atrioventricular (bicuspid) valve.

9. Slip a blunt probe into the **aorta**. Remove the probe and cut the left ventricular wall toward the aorta and then continue with a longitudinal cut along the length of the aorta. Spread the vessel open and locate the **aortic semilunar valve** at the aortic base. Examine the base and identify the right and left **coronary arteries**.

10. Upon completion of the dissection, dispose of the sheep heart as directed by the lab instructor and wash your hands and dissecting instruments.

HEART ANATOMY EXERCISE

This is a list of bold terms presented in Exercise 37. Use this list for review purposes after you have completed the laboratory activities.

SUPERFICIAL AND WALL ANATOMY

• Figures 37.1 and 37.2
apex
base
pericardium
   parietal pericardium
     pericardial cavity
       serous fluid
   visceral pericardium

three layers of heart wall:
   epicardium
   myocardium
     cardiocytes
       striations
       intercalated discs
       uninucleate
   endocardium

superficial anatomy:
coronary sulcus
anterior interventricular sulcus
posterior interventricular sulcus

CHAMBERS AND VALVES

• Figures 37.3 and 37.6
right and left auricles
right atrium
   pectinate muscle
interatrial septum
left atrium

right ventricle
   tricuspid valve
   trabeculae carneae
interventricular septum
left ventricle
   bicuspid (mitral) valve

atrioventricular valves:
tricuspid, bicuspid
chordae tendineae
papillary muscles

GREAT VESSELS OF THE HEART

• Figures 37.3, 37.4, and 37.7
pulmonary circuit
   pulmonary trunk
     pulmonary semilunar valve
   right and left pulmonary arteries
   pulmonary veins (four)

systemic circuit
   aorta
     aortic semilunar valve
   superior vena cava
   inferior vena cava

coronary circulation
   coronary arteries
     posterior interventricular artery
     circumflex artery
     anterior interventricular artery
   cardiac veins
     great cardiac vein
     middle cardiac vein
     small cardiac vein
     coronary sinus

CONDUCTION SYSTEM

• Figure 37.8
nodal cells
sinoatrial node
atrioventricular node
atrioventricular bundle
bundle branches
Purkinje fibers

FETAL HEART STRUCTURES

• Figure 37.9
fossa ovalis (fetus)
foramen ovale (after birth)
ductus arteriosus (fetus)
ligamentum arteriosum (after birth)

SHEEP HEART

• Figure 37.10 and 37.11
pericardium
epicardium
myocardium

auricles
right atrium
   pectinate muscles
tricuspid valve
   chordae tendineae
   papillary muscles

right ventricles
   carnae trabeculae
   pulmonary trunk
      pulmonary semilunar valve

left atrium
   pulmonary veins
bicuspid valve
left ventricle
interventricular septum
   aorta
      aortic semilunar valve
   coronary arteries
   brachiocephalic artery
   cardiac veins
      coronary sinus
   superior vena cava
   inferior vena cava

Name _____   Date _____

Section _____   Number _____

ANATOMY OF THE HEART

Laboratory Report Exercise 37

A. **Matching**

Match each heart structure listed on the left with the correct description on the right.

1. _____ tricuspid valve	A.	empties into left atrium
2. _____ superior vena cava	B.	left atrioventricular valve
3. _____ right ventricle	C.	muscle folds of ventricles
4. _____ aorta	D.	pumps blood to body tissues
5. _____ interventricular septum	E.	branch off left coronary artery
6. _____ left ventricle	F.	major systemic artery
7. _____ pulmonary veins	G.	muscular ridges of right atrium
8. _____ semilunar valve	H.	artery with deoxygenated blood
9. _____ bicuspid valve	I.	groove on right side of heart
10. _____ pulmonary trunk	J.	visceral pericardium
11. _____ circumflex artery	K.	drains coronary veins into heart
12. _____ trabeculae carneae	L.	wall between the ventricles
13. _____ pectinate muscle	M.	inferior tip of heart
14. _____ coronary sulcus	N.	cardiac muscle tissue
15. _____ auricle	O.	aortic or pulmonary valve
16. _____ coronary sinus	P.	empties into right atrium
17. _____ myocardium	Q.	attached to AV valves
18. _____ epicardium	R.	right atrioventricular valve
19. _____ chordae tendineae	S.	pumps blood to lungs
20. _____ apex	T.	external flap of atrium

B. **Short-Answer Questions**

1. What type of blood, oxygenated or deoxygenated, does the right ventricle pump?

2. What type of blood does the pulmonary trunk contain? Why is it an artery rather than a vein?

3. Why is the wall of the left ventricle thicker than the wall of the right ventricle?

4. What keeps the atrioventricular valves from reversing?

5. List the layers of the heart wall.

6. Describe the location of the heart.

7. List in order the spread of an impulse through the conduction system.

8. Suppose a patient has mitral valve prolapse, a weakened bicuspid valve that does not close properly. How does this affect the flow of blood in the heart?

Anatomy of Systemic Circulation

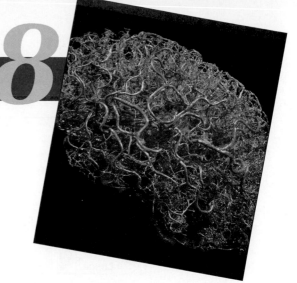

OBJECTIVES

On completion of this exercise, you should be able to:

- Compare the histology of an artery, a capillary, and a vein.
- Compare the circulation of blood to the right and left arms.
- Describe the anatomy and the importance of the circle of Willis.
- Trace a drop of blood from the ascending aorta into each abdominal organ and to the lower limbs.
- Trace a drop of blood returning to the heart from the calf.
- Discuss the unique circulatory pathways of the fetus.

WORD POWER

artery (arteri—artery)
capillary (capill—hair)
vein (ven—vein)
tunica (tunic—covering)
aorta (aort—great artery)
anastomosis (anastomos—coming together)
hepatic (hepa—liver)
portal (porta—gate or door)

MATERIALS

torso mannequin
arm model
leg model
abdominal model
head model
vascular charts
compound microscope
prepared slides:
    artery and vein

INTRODUCTION

The body has more than 60,000 miles of blood vessels to transport blood to the trillions of cells in the tissues. The basic circulatory route includes **arteries,** which distribute oxygen and nutrient-rich blood to microscopic networks of vessels called **capillaries**. At the capillaries, diffusion of nutrients, gases, wastes, and cellular products occurs between the blood and the cells. **Veins** drain deoxygenated blood from the capillaries and direct it toward the heart which then pumps it to the lungs to pick up oxygen and release carbon dioxide.

Smaller arteries, called **arterioles**, regulate blood flow into capillaries by relaxing or contracting the smooth muscle of their tunica media. Contraction of the smooth muscle causes the arteriole to narrow, or **vasoconstrict**, and results in a decrease in blood flow. Relaxation opens the vessel and promotes blood flow, a process called **vasodilation**. Bands of smooth muscle called **precapillary sphincters** are positioned where the arteriole joins a **metarteriole**, a small arteriole just upstream of the thin-walled capillary vessels. The precapillary sphincters regulate blood flow through the capillary by controlling the diameter and resistance of the metarteriole. The path of least resistance in the capillary is called the **thoroughfare channel**. This channel empties into several small veins called **venules** which empty into larger venules that join veins.

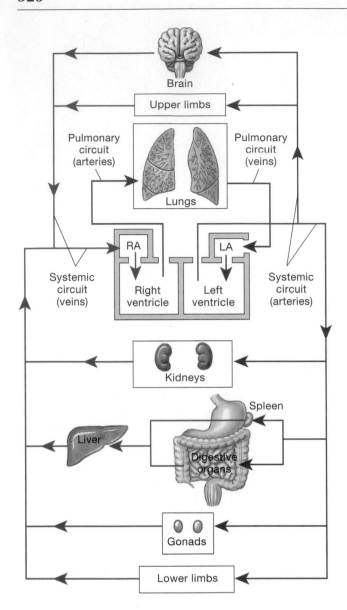

Figure 38.1
An Overview of the Pattern of Circulation.

Here RA stands for right atrium; LA, for left
atrium.

Most capillaries have several arteriole blood supplies. An **anastomosis** is where
arterioles fuse together into a single vessel to supply the same capillary. Anastomoses
ensure that a capillary receives blood. If one arteriole is pinched closed or is damaged,
the joining arteriole will provide blood to the shared capillary. Coronary circulation of
the heart has many vital anastomoses.

To accomplish the task of oxygenating and supplying blood to cells, two circulatory
pathways exist: the pulmonary circuit and the systemic circuit, illustrated in Figure
38.1. The **pulmonary circuit** takes deoxygenated blood from the right ventricle and
transports it to the lungs in pulmonary arteries. Oxygenated blood from the lungs is
returned to the left atrium by pulmonary veins. The **systemic circuit** supplies oxygenated blood to the organ systems. Oxygenated blood from the lungs is returned to the
left atrium and flows into the left ventricle. The left ventricle contracts and produces
enough pressure to pump the blood into the aorta. Systemic arteries arise off the aorta
and direct blood into the capillaries of tissues. Cells consume oxygen and nutrients
from the blood and add carbon dioxide, wastes, and regulatory molecules (enzymes and
hormones) to the blood. Veins drain the deoxygenated blood from the capillaries and
return the blood to the right atrium. The deoxygenated blood reenters the pulmonary
circuit, and the cycle repeats. Vessels of the pulmonary circuit are presented in Exercise
37. In this exercise you will study the major arteries and veins of the systemic circuit.

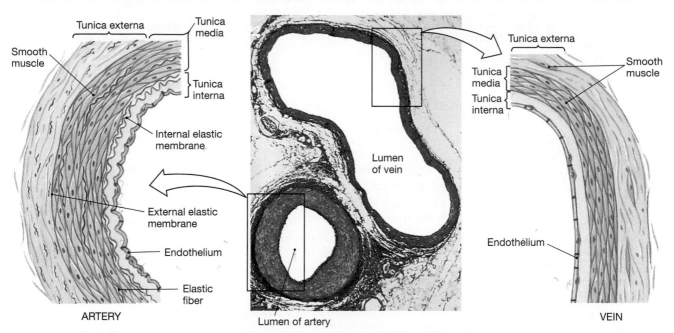

Figure 38.2
A Comparison of a Typical Artery and a Typical Vein.

Blood vessels are a continuous network of pipes, and often there is no anatomical difference between vessels within a region of the body. To facilitate identification and discussion of blood vessels, anatomists have named them after an adjacent bone or organ and changed the name of a vessel as it passes into a different region. Arteries and veins are often parallel and have the same name. For example, the large vessels in the thigh are the femoral artery and femoral vein. As the femoral artery descends the leg it is called the popliteal artery behind the knee and the posterior tibial artery of the calf.

A. Comparison of an Artery and a Vein

The walls of arteries and veins have three layers: an outer tunica externa, a middle tunica media, and an inner tunica interna. Locate these layers in Figure 38.2. The **tunica externa** is a connective tissue covering that anchors the vessel to surrounding tissues. Collagen and elastic fibers give this layer strength and flexibility. The **tunica media** is a middle layer of smooth muscle tissue. Contraction of smooth muscle in arterioles regulates blood flow into capillaries. Arteries also have elastic fibers in the tunica media which allow stretching and recoiling of the vessel in response to blood pressure changes, and veins have collagen fibers for strength. Lining the inside of the vessels is the **tunica interna**, a thin layer of **endothelium**.

Because blood pressure is much higher and fluctuates more in arteries than in veins, the walls of arteries are thicker than those of veins. Figure 38.2 compares an artery and a vein in cross section. Notice how the artery is round and has a thick tunica media with a folded tunica interna. When blood pressure increases or vasodilation occurs, the artery expands and the pleats in the tunica interna flatten as the vessel stretches.

Capillaries lack a tunica externa and tunica media. They consist of a single layer of endothelium which is continuous with the tunica interna of the artery and vein that supply and drain the capillary. Capillaries are so narrow that blood cells must line up in single file to squeeze through.

Veins have a thinner wall and have a smooth tunica interna. Their walls will collapse if emptied of blood. Blood pressure is low in veins, and to prevent backflow of blood, the peripheral veins have **valves** that function to keep blood flowing in one direction, toward the heart.

LABORATORY ACTIVITY   ARTERY AND VEIN COMPARISON

MATERIALS

  compound microscope
  prepared slide:
   artery and vein

PROCEDURES

1. Label and review the structure of arteries and veins in Figure 38.2.

2. Locate the artery and vein on the slide. Most slide preparations have one artery, an adjacent vein, and a nerve. The blood vessels are hollow and most likely have blood cells in the lumen. The nerve appears as a round, solid structure.

3. Identify the tunica externa, tunica media, and tunica interna in the artery and vein. Examine each layer at high magnification. How does the endothelial layer of the vein differ from that of the artery?

  ———————————————————————————————

  ———————————————————————————————

4. Draw and label the artery and vein in the space provided. Include enough detail in your drawing to show the anatomical differences between the vessels.

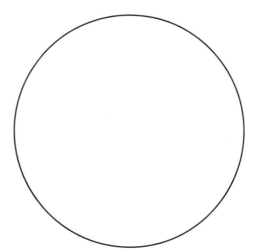

B. The Arterial System

1. Arteries of the head, neck, and upper limb

The main artery is the aorta, which receives oxygenated blood from the left ventricle of the heart. Figure 38.3 illustrates the major vessels of the systemic arterial system. Oxygenated blood is pumped out of the left ventricle of the heart and into the aorta for distribution to the organ systems. The aorta is curved like a question mark, it exits the base of the heart, curves upward and to the left, and then descends behind the heart to the abdominal cavity. Arteries that branch off the aortic arch serve the head, neck, and upper limb. Branches off the abdominal aorta serve the abdominal organs. The abdominal aorta enters the pelvic cavity and divides to send a branch into each lower limb.

  Figure 38.4 illustrates the arteries of the chest and upper limb. The **ascending aorta** begins at the base of the heart and extends to the curved **aortic arch** which descends through the thorax as the **thoracic aorta**. The aorta pierces through the diaphragm and becomes the **abdominal aorta**.

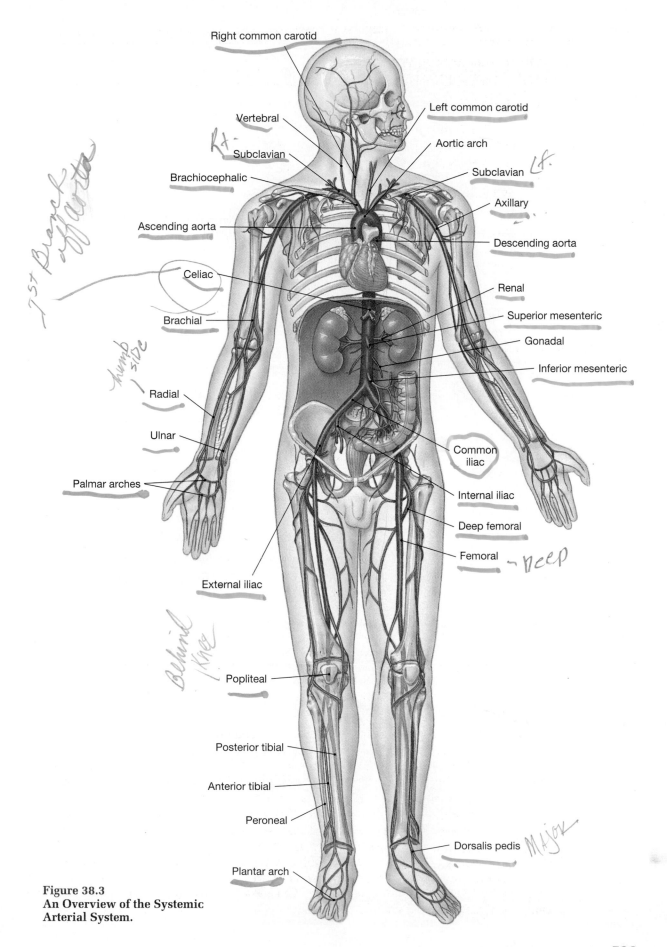

Right common carotid

Left common carotid

Vertebral

Rt.

Subclavian

Brachiocephalic

Aortic arch

Subclavian Lt.

Axillary

Ascending aorta

Descending aorta

1st Branch off aorta

Celiac

Renal

Brachial

Superior mesenteric

Thumb side

Gonadal

Inferior mesenteric

Radial

Ulnar

Palmar arches

Common iliac

Internal iliac

Deep femoral

Femoral ~ DEEP

External iliac

Behind Knee

Popliteal

Posterior tibial

Anterior tibial

Peroneal

MAJOR

Dorsalis pedis

Plantar arch

Figure 38.3
An Overview of the Systemic
Arterial System.

523

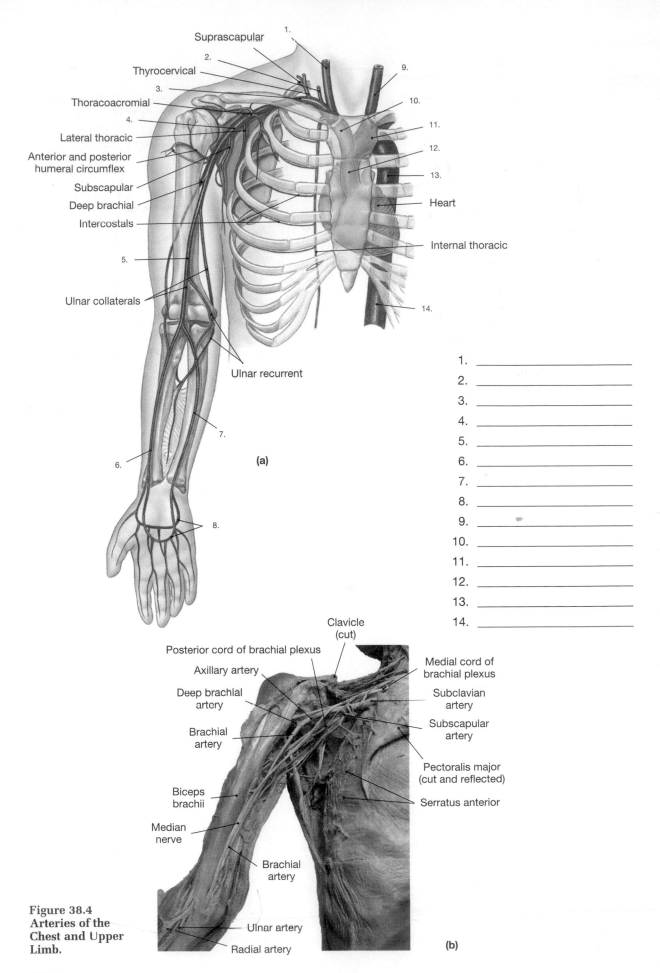

Suprascapular

Thyrocervical

Thoracoacromial

Lateral thoracic

Anterior and posterior
humeral circumflex

Subscapular

Deep brachial

Intercostals

Ulnar collaterals

Ulnar recurrent

1.

2.

3.

4.

5.

6.

7.

8.

9.

10.

11.

12.

13.

14.

Heart

Internal thoracic

(a)

1. _____
2. _____
3. _____
4. _____
5. _____
6. _____
7. _____
8. _____
9. _____
10. _____
11. _____
12. _____
13. _____
14. _____

Clavicle
(cut)

Posterior cord of brachial plexus

Axillary artery

Deep brachial
artery

Brachial
artery

Biceps
brachii

Median
nerve

Brachial
artery

Ulnar artery

Radial artery

Medial cord of
brachial plexus

Subclavian
artery

Subscapular
artery

Pectoralis major
(cut and reflected)

Serratus anterior

(b)

**Figure 38.4
Arteries of the
Chest and Upper
Limb.**

524

The aortic arch has three major arteries serving the head, neck, and arms. The first branch, the **brachiocephalic (brā-kē-ō-se-FAL-ik)** or **innominate artery**, is short and divides into the **right common carotid artery** and the **right subclavian artery**. These arteries supply blood to the right side of the head and neck and to the right arm, respectively. The left common carotid artery and left subclavian artery originate directly off the aorta. Only the right common carotid artery and right subclavian artery are derived from a brachiocephalic artery. The vertebral artery branches off each subclavian artery and supplies blood to the brain and spinal cord. Each subclavian artery passes under the clavicle and crosses the armpit as the **axillary artery** and continues into the arm as the **brachial artery**. Blood pressure is usually taken at the brachial artery. At the antecubitis, the elbow, the brachial artery divides into the lateral **radial artery** and the medial **ulnar artery**, each named after the bone it follows. In the palm of the hand, these arteries are interconnected by the superficial and deep **palmar arches** which send small **digital arteries** to the fingers. Except for the brachiocephalic artery on the right side, the vascular anatomy is symmetrical in the right and left upper limbs.

The right and left carotid arteries supply blood to the structures of the neck, face, and brain, as shown in Figure 38.5. The **right common carotid artery** originates off the brachiocephalic artery, whereas the **left common carotid artery** branches directly off the peak of the aortic arch. The term "common" suggests that the vessel joins external and internal branches. Each common carotid artery ascends deep in the neck and divides at the larynx into an **external carotid** artery and an **internal carotid** artery. The

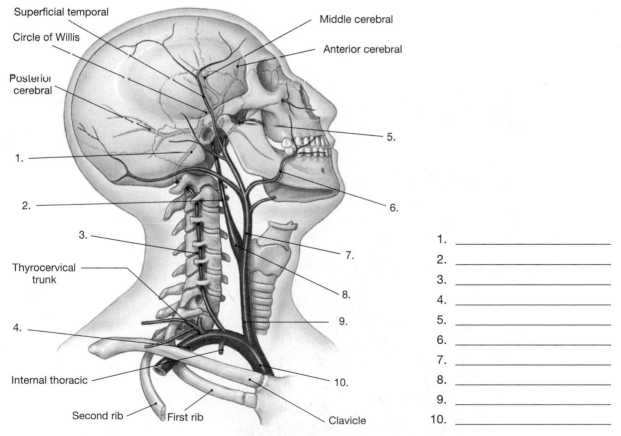

1. _____

2. _____

3. _____

4. _____

5. _____

6. _____

7. _____

8. _____

9. _____

10. _____

Figure 38.5
Arteries of the Head and Neck.

General distribution of arteries supplying the neck and superficial structures of the head.

base of the internal carotid swells as the **carotid sinus** and contains baroreceptors to monitor blood pressure. The external carotid artery branches to supply blood to the structures of the neck and face. The pulse in the external carotid artery can be felt by placing your fingers lateral to your thyroid cartilage, the Adam's apple. The external carotid artery branches into the **facial**, **maxillary**, and **superficial temporal arteries** to serve the external structures of the head.

STUDY TIP

Arteries and veins with the term "common" as part of their name always divide into an external and an internal vessel. The common carotid artery, for example, branches into an external carotid artery and an internal carotid artery.

The internal carotid artery ascends to the base of the brain and divides into three arteries: an **opthalamic artery** which provides blood to the eyes, an **anterior cerebral artery**, and an **middle cerebral artery**, each supplying blood to the brain. These vessels are detailed in Figure 38.5.

The brain has a high metabolic rate and a voracious appetite for oxygen and nutrients. A reduction in blood flow to the brain may result in permanent damage to the affected area. To ensure that the brain receives a continuous supply of blood, branches of the internal carotid artery and other arteries interconnect or **anastomose** as the **cerebral arterial circle**, also called the **circle of Willis**. Follow the vascular anatomy in Figure 38.6 as the circle is described. The right and left **vertebral arteries**, which arise off the subclavian arteries, ascend in the transverse foramina of the cervical vertebrae and

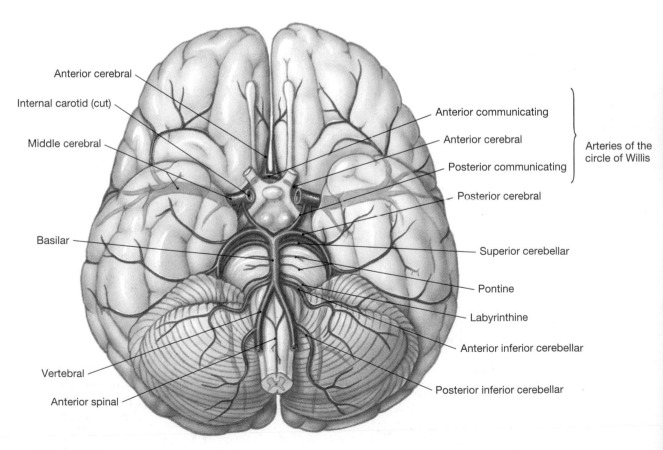

Figure 38.6
Arteries of the Brain.

enter the skull at the foramen magnum. These arteries fuse into a single **basilar artery** on the inferior surface of the brain stem. The basilar artery branches into **posterior cerebral arteries** and **posterior communicating arteries**. The posterior communicating arteries form an anastomosis with the carotid arteries. **Anterior communicating arteries** branch off the internal common carotid arteries, pass anteriorly, and meet to complete the ring of blood vessels at the base of the brain. Figure 38.6 is a photograph of a cast of the arteries of the brain. The vessels were injected with latex, and the brain tissue then chemically removed.

LABORATORY ACTIVITY   ARTERIES OF THE HEAD, NECK, AND ARMS

MATERIALS

> vascular system chart
> torso model
> head model
> arm model

PROCEDURES

1. Label and review the arteries presented in Figures 38.3–38.6.
2. Locate each artery on laboratory models, mannequins, and charts.
3. Locate the pulse in your radial and common carotid arteries. Use your index finger to feel the pulse.
4. On your body, trace the location of your arteries from the base of the aorta to your right thumb.
5. On your body, trace the location of your arteries from the base of the aorta to the left side of your face.

2. Arteries of the abdominopelvic cavity and lower limb

The descending aorta passes through the diaphragm and becomes the **abdominal aorta** in the abdominopelvic cavity. Many arteries stem from the abdominal aorta (see Figure 38.7). An easy way to identify the branches of the abdominal aorta is to distinguish between paired arteries with a right and a left branch and unpaired arteries with a single vessel.

There are three unpaired branches of the abdominal aorta. The **celiac (SĒ-lē-ak) trunk** is a short artery arising off the abdominal aorta inferior to the diaphragm. It splits into three arteries: the **common hepatic artery**, the **left gastric artery**, and the **splenic artery**. The common hepatic artery divides to supply blood to the liver, gallbladder, and part of the stomach. The left gastric artery also supports blood flow to the stomach. The splenic artery supplies blood to the spleen, stomach, and pancreas. Inferior to the celiac trunk is the next unpaired artery, the **superior mesenteric (mez-en-TER-ik) artery**. This vessel supplies blood to the large intestine, parts of the small intestine, and other abdominal organs. The third unpaired artery, the **inferior mesenteric artery**, originates before the abdominal aorta divides to enter the pelvic cavity and legs. This artery supplies parts of the large intestine and the rectum. Figure 38.8 illustrates these arteries and the organs they supply with blood.

There are four pairs of major arteries arising off the abdominal aorta. The **suprarenal arteries** arise near the superior mesenteric artery and branch into the adrenal glands located on top of the kidneys. The right and left **renal arteries**, which supply the kidneys, stem off the abdominal aorta just inferior to the suprarenal arteries. A pair of **gonadal arteries** appear near the inferior mesenteric artery. These arteries bring blood to the reproductive organs of the male and female. A pair of **lumbar arteries** originate near the terminus of the abdominal aorta and service the lower body wall.

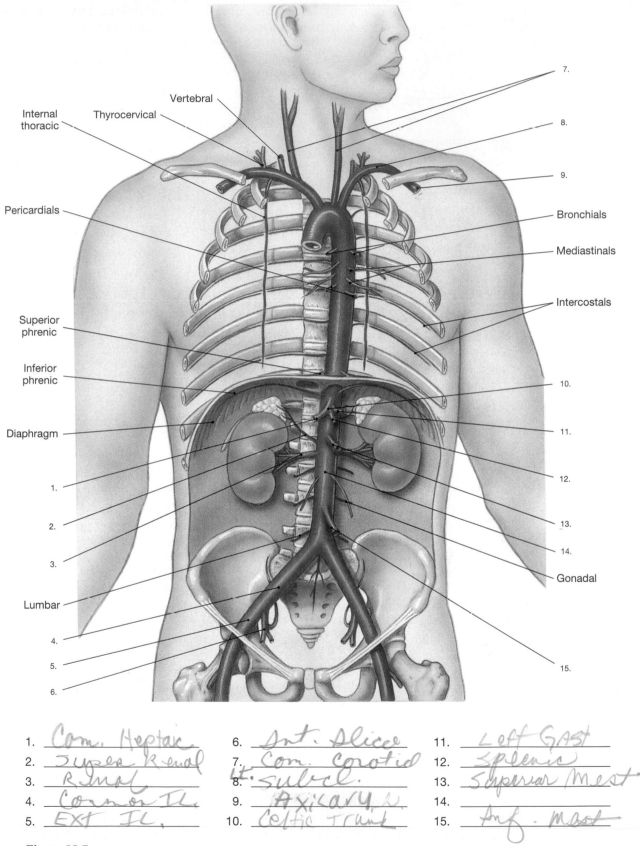

Internal thoracic
Thyrocervical
Vertebral
Pericardials
Superior phrenic
Inferior phrenic
Diaphragm
1.
2.
3.
Lumbar
4.
5.
6.

7.
8.
9.
Bronchials
Mediastinals
Intercostals
10.
11.
12.
13.
14.
Gonadal
15.

Figure 38.7
Major Arteries of the Trunk.

1. Com. Hepatic
2. Super Renal
3. Renal
4. Common IL
5. Ext IL.

6. Int. Iliac
7. Com. Carotid
8. Lt. subcl.
9. Axicary u
10. Celtic Trunk

11. Left GAst
12. Splenic
13. Superior Mest
14.
15. Inf. Mest

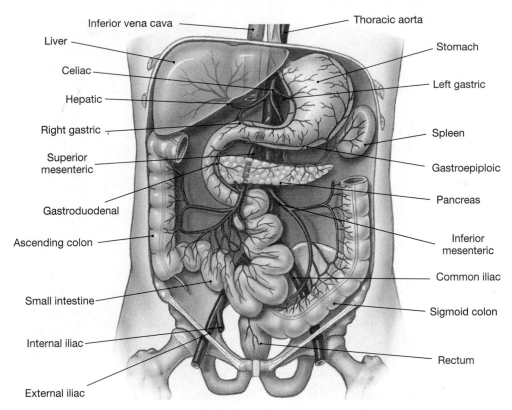

Figure 38.8
Arteries Supplying the Abdominopelvic Organs.

At the level of the hips, the abdominal aorta divides into the right and left **common iliac (IL-ē-ak) arteries** (see Figures 38.7–38.9). These arteries descend through the pelvic cavity and branch into **external iliac arteries** which enter the legs and **internal iliac arteries** which supply blood to the organs of the pelvic cavity. The external iliac artery pierces the abdominal wall and becomes the **femoral artery** of the thigh. A **deep femoral artery** arises to supply deep thigh muscles. The femoral artery passes though the posterior knee as the **popliteal (pop-LIT-ē-al) artery** and divides into the **posterior tibial artery** and the **anterior tibial artery**, each supplying blood to the lower leg. The **peroneal artery** stems laterally off the posterior tibial artery. The arteries of the lower leg branch into the foot and anastomose at the **dorsal** and **plantar arches**.

LABORATORY ACTIVITY   ARTERIES OF THE ABDOMEN AND LEG

MATERIALS

> vascular system chart
> torso model
> leg model

PROCEDURES

1. Label and review the arteries presented in Figures 38.7–38.9.
2. Locate each artery on laboratory models, mannequins, and charts.
3. On your body, trace the location of your abdominal aorta, common iliac artery, external iliac artery, femoral artery, popliteal artery, and posterior tibial artery.

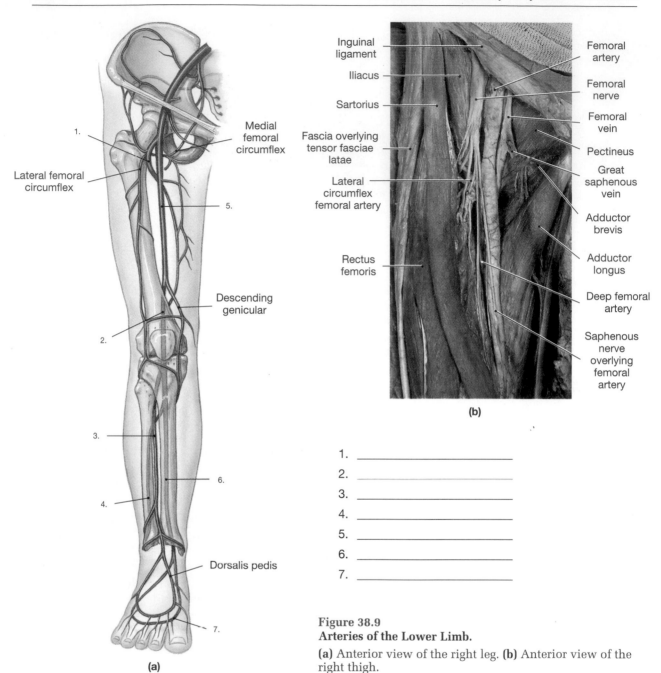

Inguinal ligament

Iliacus

Sartorius

Fascia overlying tensor fasciae latae

Lateral circumflex femoral artery

Rectus femoris

Medial femoral circumflex

Lateral femoral circumflex

Descending genicular

Dorsalis pedis

Femoral artery

Femoral nerve

Femoral vein

Pectineus

Great saphenous vein

Adductor brevis

Adductor longus

Deep femoral artery

Saphenous nerve overlying femoral artery

1.

2.

3.

4.

5.

6.

7.

(a)

(b)

1. _____

2. _____

3. _____

4. _____

5. _____

6. _____

7. _____

Figure 38.9
Arteries of the Lower Limb.

(a) Anterior view of the right leg. **(b)** Anterior view of the right thigh.

C. The Venous System

1. Veins of the head, neck, and upper limb

Once you have learned the major systemic arteries, identification of systemic veins is easy because most arteries have a corresponding vein. This similarity in distribution is seen when comparing Figures 38.3 and 38.10. Unlike arteries, many veins are superficial and easily seen under the skin.

Figure 38.10
An Overview of the Sys-
temic Venous System.

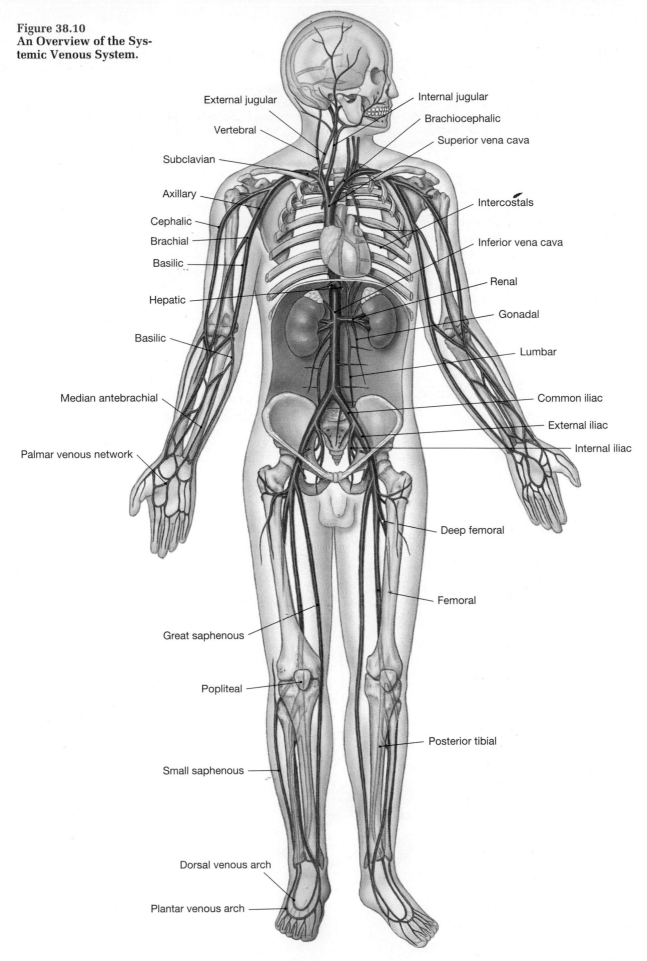

External jugular

Internal jugular

Vertebral

Brachiocephalic

Superior vena cava

Subclavian

Axillary

Cephalic

Intercostals

Brachial

Inferior vena cava

Basilic

Renal

Hepatic

Gonadal

Basilic

Lumbar

Median antebrachial

Common iliac

External iliac

Internal iliac

Palmar venous network

Deep femoral

Femoral

Great saphenous

Popliteal

Posterior tibial

Small saphenous

Dorsal venous arch

Plantar venous arch

STUDY TIP

Systemic veins are usually painted blue on vascular models and torso mannequins to indicate that they transport deoxygenated blood. When identifying veins, work in the direction of blood flow, from the periphery toward the heart.

Blood in the brain drains into large veins called sinuses, detailed in Figure 38.11. Veins deep inside the brain drain into progressively larger veins that empty into the **superior sagittal sinus**, located in the falx cerebri separating the cerebral hemispheres. Other venous sinuses and the superior sagittal sinus drain into the **internal jugular vein,** which exits the skull via the jugular foramen, descends the neck, and empties into the **brachiocephalic vein**. Superficial veins that drain the face and scalp empty into the **external jugular vein** which also descends the neck to join the subclavian vein. The right and left brachiocephalic veins merge at the **superior vena cava** and empty deoxygenated blood into the right atrium of the heart. Blood in the right atrium

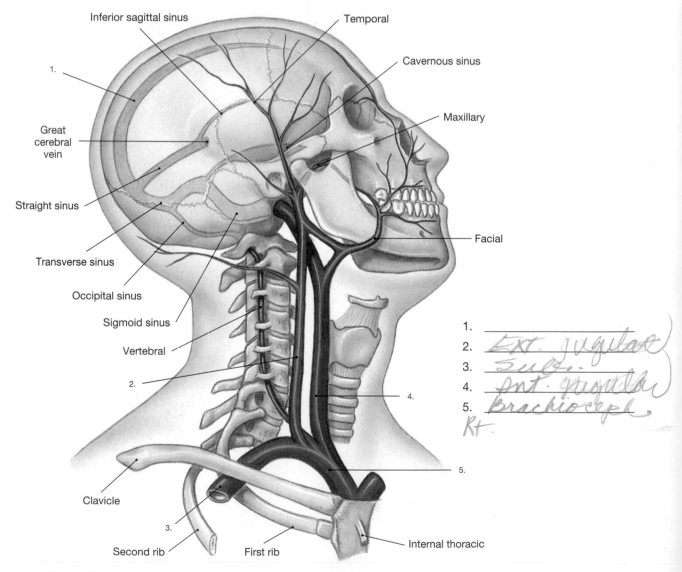

1. _____
2. ____ Ext. jugular
3. ____ Subc.
4. ____ Ant. jugular
5. ____ Brachioceph.
   Rt.

Figure 38.11
Major Veins of the Head and Neck.

enters the right ventricle, which contracts and pumps it to the lungs through the pulmonary circuit. The inferior vena cava drains veins of the abdominopelvic cavity.

STUDY TIP

The venous system has both a right and a left brachiocephalic vein, each formed by the merging of subclavian, vertebral, and internal and external jugular veins. The arterial system has a single brachiocephalic artery which branches into the right common carotid artery and right subclavian arteries. On the left side of the body, the common carotid artery and subclavian artery originate directly off the aortic arch.

Figure 38.12 illustrates the venous drainage of the arm, chest, and abdomen. The hand drains into **digital veins** which empty into a network of **palmar venous arches**. These

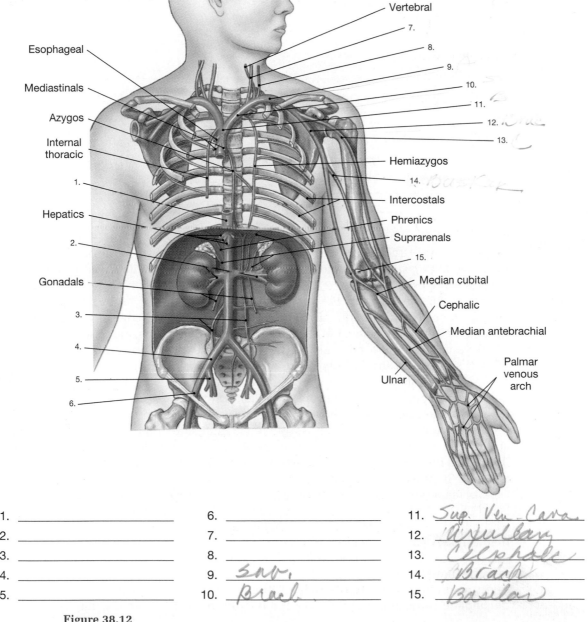

1. _____   6. _____   11. _Sup. Ven Cava_
2. _____   7. _____   12. _Axillary_
3. _____   8. _____   13. _Cephalic_
4. _____   9. _Sub._   14. _Brach_
5. _____   10. _Brach_   15. _Basilar_

Figure 38.12
Venous Drainage of the Abdomen and Chest.

vessels drain into the **cephalic vein** which ascends along the lateral margin of the arm. The **median antebrachial vein** ascends to the elbow, is joined by the **median cubital vein** which crosses over from the cephalic vein, and becomes the **basilic vein**. The median cubital vein is often used to collect blood samples from an individual. Also in the forearm are the **radial** and **ulnar veins** which fuse above the elbow into the **brachial vein**. The brachial and basilic veins meet at the armpit as the **axillary vein**, which joins the cephalic vein and becomes the **subclavian vein**. Subclavian veins and veins from the neck and head drain into the **brachiocephalic vein** which then empties into the superior vena cava. The superior vena cava returns deoxygenated blood from the head, neck, and arms to the right atrium.

LABORATORY ACTIVITY   VEINS OF THE HEAD, NECK, AND ARM

MATERIALS

> vascular system chart
> torso model
> head model
> arm model

PROCEDURES

1. Label and review the veins presented in Figures 38.10–38.12.
2. Locate each vein on laboratory models, mannequins, and charts.
3. On your body, trace your cephalic vein, subclavian vein, brachiocephalic vein, and superior vena cava.

2. Veins of the lower limb and abdomen

Veins that drain the lower limb and abdominal organs empty into the inferior vena cava, the large vein that pierces the diaphragm and empties into the right atrium of the heart (see Figure 38.12).

Figure 38.13 illustrates the venous drainage of the lower limb. The foot has many anastomoses that drain into the lateral **peroneal vein** and the **anterior tibial vein**, located on the medial aspect of the anterior shin. These veins, along with the **posterior tibial vein**, merge and become the **popliteal vein** of the posterior knee. The **small (lesser) saphenous (sa-FE-nus) vein**, which ascends from the ankle to the knee and drains blood from superficial veins, also empties into the popliteal vein. Superior to the knee, the popliteal vein becomes the **femoral vein** which ascends along the femur to the lower pelvic girdle where it joins the **deep femoral vein** at the **external iliac vein**. The **great saphenous vein** ascends from the medial side of the ankle to the upper thigh and drains into the external iliac vein. In the pelvic cavity, the external and **internal iliac veins** fuse and form the **common iliac vein**. The right and left common iliac veins merge at the **inferior vena cava**.

Six major veins from the abdominal organs drain blood into the inferior vena cava (see Figures 38.12 and 38.13). The **lumbar veins** drain the muscles of the lower body wall and the spinal cord and empty into the inferior vena cava close to the common iliac veins. A pair of **gonadal veins** empty blood from the reproductive organs into the inferior vena cava above the lumbar veins. A pair of **renal** and **suprarenal veins** drain into the inferior vena cava next to their respective organs. Before entering the thoracic cavity to drain blood into the right atrium, the inferior vena cava collects blood from the **hepatic veins** draining the liver and the **phrenic veins** from the diaphragm.

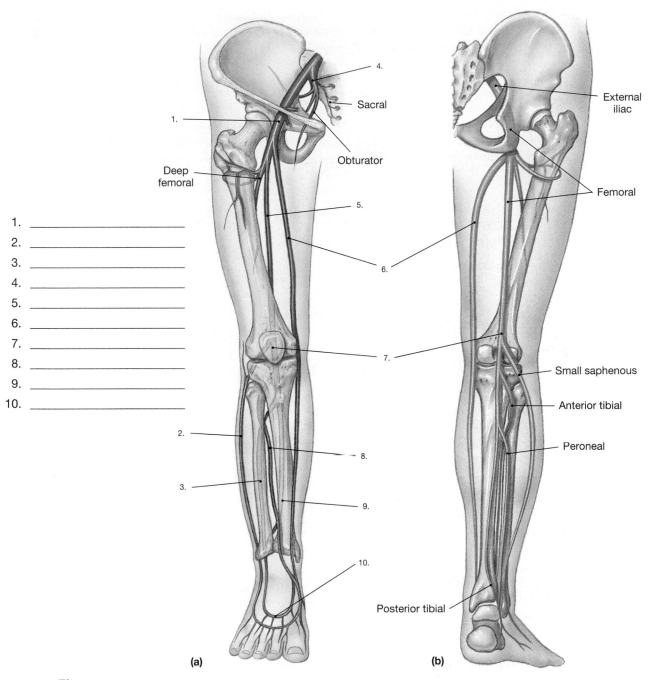

1. _____
2. _____
3. _____
4. _____
5. _____
6. _____
7. _____
8. _____
9. _____
10. _____

(a)   **(b)**

Figure 38.13
Venous Drainage from the Lower Limb.

(a) Anterior view. **(b)** Posterior view.

Figure 38.14 shows the vascular supply to the liver. The **inferior** and **superior mesenteric veins** drain the nutrient-rich blood of the digestive organs. These veins empty into the **hepatic portal vein**, which passes the blood through the liver for regulation of blood sugar concentration. Phagocytic cells consume microbes that may have entered the bloodstream through the mucous membrane of the digestive system. Blood from the hepatic arteries and hepatic portal vein mixes in the liver and is returned to the inferior vena cava by the **hepatic veins**.

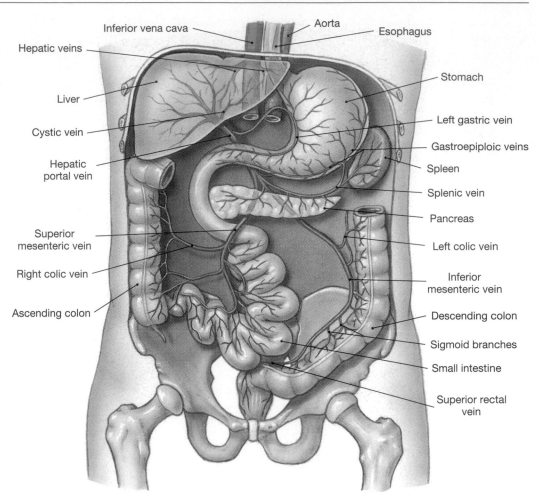

Figure 38.14
The Hepatic Portal System.
Vessels of the hepatic portal system.

LABORATORY ACTIVITY   VEINS OF THE LEG AND ABDOMEN

MATERIALS

vascular system model
torso model
leg model

PROCEDURES

1. Label and review the veins presented in Figures 38.13 and 38.14.
2. Locate each artery on laboratory models, mannequins, and charts.
3. On your body, trace the location of the veins in your lower limb.
4. Although you have studied the arterial and venous divisions separately, they are anatomically interconnected by capillaries. To reinforce this organization, trace blood through the following systemic routes:
   a. Trace a drop of blood from the heart through the vessels of the left arm and back to the heart.

   b. Trace a drop of blood from the heart through the vessels of the liver and back to the heart.

   c. Trace a drop of blood from the heart through the vessels of the right lower limb and back to the heart.

   d. Trace a drop of blood from the heart to the legs and back to the heart.

D. **Circulatory System of the Cat**

This section complements the study of human vessels. To observe the heart and blood vessels of the cat, the thoracic and abdominal cavities must be opened. Locate the large arteries leaving both the right and left ventricles which direct blood away from the heart. Trace these arteries to smaller arteries that direct blood to specific organs and tissues. Only those arteries typically injected with latex are listed for identification. Keep in mind that the cat has more arteries than listed here and that additional ones may be assigned by your instructor for you to identify.

Preparing the cat

   1. Expose the thoracic and abdominal organs of the cat by making a longitudinal midline incision through the muscles of the neck, thorax, and abdominal wall.

   2. Avoid cutting through the muscular diaphragm so that you can identify the vessels and structures passing between the thoracic and abdominal cavities. Your instructor will provide you with specific instructions for exposing these cavities and for isolating the blood vessels.

   3. Occasionally, clotted blood fills the thoracic and abdominal cavities and must be removed. If you encounter clots, check with your instructor before proceeding.

Identification of arteries

Refer to Figures 38.15–38.18 for the arteries of the cat.

Arteries of the head, neck, and thorax

Locate the heart and examine the blood vessels joining the ventricles.

   1. Deoxygenated blood is pumped from the right ventricle toward the lungs through the **pulmonary trunk**, which divides into right and left branches of the **pulmonary artery**, see Figure 38.15. It is connected to the aorta by the **ligamentum arteriosum**, a strand of connective tissue representing the obliterated ductus arteriosus. The left branch of the pulmonary artery passes ventral to the aorta to reach the left lung; the right branch passes between the aortic arch and the heart to reach the right lung.

   2. The **aorta** is the large arterial trunk that conveys oxygenated blood from the left ventricle to the body. At its origin, it makes an abrupt curve to the left, the **aortic arch**, passing dorsal to the left pulmonary artery and continuing caudally along the left side of the vertebral column to the pelvis where it divides into branches which supply the legs. The portion of the aorta anterior to the diaphragm is termed the **thoracic aorta**. Follow the aorta and locate the first branch, the **brachiocephalic** or **innominate**. It supplies the head and the right forelimb. Near the level of the second rib, the brachiocephalic divides into the **right subclavian** (which supplies the right forelimb) and the **right** and **left common carotids**.

   3. The **superior thyroid artery** branches from the common carotid artery at the cranial end of the thyroid gland; it supplies the thyroid gland, superficial laryngeal muscles, and ventral neck muscles. Just anterior to the superior thyroid artery are

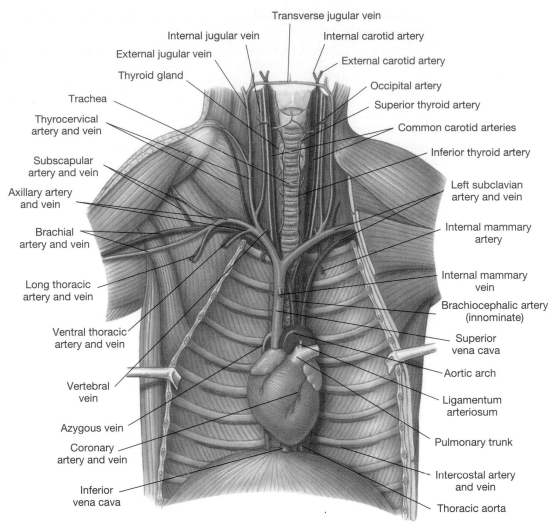

Figure 38.15
Arteries and Veins of the Chest and Neck.

the **occipital** and the **internal carotid arteries**. Sometimes they arise by a common vessel; the occipital artery supplies the deep muscles of the neck and extends to the posterior side of the skull where it runs along the superior nuchal line. The internal carotid artery enters the skull via the foramen lacerum and joins the posterior cerebral artery. After giving off the internal carotid artery, the common carotid continues as the **external carotid artery**.

4. The external carotid artery turns medially near the posterior margin of the masseter and continues as the **internal maxillary artery**. It gives off several branches and then ramifies to form the carotid plexus surrounding the maxillary branch of the trigeminal nerve near the foramen rotundum; the internal maxillary and the carotid plexus give various branches to brain, eye, and other deep structures of head.

5. The **left subclavian artery** is the second great vessel to branch off the aortic arch. It supplies the left forelimb. As the subclavian arteries approach the arms their branches are identical.

6. The **internal mammary artery** arises from the ventral surface of the subclavian at about the level of the vertebral artery and passes caudally to the ventral thoracic

wall, giving off branches to the adjacent muscles, the pericardium, the mediastinum, and the diaphragm.

7. The **vertebral artery** arises from the dorsal surface of the subclavian and passes cranially through the transverse foramen of the cervical vertebrae. It gives off branches to the deep neck muscles and to the spinal cord near the foramen magnum. Distal to the vertebral artery, the **costocervical artery** arises from the dorsal surface of the subclavian artery. It sends branches to the deep muscles of the neck and shoulder and to the first two costal interspaces. The **thyrocervical artery** arises from the cranial aspect of the subclavian (distal to the costocervical artery) and passes cranially and laterally, supplying the muscles of the neck and chest.

8. After the left subclavian artery branches from the aortic arch, paired **intercostal arteries** are given off to the interspaces between the last 11 ribs. Paired **bronchial arteries** arise from the **thoracic aorta** and supply the bronchi. The **esophageal arteries** are several small vessels of varying origin along the thoracic aorta supplying the esophagus.

Arteries of the forelimb (medial view)

1. Lateral to the first rib, the subclavian artery continues into the axilla as the **axillary artery**. From the ventral surface of the axillary artery, just lateral to the first rib, the **ventral thoracic artery** arises and passes caudally to supply the medial ends of the pectoral muscles. The **long thoracic artery** arises lateral to the ventral thoracic, passing caudally to the pectoral muscles and the latissimus dorsi.

2. The largest branch of the axillary artery is the **subscapular artery**. It passes laterally and dorsally between the long head of the triceps and the latissimus dorsi to supply the dorsal shoulder muscles. It gives off two branches, the thoracodorsal artery and the posterior humeral circumflex artery.

3. Distal to the origin of the subscapular artery, the axillary artery continues as the **brachial artery**, see Figure 38.16. Distal to the elbow, the brachial artery gives rise to the superficial **radial artery** and **ulnar artery**.

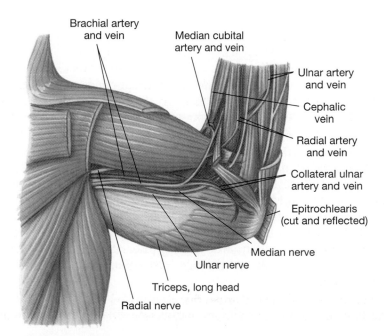

Figure 38.16
Arteries and Veins of the Fore Limb.

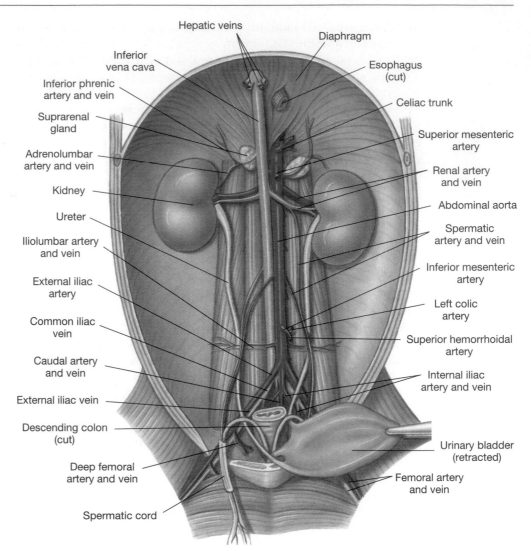

Figure 38.17
Arteries and Veins of the Abdominal Cavity.

Arteries of the abdominal cavity

1. The portion of the aorta posterior to the diaphragm is called the **abdominal aorta** or **descending aorta** after passing through the diaphragm. A single vessel, the **celiac trunk**, is the first arterial branch off the abdominal aorta. It divides into three branches: the **hepatic**, **left gastric**, and **splenic arteries**, see Figure 38.17.

2. Along the cranial border of the gastrosplenic part of the pancreas lies the **hepatic artery**. It turns cranially near the pylorus, lying in a fibrous sheath together with the portal vein and the common bile duct. Its branches include the **cystic artery** to the gallbladder and liver and the **gastroduodenal artery** near the pylorus. Along the lesser curvature of the stomach lies the **left gastric artery,** supplying many branches to both dorsal and ventral stomach walls. The **splenic artery** is the largest branch of the celiac artery. It supplies at least two branches to the dorsal surface of the stomach and divides into anterior and posterior branches to supply these portions of the spleen.

3. Just posterior to the celiac trunk, the abdominal aorta gives off the **superior mesenteric artery**. It divides into numerous intestinal branches which supply the small and large intestines.

4. The paired **adrenolumbar arteries** are just posterior to the superior mesenteric artery and supply the suprarenal glands. They pass laterally along the dorsal body wall and give rise to **phrenic** and **adrenal arteries** and then supply the muscles of the dorsal body wall. The phrenic artery branches off the adrenolumbar artery and supplies the diaphragm.

5. The paired arteries emerging from the abdominal aorta to supply the kidneys are the **renal arteries**. In some specimens, this vessel gives rise to the adrenal artery. Often double renal arteries supply each kidney.

6. The paired arteries (in the female) arising from the aorta near the caudal ends of the kidneys are the **ovarian arteries**. They pass laterally in the broad ligament to supply the ovaries. Each artery gives a branch to the cranial end of the corresponding uterine horn. In the male, the **spermatic arteries** are the paired internal arteries arising from the abdominal aorta near the caudal ends of the kidneys. They lie on the surface of the iliopsoas muscle, passing caudally to the internal inguinal ring and through the inguinal canal to the testes.

7. The **lumbar arteries** are seven pairs of arteries arising from the dorsal surface of the aorta and supplying the muscles of the dorsal abdominal wall.

8. At the level of the last lumbar vertebra, the **inferior mesenteric artery** arises from the abdominal aorta. Near its origin, it divides into the **left colic artery**, which passes anteriorly to supply the descending colon, and the **superior hemorrhoidal artery**, which passes posteriorly to supply the rectum.

9. Paired arteries, the **iliolumbar arteries,** arise near the inferior mesenteric artery and pass laterally across the iliopsoas muscle to supply the muscles of the dorsal abdominal wall.

10. Posterior to the iliolumbar arteries, near the sacrum, the abdominal aorta branches into right and left **external iliac arteries** which lead toward the hindlimbs. Posterior to branches of the external iliac arteries, the aorta gives rise to paired **internal iliac arteries**.

11. The first branch of the internal iliac artery is the **umbilical artery**, which arises near the origin of the internal iliac artery. The **caudal (medial sacral) artery** is caudal to the point where the internal iliac arteries leave the aorta. This unpaired vessel continues into the ventral aspects of the tail as the caudal artery.

Arteries of the hindlimb (medial view)

1. Just prior to leaving the abdominal cavity, the external iliac artery gives off the **deep femoral artery,** which passes between the iliopsoas and the pectineus to supply the muscles of the thigh, see Figure 38.18.

2. The external iliac artery continues outside the abdominal cavity as the **femoral artery**, lying on the medial surface of the thigh. At the posterior knee it is called the **popliteal artery** and divides into the **anterior** and **posterior tibial arteries**.

Identification of veins

Return to the heart and locate the vessels entering the atria. Trace these vessels back to the smaller veins that direct blood flow back toward the heart. In general the veins follow the pattern of the arteries. Refer to Figures 38.15–38.18 for the veins of the cat.

Veins of the head, neck, and thorax

1. To identify the **pulmonary veins**, gently pull the heart away from the lung on one side and examine the root of the lungs. From each lung lobe, a vein (usually uninjected) is seen passing toward the dorsal side of the heart to enter the right atrium.

2. The **superior vena cava (precava)** is the large vessel that drains blood from the head, neck, and forelimbs to the right atrium. Its principal tributaries include the

Figure 38.18
Arteries and Veins of the Hind Limb.

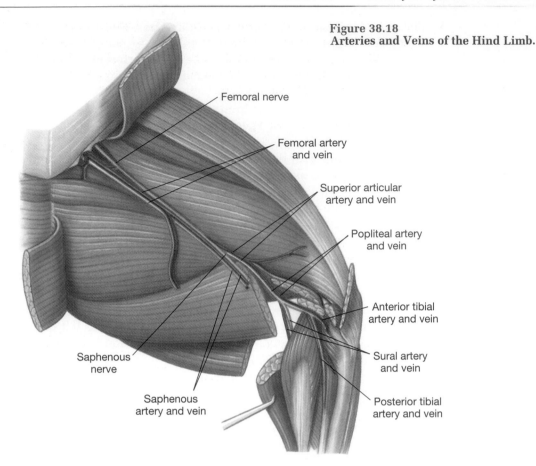

Femoral nerve

Femoral artery
and vein

Superior articular
artery and vein

Popliteal artery
and vein

Anterior tibial
artery and vein

Sural artery
and vein

Posterior tibial
artery and vein

Saphenous
nerve

Saphenous
artery and vein

internal and external jugulars, the subscapular, and the axillary veins. The **inferior vena cava (postcava)** is the large vessel that returns blood from the abdomen, hindlimbs, and pelvis to the right atrium. It drains blood from numerous vessels. The inferior vena cava pierces the diaphragm after it passes through the liver, where it receives the hepatic veins. At this point it receives drainage from the inferior phrenic veins from the diaphragm. It then passes between the heart and the caudal lobe of the right lung to enter the right atrium. Locate the venae cavae in Figure 38.15.

3. Toward the anterior end of the superior vena cava, on its ventral side, is a common stem by which the paired **internal mammary veins** enter the superior vena cava. They lie on either side of the midline to drain the ventral chest wall. If the heart is pushed to the left, the **azygous vein** can be seen arching over the root of the right lung and joining the superior vena cava near the right atrium. It lies along the right side of the vertebral column in the thorax and receives the **intercostal veins**.

4. The **brachiocephalic** or **innominate veins** are two short veins whose union forms the superior vena cava. They drain both the external jugular and subclavian veins. A short vessel within the thorax joining the brachiocephalic vein to the axillary vein is the **subclavian vein**.

5. The **costocervical** and **vertebral veins** form a common stem and dorsally connect with the brachiocephalic vein.

6. The left and right **external jugular veins** drain the head and neck. Together with the subclavian vein, each forms the brachiocephalic vein.

7. Just superior to the hyoid bone, the **transverse jugular vein** connects the left and right external jugular veins. The **internal jugular vein** drains the brain and spinal cord and joins the external jugular vein near its union with the brachiocephalic vein. At the shoulder, the external jugular vein receives the large **transverse scapular vein** that drains the dorsal surface of the scapula.

Veins of the forelimb (medial view)

1. The major vessel draining the limb in the axilla is the **axillary vein**. It forms from the brachial vein and is continuous with the subclavian vein. A small vessel emptying into the axillary vein on its ventral surface is the **ventral thoracic vein**. It is located near the point where the subscapular is found. The **long thoracic vein** is another small vessel emptying into the axillary vein. It is distal to the ventral thoracic vein.

2. The **subscapular vein** is the largest tributary of the axillary vein. It drains vessels from the shoulder. The **brachial vein** is the portion of the major vessel that drains the limb and is distal to the axillary vein in the forelimb, see Figure 38.16.

Veins of the abdominal cavity

1. Blood circulating through the stomach, intestines, spleen, and pancreas returns to the liver via the portal vein. Within the liver, the portal vein ends in a system of capillaries called the sinusoids of the liver. Blood passes through the sinusoids of the liver and enters the inferior vena cava via the **hepatic veins**, which join the inferior vena cava within the liver, see Figure 38.17.

2. The **hepatic portal vein** is formed by the union of the superior mesenteric and gastrosplenic veins. It is also joined by other veins from the gastrointestinal tract. The **superior mesenteric vein** is the largest branch of the hepatic portal vein.

3. The **adrenolumbar veins** are vessels that empty into the inferior vena cava and drain the adrenal glands. Paired **renal veins** empty into the inferior vena cava and drain the kidneys. Often, double renal veins drain each kidney.

4. Posterior to the renal vein in males, the **internal spermatic veins** are the vessels that drain the testes. Usually, the left internal spermatic vein empties into the renal vein. In females, the **ovarian veins** are posterior to the renal veins. These vessels empty into the inferior vena cava and drain the ovaries.

5. The **lumbar veins** drain the dorsal abdominal wall and empty into the inferior vena cava. The posterior dorsal abdominal wall is drained by the **iliolumbar veins**. The vessel that drains the tail and empties into the inferior vena cava is the **caudal vein**.

6. The **common iliac veins** are paired veins that join together in the posterior abdominal region to form the inferior vena cava. The inferior vena cava lies dorsal to the aorta. The large **internal iliac veins** enter the common iliac vein in the pelvic cavity and drain the rectum, bladder, and internal reproductive organs. Distal to the joining of the internal iliac vein to the common iliac vein, the **external iliac vein** is a continuation of the femoral vein from the hindlimb.

Veins of the hindlimb (medial view)

1. The **femoral vein** is a superficial vein on the anterior surface of the thigh. As it enters the abdominal cavity, it becomes the external iliac vein. The **deep femoral vein** is a medial branch entering the femoral vein.

2. The medial and lateral branches of the **popliteal vein** drain the foot and calf region, uniting in the popliteal region to form the **saphenous vein** which drains into the femoral vein. These vessels lie superficial to the muscles of the shank, see Figure 38.18.

ANATOMY OF SYSTEMIC CIRCULATION CHECKLIST

This is a list of bold terms presented in Exercise 38. Use this list for review purposes after you have completed the laboratory activities.

ARTERY AND A VEIN COMPARISON

• Figures 38.1 and 38.2
arteries, arterioles
capillaries, endothelium
precapillary sphincters
metarteriole
thoroughfare channel
veins, valves, venules
anastomosis
pulmonary circuit, systemic circuit
tunica externa, tunica media, tunica interna
vasoconstriction, vasodilation

ARTERIES OF HEAD, NECK, AND ARM

• Figures 38.4–38.6
aortic arch
ascending, thoracic, and abdominal aortas
brachiocephalic artery
common carotid arteries
internal carotid arteries, carotid sinus
external carotid arteries
ophthalmic arteries
anterior cerebral arteries
middle cerebral arteries
cerebral arterial circle
vertebral arteries, basilar arteries
posterior cerebral arteries
posterior communicating arteries
anterior communicating arteries
facial arteries, maxillary arteries
superficial temporal arteries
subclavian arteries, axillary arteries
brachial arteries
radial arteries, ulnar arteries
palmar arches, digital arteries

ARTERIES OF ABDOMEN AND LEG

• Figures 38.7 and 38.8
descending abdominal aorta
celiac trunk, common hepatic artery
left gastric artery, splenic artery
superior and inferior mesenteric arteries
suprarenal arteries, renal arteries
gonadal arteries, lumbar arteries
common iliac arteries
internal iliac arteries, external iliac arteries
femoral arteries, deep femoral arteries
popliteal, anterior, and posterior tibial arteries
peroneal arteries, dorsal and plantar arches

VEINS OF HEAD, NECK, AND ARM

• Figures 38.11 and 38.12
superior sagittal sinus
internal jugular veins, brachiocephalic veins
external jugular veins, superior vena cava
digital veins, palmar venous arches
cephalic veins, median antebrachial veins
median cubital veins, basilic veins
radial veins, ulnar veins, brachial veins
axillary veins, subclavian veins

VEINS OF ABDOMEN AND LEG

• Figures 38.12–38.14
lumbar veins, gonadal veins
renal veins, suprarenal veins
hepatic veins, phrenic veins
hepatic portal veins
superior and inferior mesenteric veins
peroneal, anterior, and posterior tibial veins
popliteal veins, small saphenous veins
femoral veins, deep femoral veins
internal and external iliac veins

great saphenous veins
common iliac veins, inferior vena cavae

ARTERIES OF THE CAT

arteries of the head, neck, and thorax
• Figure 38.15
pulmonary trunk, pulmonary artery
ligamentum arteriosum
aorta, aortic arch, thoracic aorta
brachiocephalic (innominate)
right subclavian
right and left common carotids
superior thyroid artery
occipital arteries
internal carotid arteries
external carotid arteries
internal maxillary arteries
internal mammary arteries
vertebral arteries
costocervical arteries
thyrocervical arteries
intercostal arteries
bronchial arteries, esophageal arteries

arteries of the forelimb
• Figure 38.16
axillary arteries
ventral thoracic arteries
long thoracic arteries
subscapular arteries
brachial arteries
radial arteries, ulnar arteries

arteries of the abdominal cavity
• Figure 38.17
abdominal (descending) aorta
celiac trunk
hepatic artery
left gastric artery
splenic artery
hepatic artery
cystic artery
gastroduodenal artery
superior mesenteric artery
adrenolumbar arteries
phrenic arteries
adrenal arteries, renal arteries
ovarian or spermatic arteries
lumbar arteries
inferior mesenteric artery
left colic artery
superior hemorrhoidal artery

iliolumbar arteries
external and internal iliac arteries
umbilical artery
caudal (medial sacral) artery

arteries of the hindlimb
• Figure 38.18
deep femoral arteries
femoral arteries
popliteal arteries
anterior and posterior tibial arteries

VEINS OF THE CAT

veins of the head, neck, and thorax
• Figure 38.15
pulmonary veins
superior vena cava (precava)
inferior vena cava (postcava)
internal mammary veins
azygous vein, intercostal veins
brachiocephalic (innominate) veins
subclavian veins
costocervical veins
vertebral veins
external jugular veins
transverse jugular vein
internal jugular veins
transverse scapular veins

veins of the forelimb
• Figure 38.16
axillary veins
ventral thoracic veins
long thoracic veins
subscapular veins
brachial veins

veins of the abdominal cavity
• Figure 38.17
hepatic veins, hepatic portal vein
adrenolumbar veins, renal veins
internal spermatic veins or ovarian veins
lumbar veins, iliolumbar veins
caudal veins
common iliac veins
internal iliac veins, external iliac veins

veins of the hindlimb
• Figure 38.18
femoral veins, deep femoral veins
popliteal veins, saphenous veins

ANATOMY OF SYSTEMIC CIRCULATION

Laboratory Report Exercise 38

A. Matching

Match each description on the left with the correct term on the right.

1.	_____ artery in armpit	A.	subclavian
2.	_____ artery with three branches	B.	superior mesenteric
3.	_____ vein used for blood samples	C.	popliteal
4.	_____ artery on right side only	D.	cephalic
5.	_____ long vein of leg	E.	carotid
6.	_____ deoxygenated supply to liver	F.	gonadal
7.	_____ artery to large intestine	G.	valves
8.	_____ cerebral anastomosis	H.	axillary
9.	_____ long vein of arm	I.	great saphenous
10.	_____ vein in knee	J.	median cubital
11.	_____ artery to reproductive organ	K.	hepatic portal vein
12.	_____ major artery in neck	L.	circle of Willis
13.	_____ vein under clavicle	M.	coliac
14.	_____ found only in veins	N.	brachiocephalic artery

B. Short-Answer Questions

1. Compare the wall of an artery to the wall of a vein.

2. How is the arterial anatomy to the right arm different than that to the left arm?

3. What is the function of valves in the peripheral veins?

4. How does the cerebral vascular circle ensure that the brain has a constant supply of blood?

5. Describe the major vessels that return deoxygenated blood to the right atrium of the heart.

Cardiovascular Physiology

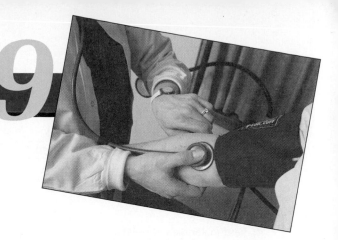

OBJECTIVES

On completion of this exercise you should be able to:

- Describe the pressure changes that occur during a cardiac cycle.
- Describe and demonstrate the steps involved in blood pressure determination.
- Explain the differences in blood pressure due to body position.
- Take a pulse rate at several locations.

WORD POWER

systole (systol—contraction)
diastole (diastol—standing apart)
sphygmomanometer (sphygm—pulse)

pulse (puls—beat)
auscultation (auscult—to listen)

MATERIALS

sphygmomanometer
stethoscope
alcohol swabs

INTRODUCTION

To fully appreciate the cardiovascular experiments in this chapter it is important to understand the anatomical features of the heart and the circulation of blood through the chambers of the heart. If necessary, review these concepts in Exercise 37 before proceeding.

One complete heartbeat is called a **cardiac cycle**. During a cardiac cycle the atria contract and relax and the ventricles contract and relax. The contraction phase of a chamber is called **systole (SIS-tō-lē)**, and the relaxation phase is termed **diastole (dī-AS-tō-lē)**. The human heart averages 75 cardiac cycles or heartbeats per minute, with each cardiac cycle occurring in 0.8 s. Figure 39.1 illustrates the events of a cardiac cycle. The cycle begins with the brief systole of the atria, lasting 0.1 s (100 ms), to fill the resting ventricles. Next, the ventricles enter their systolic phase and contract for 0.3 s to pump blood out of the heart. For the remaining 0.4 s of the cardiac cycle all chambers are in diastole and fill with blood in preparation for the next heartbeat. Most blood enters the ventricles during this resting period. Atrial systole contributes only 30% of the blood volume in the ventricle prior to ventricular systole. The remaining 70% of blood flows into the ventricles due to pressure and gravity.

Each cardiac cycle is marked with an increase and a decrease in blood pressure, both within the heart and in the arteries. When a chamber contracts, pressure increases as a result of the squeezing together of the chamber walls. The increase in blood pressure forces blood to circulate into the next chamber or out of the ventricle. As a ventricle relaxes, its walls move apart and pressure decreases. This drop in pressure draws blood into the ventricle and refills it for the next systole.

When all chambers of the heart are in diastole, the atrioventricular (AV) valves are open and blood flows from the atria into the ventricles. The semilunar (SL) valves are

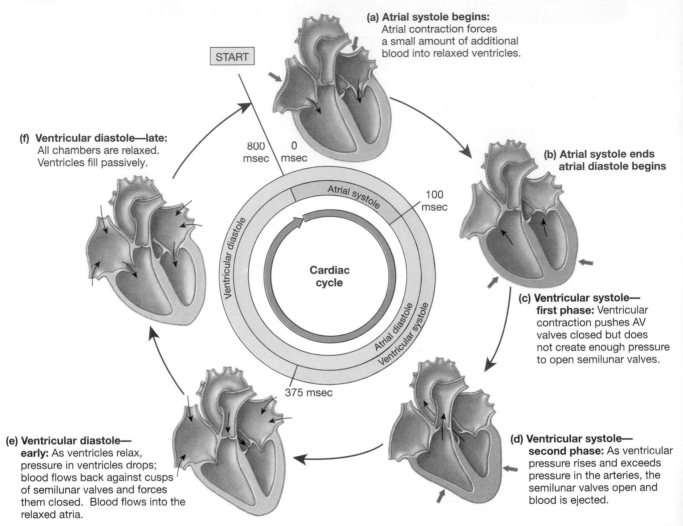

(a) Atrial systole begins: Atrial contraction forces a small amount of additional blood into relaxed ventricles.

START

(f) Ventricular diastole—late: All chambers are relaxed. Ventricles fill passively.

800 msec   0 msec

100 msec

Atrial systole

Ventricular diastole

Cardiac cycle

Atrial diastole

Ventricular systole

(b) Atrial systole ends atrial diastole begins

(c) Ventricular systole—first phase: Ventricular contraction pushes AV valves closed but does not create enough pressure to open semilunar valves.

375 msec

(e) Ventricular diastole—early: As ventricles relax, pressure in ventricles drops; blood flows back against cusps of semilunar valves and forces them closed. Blood flows into the relaxed atria.

(d) Ventricular systole—second phase: As ventricular pressure rises and exceeds pressure in the arteries, the semilunar valves open and blood is ejected.

Figure 39.1
Phases of the Cardiac Cycle.
Thin black arrows indicate blood flow, and green arrows indicate contractions.

closed to prevent backflow of blood from the aorta and pulmonary artery into the left and right ventricles, respectively. When the left ventricle contracts, pressure increases to a point where it exceeds the pressure of the blood in the aorta holding the aortic SL valve shut. The ventricle forces blood through the valve, and arterial blood pressure increases as a result of the increase in blood volume. When the ventricle relaxes, aortic and arterial pressures drop and the aortic SL valve closes. Similar events occur on the right side of the heart with the pulmonary SL valve.

In this exercise you will investigate the physiology of blood pressure and the effect of posture on blood pressure. You will also listen to heart sounds and practice taking a subject's pulse.

A. Heart Sounds

Listening to internal sounds of the body is called **auscultation**. A **stethoscope** is used to amplify the sounds to an audible level. The heart, lungs, and digestive tract are the most frequently auscultated systems. Auscultation provides the listener with valuable information concerning fluid accumulation in the lungs or blockages in the digestive

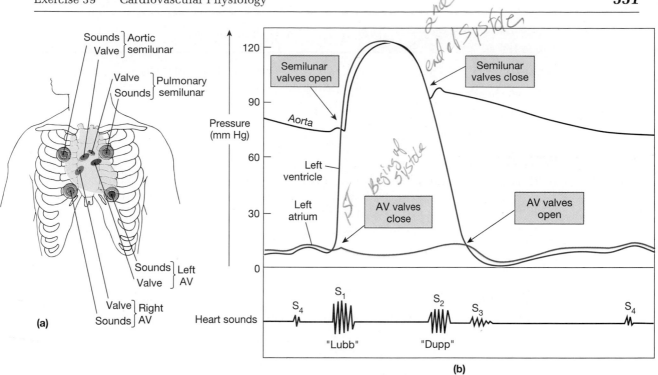

Figure 39.2
Heart Sounds.

(a) Placements of a stethoscope for listening to the different sounds produced by individual valves. **(b)** The timing of heart sounds in relation to key events in the cardiac cycle.

tract. Auscultation of the heart is used as a diagnostic tool to evaluate valve function and normal and abnormal heart sounds.

Four sounds are produced by the heart during a cardiac cycle. The first two sounds are easily heard and are the familar "lubb-dupp" of the heartbeat. The first heart sound (S_1) , the lubb, is caused by the closure of the atrioventricular valves as the ventricles begin their contraction. The second sound (S_2), the dupp, occurs as the semilunar valves close at the beginning of ventricular diastole. The third and fourth sounds are faint and difficult to hear. The third sound (S_3) is produced by blood flowing into the ventricles, and the fourth sound (S_4) is generated by the contraction of the atria. Relate each heart sound to the cardiac cycle shown in Figure 39.2.

Figure 39.2a illustrates the landmarks for proper placement of the stethoscope to listen effectively to each heart valve. Notice in the figure that the sites for auscultation do not overlie the anatomical location of the heart valves. Interference of sound conduction by soft tissue and bone deflects the sound waves to locations lateral to the valves.

CLINICAL APPLICATION—HEART MURMUR

An unusual heart sound is called a **murmur**. Not all murmurs indicate an anatomical or functional anomaly of the heart. The sound may originate from turbulent flow in a heart chamber. Some murmurs, however, are diagnostic for certain heart defects. **Septal defects** are holes in a wall between two chambers. A murmur is heard as blood passes through the hole and into the chamber on the opposite side of the heart. Abnormal operation of a valve also produces a murmur. In **mitral valve prolapse**, the left atrioventricular valve does not seal completely during ventricular systole and blood **regurgitates** upward into the left atrium.

AUSCULTATION OF THE HEART

MATERIALS

> stethoscope
> alcohol wipes
> laboratory partner

PROCEDURES

1. Auscultation of the heart is best achieved with the shirt open. Your lab partner may feel more comfortable holding the bell of the stethoscope, slipping it inside his or her shirt, and locating each valve site.

2. Clean the earpieces of the stethoscope with a sterile alcohol wipe and dispose of the used wipe in a trash can.

3. Wear the stethoscope by placing the angled earpieces facing anterior. The angle directs the earpiece into the external ear canal.

4. Use Figure 39.2a as a guide and auscultate the AV and SL valves. Can you discriminate between the "lubb" and the "dupp" sounds? _____

   _____

5. What causes the first heart sound? What causes the second heart sound?

   _____

B. Determination of Blood Pressure

Blood pressure (BP) is a measure of the force the blood exerts on the walls of the systemic arteries. Blood pressure in arteries is not constant and increases when the left ventricle contracts and pumps blood into the aorta. When the left ventricle relaxes, less blood flows into the aorta and so the pressure decreases until the next ventricular systole. Two pressures are therefore used to express blood pressure, a systolic pressure and a lower diastolic pressure. Average blood pressure is considered to be 120/80 mm for a typical male and closer to 110/70 for most females. Do not be surprised when you take your blood pressure in the following exercise and discover it is not "average." Cardiovascular physiology is a dynamic mechanism, and pressures regularly change to adjust to the demands of the body.

Blood pressure is measured using an inflatable cuff called a **sphygmomanometer (sfig-mō-ma-NOM-e-ter)** or blood pressure cuff. Figure 39.3 demonstrates the proper placement of the cuff. It is wrapped around the arm just superior to the elbow and then inflated to approximately 160 mm Hg to compress the brachial artery and block blood flow in the artery. A stethoscope is placed on the antecubital region, and pressure is gradually vented from the cuff. Once pressure in the cuff is slightly less than pressure in the brachial artery, blood spurts through the artery and the turbulent flow makes a sound, called **Korotkoff's sounds**, which are audible through the stethoscope. The pressure on the gauge when the first sound is heard is recorded as the **systolic pressure**. As more pressure is relieved from the cuff, blood flow becomes less turbulent and quieter. The sounds fade when the cuff pressure matches the **diastolic pressure** of the artery.

CLINICAL APPLICATION—HIGH BLOOD PRESSURE AND SALT INTAKE

One of the first recommendations a physician gives a patient with *hypertension* is to reduce the dietary intake of salt. A high salt diet leads to an increase in blood volume. Because the vascular system is a sealed or closed system, the additional fluid volume is "trapped" in the vessels and blood pressure increases. Think about osmosis and explain how salt increases blood volume. Also, the next time you are in the grocery store, investigate the various salt substitutes currently available.

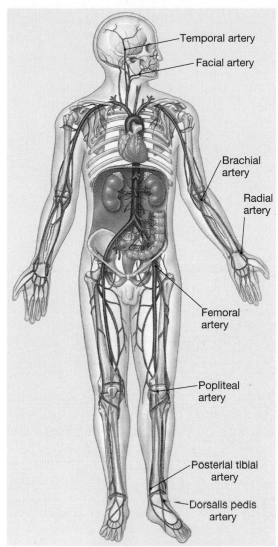

(a)

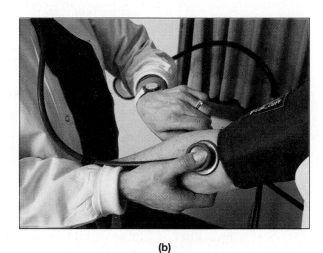

(b)

Figure 39.3
Checking the Pulse and Blood Pressure.

(a) Pressure points used to check the presence and strength of the pulse. **(b)** Use of a sphygmomanometer to check arterial blood pressure.

LABORATORY ACTIVITY   DETERMINATION OF BLOOD PRESSURE

MATERIALS

> sphygmomanometer
> stethoscope
> alcohol wipes
> laboratory partner

PROCEDURE I—RESTING BLOOD PRESSURE

1. Sit your lab partner comfortably in a chair and ask him or her to relax for several minutes. Roll up the sleeve of the right arm to expose the upper brachium. Clean the stethoscope with a sterile alcohol pad and dispose of the used pad in the trash.

2. Force all air out of the sphygmomanometer by compressing the cuff against the laboratory table. Loosely wrap the deflated cuff around the right arm of your partner so that its lower edge is just superior to the antecubital region of the elbow as in Figure 39.3.

3. If the cuff has orientation arrows, line the arrows up with the antecubitis; otherwise, position the rubber tubing over the antecubital region. Tighten the cuff so it is snug against the arm.

4. Gently close the valve on the cuff and inflate it to approximately 160 mm Hg. *Do not* leave the cuff inflated for more than 1 min. The inflated cuff prevents blood flow to the forearm, and the disruption in blood flow could lead to fainting.

5. Put on the stethoscope and place the bell below the cuff with the diaphragm over the brachial artery at the antecubitis.

6. Carefully open the valve to the sphygmomanometer and slowly deflate the cuff while listening for the Korotkoff's sounds. When the first sound is heard, note the pressure reading on the gauge.

7. Continue to vent pressure from the cuff and to listen with the stethoscope. When you hear the last faint sound, note the pressure on the gauge.

8. Open the pressure valve completely, quickly finish deflating the cuff, and remove it from your partner's arm.

9. Record the observed blood pressures in Table 39.1 under resting pressures. Repeating the above procedures and averaging the readings may yield a more accurate overall blood pressure determination.

PROCEDURE II—EFFECT OF POSTURE ON BLOOD PRESSURE

Changing posture affects the influence of gravity on blood pressure. This is readily apparent when standing on your head. In this section your partner will lie in the supine position (on the back) for approximately 5 min to allow for cardiovascular adjustments. Blood pressure will then be determined and compared to the pressure obtained in procedure I.

1. Ask your partner to lie on the lab table and relax for 5 min.

2. Wrap the sphygmomanometer around your partner's right arm and determine his or her supine blood pressure. Recore it in Table 39.1.

3. Next, ask your partner to stand still for 5 min. Take his or her blood pressure and record it in Table 39.1.

4. What was the effect of a supine posture on the blood pressure?

_____

_____

TABLE 39.1	**Blood Pressure Measurements**		
	Resting	*Supine*	*Standing*
Measurement 1	_____	_____	_____
Measurement 2	_____	_____	_____

PROCEDURE III—EFFECT OF EXERCISE ON BLOOD PRESSURE

In this section you will determine the effect of mild exercise on blood pressure. Be sure that you have determined the resting blood pressure of your subject beforehand as outlined in procedure I. The blood pressure observed in procedure I will be used as a baseline to compare with the "exercise blood pressure."

1. Secure the cuff around your partner's arm loose enough so that it is comfortable for exercise yet is in position for taking pressure readings.

2. Have your partner jog in place for 5 min without stopping if possible. This is not a stress test, so if the subject becomes excessively winded or tired, he or she should stop immediately.

3. Upon completion of the exercise, take a blood pressure reading, and record the pressures in Table 39.2. Repeat the readings once every 2 min until the pressure returns to the resting values.

TABLE 39.2	Exercise Effect on Blood Pressure					
	Start	*2 min*	*4 min*	*6 min*	*8 min*	*10 min*
PB	_____	_____	_____	_____	_____	_____
MAP	_____	_____	_____	_____	_____	_____

4. To compare your blood pressure readings, determine the **mean arterial pressure** (MAP). To calculate MAP, first determine the **pulse pressure,** the difference between the diastolic and systolic pressures (subtract diastolic P from systolic P). Next, add one-third of the pulse pressure to the diastolic pressure to get the MAP.
   Calculate the MAP for each pressure measurement, and record your data in Table 39.2.

C. Determination of Pulse

Heart rate is usually determined by measuring the **pulse** or pressure wave in an artery. During ventricular systole, blood pressure increases and stretches the walls of arteries. When the ventricle is in diastole, blood pressure decreases and the arterial walls rebound to their relaxed diameter. This change in vessel diameter is felt as a throb at a pressure point. The most commonly used pressure point is the radial artery on the lateral forearm just superior to the thumb. Other points include the common carotid artery in the neck and the popliteal artery of the posterior knee, see Figure 39.3. The number of pulses indicates the number of cardiac cycles for a given interval of time. The **pulse pressure** is the difference between the systolic and diastolic blood pressures and accounts for the pressure wave.

LABORATORY ACTIVITY   DETERMINATION OF PULSE AND PULSE PRESSURE

MATERIALS

watch or clock with a second hand
lab partner

PROCEDURE

1. Have your partner relax in a chair for several minutes.
2. Locate your subject's pulse in the right radial artery located just above the thumb. Use your index finger or index and middle fingers to palpitate and identify the pulse. Do not use your thumb for pulse measurements.
3. Gently apply light pressure to the pulse point and count the pulse rate for 15 s and then multiply this number by 4 to obtain the rate per minute. Record your data in Table 39.3.
4. Repeat the pulse determination at the carotid artery and the popliteal artery. Record your data in Table 39.3 and calculate the pulse pressure for each.
5. Is there a difference between the pulse at the various pulse points?

TABLE 39.3	Pulse Measurement
Pulse	*Pulse Rate*
Radial	_____
Carotid	_____
Popliteal	_____

CARDIOVASCULAR PHYSIOLOGY CHECKLIST

This is a list of bold terms presented in Exercise 39. Use this list for review purposes after you have completed the laboratory activities.

CARDIAC CYCLE

• Figure 39.1
cardiac cycle
diastole
systole
atrial systole
atrial diastole
ventricular systole
ventricular diastole

HEART SOUNDS

• Figure 39.2
auscultation
stethoscope
dupp

lubb
murmur
septal defect
mitral valve prolapse
regurgitation

BLOOD PRESSURE AND PULSE

• Figure 39.3
blood pressure
sphygmomanometer
Korotkoff's sounds
systolic pressure
diastolic pressure
pulse
pulse pressure

CARDIOVASCULAR PHYSIOLOGY

Laboratory Report Exercise 39

A. Matching
Match the term listed on the left with the correct description listed on the right.

1. _____ cardiac cycle	**A.**	unusual heart sound
2. _____ diastole	**B.**	contraction phase of heart
3. _____ systole	**C.**	a single heart beat
4. _____ auscultation	**D.**	listening to body sounds
5. _____ Korotkoff's sounds	**E.**	closed during ventricular systole
6. _____ first heart sound	**F.**	relaxation phase of heart
7. _____ second heart sound	**G.**	sounds heard in stethoscope
8. _____ murmur	**H.**	sound due to closure of AV valves
9. _____ semilunar valves	**I.**	closed during ventricular diastole
10. _____ atrioventricular valves	**J.**	sound due to closure of SL valves

B. Short-Answer Questions

1. Briefly describe the events of a cardiac cycle.

2. Why does blood pressure fluctuate in arteries?

3. Describe how a sphygmomanometer is used to determine blood pressure.

4. Describe how blood pressure changes during exercise and during rest.

5. Calculate the MAP if blood pressure is 130/85.

Cardiovascular Physiology Alternate

BIOPAC
Systems, Inc.

ELECTROCARDIOGRAPHY
Components of the ECG

OBJECTIVES

On completion of this exercise, you should be able to:

- Become familiar with the electrocardiograph as a primary tool for evaluating electrical events within the heart.
- Correlate electrical events as displayed on the ECG with the mechanical events that occur during the cardiac cycle.
- Observe rate and rhythm changes in the ECG associated with body position and breathing.

INTRODUCTION

During each cardiac cycle, a sequence of electrical impulses cause the heart muscle to depolarize and then repolarize. This electrical current may be detected at the body surface with a series of electrodes or **leads.** In this investigation you will use the **Lead II** configuration with a positive electrode on the left ankle, a negative electrode on the right wrist, and a ground electrode on the right ankle. A recording of the impulse is called an **electrocardiogram,** an **ECG** or **EKG.** For background information on cardiovascular physilolgy, it is recommended that you review Exercise 39.

The ECG is typically printed on a standard grid with seconds as the x-axis and amplitude or intensity in millivolts (mV) as the y-axis, see Figure 39A.1. For each small box on the grid, the vertical lines measure 0.04 seconds and the horizontal lines measure amplitude in 0.1 mV increments.

The basic components of an ECG are a straight baseline, the isoelectric line, and waves that indicate periods of depolarization and repolarization of the heart's chambers. The **P wave** occurs as the atria depolarize for contraction. Atrial systole occurs approximately 0.1 seconds after depolarization. The P wave is followed by a large spike called the **QRS complex** when the ventricles depolarize. After ventricular systole, the **T wave** results from ventricular repolarization. The ECG returns to the baseline and the next cardiac cycle soon occurs. Atrial repolarization occurs during the QRS complex and is undetected.

Within the ECG are intervals which include a wave and the return to the baseline. The **P-R interval** occurs between the start of the P wave and the start of the QRS complex. The P-R interval is the time required for an impulse to travel from the SA node to the ventricular muscle. The **Q-T interval** is the cycle of ventricular depolarization and repolarization. Segments are the baseline recordings between the waves. The **P-R segment** represents the time for an impulse to travel from the AV node to the ventricles. The **S-T segment** measures the delay between depolarization and repolarization of the ventricles. Table 39A.1 presents the average length and intensity of each phase of a typical ECG.

If the electrical activity of the heart changes, the ECG will reflect the changes. For example, damage to the AV node results in an extension of the P-R segment.

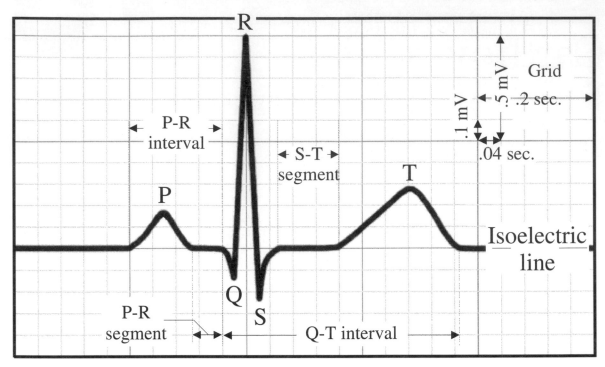

Figure 39A.1
Components of the ECG.

TABLE 39A.1	Normal Lead II ECG Values	
Phase	**Duration (second)**	**Amplitude (millivolt)**
P wave	0.06–0.11	< 0.25
PR interval	0.12–0.20	
PR segment	0.08	
QRS complex (R)	< 0.12	0.8–1.2
ST segment	0.12	
QT interval	0.36–0.44	
T wave	0.16	< 0.5

Irregularities in the heart beat are called **arrhythmias** and may indicate problems in cardiac function.

The ECG laboratory exercise is organized into four major procedures. The "**Set Up**" section describes where to plug in the electrode leads and how to apply the skin electrodes for a Lead II configuration. The "**Calibration**" section adjusts the hardware so it can collect accurate physiological data. Once the hardware is calibrated the "**Data Recording**" section describes taking ECG recordings with the Subject in four positions; lying down, sitting, while breathing deeply, and after mild exercise. After saving the ECG data to computer disc, the "**Data Analysis**" section instructs you how to use the software tools to interpret and evaluate the ECG.

The BioPac Student Lab system is safe and easy to use but be sure to read the safety notices and to carefully follow the procedures as outlined in the laboratory activities. Under no circumstances should you deviate from the experimental procedures.

LABORATORY ACTIVITY   ELECTROCARDIOGRAPHY

MATERIALS

BIOPAC electrode lead set (SS2L)

BIOPAC disposable vinyl electrodes (EL503), 3 electrodes per subject

Cot or lab table and pillow

BIOPAC electrode gel (GEL1) and abrasive pad (ELPAD) *or* Skin cleanser or alcohol prep

Computer system:

Macintosh®—minimum 68020 *or* PC running Windows 95®

Memory requirements:

The Biopac Student Lab application needs to have at least 4 MB of RAM available for its needs. This is 4 MB above and beyond the operating system needs and any other programs that are running.

BIOPAC Student Lab software V3.0

BIOPAC acquisition unit (MP30)

BIOPAC wall transformer (AC100A)

BIOPAC serial cable (CBLSERA)

PROCEDURE—SET UP

1. Turn the computer **ON.**

   The desktop should appear on the monitor. If it does not appear, ask the laboratory instructor for assistance.

2. Make sure the BIOPAC MP30 unit is turned **OFF.**

3. Plug the equipment in as follows: Electrode lead (SS2L)—CH 2 (See Figure 30A.2.)

4. Turn the BIOPAC MP30 Unit **ON.**

5. Place three electrodes on the subject, as shown in Figure 39A.3.

   The Subject should not be in contact with nearby metal objects (faucets, pipes, etc.) and should remove any wrist or ankle bracelets.

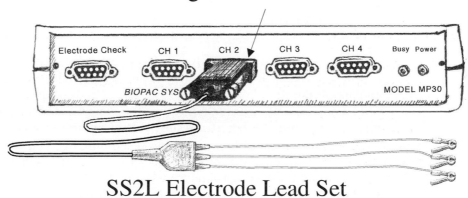

Figure 39A.2
Biopac MP 30 Set Up.

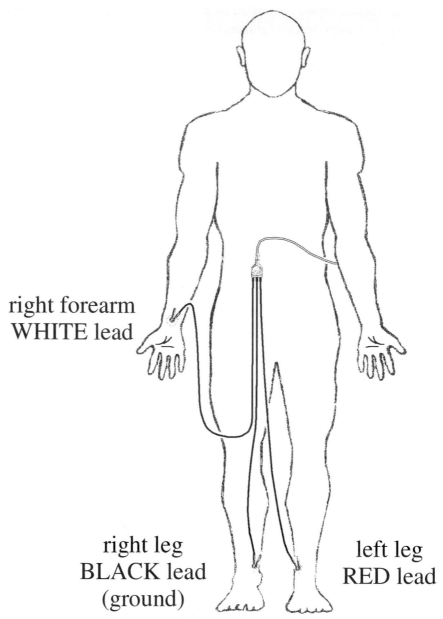

right forearm
WHITE lead

right leg
BLACK lead
(ground)

left leg
RED lead

Figure 39A.3
Lead II Electrode Placement.

Place one electrode on the medial surface of each leg, just above the ankle. Place the third electrode on the right anterior forearm at the wrist (same side of arm as the palm of hand).

Note: For optimal electrode adhesion, the electrodes should be placed on the skin at least 5 minutes before the start of the Calibration procedure.

6. Attach the electrode cable to the electrodes, following Figure 39A.3.

Each of the pinch connectors on the end of the electrode cable needs to be attached to a specific electrode. The electrode cables are each a different color, and you should follow Figure 39A.3 to ensure that you connect each cable to the proper electrode.

The pinch connectors work like a small clothespin but will only latch onto the nipple of the electrode from one side of the connector.

When the electrode cable is connected properly, the LEAD II electrode configuration will be established.

7. Have the Subject lie down and relax.

   Position the electrode cables such that they are not pulling on the electrodes. Connect the electrode cable clip (where the cable meets the three individual colored wires) to a convenient location (can be on the Subject's clothes). This will relieve cable strain.

8. Start the BIOPAC *Student Lab* Program.

9. Choose **Lesson L05-ECG.1.**

10. **Type in your filename.**

    Type in a unique identifier

11. Click **OK.**

    This ends the Set Up procedure.

PROCEDURE—CALIBRATION

The Calibration procedure establishes the hardware's internal parameters (such as gain, offset, and scaling) and is critical for optimum performance. **Pay close attention to the following Calibration steps.**

1. Double check the electrodes, and make sure the Subject is relaxed and lying down.

   Make sure the electrodes adhere securely to the skin. If they are being pulled up, you will not get a good ECG signal.

   The Subject must be relaxed and as still as possible during the Calibration procedure. The electrocardiograph is very sensitive to small changes in voltage caused by contraction of skeletal muscles, and the Subject's arms and legs need to be relaxed so that the muscle (EMG) signal does not corrupt the ECG signal.

2. Click on **Calibrate.**

   The **Calibrate** button is in the upper left corner of the **Setup** window. This will start the Calibration recording.

   The Subject needs to remain relaxed throughout Calibration.

3. **Wait** for the Calibration procedure to stop.

   The Calibration procedure will stop automatically after 8 seconds.

4. **Check** the Calibration data:

   At the end of the 8-second Calibration recording, the screen should resemble Figure 39A.4

   • If *correct,* proceed to Data Recording.

   • If *incorrect,* **Redo Calibration.**

   There should be a greatly reduced ECG waveform with a relatively flat baseline. If your data resembles Figure 39A.4, proceed to the Data Recording section.

   If the data show any large spikes, jitter, or large baseline drifts, then you should redo the Calibration by clicking on the **Redo Calibration** button and repeating the entire Calibration sequence.

PROCEDURE—RECORDING LESSON DATA

1. Prepare for the recording, and have the Subject lie down and relax.

   You will record the Subject in four conditions: lying down, after sitting up, breathing deeply, and after exercise. The Subject will perform tasks in the intervals between recordings.

   In order to work efficiently, read this entire section so you will know what to do for each recording segment.

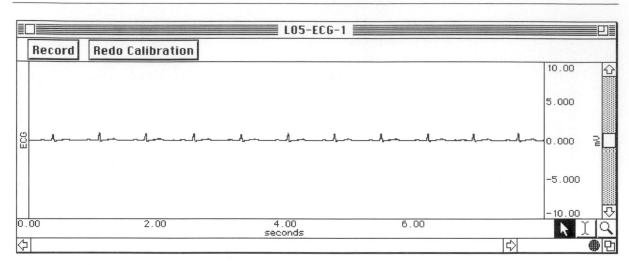

Figure 39A.4
Calibration ECG.

The Subject should remain in a supine position and should continue to relax while you review the lesson.

Check the last line of the journal and note the total amount of time available for the recording. Stop each recording segment as soon as possible so you don't use an excessive amount of time (time is memory).

Hints for obtaining optimal data:

a. The Subject should not talk or laugh during any of the recording segments.

b. The Subject should be in a relaxed state for each recording segment and in the position noted for each segment.

c. When the Subject is asked to sit up, he or she should do so in a chair, with arms relaxed on the armrest (if available).

d. For Steps 6 and 7: Hit Resume as soon as possible after the Subject sits up in order to capture the heart rate variation but not while the subject is in the process of sitting up or there will be excessive motion artifact.

e. The Subject should be as still as possible during the recording segment. The electrocardiograph is very sensitive to small changes in voltage caused by contraction of skeletal muscles, and the Subject's arms and legs need to be relaxed so that the muscle (EMG) signal does not corrupt the ECG signal.

Segment 1

2. Click on **Record.**

   The recording will begin.

3. Record for 20 seconds.

   Subject is lying down (seconds 0–20).

4. Click on **Suspend.**

   The recording should halt, giving you time to review the data and prepare for the next recording segment.

5. Review the data on the screen.

   If all went well, your data should look similar to Figure 39A.5, and you can proceed to Step 6.

   • If *correct,* go to Step 6.

   • If *incorrect,* click on **Redo.**

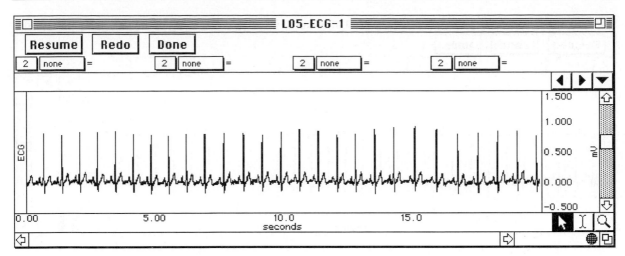

Figure 39A.5
Segment 1 ECG: Lying Down.

The data would be *incorrect* if:

a. The suspend button was pressed prematurely.

b. An electrode peeled up, causing a large baseline drift, spike, or loss of signal.

c. The Subject had too much muscle (EMG) artifact.

In this case, you should redo the recording by clicking on **Redo** and repeating Steps 2–5. Note that once you press **Redo,** the data you have just recorded will be erased.

Segment 2

6. Have the subject quickly sit up.

   In order to capture the heart rate variation, it is important that you **Resume** recording as quickly as possible after the subject sits up. However, it is also important that you do not click **Resume** while the subject is in the process of sitting up or you will capture motion artifact.

7. Click on **Resume.**

   The recording will continue from the point where it last stopped, and a marker labeled "After sitting up" will automatically come up when **Resume** is pressed.

8. Record for 20 seconds.

   Subject is seated (seconds 21–40).

9. Click on **Suspend.**

   The recording should halt, giving you time to review the data and prepare for the next recording segment.

10. Review the data on the screen.

    If all went well, your data should look similar to Figure 39A.6, and you can proceed to Step 11.

    • If *correct* and more segments are required, go to Step 11.

    • If *incorrect,* click on **Redo.**

    The data would be *incorrect* for the reasons in Step 5.

    If incorrect, you should redo the recording by clicking on **Redo** and repeating Steps 6–10. Note that once you press **Redo,** the data you have just recorded will be erased.

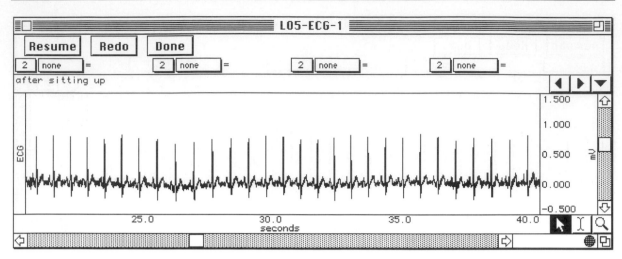

Figure 39A.6
Segment 2 ECG: After Sitting Up.

Segment 3

11. Click on **Resume.**

The recording will continue from the point where it last stopped, and a marker labeled "Deep Breathing" will automatically come up when **Resume** is pressed.

Subject is seated (seconds 41–60).

After the recording begins, the **Subject** should start a series of slow, prolonged breathing and should continue for 5 cycles.

Note: It is important to breathe with long, slow, deep breaths in order to minimize the EMG artifact.

12. Record for 20 seconds and have **Subject** take in 5 deep breaths during recording.

Recorder should insert markers at a corresponding inhale and exhale.

The **Recorder** should label these markers "Inhale" and "Exhale."

∇ "Inhale"

∇ "Exhale"

13. Click on **Suspend.**

The recording should halt, giving you time to prepare for the next recording segment.

14. Review the data on the screen.

If all went well, your data should look similar to Figure 39A.7, and you can proceed to Step 15.

• If *correct,* go to Step 15.

• If *incorrect,* click on **Redo.**

The data would be incorrect for the reasons in Step 5.

Note: The Deep Breathing recording may have some baseline drift as shown in Figure 39A.7. Baseline drift is fairly normal, and unless it is excessive, it does not necessitate redoing the recording.

If incorrect, you should redo the recording by clicking on **Redo** and repeating Steps 11–14. Note that once you press **Redo,** the data you have just recorded will be erased.

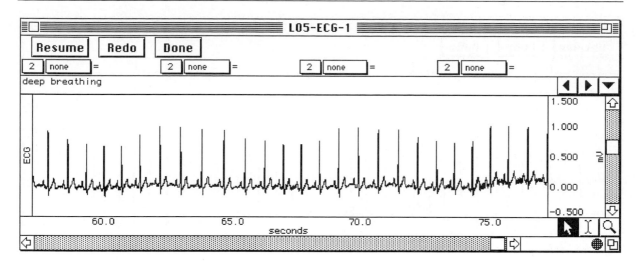

Figure 39A.7
Segment 3 ECG: Deep Breathing.

Segment 4

15. Have the **Subject** perform an exercise to elevate his or her heart rate.

Subject should perform an exercise to elevate his or her heart rate fairly rapidly, such as push-ups or jumping jacks.

Note: You may remove the electrode cable pinch connectors so that the Subject can move about freely but **do not remove the electrodes.**

If you do remove the cable pinch connectors, you must reattach them following the precise color placement in Figure 39A.3 prior to clicking **Resume.**

In order to capture the heart rate variation, it is important that you **Resume** recording as quickly as possible after the subject has performed the exercise. However, it is also important that you do not click **Resume** while the Subject is exercising or you will capture motion artifact.

16. Click on **Resume.**

The recording will continue from the point where it last stopped, and a marker labeled "After Exercise" will automatically come up when **Resume** is pressed.

17. Record for 60 seconds.

Subject is seated in a relaxed state, recovering from exercise (seconds 61–120).

18. Click on **Suspend.**

The recording should halt.

19. Review the data on the screen.

If all went well, your data should look similar to Figure 39A.8, and you can proceed to Step 20.

- If *correct,* go to Step 20.
- If *incorrect,* click on **Redo.**

The data would be incorrect for the reasons in Step 5.

Note: The After Exercise recording may have some baseline drift (as shown in Figure 39A.2). Baseline drift is fairly normal, and unless it is excessive, it does not necessitate redoing the recording.

If incorrect, you should redo the recording by clicking on **Redo** and repeating Steps 2–5. Note that once you press **Redo,** the data you have just recorded will be erased.

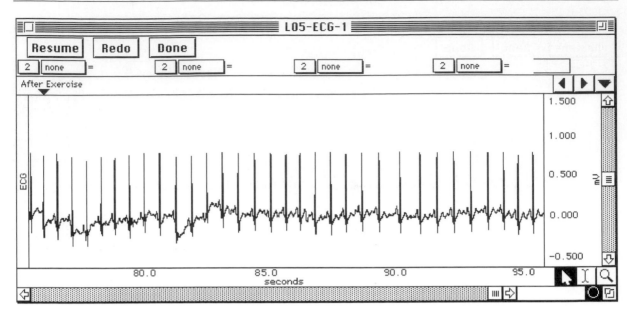

Figure 39A.8
Segment 4 ECG: After Exercise.

20. Click on **Done.**

    A pop-up window with four options will appear. Make your choice, and continue as directed.

    If choosing the "Record from another subject" option:

    a. Attach electrodes per Set Up Step 5 and continue the entire lesson from Set Up Step 8.

    b. Each person will need to *use a unique filename.*

21. Remove the electrodes.

    Remove the electrode cable pinch connectors, and peel off the electrodes. Throw out the electrodes (BIOPAC electrodes are not reusable). Wash the electrode gel residue from the skin, using soap and water. The electrodes may leave a slight ring on the skin for a few hours. This is normal and does not indicate that anything is wrong.

PROCEDURE—DATA ANALYSIS

In this section, you will examine ECG components of cardiac cycles and measure amplitudes (mV) and durations (msecs) of the ECG components.

Note: Interpreting ECGs is a skill that requires practice to distinguish between normal variation and those arising from medical conditions. Do not be alarmed if your ECG is different than the examples shown or from the tables and figures in the Introduction.

1. Enter the **Review Saved Data** mode.

   Enter the Review Saved Data mode from the Lessons menu.

   The data window should come up the same as shown in Figure 39A.9.

   Note Channel Number (CH) designation.

   CH 2   **ECG Lead II**

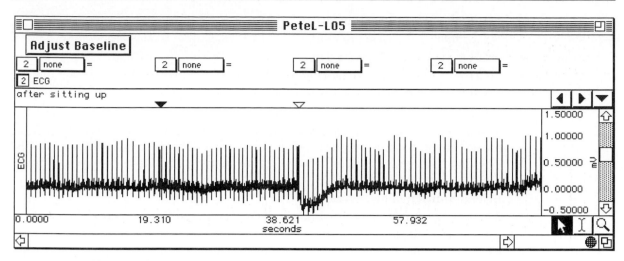

Figure 39A.9
Sample ECG Data.

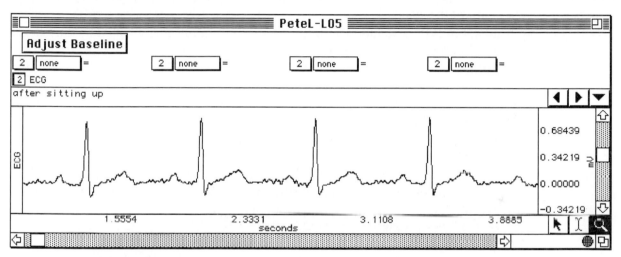

Figure 39A.10
Sample Data from Segment 1.

2. Set up your display window for optimal viewing of four successive beats from
   Segment 1 (see Figure 39A.10).

   The following tools help you adjust the data window:

Autoscale horizontal	Horizontal (Time) Scroll Bar
Autoscale waveforms	Vertical (Amplitude) Scroll Bar
Zoom Tool	Zoom Previous

   Turn **Grids ON and OFF** by choosing **Display: Preferences** from the **File** menu.

   Adjust Baseline Button: Allows you to position the waveform up or down in
   small increments so that the baseline can be exactly zero. This is not needed to
   get accurate amplitude measurements but may be desired before making a print-
   out or when using grids.

   Once the Baseline button is pressed, two more buttons, **Up** and **Down,** will be
   displayed. Simply click on these to move the waveform up or down.

3. Set up the measurement boxes as follows:

Channel	Measurement
CH 2	ΔT (Delta Time)
CH 2	BPM (Beats Per Minute)
CH 2	Δ (Delta Amplitude)
CH 2	max (Maximum Amplitude)

The measurement boxes are above the marker region in the data window. Each measurement has three sections: channel number, measurement type, and result. The first two sections are pull-down menus that are activated when you click on them.

Brief definition of measurements:

ΔT: The Delta Time measurement is the difference in time between the end and beginning of the selected area.

BPM: The Beats Per Minute measurement first calculates the difference in time between the end and beginning of the area selected by the I-Beam tool (same as ΔT) and divides this value into 60 seconds/minute.

Δ: Computes the difference in amplitude between the last point and the first point of the selected area.

max: Finds the maximum amplitude value within the area selected by the I-Beam tool (including the endpoints).

Note: The "selected area" is the area selected by the **I-beam** tool (including the endpoints).

4. Using the **I-Beam** cursor, select the area between two successive R waves.

Try to go from P wave peak to R wave peak as precisely as possible. Figure 39A.11 shows an example of the selected area.

Record your data in Section A of the lab report.

5. Take measurements at two other intervals in the current waveform display.

Record your data in Section A of the lab report.

6. **Zoom** in on a single cardiac cycle from Segment 1.

Be sure to stay in the first recorded segment (Segment 1) when you select the cardiac cycle.

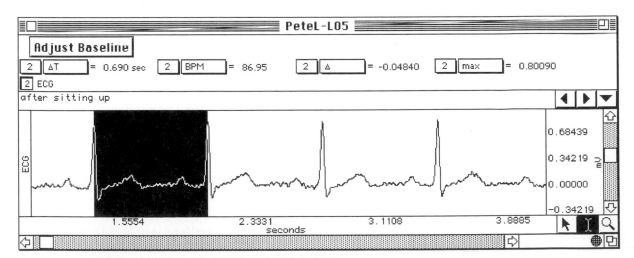

Figure 39A.11
Selection of ECG Waves.

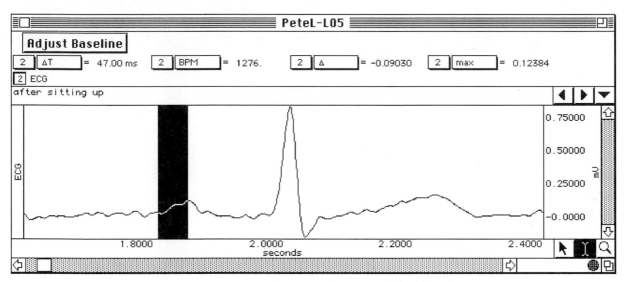

Figure 39A.12
Measurement of P Wave Amplitude.

7. Use the I-Beam cursor and measurement box values (and refer to Figure 39A.1 as necessary) to record the following amplitudes and durations.

Figure 39A.12 shows a sample setup for measuring P wave amplitude. Note that the Δ measurement shows the amplitude difference between endpoints in the selected area.

Note: Figure 39A.1 in the Introduction details components of the ECG.

Durations

P wave	QT interval
PR interval	ST segment
QRS interval	T wave

Amplitudes

P wave	T wave
QRS complex	

Record your data in part A of the lab report.

8. Repeat measurements as required for the Data Report.

Follow the examples shown in Figure 39A.12 to complete all the measurements required for your Data Report.

9. Save or print the data file.

You may save the data to a floppy drive, save notes that are in the journal, or print the data file.

10. Exit the program.

Name _____   Date _____

Section _____   Number _____

ELECTROCARDIOGRAPHY I: ECG I

Laboratory Report Exercise 39 Alternate

A. Data and Calculations

Subject Profile

Name _____   Height _____

Age _____   Weight _____

Gender: Male / Female

1. Supine, Resting, Regular Breathing (using *Segment 1* data)

Complete the following tables with the lesson data indicated, and calculate the Mean and Range as appropriate.

TABLE 39A.2

Measurement	From Channel	Cardiac Cycle 1	Cardiac Cycle 2	3	Mean	Range
ΔT	CH 2	_____	_____	_____	_____	_____
BPM	CH 2	_____	_____	_____	_____	_____

TABLE 39A.3

ECG Component	Duration ΔT [CH 2] Cycle 1	Cycle 2	Cycle 3	Mean	Amplitude (mV) Δ [CH 2] Cycle 1	Cycle 2	Cycle 3	Mean
P wave	___	___	___	___	___	___	___	___
PR interval	___	___	___	___	___	___	___	___
PR segment	___	___	___	___	___	___	___	___
QRS complex	___	___	___	___	___	___	___	___
QT interval	___	___	___	___	___	___	___	___
ST segment	___	___	___	___	___	___	___	___
T wave	___	___	___	___	___	___	___	___

TABLE 39A.4

Ventricular Readings	CH 2 ΔT Cycle 1	Cycle 2	Cycle 3	Mean
QT interval (*corresponds to Ventricular Systole*)	_____	_____	_____	_____
End of T wave to subsequent R wave (*corresponds to Ventricular Diastole*)	_____	_____	_____	_____

2. Supine, Deep Breathing

TABLE 39A.5

Rhythm	CH #	Cycle 1	Cycle 2	Cycle 3	Mean
Inspiration					
ΔT	CH 2	_____	_____	_____	_____
BPM	CH 2	_____	_____	_____	_____
Expiration					
ΔT	CH 2	_____	_____	_____	_____
BPM	CH 2	_____	_____	_____	_____

3. Sitting

TABLE 39A.6

Heart Rate	CH #	Cycle 1	Cycle 2	Cycle 3	Mean
ΔT	CH 2	_____	_____	_____	_____
BPM	CH 2	_____	_____	_____	_____

4. After Exercise

TABLE 39A.7

		CH 2 ΔT			
Ventricular Readings	Cycle 1	Cycle 2	Cycle 3	Mean	
QT interval (*corresponds to Ventricular Systole*)	_____	_____	_____	_____	
End of T wave to subsequent R wave (*corresponds to Ventricular Diastole*)	_____	_____	_____	_____	

B. Data Summary and Questions

1. Heart Rate (BPM)

Condition	Mean	Range
Supine, regular breathing	_____	_____
Supine, deep breathing, inhalation	_____	_____
Supine, deep breathing, exhalation	_____	_____
Sitting, regular breathing	_____	_____
After exercise—start of recording	_____	_____
After exercise—end of recording	_____	_____

Explain the changes in heart rate between conditions. Describe the physiological mechanisms causing these changes._____

_____

_____

_____

_____

_____

_____

_____

_____

2. Duration (ΔT)

Rhythm

Measurement	Mean	Range
Supine, regular breathing		
Inhalation	_____	_____
Exhalation	_____	_____
Supine, deep breathing		
Inhalation	_____	_____
Exhalation	_____	_____

Are there differences in the cardiac cycle with the respiratory cycle?

_____

_____

_____

Measurement	Mean	Range
Supine, regular breathing		
Ventricular systole	_____	_____
Ventricular diastole	_____	_____
After exercise		
Ventricular systole	_____	_____
Ventricular diastole	_____	_____

What changes occurred in the duration of systole and diastole between resting and post-exercise?

_____

_____

_____

3. Review Your Data

   1. Is there always one P wave for every QRS complex?   Yes   No

   2. Describe the P wave and T wave shapes:_____

_____

   3. Do the wave durations and amplitudes for all subjects fall within the normal ranges listed in Table 39A.1?   Yes   No

   4. Do the ST segments mainly measure between −0.1 mV and 0.1 mV?   Yes   No

   5. Is there baseline "drift" in the recording?   Yes   No

   6. Is there baseline "noise" in the recording?   Yes   No

Cardiovascular Physiology Alternate

ECG AND PULSE
Mechanical Action of the Heart
Peripheral Pressure Pulse
Plethysmography

OBJECTIVES

On completion of this exercise, you should be able to:

- Become familiar with the principle of plethysmography and its usefulness in qualitatively assessing peripheral changes in blood volume.
- Observe and record changes in peripheral blood volume and pressure pulse under a variety of both experimental and physiological conditions.
- Determine the approximate speed of the pressure pulse wave traveling between the heart and the finger.
- Illustrate the electrical activity associated with normal cardiac activity and how it relates to the flow of blood throughout the body.

INTRODUCTION

Each cardiac cycle pumps pressurized blood into the vascular system. This surge of ejected blood creates a pressure wave, measured as the pulse. A **plethysmogram** is a recording of the volume changes in the blood during a pulse wave. As the pulse wave flows along the artery, it causes expansion followed by the rebound of the vessel wall. A photoelectric transducer on the finger tip passes light into the skin and measures how much light is reflected back. Blood absorbs light and during a pulse wave, the increased blood volume absorbs proportionally more light. The Biopac Student Lab System converts the reflected light signals into electrical signals representing the pulse wave.

In this exercise you will collect ECG and pulse data from a Subject in different physiological conditions. For background information on cardiovascular physiology, it is recommended that you review Exercise 39.

The ECG and pulse laboratory exercise is organized into four major procedures. The "**Set Up**" section describes where to plug in the electrode leads and how to apply the skin electrodes for a Lead II configuration. A photoelectric transducer is wrapped around a finger tip to detect the pulse. The "**Calibration**" section adjusts the hardware so it can collect accurate physiological data. Once the hardware is calibrated the "**Data Recording**" section describes taking ECG and pulse recordings with the Subject in three positions; arm relaxed, hand in plastic bucket of cold or warm water, and arm raised. After saving the ECG data to computer disc, the "**Data Analysis**" section instructs you how to use the software tools to interpret and evaluate the ECG and pulse.

The Biopac Student Lab system is safe and easy to use but be sure to read the safety notices and to carefully follow the procedures as outlined in the laboratory activities. Under no circumstances should you deviate from the experimental procedures.

Laboratory Activity	ECG AND PULSE

MATERIALS

BIOPAC electrode lead set (SS2L)

BIOPAC disposable vinyl electrodes (EL503), 3 electrodes per subject

BIOPAC pulse plethysmograph (SS4LA or SS4L)

Ruler or measuring tape

Ice water or warm water in plastic bucket

BIOPAC electrode gel (GEL1) and abrasive pad (ELPAD) *or* Skin cleanser or alcohol prep

Computer system:

Macintosh®; minimum 68020 *or* PC running Windows 95®

Memory requirements: The Biopac Student Lab application needs to have at least 4 MB of RAM available for its needs. This is 4 MB above and beyond the operating system needs and any other programs that are running.

BIOPAC Student Lab software V3.0

BIOPAC acquisition unit (MP30)

BIOPAC wall transformer (AC100A)

BIOPAC serial cable (CBLSERA)

PROCEDURE—SET UP

1. Turn the computer **ON.**

   The desktop should appear on the monitor. If it does not appear, ask the laboratory instructor for assistance.

2. Make sure the BIOPAC MP30 unit is turned **OFF.**

3. Plug the equipment in as follows:

   Electrode lead (SS2L)—CH 1

   Pulse transducer (SS4LA) or

   Pulse transducer SS4L—CH 2

   (See Figure 39B.1)

4. Turn the MP30 Data Acquisition Unit **ON.**

5. Place three electrodes on the Subject (Figure 39B.2).

   • One electrode on the medial surface of the right leg, just above the ankle

   • One electrode on the medial surface of the left leg, just above the ankle

   • One electrode on the right anterior forearm just above the wrist (same side of arm as the palm of hand)

   Note: For optimal electrode adhesion, the electrodes should be placed on the skin at least 5 minutes before the start of the Calibration procedure.

6. Attach the first electrode lead set (SS2L) to the electrodes, following the color code (Figure 39B.3).

   Each of the pinch connectors on the end of the electrode cable needs to be attached to a specific electrode. The electrode cables are each a different color, and you should follow Figure 39B.3 to ensure that you connect each cable to the proper electrode.

   The pinch connectors work like a small clothespin but will only latch onto the nipple of the electrode from one side of the connector.

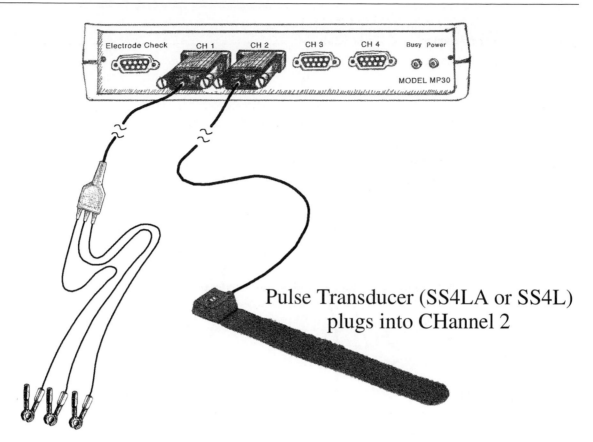

Pulse Transducer (SS4LA or SS4L)
plugs into CHannel 2

Electrode Lead (SS2L)
plugs into CHannel 1

Figure 39B.1
Biopac MP 30 Set Up.

When the electrode cable is connected properly, the LEAD II electrode configuration will be established.

7. Clean the window of the sensor.

    It is a good idea to clean the window of the sensor before each use. This will prevent any oil or dirt on the window from interfering with the signal. Use a soft cloth or other nonabrasive material to wipe it clean.

8. Wrap the pulse transducer (**SS4L**) around the tip of your index finger (Figure 39B.4).

    Position the transducer so that the sensor is on the bottom of your fingertip (the part without the fingernail), and wrap the Velcro® tape around your finger so the transducer fits snugly but not so that blood circulation is cut off. (It's a fine line between tight and too tight.)

9. Have the **Subject** sit down with forearms supported and relax.

    Position the electrode cables such that they are not pulling on the electrodes or the transducer.

    Connect the electrode cable clip (where the cable meets the three individual colored wires) to a convenient location (can be on the Subject's clothes). This will relieve cable strain.

    The **Subject** should not be in contact with nearby metal objects (faucets, pipes, and so on) and should remove any wrist or ankle bracelets.

Figure 39B.2
Lead II Electrode Placement.

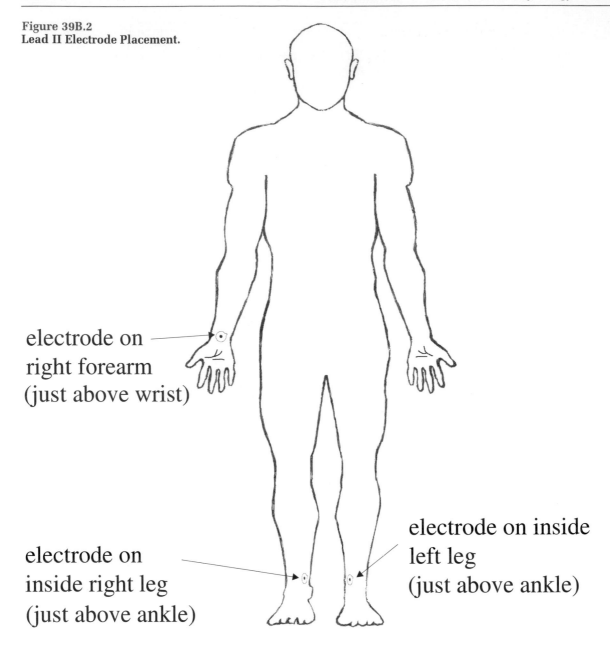

electrode on
right forearm
(just above wrist)

electrode on inside
left leg
(just above ankle)

electrode on
inside right leg
(just above ankle)

10. Start the BIOPAC Student Lab Program.
11. Choose **Lesson 7 (L07-ECG&P-1)**.
12. Type in your filename.
    Use a unique identifier.
13. Click **OK**.
    This ends the Set Up procedure.

PROCEDURE—CALIBRATION

The Calibration procedure establishes the hardware's internal parameters (such as gain, offset, and scaling) and is critical for optimum performance. **Pay close attention to the Calibration procedure.**

1. Double check the electrodes, and make sure the **Subject** is relaxed.

   Make sure the electrodes adhere securely to the skin. If they are being pulled up, you will not get a good ECG signal.

Figure 39B.3
Lead II Configuration Set Up.

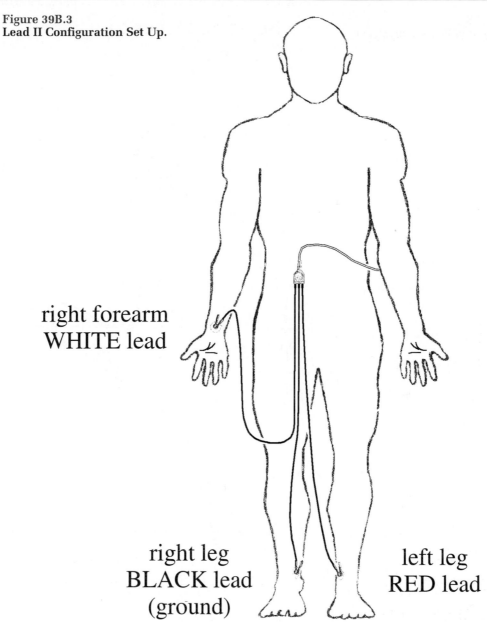

right forearm
WHITE lead

right leg
BLACK lead
(ground)

left leg
RED lead

The **Subject** must be relaxed during the Calibration procedure. The **Subject's** arms and legs need to be relaxed so that the muscle (EMG) signal does not corrupt the ECG signal.

2. Click on **Calibrate.**

   The **Calibrate** button is in the upper left corner of the **Setup** window. This will start the Calibration recording.

   The **Subject** needs to remain relaxed throughout Calibration.

3. **Wait** for the Calibration procedure to stop.

   The Calibration procedure will stop automatically after 8 seconds.

4. **Check** the Calibration data.

   At the end of the eight-second calibration recording, the screen should resemble Figure 39B.5.

   • If *similar*, proceed to Data Recording.

   • If *different*, **Redo the Calibration.**

Figure 39B.4
Placement of Pulse Transducer.

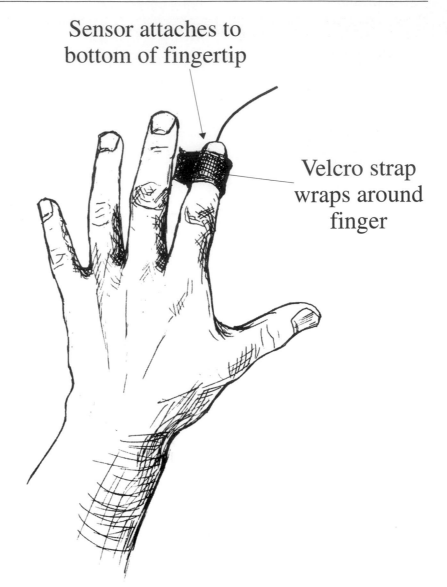

Sensor attaches to
bottom of fingertip

Velcro strap
wraps around
finger

There should be a greatly reduced ECG waveform with a relatively flat baseline in
the upper band. There should be wavelike forms in the pulse band. If your data
resemble Figure 39B.5, proceed to the Data Recording section.

If the data show any large spikes, jitter, or large baseline drifts or if you have a flat
line instead of a clear signal on the pulse channel, then you should redo the Calibration by clicking on the **Redo Calibration** button and repeating the entire Calibration sequence.

PROCEDURE—RECORDING LESSON DATA

1. Prepare for the recording, and have the **Subject** sit down in a chair and relax,
   with arms on the arm rest.

   You will record ECG on one channel and indirect pulse pressure on another
   channel with the **Subject** in three conditions: arm relaxed, temperature change,
   and arm up. The Subject will perform tasks in the intervals between recordings.

   In order to work efficiently, read this entire section so you will know what to do
   for each recording segment.

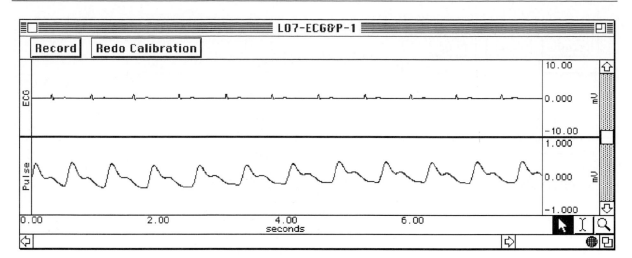

Figure 39B.5
Calibration ECG and Pulse.

The **Subject** should remain in a seated position and should continue to relax while you review the lesson.

Check the last line of the journal and note the total amount of time available for the recording. Stop each recording segment as soon as possible so you don't use an excessive amount of time (time is memory).

Hints for obtaining optimal data:

To minimize muscle (EMG) corruption of the ECG signal and baseline drift:

a. The **Subject** should keep still during all of the recording segments because the recording from the pulse transducer is sensitive to motion and the ECG recording is sensitive to EMG artifact.

b. The **Subject** should be in a relaxed state for each recording segment.

c. Initially, the **Subject's** forearms should be supported on an armrest.

d. After the **Subject** has repositioned his or her forearm, check to make sure that the cable is not pulling on the pulse transducer.

e. The recording should be suspended before the **Subject** prepares for the next recording segment.

f. Make sure electrodes do not "peel up."

Segment 1

2. Click on **Record.**

   The recording will begin.

3. Record for 15 seconds.

   Subject is seated in chair, with arms relaxed on the arm rest (seconds 0–15).

4. Click on **Suspend.**

   The recording should halt, giving you time to review the data and prepare for the next recording segment.

5. Review the data on the screen.

   If all went well, your data should look similar to Figure 39B.6, and you can proceed to Step 6.

   • If *correct,* go to Step 6.

   • If *incorrect,* click on **Redo.**

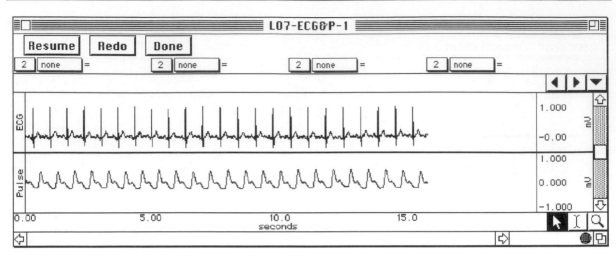

Figure 39B.6
Segment 1: Arm Relaxed.

The data would be *incorrect* if:

a. The suspend button was pressed prematurely.

b. An electrode peeled up, causing a large baseline drift, spike, or loss of signal.

c. The **Subject** has too much muscle (EMG) artifact.

If incorrect, you should redo the recording by clicking on **Redo** and repeating Steps 2–5. Note that once you press **Redo,** the data you have just recorded will be erased.

Segment 2

6. **Subject** remains seated and sticks the nonrecording hand in a plastic bucket filled with warm or cold water.

   Warning: The container for the water cannot be metal, as this is a potential danger of bypassing the electrical isolation of the BIOPAC Student Lab.

7. Click on **Resume.**

   The recording will continue from the point where it last stopped, and a marker labeled "Seated, one hand in water" will automatically come up when **Resume** is pressed.

8. Record for 30 seconds.

   Subject remains sitting with hand in water (seconds 16–45).

9. Click on **Suspend.**

   The recording should halt, giving you time to review the data and prepare for the next recording segment.

10. Review the data on the screen.

    If all went well, your data should look similar to Figure 39B.7, and you can proceed to Step 11.

    • If *correct,* and more segments are required, go to Step 11.

    • If *incorrect,* click on **Redo.**

    The data would be *incorrect* for the reasons in Step 5.

    If incorrect, you should redo the recording by clicking on **Redo** and repeating Steps 6–10. Note that once you press **Redo,** the data you have just recorded will be erased.

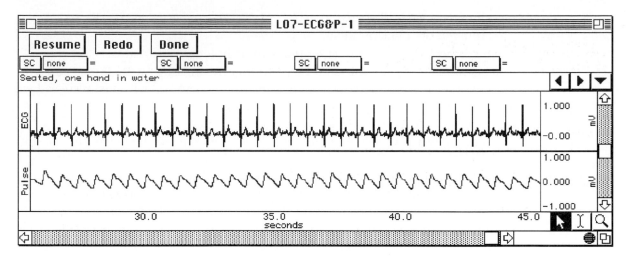

Figure 39B.7
Segment 2: One Hand in Water.

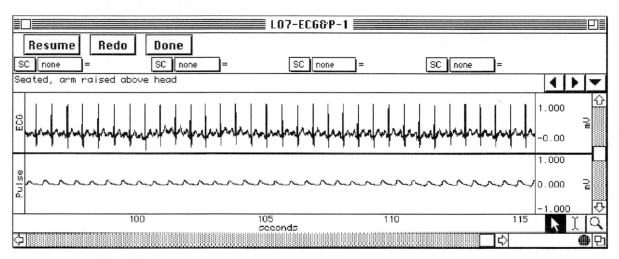

Figure 39B.8
Segment 3: Arm Raised Above Head.

Segment 3

11. **Subject** remains seated, **raises hand** to extend the arm above the head, and holds that position for the duration of the recording.

12. Click on **Resume.**

    The recording will continue from the point where it last stopped, and a marker labeled "Seated, arm raised above head" will automatically come up when Resume is pressed.

13. Record for 60 seconds.

    Subject remains seated with arm extended above head (seconds 46–105).

14. Click on **Suspend.**

    The recording should halt.

15. Review the data on the screen.

    If all went well, your data should look similar to Figure 39B.8, and you can proceed to Step 16.

- If *correct,* go to Step 16.
- If *incorrect,* click on **Redo.**

The data would be incorrect for the reasons in Step 5.

If incorrect, you should redo the recording by clicking on **Redo** and repeating Steps 11–15. Note that once you press **Redo,** the data you have just recorded will be erased.

16. Click on **Done.**

A pop-up window with four options will appear. Make your choice, and continue as directed.

If choosing the "Record from another subject" option:

   a. Attach electrodes and transducer per Set Up Steps 5–7, have the new subject sit and relax, and continue the entire lesson from Set Up Step 10.

   b. Each person will need to *use a unique filename.*

17. Remove the electrodes and the transducer.

Remove the electrode cable pinch connectors, and peel off the electrodes. Throw out the electrodes (BIOPAC electrodes are not reusable). Wash the electrode gel residue from the skin, using soap and water. The electrodes may leave a slight ring on the skin for a few hours. This is normal and does not indicate that anything is wrong.

PROCEDURE—DATA ANALYSIS

1. Enter the **Review Saved Data** mode.

Enter the Review Saved Data mode from the Lessons menu.

The data window should come up the same as shown in Figure 39B.9.

Note Channel Number (CH) designation:

Channel	Displays
CH 1	**ECG**
CH 40	**Pulse**

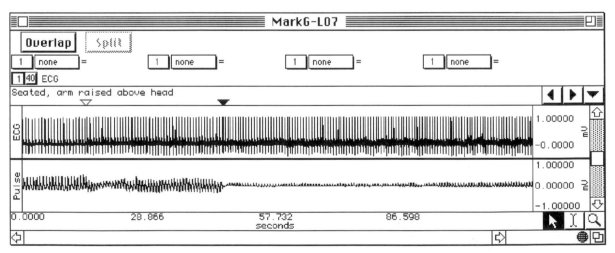

Figure 39B.9
Sample ECG and Pulse Data.

2. Set up your display window for optimal viewing of the entire recording.

   The following tools help you adjust the data window:

   Autoscale horizontal   Horizontal (Time) Scroll Bar

   Autoscale waveforms   Vertical (Amplitude) Scroll Bar

   Zoom Tool   Zoom Previous

   Grids—turn Grids ON and OFF by choosing **Preferences** from the **File** menu.

3. Set up the measurement boxes as follows:

Channel	Measurement
CH 1	ΔT
CH 1	BPM
CH 1	p-p
CH 40	p-p

   The measurement boxes are above the marker region in the data window. Each measurement has three sections: channel number, measurement type, and result. The first two sections are pull-down menus that are activated when you click on them.

   Brief definition of measurements:

   ΔT: The Delta Time measurement is the difference in time between the end and beginning of the selected area.

   BPM: The Beats Per Minute measurement first calculates the difference in time between the end and beginning of the area selected by the I-Beam tool (same as ΔT), and divides this value into 60 seconds/minute. Because the BPM only uses the time measurement of the selected area for its calculation, the BPM value is not specific to a particular channel.

   p-p: finds the maximum value in the selected area and subtracts the minimum value found in the selected area.

   Note: The "selected area" is the area selected by the **I-beam** tool (including the endpoints).

4. **Zoom** in on a small section of the Segment 1 data.

   Be sure to zoom in far enough so that you can easily measure the intervals between peaks approximately 4 cardiac cycles.

5. Using the **I-Beam** cursor, select the area between two successive R waves (one cardiac cycle).

   Try to go from R wave peak to R wave peak as precisely as possible (see Figure 39B.10).

6. Repeat the above measurements for each of the data segments.

   Record your data in the data section of the lab report.

7. Using the **I-Beam** cursor, select the area between two successive pulse peaks (one cardiac cycle) (see Figure 39B.11).

   Record your data in the data section of the lab report.

8. Repeat the above measurements for each of the data segments.

   Record your data in the data section of the lab report.

9. Select individual pulse peaks for each segment and determine their amplitudes (see Figure 39B.12).

   Use the **p-p** [CH 40] measurements.

   Record your data in the lab report.

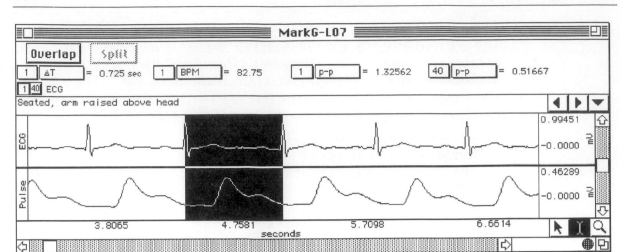

Figure 39B.10
Measurement Between R Wave Peaks.

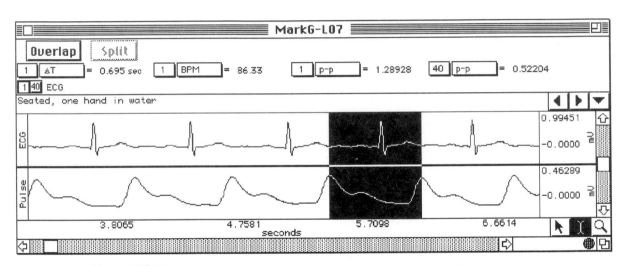

Figure 39B.11
Measurement Between Two Pulse Peaks.

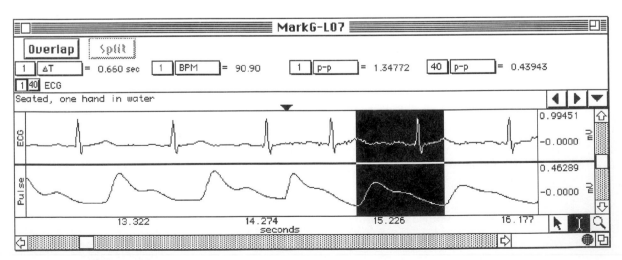

Figure 39B.12
Sample Data from Segment 1.

Important: Measure the first pulse peak after the recording is resumed. The body's homeostatic regulation of blood pressure and volume occurs quickly. The increase or decrease in your results will be dependent on the timing of your data relative to the speed of physiological adjustments.

10. Using the I-beam cursor, select the interval between the R wave and pulse peak.

    Record the time interval (ΔT) between the two peaks.

    Record your data in the lab report.

11. Save or print the data file.

    You may save the data to a floppy drive, save notes that are in the journal, or print the data file.

12. Exit the program.

LABORATORY REPORT EXERCISE 39: ECG AND PULSE

Mechanical Action of the Heart, Peripheral Pressure Pulse, and Plethysmography

A. **Data and Calculations**

Subject Profile

Name _____   Height _____

Age _____   Weight _____

Gender: Male / Female

1. Comparison of ECG with Pulse Plethysmogram (Segments 1–3)

Complete Table 39B.1 with data from three cycles from each segment and calculate the average.

TABLE 39B.1

Condition	Measurement	Channel		Cycle 1	Cycle 2	Cycle 3	Average
Arm Relaxed	R-R Interval	**ΔT**	CH 1	____	____	____	____
Segment 1	Heart Rate	**BPM**	CH 1	____	____	____	____
	Pulse Interval	**ΔT**	CH 1	____	____	____	____
	Pulse Rate	**BPM**	CH 1	____	____	____	____
Temp. Change	R-R Interval	**ΔT**	CH 1	____	____	____	____
Segment 2	Heart Rate	**BPM**	CH 1	____	____	____	____
	Pulse Interval	**ΔT**	CH 1	____	____	____	____
	Pulse Rate	**BPM**	CH 1	____	____	____	____
Arm Up	R-R Interval	**ΔT**	CH 1	____	____	____	____
Segment 3	Heart Rate	**BPM**	CH 1	____	____	____	____
	Pulse Interval	**ΔT**	CH 1	____	____	____	____
	Pulse Rate	**BPM**	CH 1	____	____	____	____

2. Relative Volume Changes (Segments 1–3)

Complete Table 39B.2 with data from each recording segment.

TABLE 39B.2

Measurement	Arm Resting Segment 1	Temperature Segment 2	Arm Up Segment 3
QRS Amplitude	_____	_____	_____
Relative Pulse Amplitude (mV)	_____	_____	_____

3. Calculation of Pulse Speed

Distance between Subject's sternum and shoulder? _____cm

Distance between Subject's shoulder and fingertip? _____cm

Total distance? _____cm

Segment 1 Data

Time between R wave and Pulse peak? _____secs

Speed? _____cm/sec

Segment 3 Data

Time between R wave and Pulse peak? _____secs

Speed? _____cm/sec

B. Questions

1. Referring to data in Table 39B.1, are the values of heart rate and pulse rate similar for each condition?

Yes   No

Explain why the values might differ or be similar.

_____

_____

_____

2. Referring to Table 39B.2 data, how much did the amplitude of the QRS complex change between conditions?

Extreme Temp − Arm Resting? _____ mV

Arm Up − Arm Resting? _____ mV

3. Referring to Table 39B.2 data, how much did the pulse volume change between arm positions?

Extreme Temp − Arm Resting? _____ mV

Arm Up − Arm Resting? _____ mV

4. Referring to Table 39B.2 data, does the amplitude of the QRS complex change with the pulse amplitudes? Why or why not?

_____

_____

_____

5. Describe one mechanism that causes changes in blood volume to your fingertip.

_____

_____

_____

6. Referring to data from section 3 of this report, how would you explain the difference in speed, if any?

_____

_____

7. Which components of the cardiac cycle (atrial systole and diastole, ventricular systole and diastole) are discernible in the pulse tracing?

_____

_____

_____

8. Would you expect the calculated pulse wave velocities of other students to be very close to, if not the same as, yours? Why or why not?

_____

_____

_____

9. Explain any amplitude or frequency changes that occurred with arm position.

_____

_____

_____

Lymphatic System

OBJECTIVES

On completion of this exercise, you should be able to:

- List the functions of the lymphatic system.
- Discuss exchange of blood plasma, interstitial fluid, and lymph.
- Describe the structure of a lymph node.
- Explain how the lymphatic system drains into the vascular system.
- Describe the gross anatomy and basic histology of the spleen.

WORD POWER

lymph (lymph—water)
cisterna chyli (cistern—storage
well; chyl—juice)
node (nod—swelling)
trabeculae (trabeculate—containing cross bars)

MATERIALS

torso models
lymphatic system chart
spleen model
compound microscope

prepared slides
lymph node
spleen

INTRODUCTION

The lymphatic system, Figure 40.1, includes lymphatic vessels and lymph nodes, tonsils, the spleen, and the thymus gland. Lymphatic vessels transport a fluid called lymph from the tissue spaces to the veins of the cardiovascular system. Scattered along a lymphatic vessel are lymph nodes that contain lymphocytes and phagocytic macrophages. The macrophages remove invading bacteria and other substances from the lymph before the lymph is returned to the blood. Although lymphocytes are a formed element of the blood, they are the main cells of the lymphatic system and colonize dense populations in lymph nodes and the spleen. **Antigens** are substances capable of causing the production of antibodies. As macrophages capture antigens in the lymph, lymphocytes are exposed to the antigens and activate the immune system to respond to the intruding cells. Some lymphocytes, the B cells, produce **antibodies** that chemically combine with and destroy specific antigens. The thymus gland is involved in the development of the functional immune system in infants. In adults, the thymus gland controls the maturation of lymphocytes. We covered the thymus gland with the endocrine system in Exercise 30 of the lab manual.

Figure 40.2 illustrates fluid circulation in the body. Pressure in blood capillaries forces **plasma** fluids and solutes out of the capillary and into the interstitial spaces. This filtrate, called **interstitial fluid**, bathes the cells with nutrients, dissolved gases, hormones, and other materials. Most of the interstitial fluid reenters the capillary owing to osmotic pressure. The remaining interstitial fluid is returned to the blood by the lymphatic system. Interstitial fluid slowly enters lymphatic vessels and becomes **lymph**. Lymphatic vessels join lymph nodes where phagocytes remove abnormal cells

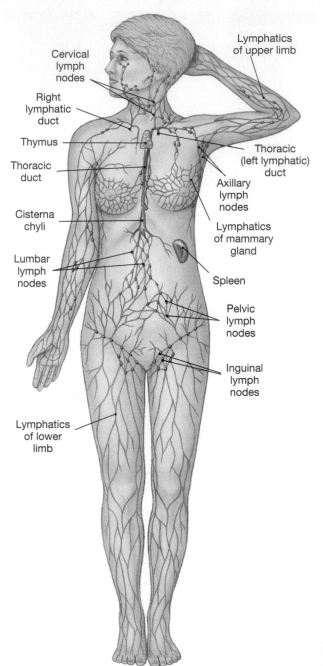

Cervical
lymph
nodes

Right
lymphatic
duct

Thymus

Thoracic
duct

Cisterna
chyli

Lumbar
lymph
nodes

Lymphatics
of lower
limb

Lymphatics
of upper limb

Thoracic
(left lymphatic)
duct

Axillary
lymph
nodes

Lymphatics
of mammary
gland

Spleen

Pelvic
lymph
nodes

Inguinal
lymph
nodes

Figure 40.1
**Components of the Lymphatic
System.**

and microbes from the lymph. A pair of lymphatic ducts join with veins near the heart
and return the lymph to the blood. Approximately three liters of fluid per day are
forced out of the capillaries and flow through lymphatic vessels as lymph.

A. Lymphatic Vessels

Lymphatic vessels occur next to the vessels of the vascular system. The major components of the lymphatic vessels and related structures are illustrated in Figure 40.3.
Lymphatic vessels, or simply lymphatics, collect fluid lost from blood capillaries, filter
the fluid in lymph nodes, and return it to the blood circulation. Lymphatic vessels are
structurally similar to veins. The vessel wall has similar layers and valves to prevent

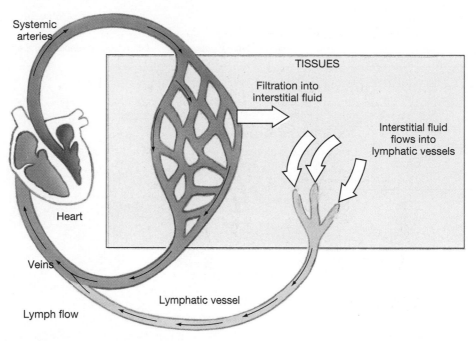

Figure 40.2
Fluid Connective Tissues.

The functional relationships between blood and lymph. Blood travels through the circulatory system, pushed by the contractions of the heart. In capillaries, hydrostatic (blood) pressure forces fluid and dissolved solutes out of the circulatory system. This fluid mixes with the interstitial fluid already in the tissue. Interstitial fluid slowly enters lymphatic vessels; now called lymph, it travels along the lymphatics and re-enters the circulatory system at one of the veins that returns blood to the heart.

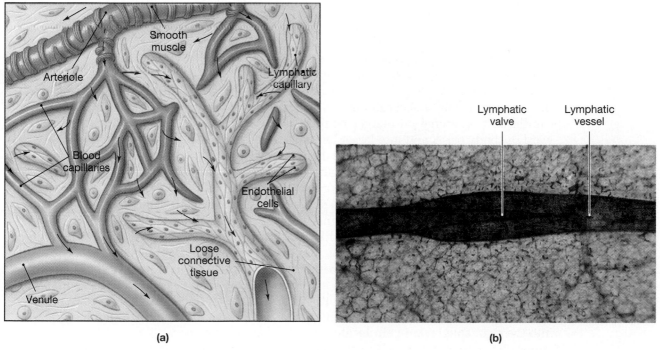

Figure 40.3
Lymphatic Capillaries.

(a) A three dimensional view of the association of blood capillaries, tissue, interstitial fluid, and lymphatic capillaries. Arrows show the direction of interstitial fluid and lymph movement. **(b)** A valve within a small lymphatic vessel. (LM ×43)

597

1. _____
2. _____
3. _____
4. _____

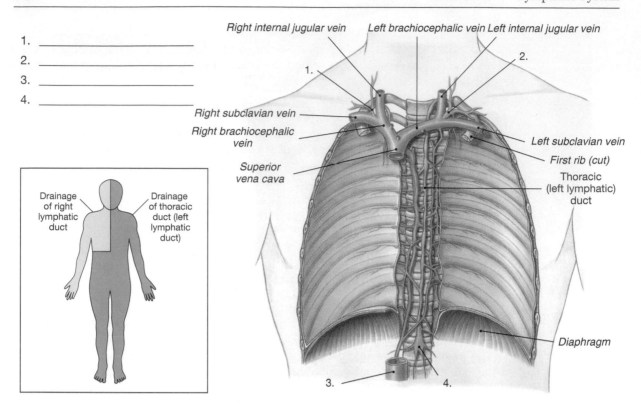

Figure 40.4
The Lymphatic Ducts and the Venous System.
The thoracic duct carries lymph originating in tissues inferior to the diaphragm and from the left side of the upper body. It empties into the left subclavian vein. The right lymphatic duct drains the right half of the body superior to the diaphragm. It empties into the right subclavian vein.

back flow of fluid. Lymphatic pressures are very low, and the lymphatic valves are close together to keep the lymph circulating toward the body trunk. Unlike blood capillaries, lymphatic capillaries are not continuous with other vessels at both ends. Lymphatic capillaries originate as closed tubules near the blood capillaries. The tubules elongate into capillaries that merge into larger lymphatic vessels. These vessels conduct lymph toward the body trunk and into larger lymphatics that empty into veins near the heart.

Two large lymphatic vessels, the **thoracic duct** and the **right lymphatic duct**, return lymph to venous circulation. Use Figures 40.1 and 40.4 to trace the drainage of the lymphatic system. Most of the lymph is returned to circulation by the thoracic duct. The thoracic duct commences at the level of the second lumbar vertebra, L_2, on the posterior abdominal wall behind the abdominal aorta. Lymphatics from the legs, pelvis, and abdomen drain into an inferior saclike portion of the duct called the **cisterna chyli (KĬ-lē)**. Lymph flows into the thoracic duct that pierces the diaphragm and ascends to the base of the heart, where it joins with the left subclavian vein to return the lymph to the blood. The only lymph that does not drain into the thoracic duct is from lymphatic vessels in the right arm and the right side of the chest, neck, and head. These areas drain into the right lymphatic duct near the right clavicle. This duct empties lymph into the right subclavian vein located near the base of the heart.

LABORATORY ACTIVITY   LYMPHATIC VESSELS

MATERIALS

torso model or lymphatic system chart

PROCEDURES

1. Locate the thoracic duct and the cisterna chyli on a chart or lab model, and label them in Figure 40.4. What areas of the body drain lymph into this duct? Where does this duct return lymph to the blood?

_____

_____

2. Locate the right lymphatic duct on a chart or lab model. Where does this vessel join the vascular system? What regions of the body drain lymph into this duct?

_____

_____

B. Lymphoid Tissues and Lymph Nodes

Two major groups of lymphatic structures occur in connective tissues: **encapsulated lymph organs** and **diffuse lymphoid tissues**. The encapsulated lymph organs include lymph nodes, the thymus gland, and the spleen. Each encapsulated organ is separated from the surrounding connective tissue by a fibrous capsule. Diffuse lymphoid tissue, such as tonsils and lymphoid nodules, do not have a defined boundary from the connective tissue.

A **lymph node** is an oval nodular organ that functions like a filter cartridge. As lymph passes though the node, phagocytes remove microbes, debris, and other antigens from the lymph. Lymph nodes are scattered throughout the lymphatic system, as depicted in Figure 40.1. Lymphatics from the lower limbs pass through a network of **inguinal** lymph nodes. **Pelvic** and **lumbar** nodes filter lymph from the pelvic and abdominal lymphatics. Many lymph nodes occur in the upper limbs and in the **axillary** and **cervical** regions. The breasts in women also have many lymphatic vessels and nodes. Infections often occur in a lymph node before they spread systemically. A swollen or painful lymph node suggests an increase in lymphocyte abundance and general immunological activity in response to antigens in the lymph nodes.

Figure 40.5 details the organization of a lymph node. Each lymph node is encased in a dense connective tissue **capsule**. Collagen fibers from the capsule extend as partitions called **trabeculae** into the interior of the lymph node. The outer **cortex** of the node contains **sinuses** full of lymph fluid. Masses of lymphocytes are produced in and surround many pale-staining **germinal centers**. Deep to the cortex is the **medulla**, where

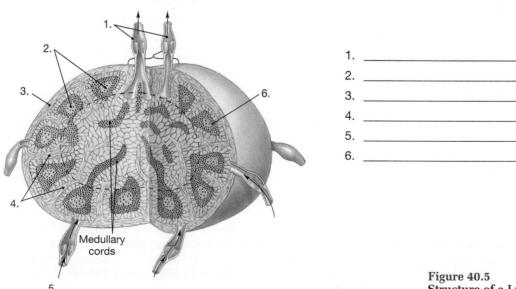

1. _____

2. _____

3. _____

4. _____

5. _____

6. _____

Figure 40.5
Structure of a Lymph Node.

medullary cords of lymphocytes extend into sinuses. Lymph enters a node in **afferent lymphatic vessels**. As lymph flows through the cortical and medullary sinuses, macrophages phagocytize abnormal cells, pathogens, and debris. Draining the lymph node are **efferent lymphatic vessels** that exit the node at a slit called the **hilus**.

Lymphocytes are the most abundant cells in the lymphatic system. This diverse group of white blood cells protects against infection. Some lymphocytes produce **antibody** proteins to destroy microbes. When lymphocytes are exposed to an antigen, the immune system responds with chemical and cellular defenses to eradicate the antigen.

Lymphoid **nodules** are found in connective tissues under the lining of the digestive, urinary, and respiratory systems. Microbes that penetrate through the exposed epithelial surface pass through lymphoid nodules where lymphocytes and macrophages destroy and remove the foreign cells from the lymph. Some nodules have a germinal center where lymphocytes are produced by cell division.

Tonsils are lymphoid nodules in the pharynx. A pair of **lingual tonsils** occurs at the posterior base of the tongue. The **palatine tonsils** are easily viewed hanging off the posterior arches of the oral cavity. A single **pharyngeal tonsil**, or **adenoids**, is located in the upper pharynx near the opening to the nasal cavity.

CLINICAL APPLICATION

The lymphatic system usually has the "upper hand" in the immunological battle against invading bacteria and viruses. Occasionally, microbes manage to populate a lymphoid nodule. When tonsils are infected, they swell and become irritated. This condition is called tonsillitis and is treated with antibiotics to control the infection. If the problem is recurrent, the tonsils are removed during a surgical procedure called a tonsillectomy. Usually, the palatine tonsils are removed. If the pharyngeal tonsils are also infected or are abnormally large, they are removed, too.

LABORATORY ACTIVITY | LYMPH NODES

MATERIALS

compound microscope
prepared slide of a lymph node

PROCEDURES

1. Label and review the structure of a lymph node in Figure 40.5.
2. Obtain a prepared slide of a lymph node. Examine the slide at low magnification, and identify the capsule and trabeculae. Within the cortex of the node, examine a germinal center at a higher magnification. What types of cells are produced in this region?

_____

_____

C. The Spleen

The **spleen** is the largest lymphatic organ in your body. It is located lateral to the stomach along the greater curvature, shown in Figure 40.6.

Histologically, the spleen is surrounded by a capsule that protects the underlying tissue called pulp. The pulp consists of two types of tissues, **red pulp** and **white pulp**. Branches of the splenic artery, called **trabecular arteries**, are surrounded by white pulp that contains large populations of lymphocytes. Capillaries of the trabecular arteries open into the red pulp. As blood flows through the red pulp, free and fixed phagocytes remove abnormal red blood cells and other antigens.

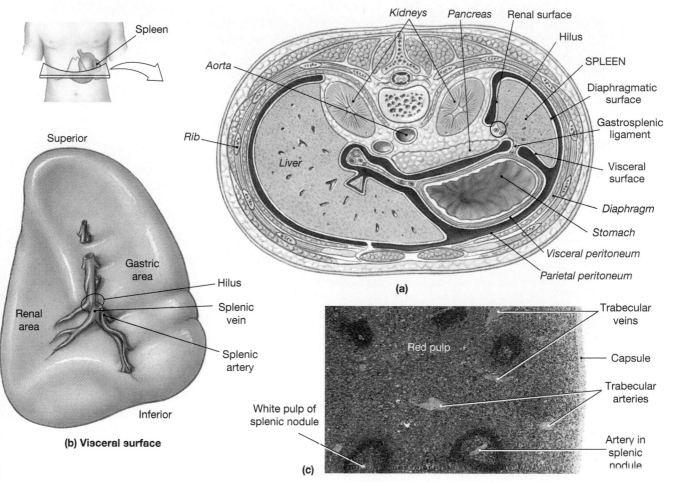

**Figure 40.6
The Spleen.**

(a) A transverse section through the trunk, showing the typical position of the spleen within the abdominopelvic cavity. The shape of the spleen roughly conforms to the shapes of adjacent organs. (b) The external appearance of the intact spleen, showing major anatomical landmarks. Compare this view with that of part (a). (c) The histological appearance of the spleen. White pulp is dominated by lymphocytes; it appears blue because the nuclei of lymphocytes stain very darkly. Red pulp contains a preponderance of red blood cells. (LM × 38)

LABORATORY ACTIVITY   THE SPLEEN

MATERIALS

torso model or chart with spleen
compound microscope
prepared slide of spleen

PROCEDURES

1. Locate the spleen on laboratory models and charts. Identify the hilus, splenic artery, and splenic vein. On the visceral surface, locate the gastric and renal areas that are in contact with the stomach and kidneys, respectively.

2. Obtain a prepared slide of the spleen. Examine the slide at low magnification, and identify the dark-stained regions of white pulp and the lighter regions of red pulp. Examine several white pulp masses for the presence of an artery.

LYMPHATIC SYSTEM CHECKLIST

This is a list of bold terms presented in Exercise 40. Use this list for review purposes after you have completed the laboratory activities.

LYMPHATIC VESSELS

• Figures 40.1 and 40.2
antigen
antibody
plasma
interstitial fluid
lymph
lymphatic vessels
   thoracic duct
   right lymphatic duct
   cisterna chyli

LYMPH TISSUES AND NODES

• Figures 40.1, 40.3, 40.4, and 40.5
encapsulated lymph organs
diffuse lymph organs
inguinal, pelvic, and lumbar nodes
axillary and cervical nodes
lymph node
   capsule
   trabeculae

   cortex
   sinuses
   germinal center
medulla
   medullary cords
afferent lymphatic vessel
efferent lymphatic vessel
hilus
lymphocytes
tonsils
   lingual tonsils
   palatine tonsils
   pharyngeal tonsil (adenoids)

SPLEEN

• Figures 40.1 and 40.6
spleen
red pulp
white pulp
trabecular arteries

LYMPHATIC SYSTEM

Laboratory Report Exercise 40

A. Matching

Match each spinal cord structure listed on the left with the correct description on the right.

1. _____ efferent vessel
2. _____ medullary cords
3. _____ cisterna chyli
4. _____ right lymphatic duct
5. _____ red pulp
6. _____ lymph node
7. _____ thoracic duct
8. _____ white pulp
9. _____ lymph
10. _____ afferent vessel

A. empties into right subclavian vein
B. empties into lymph node
C. splenic tissue with RBCs
D. fluid in lymphatic vessels
E. lymphocytes deep in node
F. empties into left subclavian vein
G. full of macrophages and lymphocytes
H. drains lymph node
I. lymphocytes surrounding trabecular artery
J. sac of thoracic duct

B. Short-Answer Questions

1. Describe the organization of a lymph node.

2. Discuss the major functions of the lymphatic system.

3. How are blood plasma, interstitial fluid, and lymph interrelated?

4. Explain how lymph is returned to the blood.

5. Describe the anatomy of the spleen.

Anatomy of the Respiratory System

OBJECTIVES

On completion of this exercise, you should be able to:

- Identify and describe the structures of the nasal cavity.
- Identify and describe the three regions of the pharynx.
- Identify and describe the cartilages and folds of the larynx.
- Describe the lining epithelium of the upper respiratory system.
- Identify and describe the structures of the trachea and the branching of the bronchial tree.
- Identify and describe the gross anatomy of the lungs.
- Identify the histological organization of the lung on a prepared slide.

WORD POWER

vestibule (vestibulum—chamber)
septum (septum—partition)
concha (concha—shell)
meatus (meatus—passageway)
pharynx (pharynx—throat)
cricoid (cricoid—ring)
glottis (glottis—aperture)
pleura (pleura—rib, side)

MATERIALS

compound microscope
prepared slides:
   nasal epithelium
   trachea
   lung tissue
lung models
human torso
hand mirror

INTRODUCTION

All cells require a constant supply of oxygen (O_2) to maintain cellular respiration. A major waste product of cellular metabolism is carbon dioxide (CO_2). The respiratory system is responsible for oxygenating the blood and removing carbon dioxide. In this exercise you will study the anatomy of the respiratory system.

The respiratory system, shown in Figure 41.1, consists of the nose, the nasal cavity and sinuses, the pharynx, the larynx, the trachea, the bronchi, and the lungs. The upper respiratory system includes the nostrils, nasal cavity, paranasal cavities, and pharynx. These structures filter, warm, and moisten air before it enters the lower respiratory system. The larynx, trachea, bronchi, and lungs comprise the lower respiratory system. The larynx regulates the opening into the lower respiratory tract and produces speech sounds. The trachea and bronchi maintain an open airway to the lungs. In the lungs, an exchange of gases occurs between millions of alveolar sacs and the blood in pulmonary capillaries.

Most of the respiratory system is lined with **pseudostratified ciliated columnar epithelium** containing many **goblet cells**. The goblet cells secrete a sticky mucus that covers the ciliated epithelium with a sticky layer to trap particles in the inspired air. The **cilia** beat to move the mucus toward the pharynx where it is swallowed. This process of mucus movement is often called the *mucus escalator*.

Figure 41.1
Components of the Respiratory System.

upper respiratory system

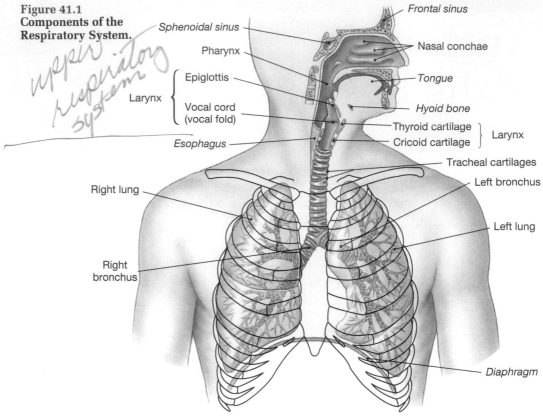

A. **The Nose and Pharynx**

The primary route for air entering the respiratory system is through two openings, the **external nares (NĀR-ēz)** or nostrils (Figure 41.2). Just inside each external naris is an expanded **vestibule (VES-ti-bul)** containing coarse hairs. The hairs help to prevent large airborne materials like dirt and insects from entering the respiratory system. A midsagittal **nasal septum** divides the **nasal cavity**. This bony septum is formed by union of the **perpendicular plate of the ethmoid bone** with the **vomer**. The external portion of the nose is composed of cartilage that forms the bridge and tip of the nose and part of the nasal septum. The nasal cavity is a vibrating or resonating chamber for the voice.

The **superior**, **middle**, and **inferior nasal conchae** are bony shelves that project from the lateral walls of the nasal cavity. The bone of each concha curls underneath and forms a tube or **meatus** that causes air to swirl in the nasal cavity. This turbulence moves inhaled air across the sticky epithelial lining where dust and debris are removed. The floor of the nasal cavity is the superior portion of the **hard palate**, formed by the maxillae, the palatine bones, and the muscular **soft palate**. Hanging off the posterior edge of the soft palate is the conical **uvula (Ū-vū-luh)**. Two openings, the **internal nares**, lead from the nasal cavity into the uppermost portion of the throat.

The throat or **pharynx (FAR-inks)** is divided into three regions, the upper nasopharynx, the middle oropharynx, and the lower laryngopharynx. The **nasopharynx (nā-zō-FAR-inks)** is located above the soft palate and serves as a passageway for airflow from the nasal cavity. A pseudostratified ciliated columnar epithelium lines the nasopharynx to warm, moisten, and clean inspired air. During swallowing, food pushes past the uvula, and the soft palate raises to prevent food from entering the nasal cavity.

Located on the posterior wall of the nasopharynx is a single **pharyngeal tonsil**, also called the adenoids. On the lateral walls are the openings of the **auditory** or

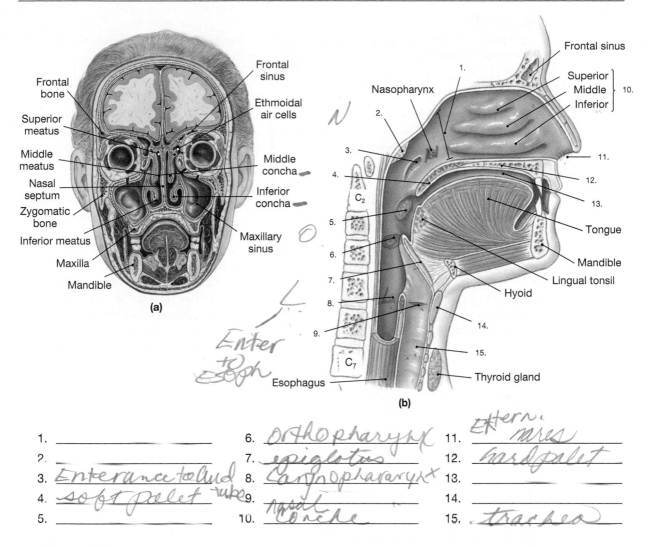

Figure 41.2
The Nose, Nasal Cavity, and Pharynx.

(a) The meatuses and the maxillary and ethmoidal sinuses. (b) The nasal cavity and pharynx as seen in sagittal section, with the nasal septum removed.

1. _____
2. _____
3. _Enterance to lud_
4. _soft palet_ tube
5. _____
6. _Orthopharynx_
7. _epiglotus_
8. _Carynopharynx_
9. _nasal_
10. _conche_
11. _Extern. nares_
12. _hard palet_
13. _____
14. _____
15. _trachea_

pharyngotympanic tubes. These tubes allow for equalization of pressure between the middle chamber of the ears and the atmosphere.

The oral cavity joins the throat at an opening called the **fauces (FAW-sēz)**. The **oropharynx** extends from the soft palate down to the epiglottis. It contains the **palatine** and **lingual** tonsils. The **laryngopharynx (la-rin-gō-FAR-inks)** is located between the hyoid bone and the entrance to the esophagus. Because food passes through the oropharynx and laryngopharynx, these areas are lined with a **stratified squamous epithelium**.

LABORATORY ACTIVITY   THE NOSE AND PHARYNX

MATERIALS

head model
respiratory system chart
hand mirror

PROCEDURES

1. Label and review the gross anatomy of the nose in Figure 41.2. Locate these structures on the models and charts available in the laboratory.

2. Label and review the three regions of the pharynx in Figure 41.2. Locate each region on the laboratory models.

3. Using a hand mirror, examine the inside of your mouth. Locate your hard and soft palates, uvula, palatine tonsils, and oropharynx.

B. **The Larynx**

The **larynx (LAR-inks)** or voice box joins the laryngopharynx with the trachea. It consists of nine cartilages (see Figure 41.3). Three of the cartilages—the thyroid cartilage, the epiglottis, and the cricoid cartilage—form the body of the larynx. The **thyroid cartilage** or Adam's apple is composed of hyaline cartilage. It is visible under the skin on the anterior neck, especially in males. The **cricoid (KRĪ-kyod) cartilage** is a ring of hyaline cartilage connecting the trachea to the base of the larynx. The **epiglottis (ep-i-GLOT-is)** is a tongue-shaped piece of elastic cartilage that falls over the opening or **glottis** during swallowing to prevent ingested food from entering the respiratory tract. The **arytenoid (ar-i-TĒ-noyd)** cartilages articulate with the superior border of the cricoid cartilage. The **corniculate (kor-NIK-ū-lāt)** cartilages articulate with the arytenoid cartilages and are involved in the opening and closing of the glottis and in the production of sound. The **cuneiform (kū-NĒ-i-form)** cartilages are club-shaped cartilages anterior to the corniculate cartilages.

Spanning the glottis from the thyroid cartilage are two pairs of folds. The upper pair of folds are the **vestibular folds**, the **false vocal cords**. These folds prevent foreign materials from entering the glottis and tightly close the glottis during coughing and sneezing. The lower pair of the folds are the **vocal cords**, also called the **true vocal cords**. To produce sounds, air is exhaled over the true vocal cords. Laryngeal muscles adjust the tension on the cords to change the pitch or frequency of the sound. Tightly stretched cords produce a high-pitched sound.

LABORATORY ACTIVITY   THE LARYNX

MATERIALS

head model
torso model
respiratory system chart

PROCEDURES

1. Review the gross anatomy of the larynx in Figure 41.3. Locate these structures on the models and charts available in the laboratory.

2. Put your finger on your thyroid cartilage and swallow. How does your thyroid cartilage move when you swallow? Is it possible to swallow and make a sound simultaneously?

3. While holding your thyroid cartilage, make a high-pitched and a low-pitched sound. Describe the tension of the true vocal cords for each sound.

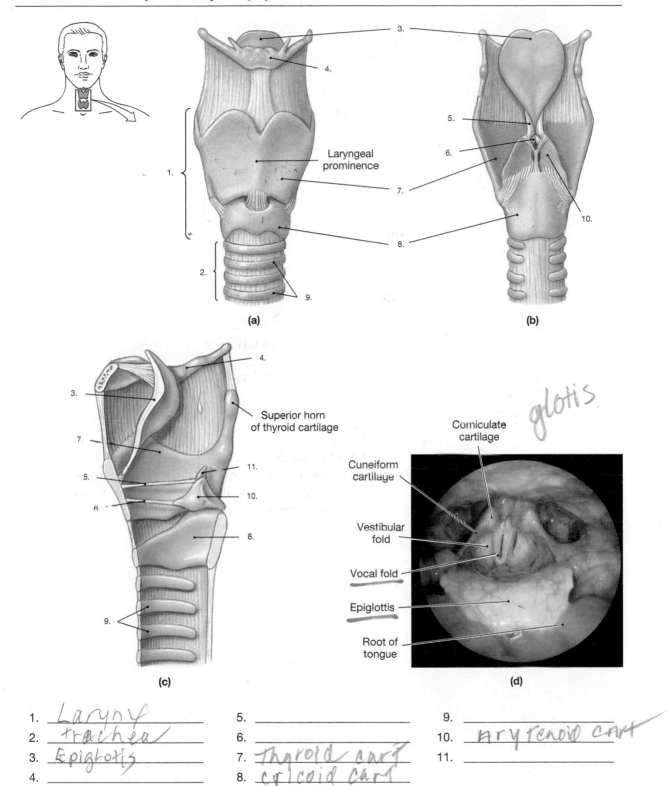

Laryngeal prominence

Superior horn of thyroid cartilage

Corniculate cartilage

glotis.

Cuneiform cartilage

Vestibular fold

Vocal fold

Epiglottis

Root of tongue

(a)

(b)

(c)

(d)

1. *Larynx*
2. *trachea*
3. *Epiglotis*
4. _____
5. _____
6. _____
7. *Thyroid cart*
8. *cricoid cart*
9. _____
10. *Arytenoid cart*
11. _____

Figure 41.3
Anatomy of the Larynx.

(a) Anterior view of the intact larynx. **(b)** Posterior view of the intact larynx. **(c)** Sagittal section through the larynx. **(d)** A fiber-optic view of the larynx with the glottis almost completely closed by the vocal folds.

C. The Trachea and the Bronchial Tree

The **trachea (TRĀ-kē-uh)** or windpipe, shown in Figure 41.4, is a tubular structure approximately 11 cm (4.25 in.) in length and 2.5 cm (1 in.) in diameter. It lies anterior to the esophagus and can be felt on the front of the neck below the thyroid cartilage of the larynx. Along the length of the trachea are 15 to 20 C-shaped pieces of hyaline cartilage, called the **tracheal cartilage**, that keep the airway open. The **trachealis** muscle holds the two sides of the C-shaped cartilage together posteriorly. This muscle allows the esophagus to expand into the trachea during swallowing. The trachea is lined with pseudostratified ciliated columnar epithelium. Palpate your trachea for the tracheal cartilage rings.

The trachea divides into the left and right **primary bronchi (BRONG-kī)**. Each bronchus branches into increasingly smaller passageways to conduct air into the lungs (see Figure 41.4). Objects that are accidentally aspirated (inhaled) often enter the right primary bronchus because it is wider and more vertical than the left bronchus. The primary bronchi branch into as many **secondary bronchi** as there are lobes of each lung. The right lung has three lobes, and each lobe receives a secondary bronchus to supply

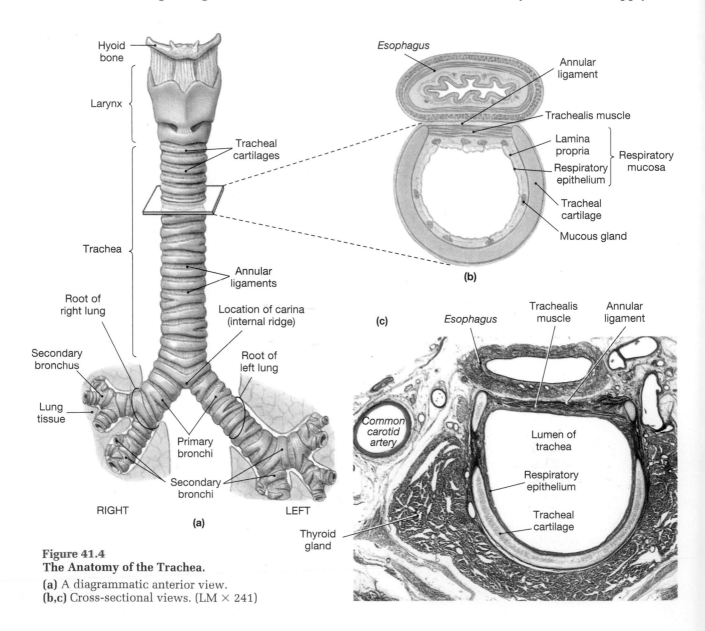

Figure 41.4
The Anatomy of the Trachea.

(a) A diagrammatic anterior view.
(b,c) Cross-sectional views. (LM × 241)

it with air. The left lung has two lobes, and two secondary bronchi branch off the left primary bronchus. This branching pattern formed by the divisions of the bronchial structures is called the **bronchial tree**.

The secondary bronchi divide into **tertiary bronchi**, also called segmental bronchi, that enter a bronchopulmonary segment of a lobe (see Figure 41.7). Smaller divisions called **bronchioles** enter lobules and branch into **respiratory bronchioles**. These narrow tubules divide into **alveolar ducts** which connect to clusters of **alveoli (al-VĒ-o-lī)** called **alveolar sacs**. Exchanges of gases occur across the membranes of the alveoli and pulmonary capillaries.

As the bronchial tree branches from the primary bronchi to the bronchioles, cartilage is gradually replaced with smooth muscle tissue. The epithelial lining of the bronchial tree also changes from pseudostratified ciliated columnar in the primary bronchus to simple squamous epithelium in the alveoli.

LABORATORY ACTIVITY   THE TRACHEA AND BRONCHIAL TREE

MATERIALS

compound microscope
prepared slide: trachea
head model
torso model
lung model
respiratory system chart

PROCEDURES

1. Review the gross anatomy of the trachea in Figure 41.4. Locate these structures on the models and charts available in the laboratory.
2. Study the bronchial tree on the laboratory models and identify the primary bronchi, secondary bronchi, tertiary bronchi, bronchioles, and respiratory bronchioles.
3. Observe a prepared slide of the trachea and locate the structures labeled in Figure 41.4.
   a. Identify the tracheal rings of hyaline cartilage.
   b. Locate the respiratory epithelium at the lumen of the trachea. What type of epithelium occurs here?

   _____

   _____

D. The Lungs

The lungs are a pair of cone-shaped organs lying in the pleural cavities (Figure 41.5). The **apex** is the conical top of each lung, and the broad inferior portion is the **base**. The heart lies on a medial concavity of the left lung called the **cardiac notch**. Each lung has a **hilus**, a medial slit where the bronchial tubes, vascularization, lymphatics, and nerves reach the lung. Each lung is divided into lobes by deep fissures. The left lung is divided by an **oblique fissure** into the **superior** and **inferior lobes**. The right lung is divided into three lobes, the **superior**, **middle**, and **inferior lobes**. The superior and middle lobes are separated by a **horizontal fissure**, and the **oblique fissure** separates the inferior lobe.

Each lung is isolated in its own serous membrane (see Figure 41.6). The **mediastinum** divides the thoracic cavity into two **pleural cavities**, each containing a lung. The pleural cavity is lined with a serous membrane, the **pleura**. The **parietal pleura**

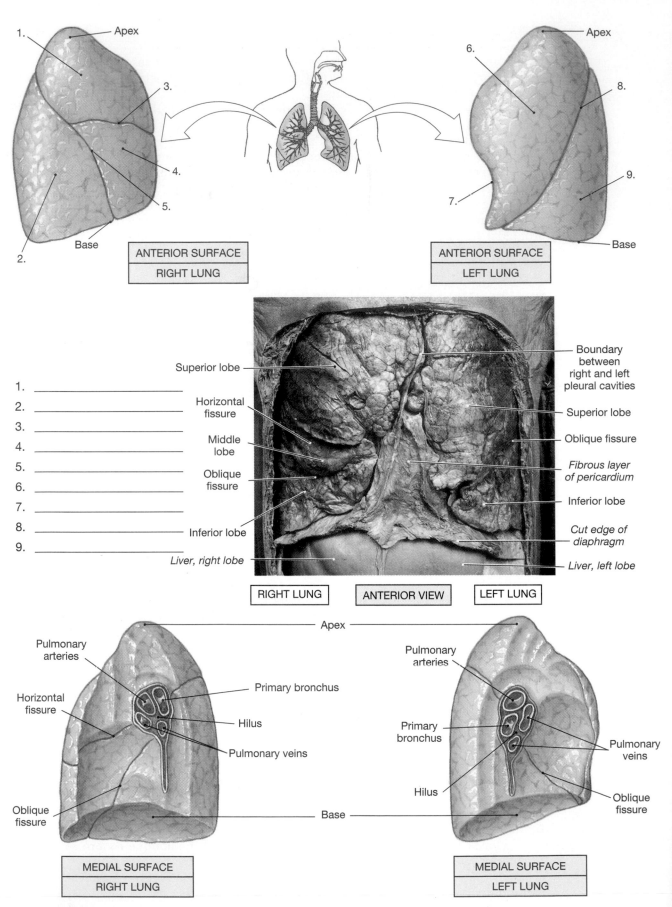

1.
2.
3.
4.
5.
6.
7.
8.
9.

Apex

Base

ANTERIOR SURFACE

RIGHT LUNG

Apex

Base

ANTERIOR SURFACE

LEFT LUNG

Superior lobe

Horizontal fissure

Middle lobe

Oblique fissure

Inferior lobe

Liver, right lobe

Boundary between right and left pleural cavities

Superior lobe

Oblique fissure

Fibrous layer of pericardium

Inferior lobe

Cut edge of diaphragm

Liver, left lobe

RIGHT LUNG   ANTERIOR VIEW   LEFT LUNG

Apex

Pulmonary arteries

Horizontal fissure

Oblique fissure

Primary bronchus

Hilus

Pulmonary veins

Base

MEDIAL SURFACE

RIGHT LUNG

Pulmonary arteries

Primary bronchus

Hilus

Pulmonary veins

Oblique fissure

MEDIAL SURFACE

LEFT LUNG

Figure 41.5
The Gross Anatomy of the Lungs.

612

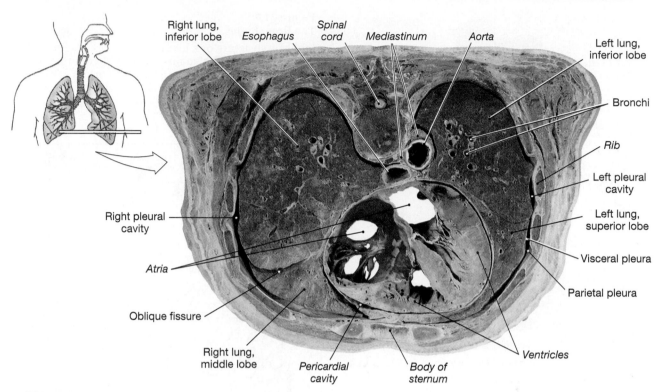

Figure 41.6
The Relationship between the Lungs and Heart.
This transverse section was taken at the level of the cardiac notch.

lines the thoracic wall, and the **visceral pleura** covers the superficial surface of the lung. The pleurae produce a slippery **serous fluid** that reduces friction and adhesion between the lungs and the thoracic wall during breathing.

The alveolar walls are constructed of **simple squamous pulmonary epithelium** (see Figure 41.7). Scattered among the epithelium are **surfactant (sur-FAK-tant) cells** that secrete an oily coating to prevent the alveoli from sticking together after exhalation. Also in the alveolar wall are macrophages that phagocytize debris. Pulmonary capillaries cover the exterior of the alveoli, and gas exchange occurs across the thin walls. Oxygen from inspired air diffuses through the simple pulmonary epithelium of the alveolar wall, moves across the basement membrane and the endothelium of the capillary, and enters the bloodstream. This membrane is about 0.5 mm thick and permits rapid gas exchange between the alveoli and blood.

LABORATORY ACTIVITY   THE LUNGS

MATERIALS

compound microscope
prepared slide: lung
head model
torso model
respiratory system chart

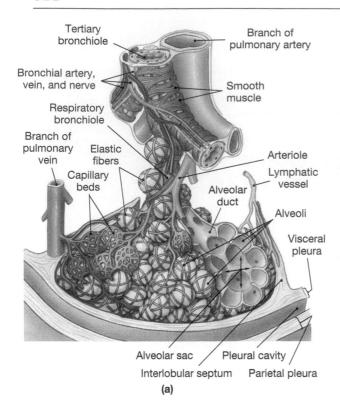

Tertiary bronchiole

Branch of pulmonary artery

Bronchial artery, vein, and nerve

Smooth muscle

Respiratory bronchiole

Branch of pulmonary vein

Elastic fibers

Capillary beds

Arteriole

Lymphatic vessel

Alveolar duct

Alveoli

Visceral pleura

Alveolar sac   Pleural cavity

Interlobular septum   Parietal pleura

(a)

Figure 41.7
The Bronchioles.

(a) Basic structure of a lobule. A network of capillaries surrounds each alveolus. **(b)** The distribution of a respiratory bronchiole supplying a portion of a lobule. **(c)** Alveolar sacs and alveoli. (LM × 42) **(d)** An SEM of the lung. Notice the open, spongy appearance of the lung tissue.

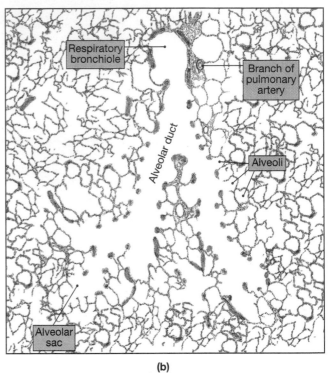

Respiratory bronchiole

Branch of pulmonary artery

Alveolar duct

Alveoli

Alveolar sac

(b)

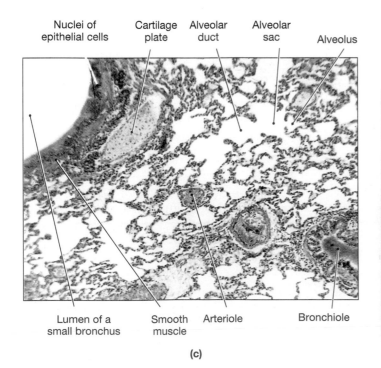

Nuclei of epithelial cells

Cartilage plate

Alveolar duct

Alveolar sac

Alveolus

Lumen of a small bronchus

Smooth muscle

Arteriole

Bronchiole

(c)

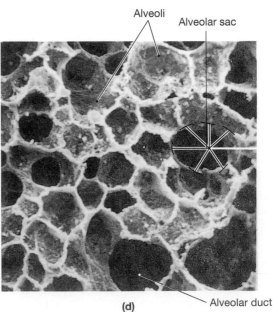

Alveoli

Alveolar sac

Alveolar duct

(d)

PROCEDURES

1. Review the gross anatomy of the lungs in Figure 41.5. Locate these structures on the models and charts available in the laboratory.
2. Observe a prepared slide of lung.
   a. Notice the thin, fragile appearance of the walls of the alveoli.
   b. Identify the alveoli and an alveolar duct using Figure 41.7 as a reference.

E. Dissection of the Respiratory System of the Cat

The dissection of the cat respiratory system will cover the nose, pharynx, larynx, trachea, bronchi, and lungs. Exercise care at all times during the dissection. Wear gloves and always be aware of your hand placement while cutting. Use Figure 41.8 as a guide to locate the described anatomy.

LABORATORY ACTIVITY   CAT RESPIRATORY SYSTEM DISSECTION

MATERIALS

cat

dissecting tools

gloves, safety glasses

PROCEDURES

1. Observe the paired **external nares**. Air enters the **nasal cavity** via these openings and passes through it to the **internal nares (choanae)**. It continues into the **pharynx,** the cavity dorsal to the soft palate and the larynx, which is partitioned into three regions **(nasopharynx, oropharynx,** and **laryngopharynx)** before it reaches the cartilaginous **larynx**. Air must pass through an opening, the **glottis,** prior to entering the larynx.
2. To observe the regions of the pharynx, use bone cutters to cut through the angle of the mandible on one side only. Pull the jaw toward the other side, leaving the salivary glands intact on the uncut side. Carefully use a scalpel to cut through soft tissues (connective tissue and muscle) until you reach the pharynx.
3. If the thoracic cavity is not open on your specimen, make a midventral incision from the caudal sternum to the jaw. Carefully cut through the chest and the neck muscles and do not penetrate the respiratory tract with your blade. Cut completely through the neck muscles to the body of a cervical vertebra. Carefully cut through any remaining connective tissue that may still be securing the larynx. Care must be taken not to cut the common carotid arteries or vagus nerves.
4. Locate the four cartilages of the larynx. The **thyroid cartilage** is the large prominent ventral cartilage deep to the ventral neck muscles. Caudal to the thyroid cartilage is the **cricoid cartilage**, which is the only complete ring of cartilage in the respiratory tract. The paired **arytenoid cartilages** occupy the dorsal surface of the larynx anterior to the cricoid cartilage. The passageway through the larynx is the glottis. A flap of cartilage called the **epiglottis** covers the glottis during swallowing and keeps food from entering the lower respiratory tract.
5. Posteriorly, the larynx is continuous with the **trachea**. Observe the **tracheal rings**, the incomplete, C-shaped rings that keep the trachea open. On the dorsal side of the trachea is the food tube, the esophagus. Laterally, the common carotid arteries, internal jugular veins, and vagus nerve pass through the neck.

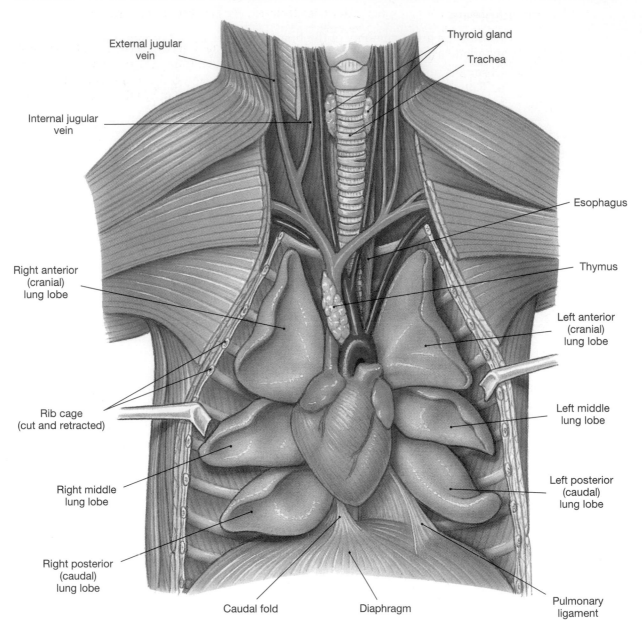

External jugular vein

Thyroid gland

Trachea

Internal jugular vein

Esophagus

Right anterior (cranial) lung lobe

Thymus

Left anterior (cranial) lung lobe

Rib cage (cut and retracted)

Left middle lung lobe

Right middle lung lobe

Left posterior (caudal) lung lobe

Right posterior (caudal) lung lobe

Caudal fold

Diaphragm

Pulmonary ligament

Figure 41.8
Respiratory System of the Cat.
Right accessory lobe posterior to heart not shown.

6. Expose and remove the larynx from the neck. Make a median incision on the dorsal surface of the larynx. Open the larynx to expose the vocal cords between the thyroid and arytenoid cartilages. The **false vocal cords** are the anterior pair of mucous membranes; the **true vocal cords** are the posterior pair. Probe the trachea through the glottis.

7. Trace the trachea to the **thoracic cavity** where it bifurcates into left and right **primary bronchi**. The bronchi penetrate the lungs and branch repeatedly to supply the alveoli with air. Notice the many lobes of the cat lungs. The right lung has four lobes, and the left has three. Use Figure 41.8 as a guide and identify each lobe. A lung is attached to other structures at its root, which consists of a

bronchus, pulmonary artery, pulmonary veins, bronchial arteries and veins, nerves, and lymphatic vessels.

8. The **pleura** surrounds each lung. This serous membrane consists of the **visceral pleura** on the lung surface and the **parietal pleura** lining the thoracic wall. These membranes can be observed as the glistening surface of the lungs and interior surface of the thoracic cavity. Between these layers is a potential space, the **pleural cavity**, which contains pleural fluid secreted by the pleura.

9. The floor of the thoracic cavity is formed by a skeletal muscle sheet, the **diaphragm**—a major muscle of respiration.

RESPIRATORY SYSTEM CHECKLIST

This is a list of bold terms presented in Exercise 41. Use this list for review purposes after you have completed the laboratory activities.

NOSE AND PHARYNX

• Figures 41.1 and 41.2
external nares, vestibule
nasal septum
nasal cavity
perpendicular plate of ethmoid
vomer
conchae, meatuses
hard and soft palates, uvula
internal nares
nasopharynx
pharyngeal tonsils
auditory tubes
pharynx
oropharynx, fauces
palatine tonsils
lingual tonsils
stratified squamous epithelium
laryngopharynx

LARYNX

• Figures 41.1 and 41.3
larynx
thyroid cartilage
cricoid cartilage
epiglottis, glottis
arytenoid cartilages
corniculate cartilages
cuneiform cartilages
vestibular folds
false vocal cords
vocal cords
true vocal cords

TRACHEA AND BRONCHIAL TREE

• Figures 41.1 and 41.4
trachea
tracheal cartilage
trachealis muscle
left and right primary bronchi
secondary and tertiary bronchi
bronchioles
respiratory bronchioles
alveolar ducts, alveolar sacs

LUNGS

• Figures 41.1, 41.5–41.7
mediastinum, pleural cavities
parietal and visceral pleura
serous fluid
apex, base, cardiac notch, hilus
horizontal and oblique fissures
superior, middle, and inferior lobes
simple squamous pulmonary
    epithelium
surfactant cells

RESPIRATORY SYSTEM OF THE CAT

• Figure 41.8
external nares
nasal cavity
internal nares (choanae)
pharynx
nasopharynx, oropharynx,
    laryngopharynx
larynx
glottis
thyroid cartilage
cricoid cartilage
arytenoid cartilages
epiglottis
trachea, tracheal rings
false vocal cords
true vocal cords
thoracic cavity
primary bronchi
pleura
visceral pleura
parietal pleura
pleura cavity
right and left cranial lobes
right and left middle lobes
right and left caudal lobes
right accessory lobe
diaphragm

Name _____   Date _____

Section _____   Number _____

ANATOMY OF THE RESPIRATORY SYSTEM
Laboratory Report Exercise 41

A. **Matching**

   Match the answer on the right with the short definition on the left.

   1. _____ C rings   **A.** voice box
   2. _____ internal intercostals   **B.** tissue of epiglottis
   3. _____ cricoid cartilage   **C.** serous membrane of lung
   4. _____ pleura   **D.** left lung
   5. _____ elastic cartilage   **E.** elevate rib cage
   6. _____ larynx   **F.** tracheal cartilage
   7. _____ diaphragm   **G.** right lung
   8. _____ cardiac notch   **H.** muscular thoracic floor
   9. _____ external intercostals   **I.** depress ribs cage
   10. _____ has three lobes   **J.** base of larynx

B. **Short-Answer Questions**

   1. List the parts of the upper and lower respiratory system.

   2. What are the functions of the superior, middle, and inferior meatuses?

   3. Where is the pharyngeal tonsil located?

   4. What is the function of goblet cells?

5. Why are the oropharynx and laryngopharynx lined with stratified squamous epithelium?

6. Which type of epithelium lines the trachea and bronchial tubes? What specialized function does this tissue perform?

EXERCISE 42

Physiology of the Respiratory System

OBJECTIVES

On completion of this exercise, you should be able to:

- Discuss the processes of pulmonary ventilation, internal respiration, and external respiration.
- Describe the activities of the respiratory muscles during inspiration and expiration.
- Define the various lung capacities and explain how they are measured.
- Show how to use a dry spirometer.

WORD POWER

spirometer (spiro—to breathe)
emphysema (en—in, physema—blowing)
apnea (apnoia—want of breath)

MATERIALS

dry spirometers
disposable mouthpieces

INTRODUCTION

The respiratory process involves three distinct phases, pulmonary ventilation, external respiration, and internal respiration. Breathing or **pulmonary ventilation** is the movement of air into and out of the lungs. This requires coordinated contractions of the diaphragm, the intercostal muscles, and the abdominal muscles. **External respiration** is the exchange of gases between the lungs and the blood. Inhaled air is rich in oxygen, and this gas constantly diffuses through the alveolar wall of the lungs and into the blood of the pulmonary capillaries. Simultaneously, carbon dioxide diffuses out of the blood and into the lungs where it is exhaled. The freshly oxygenated blood is pumped to the tissues to deliver the oxygen and take up carbon dioxide from the tissues. **Internal respiration** is the exchange of gases between the blood and the tissues. Cellular respiration is the process by which cells use oxygen to produce ATP, carbon dioxide, and water.

Pulmonary **ventilation** or breathing consists of two separate processes, inspiration and expiration. **Inspiration** is inhalation, the replenishing of oxygen rich air into the lungs. **Expiration**, or exhalation, involves emptying the carbon dioxide-laden air from the lungs into the atmosphere. The average respiratory rate is approximately 12 breaths per minute. This rate is modified by many factors such as exercise and stress, which increase respiration rates, and sleep and depression which slow respiration rates.

To ventilate the lungs, a pressure difference must exist between the surrounding air pressure, the **atmospheric pressure**, and the pressure in the thoracic cavity. Atmospheric pressure is normally 760 mm Hg or approximately 15 psi. To inhale, the pressure in the thoracic cavity must be lower than the surrounding air pressure. Boyle's law explains how changing the size of the thoracic cavity and the lungs modifies pulmonary pressures to create a pressure gradient for breathing. **Boyle's law** states that the pressure of a gas in a closed container is inversely proportional to the volume of the container. Simply put, if a sealed container is made smaller, the gas in the container is pushed together and therefore has a higher pressure.

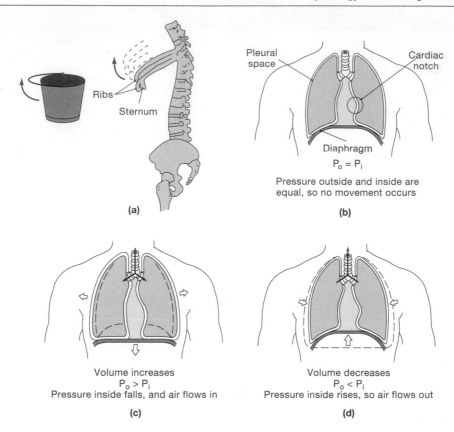

Figure 42.1
Mechanisms of Pulmonary Ventilation.

(**a**) As the ribs are elevated or the diaphragm depressed, the thoracic cavity increases in volume. (**b**) Anterior view at rest, with no air movement. (**c**) *Inhalation:* Elevation of the rib cage and contraction of the diaphragm increase the size of the thoracic cavity. Pressure decreases, and air flows into the lungs. (**d**) *Exhalation:* When the rib cage returns to its original position, the volume of the thoracic cavity decreases. Pressure rises, and air moves out of the lungs.

Figure 42.1 illustrates the mechanisms of pulmonary ventilation. When the **diaphragm** is relaxed, it is dome-shaped. Upon contraction it lowers and flattens the floor of the thoracic cavity. This results in an increase in thoracic volume and a subsequent decrease in thoracic pressure. Simultaneously, the **external intercostal muscles** contract and elevate the rib cage and further increase the thoracic volume. An expansion of the thoracic walls and a decrease in thoracic pressure cause a concurrent expansion of the lungs and a decrease in lung pressure, the **intrapulmonic (in-tra-PUL-mo-nik) pressure**. Once intrapulmonic pressure falls below atmospheric pressure, air begins to flow into the lungs.

Inspiration is an active process because it requires the contraction of several muscles to change pulmonary volumes and pressures. Expiration is essentially a passive process that occurs when the muscles for inspiration relax and the thoracic wall and elastic lung tissue recoil. During exercise, however, air may be actively exhaled by the combined contractions of the internal intercostal muscles and the abdominal muscles. The **internal intercostal muscles** depress the rib cage, and the abdominal muscles push the diaphragm higher into the thoracic cavity. Both actions decrease the thoracic volume and increase the thoracic pressure that forces more air out during exhalation.

A. LUNG VOLUMES AND CAPACITIES

The respiratory system must accommodate the oxygen demands of the body. During exercise, the respiratory system must supply the muscular system with more oxygen.

Know this

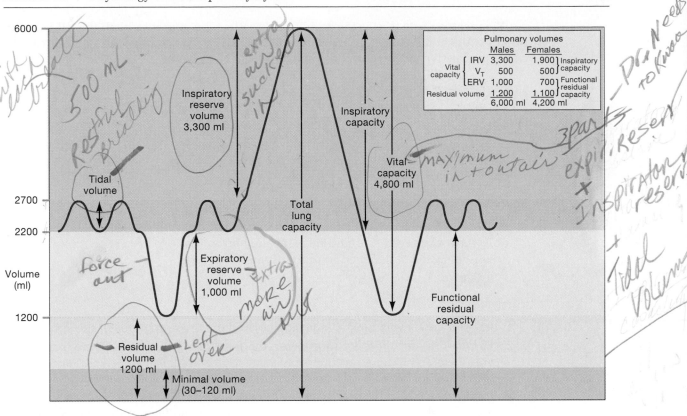

With each breath
500 mL
Restful Breathing

extra air sucked into

Dr. Needs to know

3 parts expir. resers + Inspiratory reserv + Tidal volume

Pulmonary volumes

		Males	Females	
Vital capacity	IRV	3,300	1,900	Inspiratory capacity
	V$_T$	500	500	
	ERV	1,000	700	Functional residual capacity
Residual volume		1,200	1,100	
		6,000 ml	4,200 ml	

Maximum in + out air

force out

Expiratory reserve volume 1,000 ml

extra more air out

Left over

Residual volume 1200 ml

Minimal volume (30–120 ml)

Figure 42.2
Respiratory Volumes and Capacities.

The graph diagrams the relationships between the respiratory volumes and capacities of an average male.

Go to this

The respiratory rate increases as does the volume of inhaled and exhaled gases. In this section you will measure a variety of respiratory volumes.

An instrument called a **spirometer (spī-ROM-e-ter)** is used to measure respiratory volumes. The dry spirometer is a hand-held unit with disposable mouthpieces. This unit employs a small turbine that spins to measure the amount of air you *exhale*. Keep in mind during your volume measurements that the volumes vary according to the individual's sex, height, age, and general physical condition. Refer to Figure 42.2 and compare the relationships of the following lung volumes.

Tidal volume (TV) is the amount of air one inspires and exhales during normal resting breathing. Tidal volume averages 500 ml. Additional air can be inhaled or exhaled beyond the tidal volume. The **inspiratory reserve volume** (IRV) is the amount of air that can be forcibly inspired above a normal inhalation. This volume averages 3,300 ml. The amount of air that can be forcefully exhaled after a normal exhalation is the **expiratory reserve volume** (ERV). The ERV averages 1,000 ml.

Vital capacity (VC) is the maximum amount of air that can be exhaled from the lungs after a maximum inhalation. This volume averages 4,500–5,500 ml and includes the combined volumes of the IRV, TV, and ERV. VC = IRV + TV + ERV.

The respiratory system always contains some air. The **residual volume** is the amount of air that cannot be forcefully exhaled from the lungs. Surfactant produced by the septal cells of the alveoli prevents the alveoli from collapsing completely during exhalation. Since the alveoli are not allowed to completely empty, they always maintain a resident volume of air, which averages 1,200 ml.

To calculate the **total lung capacity** (TLC), the vital capacity is added to the residual volume. TLC = VC + RV. Total lung capacity averages 5,900 ml.

Respiratory rate (RR) is the number of breaths taken per minute. If the RR is multiplied by the tidal volume, the **minute volume** (MV) is obtained, the amount of air exchanged between the lungs and the environment in 1 min. MV = TV × RR.

LABORATORY ACTIVITY   SPIROMETRY

MATERIALS

    dry spirometer
    disposable mouthpieces

Spirometry safety notices

- Do not use the spirometer if you have a cold or a communicable disease.
- Always use a clean mouthpiece on the spirometer. Do not reuse a mouthpiece which has been removed from a spirometer.
- Exhale only into the dry spirometer.

PROCEDURES

1. Insert a new, clean plastic mouthpiece onto the breathing tube of the spirometer.
2. *Exhale only into the dry spirometer;* the instrument cannot measure inspiratory volumes. To obtain as accurate a reading as possible, use a nose clip or your fingers to pinch your nose closed while exhaling into the spirometer.
3. Before using a spirometer you will need to reset the dial face to zero. This is accomplished by turning the silver ring around the face of the dial (see Figure 42.3). As you turn this ring, you will notice the scale moving also. Set the zero even with the needle of the dial.
4. You will conduct the test for each volume three times and then calculate the average volume of that respiratory volume.

Tidal Volume (TV)

1. Set the dial face to 1,000. Tidal volume is a small volume, and the scale on most dry spirometers is not graded before the 1,000-ml setting.
2. Take a normal breath and exhale normally into the spirometer. The exhalation should not be forcible, but should be more of a sigh. Repeat this test two more times and calculate your average tidal volume. Record your data in Table 42.1.

**Figure 42.3
Dry Spirometer.**

TABLE 42.1   **Spirometry Data**

Volume	Reading 1	Reading 2	Reading 3	Average
Tidal volume				
Expiratory reserve volume	1500	1000	1500	1300
Vital capacity	2500	2400	2400	2450
Respiratory rate				
Minute volume (calculated)				
Inspiratory reserve volume (calculated)				

1500
1500
1000
3900 00

3900 00

Expiratory reserve volume (ERV)

1. Set the dial face to zero.
2. After a normal exhalation:
   a. stop breathing.
   b. pinch your nose with your fingers or a nose clip.
   c. forcibly exhale all the remaining air.
3. Repeat twice and then record your readings in Table 42.1.

Vital capacity

1. Set the dial face to zero.
2. Inhale maximally once and then exhale maximally.
3. Inhale maximally and then:
   a. insert the mouthpiece into your mouth.
   b. pinch your nose with your fingers or a nose clip.
   c. forcibly exhale all the air from your lungs.
4. Repeat the measurement twice and record your data in Table 42.1.
5. Compare your vital capacity with the predicted vital capacities for females and males in Table 42.2. To use these tables, find your height across the top of the chart and your age along the side of the chart. (Multiply your height in inches by 2.54 to get your height in centimeters.) Follow the age row across and the height column down. The box where the two intersect gives the value of the average vital capacity of the population of individuals your height and sex.

   Your vital capacity might differ from the average listed in the chart. Genetics has an influence on your potential lung capacities. Damage from smoking or environmental factors such as air pollution decrease vital capacity. Cardiovascular exercises such as swimming and jogging increase lung volumes.

Respiratory rate (RR) determination

1. Have your partner observe your breathing rate while you read a textbook. Record the number of respirations that occur in 60 s.
2. Record your minute respiratory rate in Table 42.1.
3. Calculate the minute volume by multiplying your tidal volume by your respiratory rate. Enter this calculation in Table 42.1.

Inspiratory reserve volume (IRV) calculation

Although the dry spirometer cannot measure your IRV, it can be calculated from other volumes. IRV = VC − (TV + ERV). Calculate your IRV and enter it in Table 42.1.

TABLE 42.2A   Predicted Vital Components—Females

Height in centimeters and inches

Age	cm in.	152 59.8	154 60.6	156 61.4	158 62.2	160 63.0	162 63.7	164 64.6	166 65.4	168 66.1	170 66.9	172 67.7	174 68.5	176 69.3	178 70.1	180 70.9	182 71.7	184 72.4	186 73.2	188 74.0
16		3,070	3,110	3,150	3,190	3,230	3,270	3,310	3,350	3,390	3,430	3,470	3,510	3,550	3,590	3,630	3,670	3,715	3,755	3,800
17		3,055	3,095	3,135	3,175	3,215	3,255	3,295	3,335	3,375	3,415	3,455	3,495	3,535	3,575	3,615	3,655	3,695	3,740	3,780
18		3,040	3,080	3,120	3,160	3,200	3,240	3,280	3,320	3,360	3,400	3,440	3,480	3,520	3,560	3,600	3,640	3,680	3,720	3,760
20		3,010	3,050	3,090	3,130	3,170	3,210	3,250	3,290	3,330	3,370	3,410	3,450	3,490	3,525	3,565	3,605	3,645	3,695	3,720
22		2,980	3,020	3,060	3,095	3,135	3,175	3,215	3,255	3,290	3,330	3,370	3,410	3,450	3,490	3,530	3,570	3,610	3,650	3,685
24		2,950	2,985	3,025	3,065	3,100	3,140	3,180	3,220	3,260	3,300	3,335	3,375	3,415	3,455	3,490	3,530	3,570	3,610	3,650
26		2,920	2,960	3,000	3,035	3,070	3,110	3,150	3,190	3,230	3,265	3,300	3,340	3,380	3,420	3,455	3,495	3,530	3,570	3,610
28		2,890	2,930	2,965	3,000	3,040	3,070	3,115	3,155	3,190	3,230	3,270	3,305	3,345	3,380	3,420	3,460	3,495	3,535	3,570
30		2,860	2,895	2,935	2,970	3,010	3,045	3,085	3,120	3,160	3,195	3,235	3,270	3,310	3,345	3,385	3,420	3,460	3,495	3,535
32		2,825	2,865	2,900	2,940	2,975	3,015	3,050	3,090	3,125	3,160	3,200	3,235	3,275	3,310	3,350	3,385	3,425	3,460	3,495
34		2,795	2,835	2,870	2,910	2,945	2,980	3,020	3,055	3,090	3,130	3,165	3,200	3,240	3,275	3,310	3,350	3,385	3,425	3,460
36		2,765	2,805	2,840	2,875	2,910	2,950	2,985	3,020	3,060	3,095	3,130	3,165	3,205	3,240	3,275	3,310	3,350	3,385	3,420
38		2,735	2,770	2,810	2,845	2,880	2,915	2,950	2,990	3,025	3,060	3,095	3,130	3,170	3,205	3,240	3,275	3,310	3,350	3,385
40		2,705	2,740	2,775	2,810	2,850	2,885	2,920	2,955	2,990	3,025	3,060	3,095	3,135	3,170	3,205	3,240	3,275	3,310	3,345
42		2,675	2,710	2,745	2,780	2,815	2,850	2,885	2,920	2,955	2,990	3,025	3,060	3,100	3,135	3,170	3,205	3,240	3,275	3,310
44		2,645	2,680	2,715	2,750	2,785	2,820	2,855	2,890	2,925	2,960	2,995	3,030	3,060	3,095	3,130	3,165	3,200	3,235	3,270
46		2,615	2,650	2,685	2,715	2,750	2,785	2,820	2,855	2,890	2,925	2,960	2,995	3,030	3,060	3,095	3,130	3,165	3,200	3,235
48		2,585	2,620	2,650	2,685	2,715	2,750	2,785	2,820	2,855	2,890	2,925	2,960	2,995	3,030	3,060	3,095	3,130	3,160	3,195
50		2,555	2,590	2,625	2,655	2,690	2,720	2,755	2,785	2,820	2,855	2,890	2,925	2,955	2,990	3,025	3,060	3,090	3,125	3,155
52		2,525	2,555	2,590	2,625	2,655	2,690	2,720	2,755	2,790	2,820	2,855	2,890	2,925	2,955	2,990	3,020	3,055	3,090	3,125
54		2,495	2,530	2,560	2,590	2,625	2,655	2,690	2,720	2,755	2,790	2,820	2,855	2,885	2,920	2,950	2,985	3,020	3,050	3,085
56		2,460	2,495	2,525	2,560	2,590	2,625	2,655	2,690	2,720	2,755	2,790	2,820	2,855	2,885	2,920	2,950	2,980	3,015	3,045
58		2,430	2,460	2,495	2,525	2,560	2,590	2,625	2,655	2,690	2,720	2,750	2,785	2,815	2,850	2,880	2,920	2,945	2,975	3,010
60		2,400	2,430	2,460	2,495	2,525	2,560	2,590	2,625	2,655	2,685	2,720	2,750	2,780	2,810	2,845	2,875	2,915	2,940	2,970
62		2,370	2,405	2,435	2,465	2,495	2,525	2,560	2,590	2,620	2,655	2,685	2,715	2,745	2,775	2,810	2,840	2,870	2,900	2,935
64		2,340	2,370	2,400	2,430	2,465	2,495	2,525	2,555	2,585	2,620	2,650	2,680	2,710	2,740	2,770	2,805	2,835	2,865	2,895
66		2,310	2,340	2,370	2,400	2,430	2,460	2,495	2,525	2,555	2,585	2,615	2,645	2,675	2,705	2,735	2,765	2,800	2,825	2,860
68		2,280	2,310	2,340	2,370	2,400	2,430	2,460	2,490	2,520	2,550	2,580	2,610	2,640	2,670	2,700	2,730	2,760	2,795	2,820
70		2,250	2,280	2,310	2,340	2,370	2,400	2,425	2,455	2,485	2,515	2,545	2,575	2,605	2,635	2,665	2,695	2,725	2,755	2,780
72		2,220	2,250	2,280	2,310	2,335	2,365	2,395	2,425	2,455	2,480	2,510	2,540	2,570	2,600	2,630	2,660	2,685	2,715	2,745
74		2,190	2,220	2,245	2,275	2,305	2,335	2,360	2,390	2,420	2,450	2,475	2,505	2,535	2,565	2,590	2,620	2,650	2,680	2,710

TABLE 42.2B   **Predicted Vital Capacities—Males**

								Height in centimeters and inches											
	cm	152	154	156	158	160	162	164	166	168	170	172	174	176	178	180	182	184	186	188
Age	in.	59.8	60.6	61.4	62.2	63.0	63.7	64.6	65.4	66.1	66.9	67.7	68.5	69.3	70.1	70.9	71.7	72.4	73.2	74.0
16		3,920	3,975	4,025	4,075	4,130	4,180	4,230	4,285	4,335	4,385	4,440	4,490	4,540	4,590	4,645	4,695	4,745	4,800	4,850
18		3,890	3,940	3,995	4,045	4,095	4,145	4,200	4,250	4,300	4,350	4,405	4,455	4,505	4,555	4,610	4,660	4,710	4,760	4,815
20		3,860	3,910	3,960	4,015	4,065	4,115	4,165	4,215	4,265	4,320	4,370	4,420	4,470	4,520	4,570	4,625	4,675	4,725	4,775
22		3,830	3,880	3,930	3,980	4,030	4,080	4,135	4,185	4,235	4,285	4,335	4,385	4,435	4,485	4,535	4,585	4,635	4,685	4,735
24		3,785	3,835	3,885	3,935	3,985	4,035	4,085	4,135	4,185	4,235	4,285	4,330	4,380	4,430	4,480	4,530	4,580	4,630	4,680
26		3,755	3,805	3,855	3,905	3,955	4,000	4,050	4,100	4,150	4,200	4,250	4,300	4,350	4,395	4,445	4,495	4,545	4,595	4,645
28		3,725	3,775	3,820	3,870	3,920	3,970	4,020	4,070	4,115	4,165	4,215	4,265	4,310	4,360	4,410	4,460	4,510	4,555	4,605
30		3,695	3,740	3,790	3,840	3,890	3,935	3,985	4,035	4,080	4,130	4,180	4,230	4,275	4,325	4,375	4,425	4,470	4,520	4,570
32		3,665	3,710	3,760	3,810	3,855	3,905	3,950	4,000	4,050	4,095	4,145	4,195	4,240	4,290	4,340	4,385	4,435	4,485	4,530
34		3,620	3,665	3,715	3,760	3,810	3,855	3,905	3,950	4,000	4,045	4,095	4,140	4,190	4,225	4,285	4,330	4,380	4,425	4,475
36		3,585	3,635	3,680	3,730	3,775	3,825	3,870	3,920	3,965	4,010	4,060	4,105	4,155	4,200	4,250	4,295	4,340	4,390	4,435
38		3,555	3,605	3,650	3,695	3,745	3,790	3,840	3,885	3,930	3,980	4,025	4,070	4,120	4,165	4,210	4,260	4,305	4,350	4,400
40		3,525	3,575	3,620	3,665	3,710	3,760	3,805	3,850	3,900	3,945	3,990	4,035	4,085	4,130	4,175	4,220	4,270	4,315	4,360
42		3,495	3,540	3,590	3,635	3,680	3,725	3,770	3,820	3,865	3,910	3,955	4,000	4,050	4,095	4,140	4,185	4,230	4,280	4,325
44		3,450	3,495	3,540	3,585	3,630	3,675	3,725	3,770	3,815	3,860	3,905	3,950	3,995	4,040	4,085	4,130	4,175	4,220	4,270
46		3,420	3,465	3,510	3,555	3,600	3,645	3,690	3,735	3,780	3,825	3,870	3,915	3,960	4,005	4,050	4,095	4,140	4,185	4,230
48		3,390	3,435	3,480	3,525	3,570	3,615	3,655	3,700	3,745	3,790	3,835	3,880	3,925	3,970	4,015	4,060	4,105	4,150	4,190
50		3,345	3,390	3,430	3,475	3,520	3,565	3,610	3,650	3,695	3,740	3,785	3,830	3,870	3,915	3,960	4,005	4,050	4,090	4,135
52		3,315	3,353	3,400	3,445	3,490	3,530	3,575	3,620	3,660	3,705	3,750	3,795	3,835	3,880	3,925	3,970	4,010	4,055	4,100
54		3,285	3,325	3,370	3,415	3,455	3,500	3,540	3,585	3,630	3,670	3,715	3,760	3,800	3,845	3,890	3,930	3,975	4,020	4,060
56		3,255	3,295	3,340	3,380	3,425	3,465	3,510	3,550	3,595	3,640	3,680	3,725	3,765	3,810	3,850	3,895	3,940	3,980	4,025
58		3,210	3,250	3,290	3,335	3,375	3,420	3,460	3,500	3,545	3,585	3,630	3,670	3,715	3,755	3,800	3,840	3,880	3,925	3,965
60		3,175	3,220	3,260	3,300	3,345	3,385	3,430	3,470	3,500	3,555	3,595	3,635	3,680	3,720	3,760	3,805	3,845	3,885	3,930
62		3,150	3,190	3,230	3,270	3,310	3,350	3,390	3,440	3,480	3,520	3,560	3,600	3,640	3,680	3,730	3,770	3,810	3,850	3,890
64		3,120	3,160	3,200	3,240	3,280	3,320	3,360	3,400	3,440	3,490	3,530	3,570	3,610	3,650	3,690	3,730	3,770	3,810	3,850
66		3,070	3,110	3,150	3,190	3,230	3,270	3,310	3,350	3,390	3,430	3,470	3,510	3,550	3,600	3,640	3,680	3,720	3,760	3,800
68		3,040	3,080	3,120	3,160	3,200	3,240	3,280	3,320	3,360	3,400	3,440	3,480	3,520	3,560	3,600	3,640	3,680	3,720	3,760
70		3,010	3,050	3,090	3,130	3,170	3,210	3,250	3,290	3,330	3,370	3,410	3,450	3,480	3,520	3,560	3,600	3,640	3,680	3,720
72		2,980	3,020	3,060	3,100	3,140	3,180	3,210	3,250	3,290	3,330	3,370	3,410	3,450	3,490	3,530	3,570	3,610	3,650	3,680
74		2,930	2,970	3,010	3,050	3,090	3,130	3,170	3,200	3,240	3,280	3,320	3,360	3,400	3,440	3,470	3,510	3,550	3,590	3,630

Predicted Vital Capacities (a) Female (b) Male. From Archives of Environmental Health, Volume 12, pp. 146–189, February 1966. Reprinted with permission of the Helen Dwight Reid Educational Foundation. Published by Heldref Publications, 4000 Albemarle St., N.W., Washington, D.C. 20016. Copyright © 1966.

RESPIRATORY PHYSIOLOGY CHECKLIST

This is a list of bold terms presented in Exercise 42. Use this list for review purposes after you have completed the laboratory activities.

RESPIRATORY PHYSIOLOGY

- Figures 42.1 and 42.2
pulmonary ventilation
inspiration
expiration
Boyle's law
external intercostal muscles
diaphragm
internal intercostal muscles
intrapulmonic pressure
atmospheric pressure

spirometer
tidal volume (TV)
inspiratory reserve volume (IRV)
$IRV = VC - (TC + ERV)$
expiratory reserve volume (ERV)
vital capacity (VC)
residual volume
reserve volume
total lung capacity
respiratory rate, minute volume
external and internal respiration

RESPIRATORY PHYSIOLOGY

Laboratory Report Exercise 42

A. Matching

Match each term listed on the left with the correct description on the right.

1. _____ vital capacity
2. _____ tidal volume
3. _____ IRV
4. _____ ERV
5. _____ residual volume
6. _____ total lung volume
7. _____ respiratory rate
8. _____ minute volume
9. _____ spirometer
10. _____ pulmonary ventilation

A. respiratory rate multiplied by tidal volume

B. volume of air that can be forcefully exhaled after a normal exhalation

C. instrument used to measure respiratory volumes

D. amount of air normally inhaled or exhaled

E. inspiration and expiration

F. IRV + TV + ERV

G. number of breaths per minute

H. volume of air that can be forcefully inhaled after a normal inhalation

I. vital capacity + residual volume

J. volume of air that cannot be forcefully exhaled

B. Short-Answer Questions

1. Define the following:
   a. pulmonary ventilation

   b. intrapulmonary pressure

   c. intrapleural pressure

    d. inspiration

    e. expiration

2. Describe how to calculate the inspiratory reserve volume.

3. Use Boyle's law to explain the process of pulmonary ventilation.

4. How do external and internal respiration differ?

Anatomy of the Digestive System

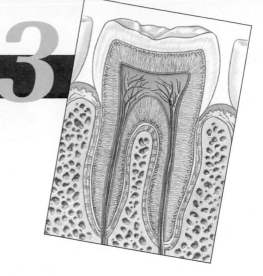

OBJECTIVES

On completion of this exercise, you should be able to:

- Identify the major layers and tissues of the digestive tract.
- Identify all digestive anatomy on lab models, charts, and exhibits.
- Identify the histological structure of the various digestive organs.
- Trace the secretion of bile from the liver to the duodenum.
- List the organs of the digestive tract and the accessory organs that empty secretions into them.

WORD POWER

mucosa (muco—mucus)
deciduous (decid—falling off)
omentum (oment—fat skin)
plicae (plico—fold)
cecum (cec—blind)

MATERIALS

torso model
anatomical charts
digestive tract (ileum) model
head model
dentition model set
stomach model
intestinal model
liver and gallbladder model
pancreas model

compound microscope
microscope slides:
    stomach
    duodenum
    ileum
    liver
    pancreas
dissected cat for demonstrations

INTRODUCTION

The digestive system functions in processing food into usable forms for cellular respiration. The major processes include (1) ingestion of food into the mouth, (2) movement of food through the digestive tract, (3) digestion of food by mechanical and enzymatic activities, (4) absorption of nutrients into the blood, and (5) formation and elimination of indigestible material and waste. To accomplish these functions the digestive system is highly specialized. Food passes through a tubular **digestive tract** extending from the mouth to the anus, and each organ of the tract has a specific function in the digestive process. *Accessory organs* occur outside the digestive tract. The salivary glands, liver, gallbladder, and pancreas manufacture and squirt enzymes, hormones, and other compounds onto the inner lining of the digestive tract. Food does not pass through the accessory organs of the digestive system. Figure 43.1 illustrates the organs of the digestive system.

A. Anatomy of the Digestive Tract

Examine the inside of your cheek with your tongue. What do you feel? The mucosal layer lining your mouth and the rest of your digestive tract is a mucous membrane that is kept wet. Glands drench the lining epithelium with enzymes, mucus, hormones, pH buffers, and other compounds to orchestrate the step-by-step breakdown of food as it

631

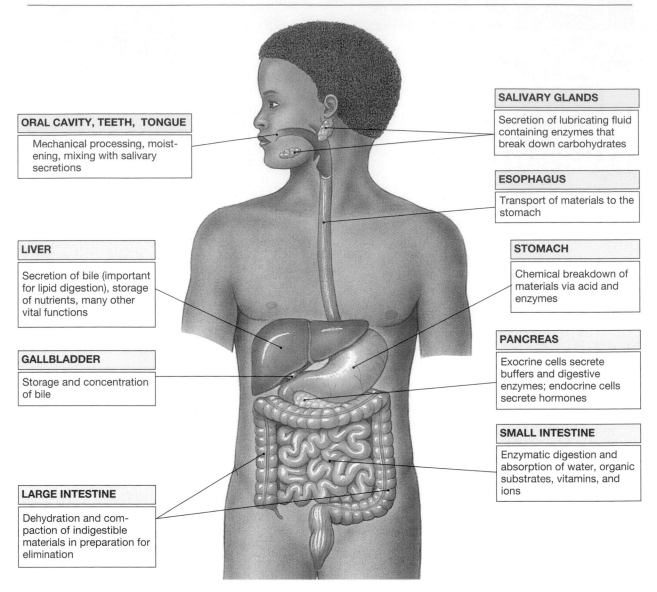

ORAL CAVITY, TEETH, TONGUE

Mechanical processing, moistening, mixing with salivary secretions

LIVER

Secretion of bile (important for lipid digestion), storage of nutrients, many other vital functions

GALLBLADDER

Storage and concentration of bile

LARGE INTESTINE

Dehydration and compaction of indigestible materials in preparation for elimination

SALIVARY GLANDS

Secretion of lubricating fluid containing enzymes that break down carbohydrates

ESOPHAGUS

Transport of materials to the stomach

STOMACH

Chemical breakdown of materials via acid and enzymes

PANCREAS

Exocrine cells secrete buffers and digestive enzymes; endocrine cells secrete hormones

SMALL INTESTINE

Enzymatic digestion and absorption of water, organic substrates, vitamins, and ions

Figure 43.1
Components of the Digestive System.

This figure introduces the accessory organs and major regions of the digestive tract together with their primary functions.

passes through your mouth, pharynx, esophagus, stomach, small intestine, large intestine, and finally, rectum.

Four major tissue layers compose the digestive tract, the **mucosa**, **submucosa**, **muscularis externa**, and **serosa**, see Figure 43.2. Although the basic histological structure of the tract is similar along its entire length, each region has anatomical specializations reflecting that region's role in the digestive process. Figure 43.2 illustrates the anatomy of the digestive tract. Figure 43.3 is a micrograph of the ileum of the small intestine in transverse section. Locate each layer of the digestive tract in this figure.

The **mucosa** lines the digestive **lumen** and is in contact with the food being pushed through the tract. Three distinct mucosal layers can be identified: the **digestive epithelium**, a sheet of connective tissue called the **lamina propria (LA-mi-nuh PRO-prē-uh)**, and a thin muscle layer, the **muscularis (mus-kū-LAR-is) mucosae**. The digestive

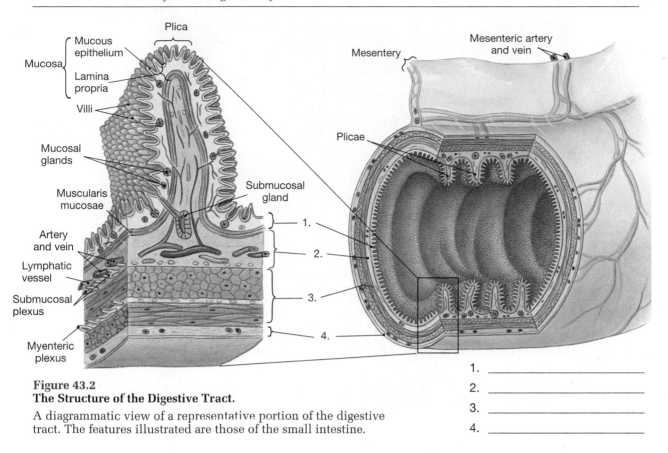

Figure 43.2
The Structure of the Digestive Tract.

A diagrammatic view of a representative portion of the digestive
tract. The features illustrated are those of the small intestine.

1. _____

2. _____

3. _____

4. _____

epithelium is the only layer exposed to the lumen of the tract. From the mouth to the
esophagus the digestive epithelium is **stratified squamous epithelium,** protecting the
mucosa from abrasion during swallowing. The stomach and small and large intestines
are lined with **simple columnar epithelium.** Food in these parts of the tract is liquid
and less abrasive to the digestive epithelium. Beneath the digestive epithelium is the
lamina propria, a layer of connective tissue that attaches the epithelium and contains
blood vessels, lymphatic vessels, and nerves. A thin layer of smooth muscle, the muscularis mucosae, is the outermost layer of the mucosa.

Superficial to the mucosa is the **submucosa,** a loose connective tissue layer containing large blood and lymphatic vessels and numerous nerves. Between the submucosa and the adjoining superficial layer of the muscularis externa is a network of
sensory and autonomic nerves, the **submucosa plexus,** that controls the tone of the
muscularis mucosae.

Surrounding the submucosa are several layers of smooth muscle tissue in the **muscularis externa.** The inner layer is **circular** muscle that wraps around the digestive
tract, causing the tract to narrow or constrict upon contraction. The outer layer is
longitudinal muscle with fibers oriented parallel to the length of the tract. Contraction
of this muscle layer shortens the tract. The layers of the muscularis externa produce
waves of contraction called **peristalsis (per-i-STAL-sis),** which move materials along
the digestive tract. Between the muscle layers is the **myenteric (mī-en-TER-ik) plexus,**
nerves that control the activity of the muscularis externa.

The superficial layer of the digestive tract is the **serosa** or **adventitia,** a loose connective tissue covering that attaches and holds the tract in position. In the abdominal
cavity the adventitia is modified as the **peritoneum,** a serous membrane covering of the
stomach and intestines.

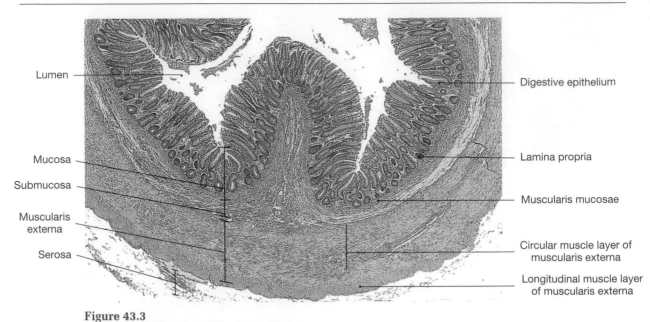

Figure 43.3
Histological Structure of the Digestive Tract.

Photomicrograph of ileum showing general histological organization. [LM ×160]

LABORATORY ACTIVITY   Anatomy of the Digestive Tract

MATERIALS

compound microscope
prepared slide: ileum
intestinal wall model

PROCEDURES

1. Label and review the anatomy presented in Figure 43.2.
2. Identify the major layers of the digestive tract on an intestinal wall model.
3. Identify the four layers of the digestive tract from a cross section of intestine.
   a. Set up the microscope and focus on the slide using low magnification.
   b. Use Figure 43.3 as a guide and locate the lumen and the **epithelium** which would be in contact with the digestive contents. Locate the **lamina propria** and the entire mucosa layer.
   c. Locate the **submucosa**, identified by its loose connective tissue and numerous blood and lymphatic vessels.
   d. Identify the **circular** and **longitudinal** muscles of the **muscularis externa.**
   e. Locate the outer **adventitia** layer.
   f. In the space provided, draw a section of the microscope field, including all four main layers of the tract.

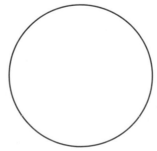

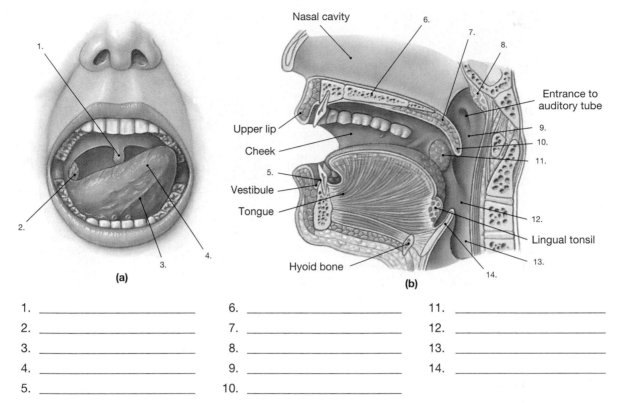

Figure 43.4
The Oral Cavity.

(a) An anterior view of the oral cavity as seen through the open mouth. (b) The oral cavity as seen in sagittal section.

1. _____   6. _____   11. _____
2. _____   7. _____   12. _____
3. _____   8. _____   13. _____
4. _____   9. _____   14. _____
5. _____   10. _____

B. The Mouth

The mouth is the site of food ingestion, the most enjoyable process of digestion for most individuals. Figure 43.4 presents the anatomy of the oral cavity in sagittal and anterior views. The mouth is the **oral cavity** or the **buccal (BUK-al) cavity** and is defined by the space to the lips or **labia,** posterior to the **fauces (FAW-sēz),** the opening between the mouth and the throat. A cone-shaped **uvula (Ū-vū-luh)** is suspended from the posterior soft palate just anterior to the fauces. The lateral walls are composed of the **cheeks,** and the roof is the **hard** and **soft palates** of the maxillae and palatine bones. Glide your tongue around in the **vestibule** between your cheek and teeth. The floor of the mouth is muscular, mostly because of the muscles of the **tongue.** A fold of tissue, the **lingual frenulum (FREN-ū-lum),** anchors the tongue yet allows free movement for food processing and speech. Between the posterior base of the tongue and the roof of the mouth is the **palatoglossal (pal-a-tō-GLOS-al) arch.** At the fauces is the **palatopharyngeal arch.**

Accessory structures of the mouth include the salivary glands and the teeth. Three major pairs of salivary glands, illustrated in Figure 43.5, produce the majority of the saliva, enzymes, and mucus of the oral cavity. The largest, the **parotid (pa-ROT-id) gland,** is in front of your ear between the skin and the masseter muscle. The **parotid duct** (Stenson's duct) pierces the masseter and enters the oral cavity to secrete saliva at the upper second molar. Saliva from this gland is a thick serous secretion and contains the enzyme salivary amylase. Salivary amylase starts the digestive process by splitting complex carbohydrates like starch into shorter polysaccharides and some disaccharide sugar molecules.

The **submandibular gland,** as its name implies, is under the mandible from the mandibular arch posterior to the ramus. The **submandibular duct** (Wharton's duct) passes through the **lingual frenulum** and opens at the swelling on the central margin of

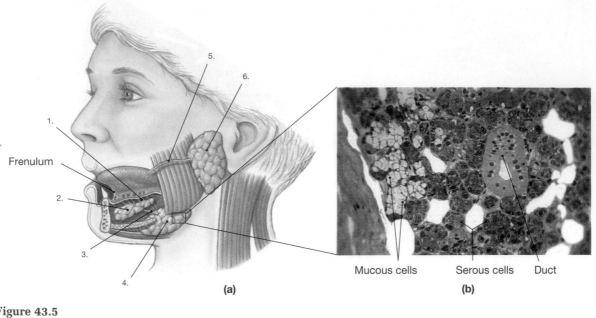

Frenulum

Mucous cells   Serous cells   Duct

(a)   (b)

Figure 43.5
The Salivary Glands.

Lateral view, showing the relative
positions of the salivary glands
and ducts on the left side of the
head.

1. _____   4. _____

2. _____   5. _____

3. _____   6. _____

this tissue. Submandibular secretions are thicker than that of the parotid glands because of the presence of **mucin**, a thick mucus that helps to keep food in a *bolus* or ball for swallowing. Notice the mucous cells in the micrograph in Figure 43.5.

The **sublingual (sub-LING-gual) glands** are located under the tongue at the floor of the mouth. A mucous saliva is secreted into numerous **lesser sublingual ducts** (ducts of Rivinus) which open along the base of the tongue.

The **teeth** are accessory digestive structures for chewing or mastication (mas-ti-KĀ-shun). Figure 43.6a details the anatomy of a typical adult tooth. The tooth is anchored in the alveolar bone of the jaw by a strong **periodontal ligament** which lines the embedded part of the tooth, the **root**. The **crown** is the portion of the tooth above the **gingiva (JIN-ji-va)** or gum. The crown and the root meet at the **neck** where the gum or **gingiva** forms the **gingival sulcus**, a tight seal around the tooth. Although a tooth is comprised of many distinct layers, only the inner **pulp cavity** is filled with living tissue, the **pulp**. Supplying the pulp tissue are blood vessels, lymphatic vessels, and nerves, all of which enter the pulp cavity through the **root canal**. Surrounding the pulp cavity is **dentin (DEN-tin)**, a hard, nonliving solid similar to bone matrix. Dentin comprises most of the structural mass of a tooth. At the root, the dentin is covered by **cementum** which provides attachment for the periodontal ligament. The exposed crown is covered with **enamel**, the hardest substance produced by living organisms. Because of the hard enamel of teeth, they are often used to identify accident victims and skeletal remains that have no other identifying features.

Figure 43.6c shows the various dental surfaces. The **occlusal surface** is the superior area where food is ground, snipped, and torn by the tooth. The **labial surface** of the upper and lower teeth face the lips. The posterior aspect of the upper dentition is the **palatal surface**, and the lower posterior surface is the **lingual surface**.

Humans have two sets of teeth during their lifetime, the first being the **deciduous (de-SID-ū-us)** which are replaced by the **permanent** dentition starting at around the age

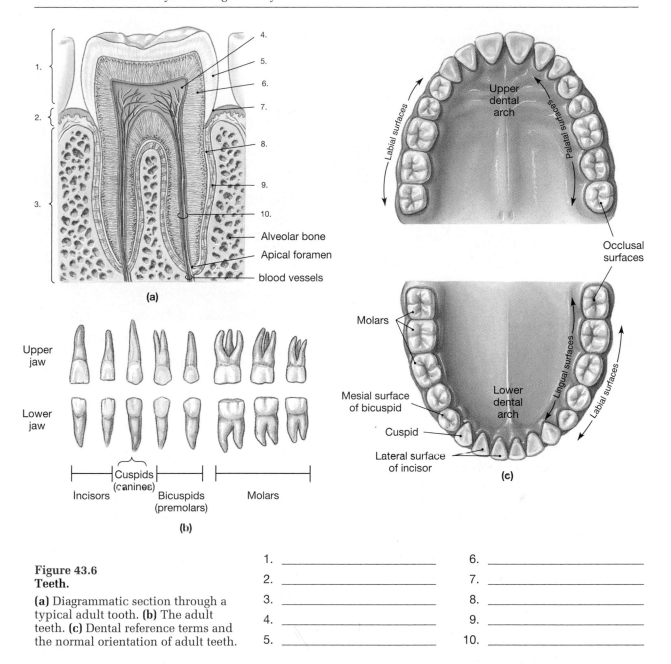

(a)

(b)

(c)

**Figure 43.6
Teeth.**

(a) Diagrammatic section through a typical adult tooth. **(b)** The adult teeth. **(c)** Dental reference terms and the normal orientation of adult teeth.

1. _____   6. _____

2. _____   7. _____

3. _____   8. _____

4. _____   9. _____

5. _____   10. _____

of 6. There are a total of 20 milk teeth, 5 in each jaw quadrant (the mouth is divided into four quadrants, upper right and left quadrants and lower right and left quadrants). These teeth begin to appear at about the age of 6 months and are replaced by permanent teeth starting at 6 years of age. From the midline of either jaw, the milk teeth are the **central incisor**, **lateral incisor**, **cuspid**, **first molar**, and **second molar**. The permanent dentition, depicted in Figure 43.6b, consist of 32 teeth, each quadrant with a **central incisor**, **lateral incisor**, **cuspid** or **canine**, **first** and **second premolars (bicuspids)**, and **first**, **second,** and **third molars**. The third molar is also called the wisdom tooth. Figure 43.7 is a photograph of the skull of an infant with the maxilla and mandible cut to reveal the permanent dentition developing at the roots of the primary dentition. During replacement of the deciduous dentition by the permanent dentition, the permanent teeth help push the deciduous teeth out of their sockets.

Figure 43.7
Developing Teeth in an Infant Skull.

Parts of the maxilla and mandible have been removed to show the presence of developing permanent teeth.

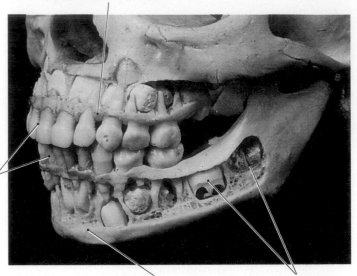

Maxilla exposed to show erupted deciduous teeth and unerupted permanent teeth

Erupted deciduous teeth

Mandible exposed to show erupted deciduous teeth and unerupted permanent teeth

First and second molars

LABORATORY ACTIVITY   THE MOUTH

MATERIALS

torso model
digestive system chart
tooth model
hand mirror

PROCEDURES

1. Label and review the anatomy presented in Figures 43.4–43.6.
2. Identify the anatomy of the mouth on all laboratory models and charts.
3. Locate the part of a sectioned tooth on laboratory materials.
4. Identify each salivary gland and duct on laboratory materials.
5. Examine your mouth in a mirror. Locate your uvula, fauces, and palatoglossal arch. Lift your tongue and examine your submandibular duct.
6. Examine your teeth in a mirror. Locate your incisors, cuspids, bicuspids, and molars. How many teeth do you have? Are you missing any teeth because of extractions? Do you have any wisdom teeth?

C. **The Pharynx**

The throat or **pharynx** is a common passageway for nutrients and air. The pharynx is divided into three anatomical regions, the nasopharynx, oropharynx, and laryngopharynx, as shown in Figure 43.4. When you swallow a **bolus** of food, it passes through the fauces into the **oropharynx** directly posterior to the oral cavity. The **nasopharynx** is superior to the oropharynx. Muscles of the soft palate contract during swallowing and close the passageway to the nasopharynx to prevent food from entering the nasal cavity. The **auditory tube** (Eustachian tube) is located near the opening between the nasopharynx and the nasal cavity. Inferior to the oropharynx is the **laryn-gopharynx**. Toward the base of this area of the pharynx the throat branches into the larynx of the respiratory system and the **esophagus** leading to the stomach. The larynx has a flap called the **epiglottis** which closes the larynx so swallowed food can enter only the esophagus and not the respiratory passageways.

LABORATORY ACTIVITY   THE PHARYNX

MATERIALS

   torso model
   digestive system chart

PROCEDURES

   1. Label the regions of the pharynx in Figure 43.4.
   2. Identify the anatomy of the pharynx on all laboratory models and charts.
   3. Examine your oropharynx in a mirror. Locate your uvula, fauces, and palato-glossal arch.
   4. Put your finger on your adam's apple, the larynx, and swallow. How does your larynx move and what is the purpose of this movement?

D. The Esophagus

The food tube or **esophagus** connects the pharynx to the stomach, illustrated in Figures 43.1 and 43.8. It is inferior to the pharynx and posterior to the trachea, the windpipe. The esophagus is approximately 25 cm (12 in.) long. It pierces the diaphragm at the **esophageal hiatus (hī-Ā-tus)** to connect with the stomach in the abdominal cavity. At the stomach the esophagus terminates in a lower esophageal sphincter, a muscular valve that prevents stomach contents from backwashing into the esophagus.

A baby is born with esophageal atresia, an incomplete connection between the esophagus and the stomach. What will most likely happen to the infant if this defect is not corrected?

LABORATORY ACTIVITY   THE ESOPHAGUS

MATERIALS

   torso model
   digestive system chart

PROCEDURES

   1. Label the esophagus in Figure 43.8.
   2. Identify the anatomy of the esophagus on all laboratory models and charts.

E. The Stomach

Figure 43.8 depicts the gross anatomy of the stomach. Your stomach is a J-shaped organ located just inferior to the diaphragm in your epigastric and lower left hypochondriac regions. Locate in Figure 43.8a the four major regions of the stomach: the **cardia (KAR-dē-uh)** where the stomach connects with the esophagus; the **fundus (FUN-dus)**, the upper rounded area of the stomach; the **body**, the middle region; and the **pylorus**

Figure 43.8
Gross Anatomy of the Stomach.

(a) Anterior view of the stomach showing superficial landmarks.
(b) Diagrammatic view of the organization of the stomach wall.

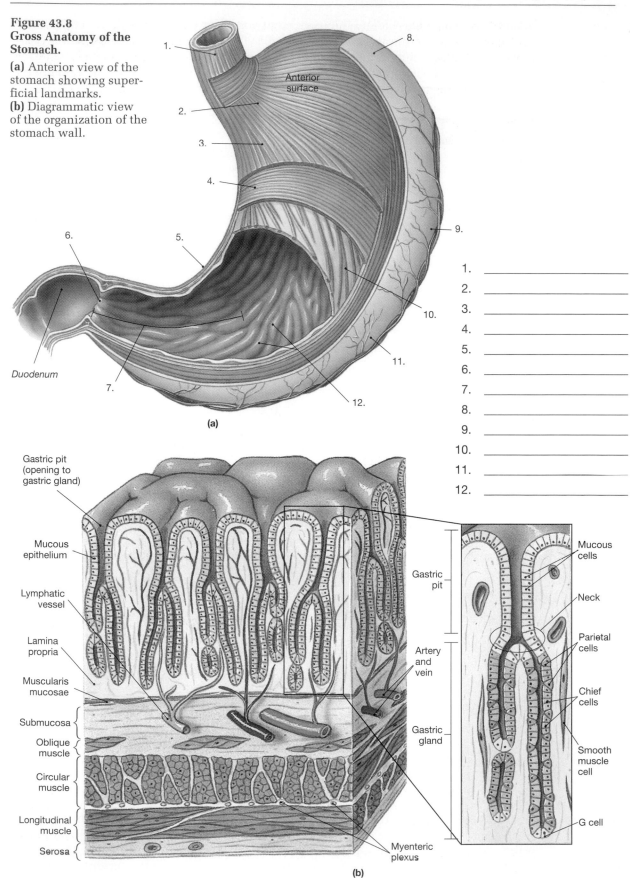

1. _____
2. _____
3. _____
4. _____
5. _____
6. _____
7. _____
8. _____
9. _____
10. _____
11. _____
12. _____

(**pī-LOR-us**), the narrowed distal end connected to the small intestine. The **pyloric sphincter** or **valve** controls movement of material from the stomach into your small intestine. The lateral, convex border of the stomach is the **greater curvature**, and the medial stomach margin is the concave **lesser curvature.** Extending from the greater curvature is the **greater omentum (ō-MEN-tum)**, commonly referred to as the fatty apron. This fatty layer is part of the outer serosa of the stomach wall and functions to protect abdominal organs and in the attachment of the transverse colon of the large intestine. The **lesser omentum**, also part of the serosa, suspends the stomach from the liver.

The inner lining of the stomach has **rugae (ROO-gē)** or folds that enable the stomach to expand as it fills with food. Unlike other regions of the digestive tract, the **muscularis externa** of the stomach contains three layers of smooth muscle instead of two. The inner layer is an oblique layer, followed by a circular layer and then an external longitudinal layer. The three muscle layers contract and churn stomach contents, mixing gastric juice and liquidifying the food into *chyme.*

LABORATORY ACTIVITY   THE STOMACH

MATERIALS

  torso model
  digestive system chart
  compound microscope
  prepared slide: stomach

PROCEDURES

1. Label the anatomy of the stomach in Figure 43.8.
2. Identify the gross anatomy of the stomach on all laboratory models and charts.
3. If specimens are available, examine the stomach of a cat or other animal. Locate the rugae, main regions of the stomach, omenta, and sphincters.
4. Histology of the stomach. Figure 43.9 details the histology of the stomach wall in cross section. At low power observe the large folds of the mucosa called **rugae.** Note the digestive epithelium of simple columnar epithelium covering the rugae. Increase the magnification and locate the numerous **gastric pits**, the invaginations along the rugae. Observe at the margin of the pits and the rugae the **simple columnar epithelium** lining the stomach mucosa. Within the pits distinguish between **parietal cells**, which are more numerous in the upper areas, and **chief cells**, those with nuclei at the basal region of the cells. Chief cells secrete pepsinogen, an inactive proteolytic enzyme, and parietal cells secrete hydrochloric acid.

F. **The Small Intestine**

The small intestine, Figure 43.10, is approximately 21 ft long and composed of three segments, the duodenum, the jejunum and the ileum. Thin sheets of membrane called **mesenteries (MEZ-en-ter-ēz)** extend from the serosa to support and attach the small intestine. The first 10 in. is the **duodenum (doo-AH-de-num)** which is attached to the distal region of the pylorus. Digestive secretions from the liver, gallbladder, and pancreas flow into ducts that merge and empty into the duodenum. This anatomy is described in the upcoming section on the liver. The **jejunum (je-JOO-num)** is approximately 12 ft long and is the site of most nutrient absorption. The last 8 ft is the **ileum (IL-ē-um)** which terminates at the **ileocecal valve** and empties into the cecum of the large intestine.

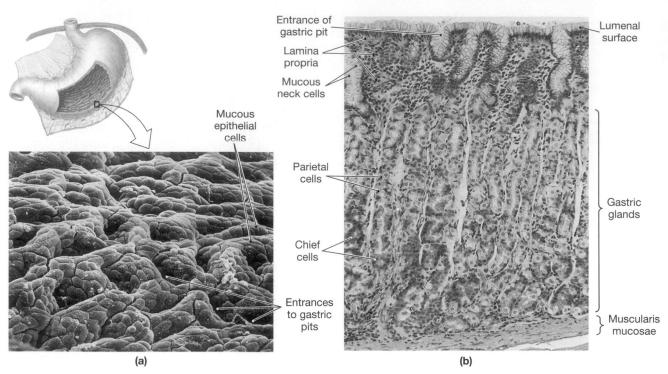

Figure 43.9
The Stomach Lining.
(a) A surface view of the gastric mucosa of the full stomach, showing the entrances to the gastric pits. (SEM × 35)
(b) A section through gastric pits and gastric glands. (LM × 300)

The small intestine is the site of most digestive and absorptive activities and has specialized folds to increase the surface area for these functions (see Figure 43.10). The submucosa and mucosa are creased together into large folds called **plicae (PLĪ-sē)**. Along the plicae, the mucosa is convoluted into fingerlike projections called **villi**. Lining each villus is simple columnar epithelium which has plasma membrane extensions called **microvilli**.

LABORATORY ACTIVITY   THE SMALL INTESTINE

MATERIALS

> torso model
> digestive system chart
> compound microscope
> prepared slide: duodenum or ileum

PROCEDURES

1. Label and review the regions and organization of the small intestine in Figure 43.10.
2. Identify the anatomy of the small intestine on all laboratory models and charts.
3. If a specimen is available, examine a segment of the small intestine of a cat or other animal.
4. Histology of the Small Intestine. Examine a slide of the small intestine and observe the folds of the mucosa at low magnification, as in Figure 43.11. Increase the magnification to examine the histology of each tissue layer. Also refer to Figure 43.3 showing the ileum wall.

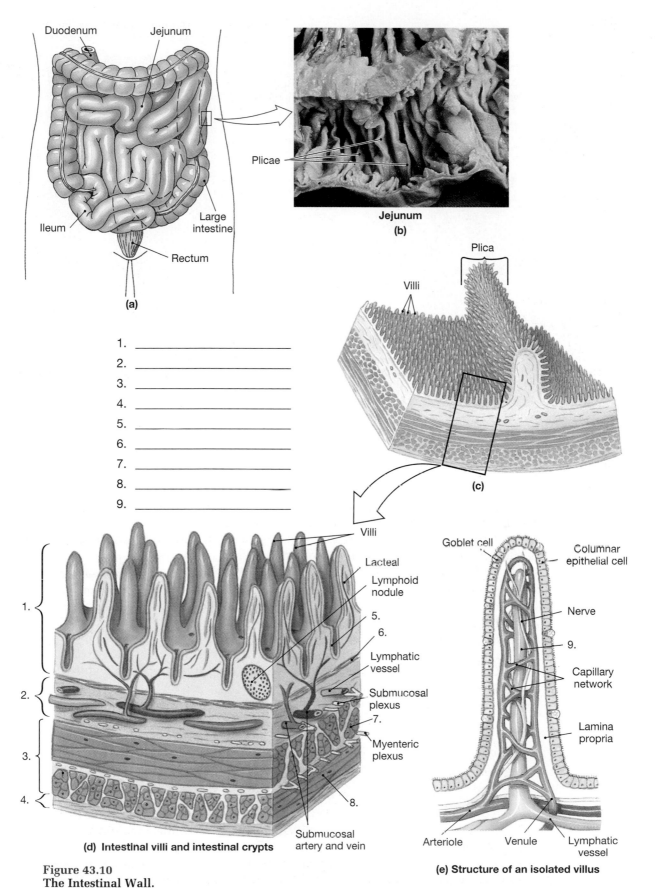

Duodenum   Jejunum

Ileum

Large
intestine

Rectum

(a)

Plicae

Jejunum
(b)

Plica

Villi

(c)

1. _____
2. _____
3. _____
4. _____
5. _____
6. _____
7. _____
8. _____
9. _____

Villi

1.

2.

3.

4.

Lacteal

Lymphoid
nodule

5.

6.

Lymphatic
vessel

Submucosal
plexus

7.

Myenteric
plexus

8.

Submucosal
artery and vein

(d) Intestinal villi and intestinal crypts

Goblet cell

Columnar
epithelial cell

Nerve

9.

Capillary
network

Lamina
propria

Arteriole   Venule   Lymphatic
vessel

(e) Structure of an isolated villus

Figure 43.10
The Intestinal Wall.

(a) Regions of the small intestine. **(b)** View of the plicae of the jejunum. **(c)** A single plica and many villi. **(d)** The organization of villi, intestinal crypts, and the intestinal wall. **(e)** Diagrammatic view of a single villus, showing the capillary and lymphatic supply.

643

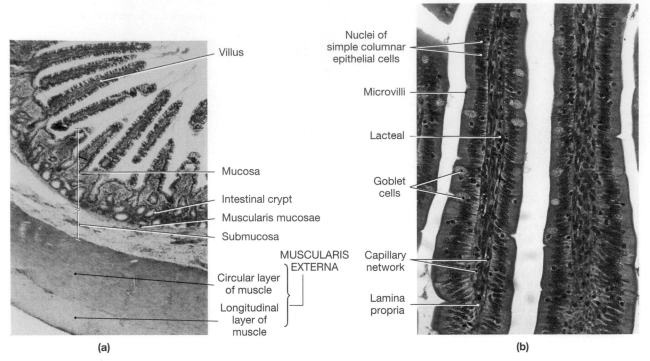

Figure 43.11
The Intestinal Wall.

(a) Panoramic view of the small intestine wall showing mucosa with characteristic villi, plica, submucosa, and muscularis layers. (LM ×49) **(b)** Photomicrograph of several villi, showing lacteals and epithelial details. (LM ×360)

5. On the duodenum slide locate the lumen of the section. Note the large folds called **plicae**. Along the plicae are smaller fingerlike extensions, the **villi**. Increase the magnification of a single villus and locate the **simple columnar epithelium** and **goblet cells** lining the mucosa and the **lamina propria** composed of loose connective tissue. In the middle of a villus locate the **lacteal (LAK-tē-al)**, a lymphatic vessel that absorbs fatty acids and monoglycerides from lipid digestion. These structures are detailed in Figure 43.11b. Also in the lamina are **intestinal crypts** (crypts of Lieberkuhn) which secrete intestinal juice. In the **submucosa** are scattered mucus-producing **submucosal** (Brunner's) **glands**. Finally, identify the **muscularis externa** and the outer **serosa**. If the intestinal slide is of the ileum, the submucosa layer will show **Peyer's patches**, large lymphatic nodules that prevent bacteria from entering the blood.

G. Large Intestine

The large intestine is the site of water absorption and waste compaction. It is approximately 5 ft long and is divided into two groups, the **colon (KŌ-lin)** and the **rectum**. Figure 43.12a details the gross anatomy of the large intestine.

The colon makes up most of the large intestine. At the **ileocecal (il-ē-ō-SĒ-kal) valve** the ileum meets the colon. The valve is illustrated in Figure 43.12a as a small crease in the colon wall just above the ileum. The first part of the colon, a pouchlike **cecum (SĒ-kum)**, is located in the right lumbar region. At the medial floor of the cecum is the wormlike **appendix**. Past the cecum the colon is divided into several regions: the **ascending colon** travels up the right side of the abdomen, bends left at the **right colic flexure**, and crosses the abdomen below the stomach as the **transverse colon**. The **left colic flexure** turns the colon downward to the **descending colon**. The S-shaped **sigmoid (SIG-moyd) colon** passes through the pelvic cavity to join the **rectum**. The rectum, detailed in

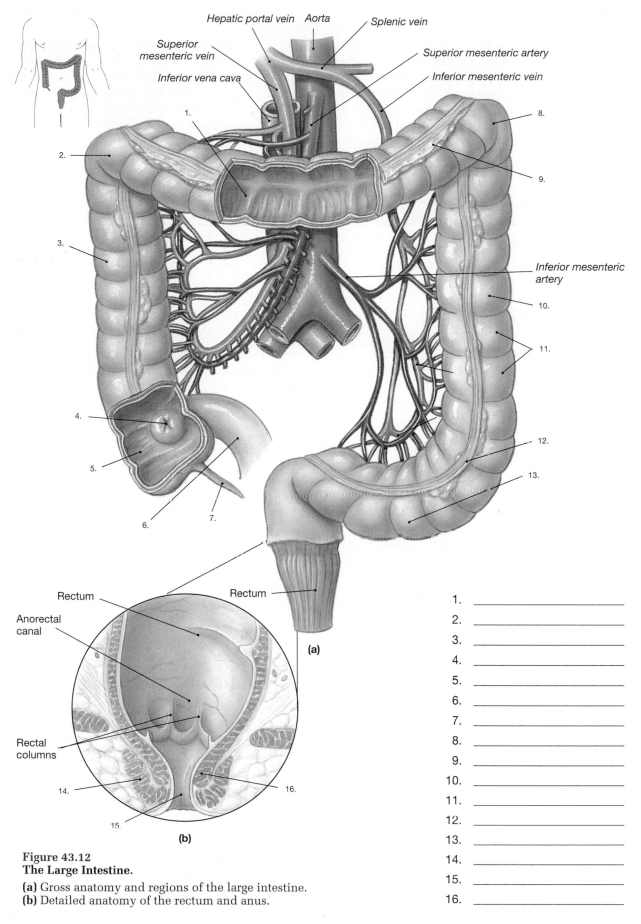

Hepatic portal vein   Aorta   Splenic vein

Superior
mesenteric vein   Superior mesenteric artery

Inferior vena cava   Inferior mesenteric vein

Inferior mesenteric
artery

Rectum   Rectum

Anorectal
canal

Rectal
columns

(a)

(b)

Figure 43.12
The Large Intestine.

(a) Gross anatomy and regions of the large intestine.
(b) Detailed anatomy of the rectum and anus.

1. _____
2. _____
3. _____
4. _____
5. _____
6. _____
7. _____
8. _____
9. _____
10. _____
11. _____
12. _____
13. _____
14. _____
15. _____
16. _____

Figure 43.12b, is the last 6 in. of the large intestine and the end of digestive tract. The opening of the rectum is the **anus** which is controlled by internal and external **anal sphincters**. Longitudinal folds called **rectal columns** occur in the rectum where the digestive epithelium changes from simple columnar epithelium to stratified squamous epithelium.

Unlike the continuous sheets of muscle in the muscularis externa extending from the mouth to the small intestine, the longitudinal muscle of the colon's muscularis externa is modified into three separate bands of muscle, the **taenia coli (TĒ-nē-a KŌ-lī)**. The muscle tone of the taenia coli constricts the colon wall into pouches called **haustrae (HAWS-truh)** which permit expansion and stretching of the colon wall.

LABORATORY ACTIVITY   THE LARGE INTESTINE

MATERIALS

torso model
digestive system chart

PROCEDURES

1. Label and review the anatomy of the large intestine in Figure 43.12.
2. Identify the gross anatomy of the large intestine on all laboratory models and charts.
3. If a specimen available, examine the colon of a cat or other animal. Locate each region, the taenia coli, and the haustrae.

H. The Liver and Gallbladder

The liver is located mostly in the right hypochondratic region, inferior to the diaphragm. It is attached to the abdominal wall by the **falciform ligament**. As shown in Figure 43.13,

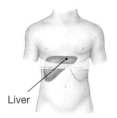
Liver

1. _____   4. _____
2. _____   5. _____
3. _____   6. _____

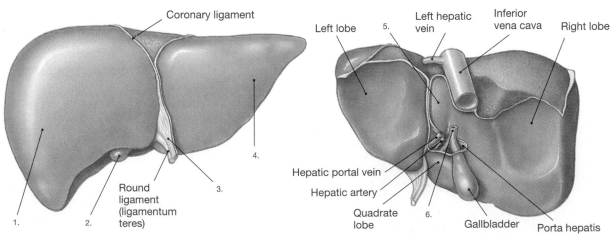

Coronary ligament

Left hepatic vein   Inferior vena cava   Right lobe

Left lobe   5.

Hepatic portal vein

Hepatic artery

Round ligament (ligamentum teres)

Quadrate lobe   6.

Gallbladder   Porta hepatis

1.   2.   3.   4.

(a) Anterior (parietal) surface   **(b) Posterior (visceral) surface**

Figure 43.13
Anatomy of the Liver.
(a) The anterior surface of the liver. **(b)** The posterior surface of the liver.

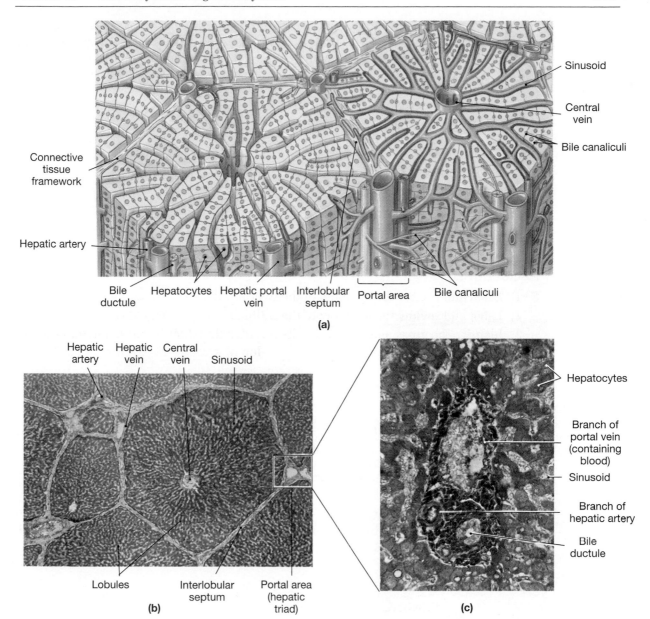

Figure 43.14
Liver Histology.

(a) Diagrammatic view of lobular organization. **(b)** Light micrograph showing a section through a liver lobule. (LM ×38) **(c)** A portal area. (LM ×31)

the liver has four major lobes. The **right** and **left lobes** are separated by the falciform ligament, the square **quadrate lobe** is located on the inferior surface of the right lobe, and the **caudate lobe** is posterior, near the site of the inferior vena cava. Each lobe is divided into thousands of smaller **lobules** which you will examine microscopically later in this exercise. Examine Figure 43.14 which highlights liver histology. Liver cells called **hepatocytes (he-PAT-ō-sits)** manufacture and secrete bile into small ducts called **bile canaliculi** which then empty into **bile ductules (DUK-tūlz)** surrounding the lobule. Progressively larger ducts drain bile into the right and left **hepatic ducts** which join a single **hepatic duct**.

The **gallbladder** is located inferior to the right lobe of the liver. It is a small, muscular sac that stores and concentrates bile salts used in the digestion of lipids. Figure 43.14 details the ducts of the liver and gallbladder. The hepatic duct from the liver meets the **cystic duct** of the gallbladder to form the **common bile duct**. This duct passes through the lesser omentum and joins the pancreatic duct of the pancreas at the **duo-denal ampulla (am-PUL-a)**. The ampulla projects into the lumen of the duodenum at the **duodenal papilla**. A band of muscle called the **pancreaticohepatic sphincter** (of Oddi) regulates the flow of pancreatic secretions and bile into the duodenum.

LABORATORY ACTIVITY   THE LIVER AND GALLBLADDER

MATERIALS

> torso model
> digestive system chart
> liver model
> compound microscope
> prepared slide: liver

PROCEDURES

1. Label and review the anatomy of the liver in Figures 43.13 and 43.14.
2. Label and review the anatomy of the gallbladder in Figures 43.13 and 43.15.
3. Identify the gross anatomy of the liver and gallbladder on all laboratory models and charts.

Figure 43.15
The Gallbladder.

(a) A view of the inferior surface of the liver, showing the position of the gallbladder and ducts that transport bile from the liver to the gallbladder and duodenum. A portion of the lesser omentum has been cut away to make it easier to see the relationship of the common bile duct to the hepatic duct and cystic duct. **(b)** Interior view of the duodenum, showing the duodenal ampulla and related structures.

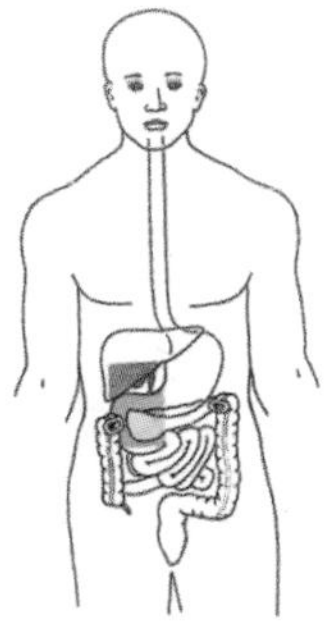

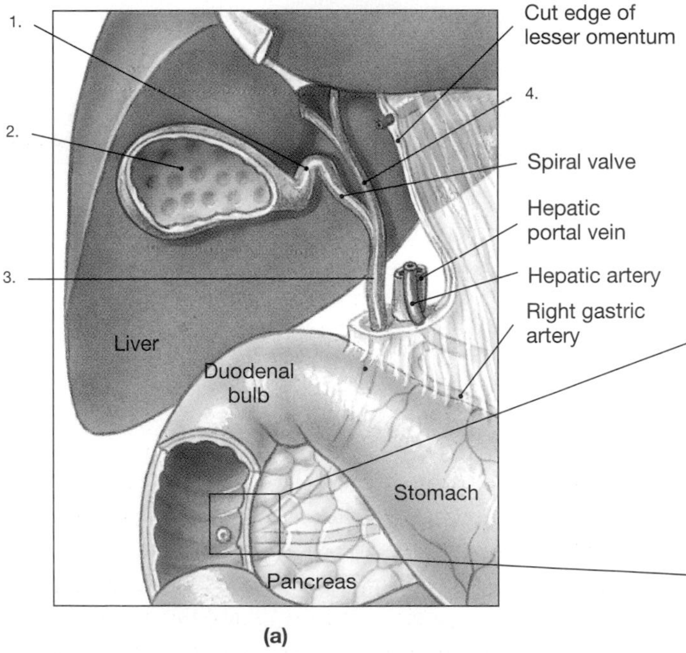

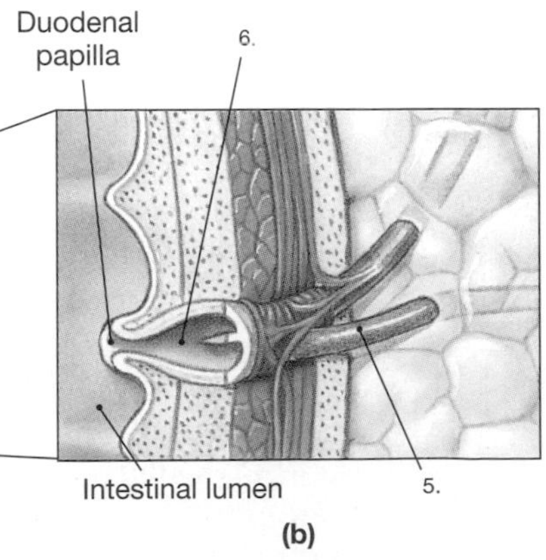

1. _____   3. _____   5. _____
2. _____   4. _____   6. _____

4. If specimens are available, examine the liver and gallbladder of a cat or other animal. Locate each liver lobe, the falciform ligament, and the hepatic, cystic, and common bile ducts.

5. Histology of the liver. The liver is composed of approximately 100,000 small lobules (see Figure 43.14). At low magnification notice the numerous oval holes in the tissue. Most of these are **central veins** that drain blood from the lobules. Pick a central vein and increase the magnification to observe a **lobule**. Depending on the quality of your slide you may be able to define the outline of an individual lobule. Note the open **sinusoids**, open spaces where blood circulates, and the **hepatocyte** cells lining the sinusoids. Hepatocytes produce bile required for fat digestion.

I. The Pancreas

The pancreas lies between the stomach and the duodenum. The **head** is adjacent to the duodenum, the **body** is the central region, and the **tail** tapers to the distal end of the gland. Locate these regions in Figure 43.16. Pancreatic exocrine or **acinar (AS-i-nar) cells** secrete pancreatic juice into small acini, the lumen of the pancreatic glands. The acini drain into progressively larger ducts that merge as the **pancreatic duct** and, in some individuals, an **accessory duct**. The pancreatic duct joins the common bile duct at the duodenal ampulla.

The pancreas is a "double" gland with both endocrine and exocrine functions. The endocrine cells occur in **pancreatic islets** and secrete hormones for sugar metabolism. Most of the glandular epithelium of the pancreas is exocrine in function. Acinar cells secrete pancreatic juice rich in enzymes and buffers into the pancreatic duct which meets the common bile duct in the duodenal ampulla.

If the duodenal ampulla were blocked, what materials could not be released into the lumen of the small intestine?

LABORATORY ACTIVITY   THE PANCREAS

MATERIALS

> torso model
> digestive system chart
> compound microscope
> prepared slide: pancreas

PROCEDURES

1. Label and review the anatomy of the pancreas in Figure 43.16.
2. Identify the anatomy of the pancreas on all laboratory models and charts.
3. If a specimen is available, examine the pancreas of a cat or other animal. Locate the three regions of the pancreas and the pancreatic duct.
4. Histology of the pancreas. The pancreas is a double gland with endocrine functions for hormone production and exocrine functions for the production of pancreatic juice containing digestive enzymes and pH buffers. Recall that exocrine glands secrete into ducts. Using Figure 43.16 as a reference, observe the pancreas slide at low magnification and locate the numerous oval ducts. The exocrine cells are the dark-stained **acinar cells**. Surrounding the acinar cells are scattered

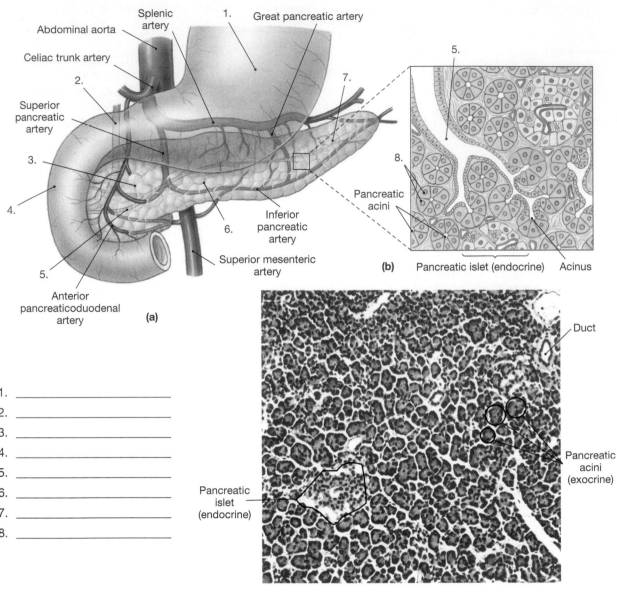

(c) Exocrine and endocrine cells (LM × 120)

1. _____
2. _____
3. _____
4. _____
5. _____
6. _____
7. _____
8. _____

Figure 43.16
The Pancreas.

(a) Gross anatomy of the pancreas. The head of the pancreas is tucked into a curve of the duodenum that begins at the pylorus of the stomach. **(b)** Diagrammatic view of the cellular organization of the pancreas, showing exocrine and endocrine regions. **(c)** Photomicrograph of the pancreas, showing exocrine and endocrine cells.

light-stained **pancreatic islets**, endocrine cells of the pancreas. The open structures in the tissue are blood vessels and branches to the **pancreatic duct**. Pancreatic juice is secreted by the acinar cells into acini, the small packets at the tips of the ducts.

J. Digestive System of the Cat

This dissection complements the study of the human digestive system. The digestive system of the cat is very similar to that of the human. During our examination of cat digestive structures, some differences will be described.

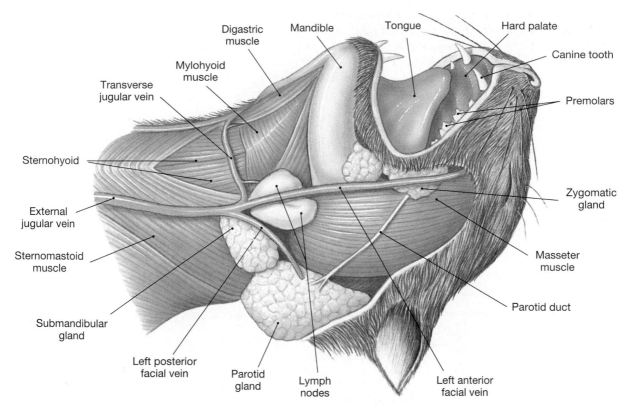

Figure 43.17
Facial Glands of the Cat.

If the thoracic and abdominal cavities have not been opened, expose the thoracic and abdominal organs by making a longitudinal midline incision through the muscles of the neck, the bones of the sternum, and the muscles of the midline of the thorax and of the abdominal wall. Avoid cutting the diaphragm in order to identify the structures that pass between the thoracic and abdominal cavities. Make lateral wall incisions in the thoracic cavity cranial to the diaphragm and in the abdominal cavity caudal to the diaphragm. In your dissection of the digestive system, trace the path of ingested materials through the body.

Use Figures 43.17 and 43.18 as guides during your dissection and observation of the cat digestive system.

1. Oral Cavity, Salivary Glands, Pharynx, and Esophagus

a. The **oral cavity** consists of the **vestibule** and the oral cavity proper. The vestibule is surrounded by the lips and cheeks. The **tongue** and **teeth** are located within the oral cavity proper. With the aid of a magnifying lens, locate the papillae on the tongue. **Filiform papillae** are anterior, numerous, and pointed. **Fungiform papillae** are small, rounded, and posterior to the filiform papillae. A few **circumvallate papillae**, which are large and rounded, are located near the back of the tongue. Also, observe that the tongue is attached to the floor of the mouth by a mucous membrane, the **lingual frenulum**.

b. The roof of the mouth consists of a bony **hard palate**, and its posterior extension, the **soft palate**. At the posterior region of the oral cavity, locate the **pharynx**. It is the cavity that occupies the space from the dorsal to the soft palate to the larynx. There are three regions of the pharynx: the **nasopharynx** is dorsal to the soft palate; the **oropharynx** is the posterior region of the oral cavity; and the

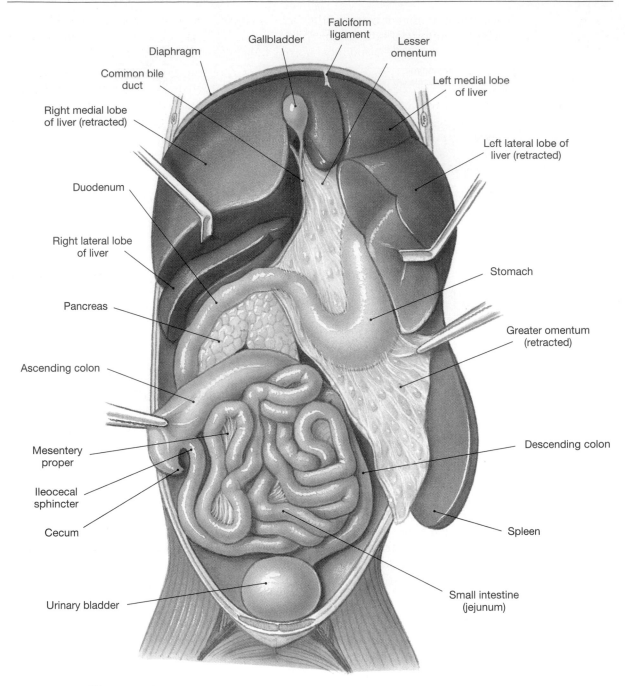

Figure 43.18
Digestive System of the Cat. (Caudate lobe of liver not shown.)

laryngopharynx is the region of the epiglottis and the opening to the esophagus. The **larynx** forms the passageway between the pharynx and the trachea.

c. The salivary glands secrete saliva into the mouth. Do not confuse the small, dark, kidney bean–shaped lymph nodes with the oatmeal-colored and textured salivary glands. Locate the large **parotid gland** inferior to the ear on the surface of the masseter muscle. The **parotid duct** passes over the surface of this muscle and enters the oral cavity. The **submandibular gland** lies inferior to the parotid and deep to the point of merger between the facial and external jugular veins. The **sublingual gland** is anterior to the submandibular gland. Ducts of both glands open onto the floor of the mouth, but typically only the **submandibular duct** can be traced.

    d. Return to the pharynx and identify the opening into the **esophagus** posterior to the epiglottis of the larynx. The esophagus is the flattened tube dorsal to the trachea. It connects the laryngopharynx to the stomach. Reflect the organs of the thoracic cavity and trace the esophagus through the **diaphragm** into the abdominal cavity where it connects with the stomach.

2. The Abdominal Cavity, Stomach, and Spleen

    a. The inside of the abdominal cavity and the outer surface of the viscera are lined by a serous membrane, the **peritoneum**. The **visceral peritoneum** covers the viscera, and the **parietal peritoneum** lines the inside of the abdominal body wall. The **peritoneal cavity** is found between these membranes. It contains serous fluid, called peritoneal fluid that permits the organs to slide past each other with a minimum amount of friction. The peritoneal sheets extending between the body wall and the viscera are termed mesenteries, omenta, and ligaments. These membranous sheets provide support for vessels, nerves, and lymphatics that supply the viscera.

    b. Locate the **stomach** on the left side of the abdominal cavity and identify its the four regions: the **cardia**, at the entrance of the esophagus; the **fundus,** the dome-shaped pouch rising above the esophagus; the **body**, the main portion of the stomach; and the **pylorus**, the posterior region of the stomach ending at the **pyloric sphincter** where the digestive tube continues as the duodenum.

    c. Make an incision through the wall of the stomach along the greater curvature, continuing this incision 2 in. past the pyloric valve. Reflect the stomach wall to observe the pyloric valve and its inner surface. Large folds of the stomach mucosa, termed **rugae**, can be observed in the empty stomach.

    d. The outer (left) margin of the stomach is convex in shape and is called the **greater curvature**. The medial (right) margin of the stomach is concave in shape and is called the **lesser curvature**. Attached to the greater curvature is a curtain of fat, the **greater omentum,** which covers the ventral surface of the abdominal organs. The **lesser omentum** connects the lesser curvature of the stomach to the liver.

    e. Posterior to the stomach observe a large, dark-brown organ, the **spleen**.

3. The Liver, Gallbladder, and Pancreas

    a. Observe the large, brown **liver** posterior to the diaphragm. It is the largest organ in the abdominal cavity and is divided into five lobes: right and left medial, right and left lateral, and caudate (posterior). Identify a dark-green sac, the **gallbladder**, within a fossa in the right medial liver lobe. The **falciform ligament** is a delicate membrane that attaches the liver superiorly to the diaphragm and abdominal wall. This ligament can be located between the right and left medial liver lobes.

    b. Identify the **common bile duct** and trace its course anteriorly to where it splits into the **cystic duct**, to the gallbladder, and the **common hepatic duct**, to the liver. Locate these ducts by teasing away connective tissue.

    c. Posterior to the stomach and within the curvature of the duodenum locate the **pancreas**, the major glandular organ of the digestive system. In the cat, the pancreas has two regions, head and tail. Identify the region within the duodenum as the **head** of the pancreas and the portion passing along the posterior surface of the stomach as the **tail** of the pancreas.

d. Locate the **pancreatic duct** (Duct of Wirsung) as it perforates the duodenum at an enlarged area, the **duodenal ampulla**. Identify the duct and ampulla by using a teasing needle probe to scrape away the pancreatic tissue of the head portion. Trace the duct to the ampulla. The pancreatic and bile ducts are adjacent to each other.

e. Return to the cut segment of the duodenum, reflect it, and secure it open with dissecting pins. Now use a hand lens to observe the numerous **villi**, the **duodenal ampulla**, and the opening of the duct.

4. The Small and Large Intestines

a. The three regions of the **small intestine** can be traced from its proximal end to its distal end. The first region, termed the **duodenum,** is the C-shaped portion distal to the pyloric valve. It is about 6 in. in length and receives the pancreatic and the common bile ducts. The **jejunum** comprises the bulk of the remaining length of the small intestine. The **ileum** is the last region of the small intestine and merges with the large intestine. The ileum terminates at the **ileocecal sphincter (valve)**, which controls the flow of chyme into the large intestine. This valve can be felt as a firm ring at the end of the ileum as it projects into the large intestine.

b. The **large intestine** is divided into cecum, colon, and rectum portions. To view the large intestine it will be necessary to pull the loops of the small intestine to the left and let them drape out of the body cavity.

c. The initial expanded region of the large intestine is the **cecum**. The small intestine enters the large intestine at about 10 cm ($\frac{1}{2}$ in.) from the base of the cecum. Cats do not have an appendix.

d. Trace the short **ascending, transverse,** and **descending colon** portions. The ascending colon lies on the right side of the abdominal cavity and begins just superior to the ileocecal valve, the transverse colon extends transversely across the abdominal cavity, and the descending colon descends on the left side of the posterior abdominal wall. It is continuous with the terminal portion of the large intestine, the **rectum**, whose opening to the exterior is the **anus**. The membranous structure that supports the colon and attaches it to the posterior body wall is the **mesocolon**.

DIGESTIVE SYSTEM ANATOMY CHECKLIST

This is a list of bold terms presented in Exercise 43. Use this list for review purposes after you have completed the laboratory activities.

THE DIGESTIVE TRACT

• Figures 43.2 and 43.3

mucosa, digestive epithelium, lumen

lamina propria, muscularis mucosae

submucosa, submucosal plexus

muscularis externa, circular and longitudinal muscle layers, myenteric plexus

adventitia, serosa, mesentery, peritoneum

stratified squamous epithelium

simple columnar epithelium

THE MOUTH AND PHARYNX

• Figures 43.4 and 43.5

oral or buccal cavity, labia, cheeks, vestibule

lingual frenulum, tongue, body, root, dorsum

hard and soft palates

palatoglossal arch, palatopharyngeal arch

parotid glands and ducts

sublingual glands and ducts

submandibular glands and ducts

uvula, fauces, pharynx, pharyngotympanic tube

nasopharynx, oropharynx, laryngopharynx

epiglottis

THE TONGUE AND TEETH

• Figures 43.6 and 43.7

deciduous and permanent dentition

dentin, pulp, pulp cavity, root canal, and root cementum, periodontal ligament

neck, crown, enamel, gingival sulcus, gingivae

incisors, cuspids (canines)

bicuspids (premolars), molars

deciduous and secondary dentition

occlusal, labial, palatal, and lingual surfaces

THE ESOPHAGUS AND STOMACH

• Figure 43.8 and 43.9

esophageal hiatus

lesser and greater curvatures

cardia, fundus, body, rugae

pylorus, pyloric sphincter

greater and lesser omentum

gastric pits and glands

parietal cells, chief cells

THE SMALL INTESTINE

• Figures 43.10 and 43.11

mesentery

duodenum, jejunum, ileum

plicae, intestinal villi, microvilli, lacteal

simple columnar epithelium

goblet cells

lamina propria

lacteal

intestinal cripts (of Lieberkuhn)

submucosa

submucosal (Brunner's) glands

muscularis externa

serosa

Peyer's patches

THE LARGE INTESTINE

• Figure 43.12

haustrae, taenia coli

cecum, ileocecal valve, appendix

ascending and transverse colon

right and left colic flexures

descending and sigmoid colon

intestinal glands, rectum, anus

rectal columns, anal sphincters

THE LIVER AND GALLBLADDER

• Figures 43.13, 43.14, and 43.15

falciform ligament

right, left, caudate and quadrate lobes

liver lobule, hepatocytes

central vein, bile canaliculi, bile ductules

cystic, hepatic and common bile duct

duodenal ampulla

duodenal papilla, pancreaticohepatic sphincter (of Oddi)

THE PANCREAS

• Figure 43.16

head, body, tail

accessory duct, pancreatic islets, pancreatic duct

acini

acinar cells

DIGESTIVE SYSTEM OF THE CAT

• Figures 43.17 and 43.18

mouth

oral cavity, vestibule, teeth, tongue, lingual frenulum, filiform papilla, fungiform papilla, circumvallate papilla, hard palate, soft palate, larynx, parotid gland, parotid duct, submandibular gland, submandibular duct, sublingual gland

pharynx and esophagus

pharynx, oropharynx, nasopharynx, laryngopharynx, esophagus, diaphragm

abdominal cavity

peritoneum, visceral peritoneum

parietal peritoneum

peritoneal cavity

stomach

cardia, fundus, body, pylorus, pyloric sphincter, rugae, greater curvature, lesser curvature, greater omentum, lesser omentum

pancreas

head, tail

pancreatic duct

duodenal ampulla

villi

liver and gallbladder

falciform ligament, right and left medial lobes, right and left lateral lobes, caudate lobe, common bile duct, cystic duct

common hepatic duct

small intestine

duodenum, jejunum, ileum

ileocecal sphincter

large intestine

cecum

ascending colon, transverse colon, descending colon, rectum, anus, mesocolon

ANATOMY OF THE DIGESTIVE SYSTEM

Laboratory Report Exercise 43

A. Matching

Match each structure listed on the left with the correct description on the right.

1. _____ pyloric sphincter	**A.**	fatty apron hanging off stomach
2. _____ greater omentum	**B.**	muscle layer of mucosa
3. _____ incisor	**C.**	muscle folds of stomach wall
4. _____ esophageal hiatus	**D.**	opening between mouth and pharynx
5. _____ taenia coli	**E.**	hard outer layer of tooth
6. _____ haustra	**F.**	pouches in colon wall
7. _____ mesentery	**G.**	folds of intestinal wall
8. _____ muscularis mucosae	**H.**	tooth used for snipping food
9. _____ muscularis externa	**I.**	longitudinal muscle of colon
10. _____ serosa	**J.**	gum surrounding tooth
11. _____ fauces	**K.**	tooth used for grinding
12. _____ lingual frenulum	**L.**	valve between stomach and duodenum
13. _____ gingiva	**M.**	swollen duct transporting bile and pancreatic juice
14. _____ enamel	**N.**	connective membrane of intestines
15. _____ rugae	**O.**	passageway for esophagus in diaphragm
16. _____ plicae	**P.**	major muscle layer superficial to submucosa
17. _____ molar	**Q.**	middle segment of small intestine
18. _____ jejunum	**R.**	duct joined by cystic duct
19. _____ common bile duct	**S.**	anchors tongue to floor of mouth
20. _____ duodenal ampulla	**T.**	also called the adventitia

B. Short-Answer Questions

1. List the three major pairs of salivary glands and the type of saliva each gland secretes.

2. List the modifications of the intestinal wall that increase surface area.

3. List the accessory organs of the digestive system.

4. How is the wall of the stomach different from the wall of the esophagus?

5. List the regions of the large intestine.

6. Trace a drop of bile from site of production to the release of bile in the intestinal lumen.

7. Discuss the anatomy of the pancreas.

8. Where is the appendix located?

C. Drawing

1. Draw a longitudinal section detailing the anatomy of a typical tooth.

2. Draw a cross section of the wall of the small intestine showing the structures that increase surface area for digestion and absorption of food.

Digestive Physiology

OBJECTIVES

On completion of this exercise, you should be able to:

- Describe why enzymes are used to digest your food.
- Discuss dehydration synthesis and hydrolysis reactions.
- Describe the chemical composition of carbohydrates, lipids, and proteins.
- List the enzymes and substrates for each digestion experiment.
- Describe each chemical test performed on the substrates and analyze the results of each experiment.
- Discuss the function of a control group in scientific experiments.

WORD POWER

catabolism (cata—to break down)
monomer (mono—one)
hydrolysis (lysis—loosen)
saccharide (sacchar—sugar)
glycogen (glyco—sweet)
disaccharide (di—two)
polysaccharide (poly—many)

MATERIALS

water bath
thermometer
test tube racks
10 test tubes
wax marker
10% starch solution
amylase enzyme solution
Lugol's solution (IKI)
Benedict's solution
squeeze bottles for reagents

litmus
whipping cream
pancreatic lipase enzyme solution
1% protein solution
Biuret's solution
hot plate
beaker
test tube holder
pH paper

INTRODUCTION

Before nutrients can be converted to usable energy by cells, the large organic **macromolecules** in food must be **catabolized** or broken down into **monomers**, the building blocks of macromolecules. These chemical reactions are controlled by enzymes produced by the digestive system. **Enzymes** are protein (organic) **catalysts** that lower **activation energy**, the energy required for a chemical reaction (see Figure 44.1). Without enzymes, the body would have to heat up to dangerous temperatures to provide the activation energy necessary to decompose ingested food. Enzymes have a narrow range of physical conditions in which they operate at maximum efficiency. Temperature and pH are two important factors of enzymatic reactions. For example, the enzymes involved in protein digestion require a different pH. The enzyme pepsin is most active in acidic conditions of the stomach. The next protein-digesting enzyme in the sequence, trypsin, requires an alkaline environment.

An enzymatic reaction involves reactants, called the **substrate**, and results in a **product** (see Figure 44.2). The enzyme has an **active site** where the substrate binds. Only a substrate that is compatible with an enzyme's active site is metabolized, and the enzyme is said to have **specificity** for compatible substrates. Upon completion of the

659

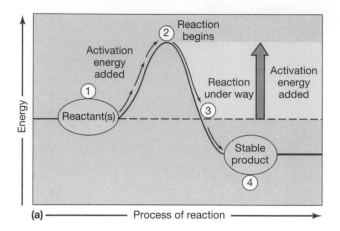

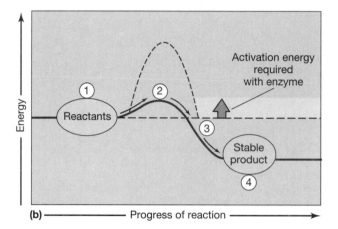

Figure 44.1
Activation Energy and Enzyme Function.
(a) Before a reaction can begin, considerable activation energy must be provided. In this diagram the activation energy represents the energy required to proceed from point 1 to point 2.
(b) The activation energy requirement of the reaction is much lower in the presence of an appropriate enzyme. The allows the reaction to take place much more rapidly, without the need for extreme conditions that would harm cells.

chemical reaction the product is released, and the enzyme, unaltered in the reaction, can bind to another substrate and repeat the reaction.

Each enzyme has a narrow range of physical conditions in which it operates at maximum efficiency. Temperature and pH are two important factors in enzymatic reactions. A good example of enzyme conditions are those required in the protein digestion experiment in this exercise. The stomach produces an inactive enzyme, a **proenzyme**, called pepsinogen. When pepsinogen is mixed with stomach acid, the proenzyme is activated into the protein-digesting enzyme **pepsin** which begins to catabolize large protein molecules in the food. When materials from the stomach move into the small intestine, the pancreas releases sodium bicarbonate which buffers the stomach acid and raises the pH to 7–8. Because pepsin is active only at an acidic pH, it becomes less active in the small intestine while other enzymes take their respective turns metabolizing the substrate. Like all protein molecules, an enzyme's function is related to its structural shape, much as the shape of a key determines which lock it fits. All enzyme experiments in this exercise will be incubated in a warm-water bath set at body temperature, 37° C. Too high a temperature **denatures** an enzyme, causing a change in shape, and destroys the enzyme.

Your digestive system catabolizes the food you ingest with a complex sequence of enzymes. If an enzyme is absent or secreted in an insufficient quantity, the substrate cannot be digested. For example, some individuals are *lactose-intolerant* and do not produce the enzyme lactase which digests lactose, commonly called milk sugar. If a lactose-intolerant individual consumes dairy products, which are high in lactose, the sugar remains in the digestive tract and is slowly digested by bacteria. This results in gas, intestinal cramps, and diarrhea.

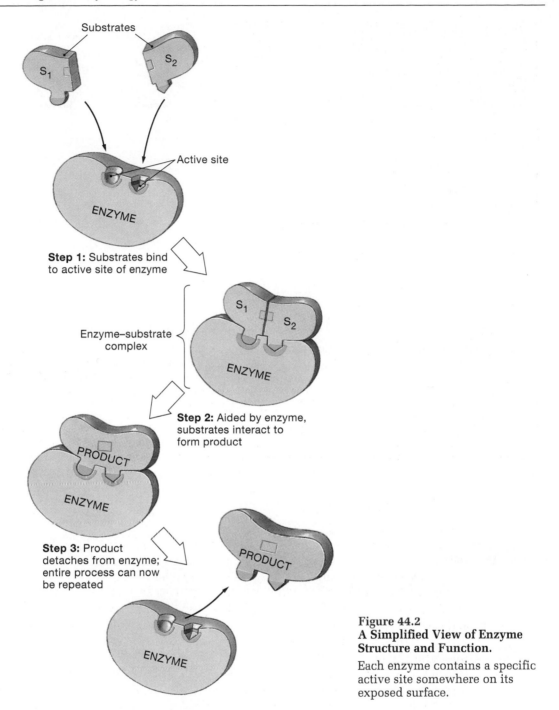

Step 1: Substrates bind to active site of enzyme

Enzyme–substrate complex

Step 2: Aided by enzyme, substrates interact to form product

Step 3: Product detaches from enzyme; entire process can now be repeated

Figure 44.2
A Simplified View of Enzyme Structure and Function.

Each enzyme contains a specific active site somewhere on its exposed surface.

Enzymes as a group are involved in both catabolic (decomposition) and anabolic (synthesis) reactions. A specific enzyme, however, functions in only one type of reaction. Enzymes of the digestive system generally cause catabolic reactions to metabolize your food into smaller molecules which can cross cell membranes and supply raw materials for cellular respiration.

Figure 44.3 details carbohydrate synthesis and decomposition reactions. Macromolecules are broken down into monomers by a process called **hydrolysis** in which water is split and added to specific areas of the macromolecule to break chemical bonds. **Dehydration synthesis** reactions remove a molecule of water between two free monomers and cause them to bond together and form a macromolecule.

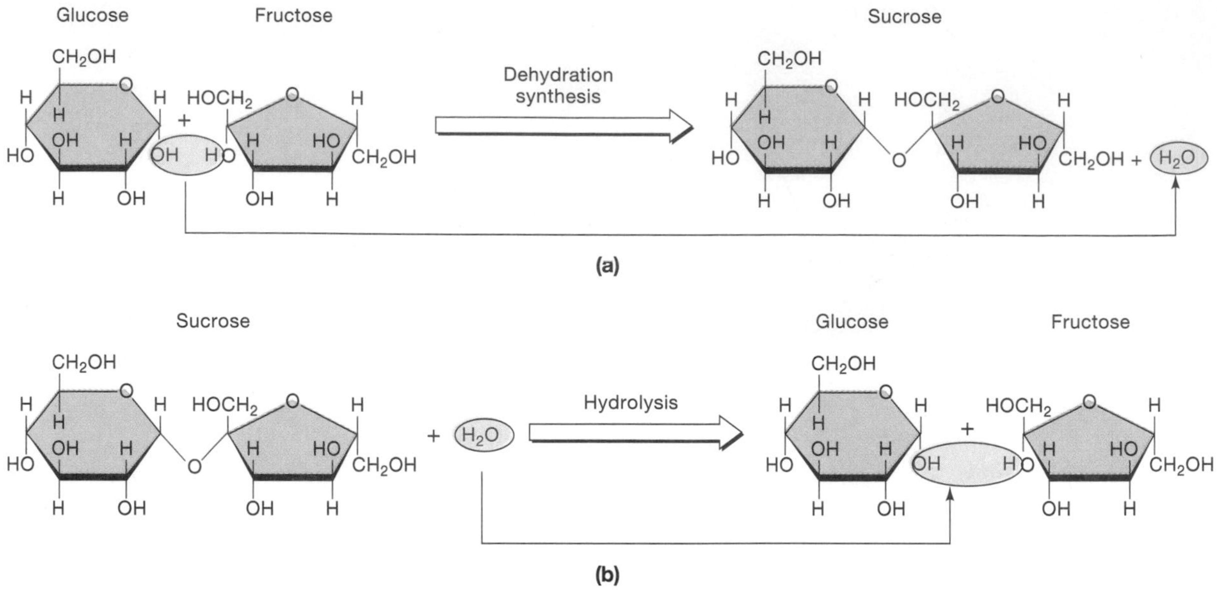

Glucose   Fructose   Sucrose

(a)

Sucrose   Glucose   Fructose

(b)

Figure 44.3
The Formation and Breakdown of Complex Sugars.
(a) During dehydration synthesis two molecules are joined by the removal of a water molecule. In this example, glucose and fructose are combined to form the disaccharide sucrose. **(b)** Hydrolysis reverses the steps of dehydration synthesis; a complex molecule is broken down by the addition of a water molecule.

In this laboratory exercise you will study the conditions necessary for the enzymatic reactions for carbohydrate, lipid, and protein digestion. Table 44.1 summarizes the time and materials required for each enzyme experiment. Use this table to help manage your laboratory time.

TABLE 44.1	Summary of Enzyme Experiments		
	Carbohydrate	*Lipid*	*Protein*
Incubation time	$\frac{1}{2}$ hr	1 hr	1 hr
Number of test tubes required	six	two	two
Solutions required	Starch solution	Litmus cream	Protein solution
	Amylase	Lipase	Pepsinogen
	Lugol's solution		Hydrochloric acid
	Benedict's reagent		Biuret's solution

A. Digestion of Carbohydrate

Starch and sugar molecules are classified as **carbohydrates (kar-bō-HĪ-dratz).** These molecules are composed of **saccharide** molecules. A carbohydrate composed of a single saccharide is a **monosaccharide (mon-ō-SAK-uh-rid)**, a simple sugar. Glucose and fructose are examples of monosaccharides. Monosaccharides are the monomers used to build larger sugar and starch molecules. A **disaccharide (dī-SAK-uh-rid)** is formed when two monosaccharides bond together by a dehydration synthesis reaction. Table sugar, sucrose, is a common disaccharide. Lactose, mentioned previously, is also a dis-

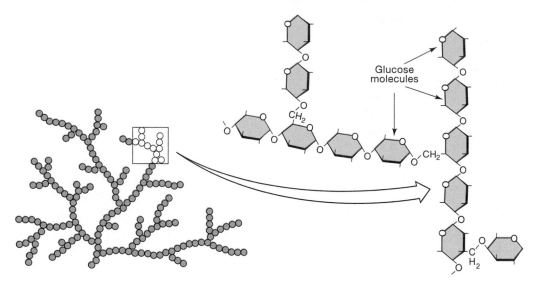

Figure 44.4
Polysaccharides.
Liver and muscle cells store glucose in glycogen molecules.

accharide. Complex carbohydrates are **polysaccharides (pol-ē-SAK-uh-ridz)** and consist of long chains of monosaccharides (see Figure 44.4). Cells store polysaccharides as future energy sources. Plant cells store carbohydrate as **starch**, the molecules found in potatoes and grains. Animal cells, such as muscle and liver cells, store polysaccharides as **glycogen (GLĪ-kō-jen)**.

In this experiment you will start with a polysaccharide substrate, a starch solution prepared by your instructor, and a carbohydrate-digesting enzyme called amylase. Although amylase is easily obtained from your saliva, your lab instructor may provide amylase produced by a nonhuman source.

LABORATORY ACTIVITY CARBOHYDRATE DIGESTION

SAFETY TIPS

- Read all procedures before doing an experiment.
- Wear gloves and safety glasses while pouring reagents and working near water baths.
- Report all spills and broken glass to the laboratory instructor.

MATERIALS

Refer to Table 44.1 for solutions and chemicals.

    wax marker
    37° C water bath
    test tube rack
    boiling water bath

I. Preparation
PROCEDURES

1. Number six test tubes with a wax marker. Fill tubes 1 and 2 with 20 ml of the starch solution. Be sure to shake the solution before pouring.

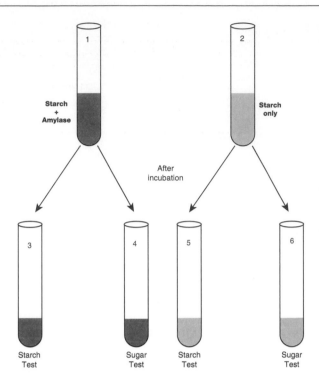

Figure 44.5
Overview of Carbohydrate Digestion Procedures.

2. Add 20 ml of amylase enzyme to test tube 1 only.

3. Place tubes 1 and 2 in a water bath set at body temperature, 37° C. Incubate the solutions for a minimum of 30 min and then remove both from the water bath.

II. Analysis
PROCEDURES

1. Divide the solution in test tube 1 equally into tubes 3 and 4 (see Figure 44.5).

2. Divide the solution in test tube 2 equally into tubes 5 and 6.

3. Use the iodine test for the presence of starch in tubes 3 and 5 as follows:
   a. Place one or two drops of **Lugol's solution (IKI)** in each test tube.
   b. A dark-blue color will show the presence of starch.
   c. Record your observations in Table 44.2.

4. Use Benedict's test for monosaccharides in tubes 4 and 6 as follows:
   a. Place 10 ml of **Benedict's reagent** in each test tube and mix by gently swirling the test tubes.
   b. Set each tube in a boiling water bath for 4–5 min. Do not point the test tubes toward anyone because the solution could splatter and cause a burn.
   c. A color change will show the presence of monosaccharides. Benedict's solution turns from light olive to dark orange, depending on the quantity of reducing sugars such as glucose and fructose.
   d. Record your observations in Table 44.2.

B. Digestion of Lipid

There are many classes of fats or lipids. Most dietary lipids are **monoglycerides** and **triglycerides**. These lipids are constructed of one or more **fatty acid** molecules bonded to a **glycerol** molecule. Figure 44.6 shows the formation and decomposition of a monoglyceride.

TABLE 44.2	Digestion of Starch by Amylase		
Benedict's Test for Sugar		*Lugol's Test for Starch*	
Tube 3	_____	Tube 4	_____
Tube 5	_____	Tube 6	_____
Analysis	_____	Analysis	_____

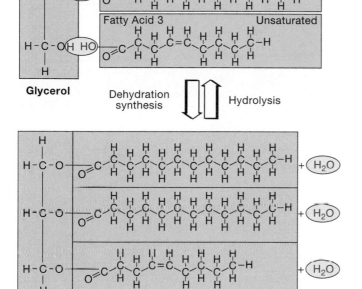

Figure 44.6
Triglyceride Formation.

The formation of a triglyceride involves the attachment of fatty acids to the carbons of a glycerol molecule. In this example, a triglyceride is formed by the attachment of one unsaturated and two saturated fatty acids to a glycerol molecule.

Pancreatic juice contains **pancreatic lipase**, a lipid-digesting enzyme that hydrolyzes triglycerides into monoglycerides and free fatty acids. In this experiment, the released fatty acids will cause a pH change in the test tube, an indication that lipid digestion has occurred. The lipid substrate you will use is whipping cream to which a pH indicator, **litmus**, has been added. Litmus is blue in basic (alkaline) conditions and pink in acidic conditions.

LABORATORY ACTIVITY   LIPID DIGESTION

MATERIALS

Refer to Table 44.1 for solutions and chemicals.

    wax marker

    37° C water bath

    test tube rack

I. Preparation
PROCEDURES

1. Number two test tubes 7 and 8. Fill each tube one-fourth full with litmus cream.
2. Add the same amount of pancreatic lipase as there is litmus cream to tube 7. Do not add the enzyme to tube 8.
3. Record the smell and color of the solution in each test tube in Table 44.3.
4. Incubate the test tubes for 1 hr in a water bath set at body temperature.

II. Analysis
PROCEDURES

1. After 1 hr, remove the test tubes from the water bath.
2. Record the color of the litmus cream in each tube in Table 44.3.
3. Carefully smell the solutions by wafting fumes from the tube up to your nose. Record the smell in Table 44.3.

TABLE 44.3	**Digestion of Lipid by Pancreatic Lipase**				
	Tube 7 **Litmus Cream + Enzyme**			**Tube 8** **Litmus Cream Only**	
	Start	End		Start	End
Smell	_____	_____		_____	_____
Color	_____	_____		_____	_____
Analysis	_____	_____		_____	_____

C. Digestion of Protein

Proteins are composed of long chains of **amino acids** bonded together by **peptide bonds**. There are 20 different amino acids, and their positions in a protein molecule determines the function of the protein. Figure 44.7 summarizes protein structure.

In this experiment, you will use an inactive proenzyme, pepsinogen. The proenzyme alone cannot cause digestion but must be activated by certain chemical conditions. By doing this experiment, you will not only learn about protein digestion but will also discover the importance of the environment in which the enzyme operates. The protein substrate you will use was obtained from gelatin, the jellylike substance in Jell-O.

LABORATORY ACTIVITY   PROTEIN DIGESTION

MATERIALS

Refer to Table 44.1 for solutions and chemicals.

    wax marker
    37° C water bath
    test tube rack
    pH paper

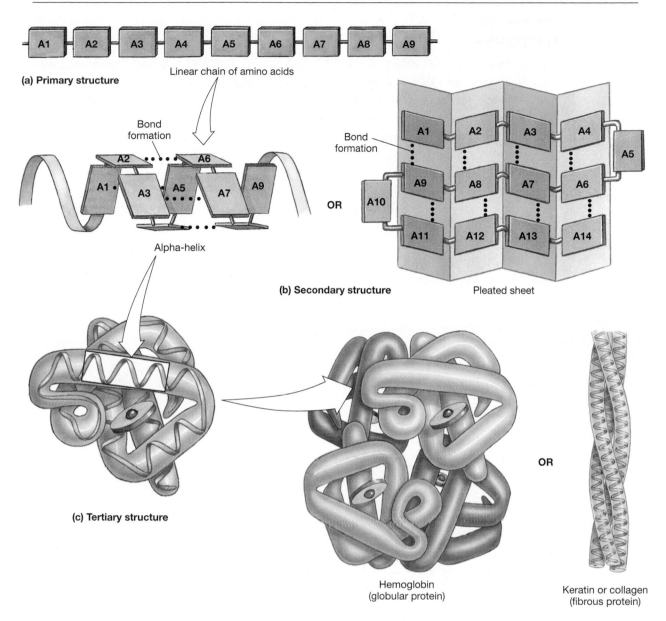

(a) Primary structure

Linear chain of amino acids

Bond formation

Alpha-helix

(b) Secondary structure

Pleated sheet

OR

Bond formation

(c) Tertiary structure

Hemoglobin (globular protein)

OR

Keratin or collagen (fibrous protein)

(d) Quaternary structure

Figure 44.7
Protein Structure.

(a) The primary structure of a polypeptide is the sequence of amino acids (A1, A2, A3, and so on) along its length. (b) Secondary structure is primarily the result of hydrogen bonding along the length of the polypeptide chain. Such bonding often produces a simple spiral (an alpha-helix) or a flattened arrangement known as a pleated sheet. (c) Tertiary structure is the coiling and folding of a polypeptide. Within the cylindrical segments of this globular protein, the polypeptide chain is arranged in an alpha-helix. (d) Quaternary structure develops when separate polypeptide subunits interact to form a larger molecule. A single hemoglobin molecule contains four globular subunits. Hemoglobin transports oxygen in the blood; the oxygen binds reversibly to the heme units. In keratin and collagen, three fibrous subunits intertwine. Keratin is a tough, water-resistant protein in skin, hair, and nails. Collagen is the principal extracellular protein in most organs.

I. Preparation
PROCEDURES

1. Number two test tubes 9 and 10. Add 10 ml of protein solution to each test tube.
2. To tube 9 add 10 ml of the proenzyme **pepsinogen**.
3. To tube 10 add 10 ml of pepsinogen and 10 ml of 0.5M **hydrochloric acid** (HCl).
4. Measure the pH in each test tube and record your data in Table 44.4.
5. Incubate both tubes for 1 hr in a 37° C water bath and then remove.

II. Analysis
PROCEDURE

1. After 1 hr, remove tubes 9 and 10 from the water bath. To test for protein digestion use **Biuret's reagent** which turns pink in the presence of free amino acids and purple in the presence of protein.
2. Add 50 ml of Biuret's reagent to each test tube.
3. Observe the color of the protein solution. Record your observations in Table 44.4. If the color is too subtle, add 20 ml more of Biuret's reagent.

TABLE 44.4   **Digestion of Protein by Pepsin**

	Tube 9 *Protein + Enzyme*	*Tube 10* *Protein + Enzyme + HCl*
pH of solution	_____	_____
Color after Biuret's Test	_____	_____
Analysis	_____	_____

DIGESTIVE PHYSIOLOGY CHECKLIST

This is a list of bold terms presented in Exercise 44. Use this list for review purposes after you have completed the laboratory activities.

INTRODUCTION

- Figures 44.1–44.3

macromolecules
catabolized
monomers
enzyme
proenzyme
substrate
active site
product
activation energy
catalyst
specificity
hydrolysis
dehydration synthesis

CARBOHYDRATE DIGESTION

- Figures 44.3–44.5

saccharide
monosaccharide
  glucose
disaccharide
polysaccharide
  glycogen
Benedict's reagent
Lugol's solution (IKI)
amylase

LIPID DIGESTION

- Figure 44.6

fatty acids
glycerol
monoglyceride
triglyceride
litmus
pancreatic lipase

PROTEIN DIGESTION

- Figure 44.7

protein
  amino acids
  peptide bonds
denature
Biuret's reagent
pepsinogen
pepsin
hydrochloric acid

DIGESTIVE PHYSIOLOGY

Laboratory Report Exercise 44

A. Fill in the Blanks

1. An organic catalyst that lowers the activation energy of a chemical reaction is called an _____.

2. The reactant in an enzymatic reaction is called the _____.

3. Enzymes have substrate specificity because of the presence of _____.

4. Large macromolecules can be catabolized into smaller units called _____.

5. The three primary nutrients are _____, _____, and _____.

6. The reagent used to test for the presence of starch was _____ (IKI).

7. The reagent used to test for the presence of sugar was _____.

B. Short-Answer Questions

1. In each digestion experiment you used a control tube that had no enzyme or other reagent added to the substrate. What was the function of the control?

2. Why were both test tubes (1 and 2) used in the starch experiment tested for starch and sugar at the conclusion of the incubation period?

3. Discuss the basic processes of dehydration synthesis and hydrolysis reactions.

4. Briefly outline the chemical composition of carbohydrates, lipids, and proteins.

5. Why was a warm-water bath used to incubate test tubes containing substrates and enzymes?

Anatomy of the Urinary System

OBJECTIVES

On completion of this exercise, you should be able to:

- Identify and describe the basic anatomy of the urinary system.
- Trace the blood flow through the kidney.
- Explain the function of the kidney.
- Identify the basic components of the nephron.
- Describe the differences between the male and female urinary tracts.

WORD POWER

hilus (hilus—a depression)
renal (ren—kidney)
rugae (rugae—wrinkle)
trigone (trigone—triangle)
arcuate (arcuate—bowed, arched)
afferent (afferent—to bring to)
glomerulus (glomus—a ball of yarn)
efferent (efferent—to bring out)
vasa (vas—vessel)
recta (recta—straight)

MATERIALS

human torso models
kidney models
model of the nephron
compound microscope
prepared slide: kidney
preserved sheep kidneys
preserved cat
dissecting instruments
latex gloves

INTRODUCTION

All cells produce waste products through their metabolic activities. Wastes such as urea, creatinine, carbon dioxide, nitrogen wastes, and excess electrolytes must be eliminated from the body to maintain homeostasis. Several organ systems eliminate wastes from the body. The lungs remove carbon dioxide during exhalation, the digestive tract eliminates undigested solid wastes, and the skin removes salts and urea in sweat. The primary function of the urinary system is to control the composition, volume, and pressure of blood by removing and restoring selected amounts of water and solutes. These eliminated products are collectively called urine. The urinary system, highlighted in Figure 45.1, is composed of a pair of kidneys and ureters, a urinary bladder, and a urethra.

A. The Kidneys

The kidneys lie on the posterior surface of the abdomen between the waist and the eleventh and twelfth pairs of ribs. The right kidney is positioned lower than the left kidney because of the position of the liver. The kidneys, adrenal glands, and most of the ureters are *retroperitoneal* because they are located behind the parietal peritoneum.

Figure 45.1
An Introduction to the Urinary System.

An anterior view of the urinary system,
showing the positions of the kidneys and
other components.

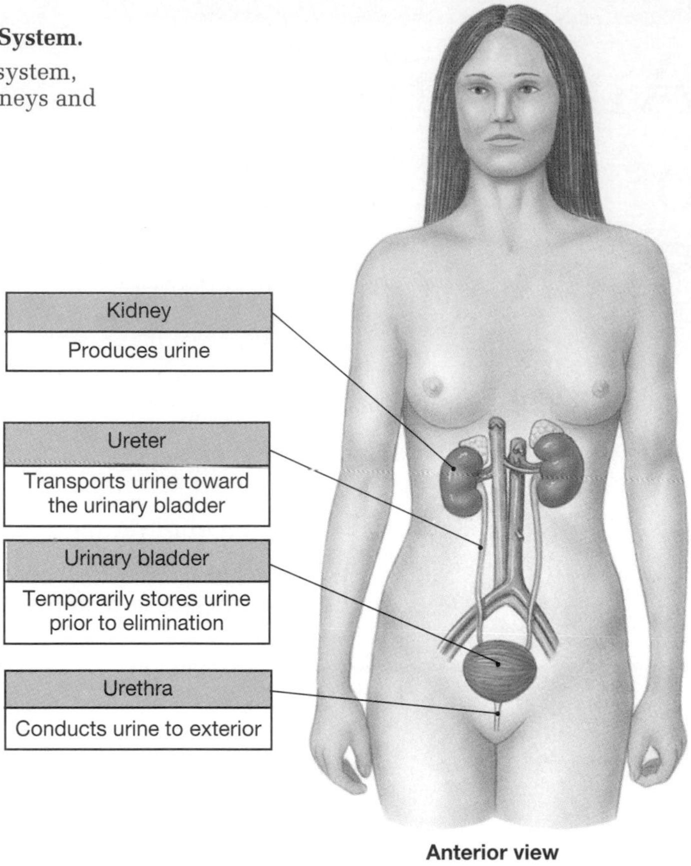

Kidney
Produces urine

Ureter
Transports urine toward the urinary bladder

Urinary bladder
Temporarily stores urine prior to elimination

Urethra
Conducts urine to exterior

Anterior view

Figure 45.2 illustrates the sectional anatomy of the kidney. Each kidney is secured in the abdominal cavity by three layers of tissue: the renal capsule, the adipose capsule, and the renal fascia. The **renal capsule** is a fibrous membrane that envelopes the external surface of each kidney. It protects the kidneys from trauma and infection. The **adipose capsule** is a mass of adipose that envelopes the kidneys and protects them from trauma as well as helping to anchor them to the abdominal wall. The outer layer, the **renal fascia**, is composed of dense irregular connective tissue used to anchor the kidneys to the abdominal wall. Nephroptosis or "floating kidneys" results when the integrity of the adipose capsule or the renal fascia is jeopardized because of excessive weight loss and there is therefore less adipose to secure the kidneys. This can result in the pinching or kinking of one or both ureters, preventing the normal flow of urine to the urinary bladder.

A kidney is about 13 cm (5.12 in.) long and 2.5 cm (1 in.) thick. The medial aspect of the kidney contains a notch called the **hilus**. Blood vessels, nerves, lymphatics, and the ureters enter and exit the kidney through the hilus. The hilus also leads to a cavity in the kidney called the **renal sinus**. The **cortex** is the outer light-red layer or area of the kidney. Deep to the cortex is a region called the **medulla** which consists of triangular **renal pyramids** projecting toward the center of the kidney. Areas of the cortex extending between the renal pyramids are **renal columns**. At the apex or tip of each renal pyramid is a **renal papilla** which empties urine into a small cuplike space called the **minor calyx (KĀ-liks)**. Several minor calyces (KĀL-i-sēz) empty into a common space, the **major calyx**. These larger calyces merge to form the **renal pelvis**. The cavity surrounding the renal pelvis is the **renal sinus**. Each kidney has a single ureter, a muscular tube that transports urine from the renal pelvis to the urinary bladder.

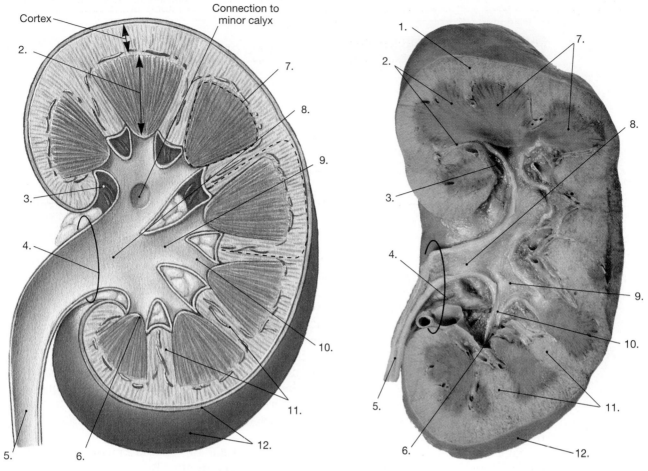

Cortex

Connection to minor calyx

1. 2. 3. 4. 5. 6. 7. 8. 9. 10. 11. 12.

(a) Frontal section of left kidney, anterior view

Minor calyx

Major calyx

Renal pelvis

Ureter

(b) Calyces and renal pelvis

1. _____
2. _____
3. _____
4. _____
5. _____
6. _____
7. _____
8. _____
9. _____
10. _____
11. _____
12. _____

Figure 45.2
Structure of the Kidney.

(a) Frontal section through the left kidney, showing major structures. The outlines of a renal lobe and a renal pyramid are indicated by dotted lines. **(b)** Shadow drawing to show the arrangement of the calyces and renal pelvis within the kidney.

LABORATORY ACTIVITY	OBSERVATIONS OF THE KIDNEY

MATERIALS

kidney models

PROCEDURES

1. Complete Figure 45.2 by labeling each structure of the kidney.
2. Examine the kidney models and charts and locate each structure shown in Figure 45.2.

B. The Nephron

The basic functional unit of the kidney is called the **nephron (NEF-ronz)**. Approximately 1.25 million nephrons occur in each kidney. In each nephron, materials are removed from the blood by a process called *filtration* to produce a fluid called *filtrate*. The filtrate circulates through a series of convoluting or twisting tubules and a U-shaped section to reclaim essential materials and remove wastes and excess ions. The remaining filtrate is excreted as urine. The process of urine formation is covered in more detail in Exercise 46.

Each nephron consists of two portions: a **renal corpuscle** and a **renal tubule**. The renal corpuscle includes a **glomerulus (glo-MER-ū-lus)** inside of a **Bowman's capsule**. Blood pressure in the glomerulus forces materials out of the blood, and the resulting filtrate flows into the **capsular space** of the Bowman's capsule. The renal corpuscle empties filtrate into the renal tubule of the nephron.

The renal tubule consists of twisted and straight tubules composed mainly of cuboidal epithelium (see Figure 45.3). The first segment after Bowman's capsule is a coiled tubule called the **proximal convoluted tubule**. The next portion of the tubule, the **loop of Henle (HEN-lē)**, consists of a **thin descending limb**, a loop of the limb, and a **thick ascending limb**. The ascending limb leads to a second convoluted section, the **distal convoluted tubule**. Surrounding the convoluted tubules are the peritubular capillaries, while the loop of Henle is covered with the vasa recta capillary. The nephron ends where the distal convoluted tubule empties into a **collecting duct**. Several nephrons drain into the same collecting duct. These ducts merge with other collecting ducts into a common **papillary duct** which drains into a minor calyx. There are between 25 and 35 papillary ducts per renal pyramid.

Two types of nephrons occur in the kidney, cortical and juxtamedullary. Approximately 85% of the nephrons are **cortical nephrons** and have their glomeruli and most of their tubules in the cortex. **Juxtamedullary (juks-ta-MED-ū-lar-ē) nephrons** have their glomeruli at the junction of the cortex and the medulla. These nephrons have long loops that extend deep into the medulla before turning back toward the cortex.

The renal corpuscle is a double walled capsule called **Bowman's (glomerular) capsule**. Inside this capsule is a capillary called the **glomerulus**. The glomerular capsule has an outer **parietal layer** and an inner **visceral layer** that wraps around the surface of the glomerulus. Between these two layers is the **capsular space**. An afferent arteriole passes into Bowman's capsule and supplies blood to the glomerulus.

The visceral layer of Bowman's capsule consists of specialized cells called **podocytes** (see Figure 45.4). These cells wrap extensions called **pedicels** around the endothelium of the glomerulus. Small gaps between the pedicels are **filtration slits** or slit pores. To be filtered out of the glomerulus a substance must be small enough to pass through the **capillary endothelium** and its **basement membrane** (lamina densa) and squeeze through the filtration slits to enter the capsular space. Any substance that can pass through these layers is removed from the blood as part of the filtrate. The filtrate, therefore, contains both essential materials and wastes.

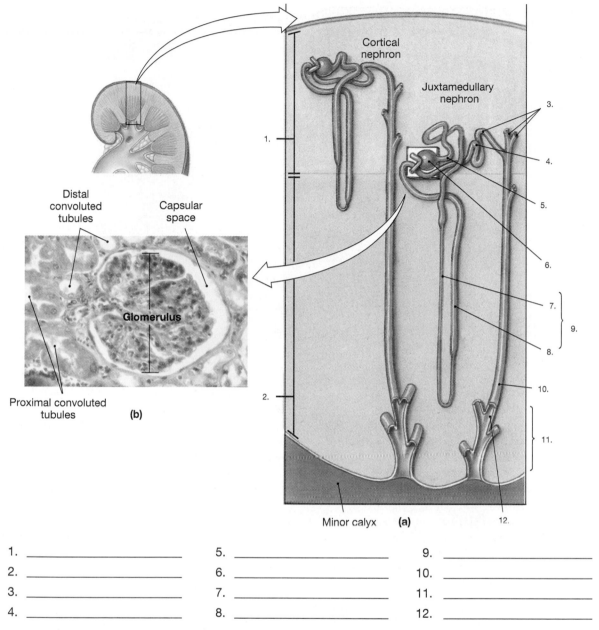

1. _____   5. _____   9. _____

2. _____   6. _____   10. _____

3. _____   7. _____   11. _____

4. _____   8. _____   12. _____

Figure 45.3
Sectional Views of the Nephron.

(a) Approximate locations of the sections in part (a). **(b)** The association of proximal and distal convoluted tubules with a glomerulus. (LM ×370)

The ascending limb of the loop of Henle twists back and comes into contact with the afferent arteriole that supplies its glomerulus. This point of contact is called the **juxtaglomerular apparatus** (see Figure 45.4). The cells of the renal tubule become tall and crowded together and form the **macula densa (MAK-ū-la DEN-sa)**. The macula densa monitors NaCl concentrations in this area of the renal tubule. The smooth muscle cells of the afferent arteriole develop round nuclei and contain an enzyme called renin, a hormone that causes vasodilation of blood vessels and, through the action of the hormone aldosterone, an increase in sodium and water reabsorption in the distal convoluted tubule.

Figure 45.4
The Renal Corpuscle.

(a) Structure and location of a juxtamedullary nephron.
(b) The renal corpuscle. Arrows show the pathway of blood flow.

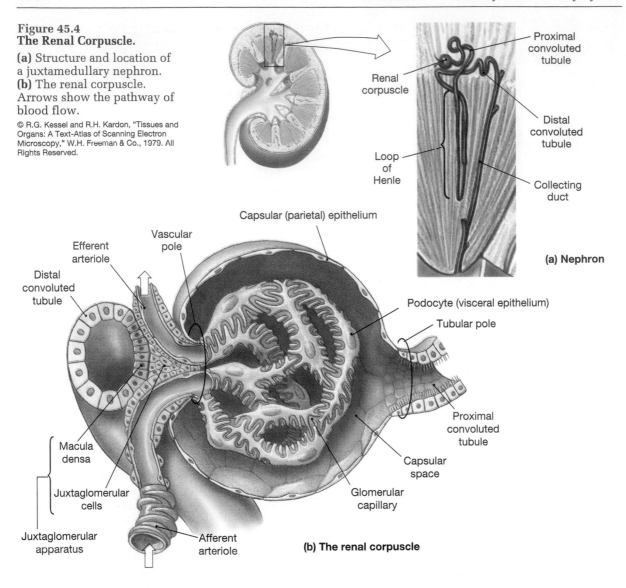

(a) Nephron

(b) The renal corpuscle

LABORATORY ACTIVITY   OBSERVATIONS OF THE NEPHRON

MATERIALS

kidney and nephron models
compound microscope
prepared slide: kidney

PROCEDURES

1. Examine the kidney models and locate all the structures of the nephron. Complete the labeling of the nephron in Figure 45.3 and of the renal corpuscle in Figure 45.4.

2. Obtain a microscope and a slide of the kidney. Use the micrographs in Figure 45.3 for reference.

   a. Use 10× and 40× magnifications and locate the renal cortex and renal medulla. Examine several renal tubules, visible as ovals on the kidney slide. Is the renal capsule visible on the outer margin of the kidney tissue?

    b. Locate a renal corpuscle which appears as a small knot in the cortex. Distinguish among the parietal layer of Bowman's capsule, the capsular space, and the glomerulus.

    c. Draw a section of the kidney slide in the space below. Label the cortex, a renal corpuscle, and a renal tubule. The medulla is usually not present on most kidney slides.

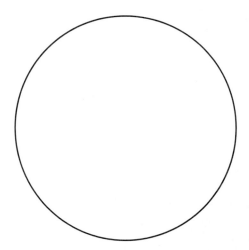

C. Blood Supply

Every minute approximately 25% of the total blood volume travels through the kidneys. This blood is delivered to a kidney by the **renal artery** that branches off the abdominal aorta (see Figure 45.5). Once it enters the hilus of the kidney it divides into five **segmental arteries** which branch into **interlobar arteries** that pass through the renal columns. The interlobar arteries divide into **arcuate (AR-kū-āt) arteries** which cross the bases of the pyramids and enter the cortex as **interlobular arteries**. These arteries branch into **afferent arterioles** that enter the nephron, the site of urine formation.

    When an afferent arteriole enters a nephron, it forms a capillary tuft called the **glomerulus**. A smaller **efferent arteriole** exits the glomerulus and forms two additional capillary beds around the tubular portion of the nephron. The capillaries surrounding the cortical nephrons are called **peritubular capillaries**. In juxtamedullary nephrons the peritubular capillaries surround only the proximal and distal convoluted tubules. The loop of Henle in juxtamedullary nephrons is surrounded by **vasa recta** capillaries. Both capillary networks are involved in the reabsorption of materials from the filtrate of the renal tubules back into the blood. The vasa recta and peritubular capillaries drain into **interlobular veins** which then drain into **arcuate veins** along the base of the renal pyramids. **Interlobar veins** pass through the renal columns and join the **renal vein** which drains into the inferior vena cava.

LABORATORY ACTIVITY   BLOOD SUPPLY TO THE KIDNEY

MATERIALS

    kidney and nephron models

PROCEDURES

    1. Label each blood vessel in Figure 45.5.

    2. On the kidney models and charts trace the blood vessels that supply and drain the kidneys.

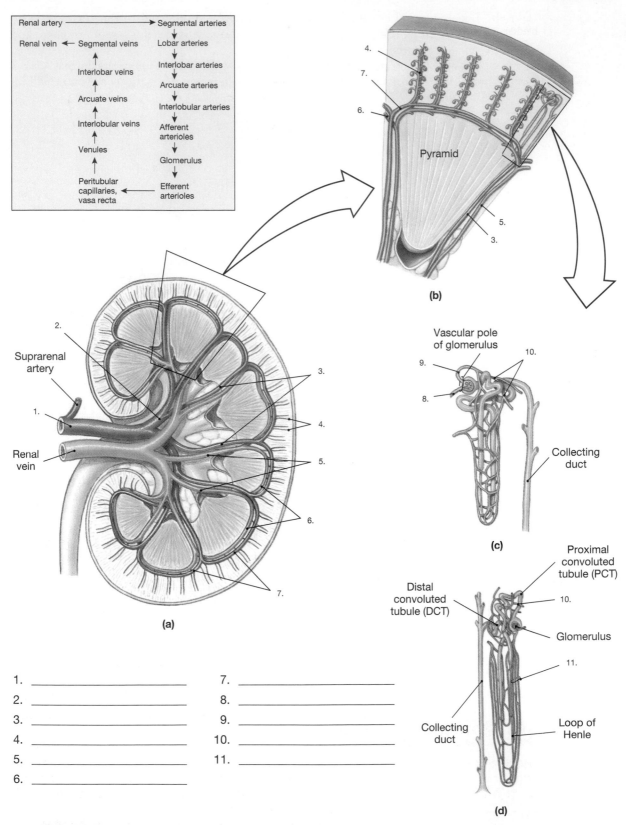

Figure 45.5
Blood Supply to the Kidneys.

(a) Sectional view, showing major arteries and veins. (b) Circulation in the cortex. (c) Circulation to a cortical nephron. (d) Circulation to a juxtamedullary nephron.

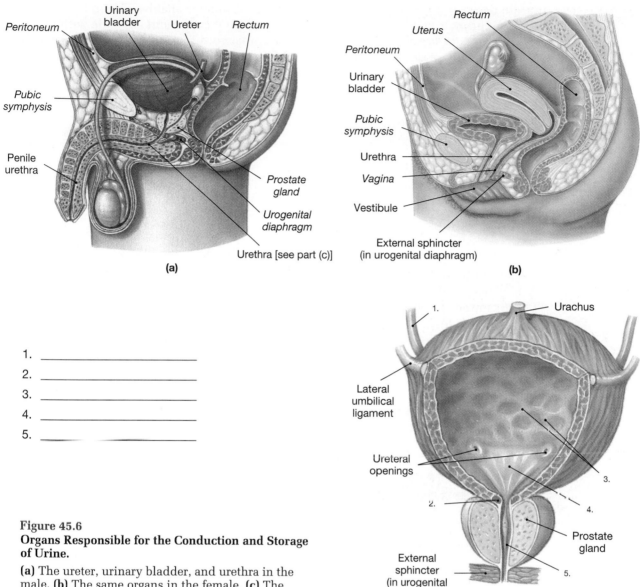

(a)

(b)

1. _____

2. _____

3. _____

4. _____

5. _____

Figure 45.6
Organs Responsible for the Conduction and Storage of Urine.

(a) The ureter, urinary bladder, and urethra in the male. **(b)** The same organs in the female. **(c)** The urinary bladder in a male.

(c)

D. The Ureters, Urinary Bladder, and Urethra

The two **ureters** conduct urine, by way of gravity and peristalsis, to the urinary bladder. The ureters, detailed in Figure 45.6, are lined with a mucus-producing epithelium to protect the ureteral walls from the acidic urine. The ureters enter the urinary bladder on the posterior surface at the top two corners of the **trigone (TRĪ-gōn)**, a smooth triangle-shaped area of the urinary bladder floor.

The **urinary bladder** is a hollow muscular organ that functions in the temporary storage of urine. In males it lies between the pubic symphysis and the rectum. In females it is posterior to the pubic symphysis, inferior to the uterus, and superior to the vagina. The inner wall contains folds called **rugae** which allow expansion and shrinkage of the bladder as it fills with and empties urine.

A single duct, the **urethra**, drains urine from the bladder out of the body. Around the opening to the urethra are two sphincter muscles, the internal and external urethral sphincters. The **internal urethral sphincter** is composed of smooth muscle and is therefore under involuntary nerve control. The **external urethral sphincter** is skeletal muscle tissue and under voluntary control. This sphincter enables you to control the release of urine from the urinary bladder. As the bladder fills and expands with urine, stretch receptors signal motor neurons in the spinal cord to relax the internal urethral sphincter and contract the **detrusor (de-TROO-sor) muscle** of the bladder wall. This reflex automatically causes closing of the external urethral sphincter. When convenient, the external urethral sphincter is relaxed and urination, or *micturition*, occurs.

LABORATORY ACTIVITY   THE URETERS, URINARY BLADDER, AND URETHRA

MATERIALS

urinary system models and charts

PROCEDURES

1. Locate the ureters on a urinary tract model. Trace the transport of urine into the ureter from the renal papilla. Label all structures in Figure 45.6.
2. Examine the wall of the urinary bladder. Identify the trigone and the rugae. Which structures control emptying of the urinary bladder?
3. Examine the urethra on the models. How does this structure differ in males and females?

E. Dissection of the Sheep Kidney

The sheep kidney is very similar to the human kidney in both size and anatomy. Dissection of a sheep kidney will reinforce your observations of kidney models in the laboratory. Use the cadaver photograph in Figure 45.2 for reference during your dissection.

LABORATORY ACTIVITY   DISSECTION OF THE SHEEP KIDNEY

MATERIALS

preserved sheep kidneys
dissection kit and dissection pan
disposable gloves, safety glasses

PROCEDURES

1. Wear latex gloves while dissecting to protect your skin from the fixatives used to preserve the kidneys. Also, wear safety glasses to prevent preservatives from entering your eyes. Rinse the kidney with water to remove excess preservative.
2. Examine the external features of the kidney. Use Figure 45.2 as a guide and locate the **hilus**, a concave slit on the medial border where the renal vessels and the ureters pass through the kidney. Locate the outer renal capsule and gently lift it with a teasing needle. Below this capsule is a light-pink area called the **cortex**.
3. With a scalpel, make a longitudinal cut to divide the kidney into anterior and posterior portions. A single long, smooth cut is less damaging to the internal anatomy than a sawing motion with the scalpel.
4. Distinguish between the cortex and the darker **medulla** which is organized into many triangular **renal pyramids**. The bases of the pyramids face the cortex, while the tips or apices narrow into **renal papillae**.

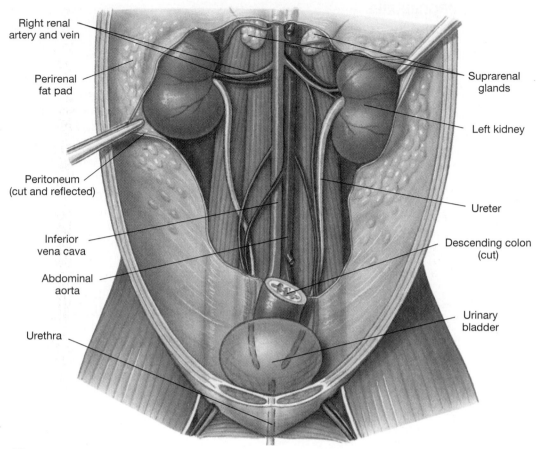

Right renal
artery and vein

Perirenal
fat pad

Peritoneum
(cut and reflected)

Inferior
vena cava

Abdominal
aorta

Urethra

Suprarenal
glands

Left kidney

Ureter

Descending colon
(cut)

Urinary
bladder

Figure 45.7
Urinary System of the Cat.

5. The **renal pelvis** is the large, expanded end of the ureter. Extending from this area are structures called **major calyces** which have smaller channels called **minor calyces** into which the renal papillae project. The minor calyces collect urine from the renal papillae and channel it into the major calyx which then delivers it to the renal pelvis. From there the urine leaves the kidney by way of the **ureter**.

F.   Dissection of the Cat Urinary System

The urinary system of the cat is similar to that of the human. In your dissection of the cat urinary system, trace the pathway of urine from its site of formation in the kidney, through its passage to the urinary bladder, and through the urethra to the exterior of the body. Use Figure 45.7 as a guide during your observation of the cat's urinary system.

LABORATORY ACTIVITY   DISSECTION OF THE CAT URINARY SYSTEM

MATERIALS

preserved cat
dissection kit and dissection pan
disposable gloves, safety glasses

PROCEDURES

1. Wear latex gloves while dissecting to protect your skin from the fixatives used to preserve the cat. Also, wear safety glasses to prevent preservatives from entering your eyes.

2. Reflect the abdominal viscera to the opposite side of the abdominal cavity to locate the large, bean-shaped **kidneys**. They are retroperitoneal and do not appear symmetrical on the dorsal body wall. Both kidneys are surrounded by pads of perirenal fat, the **adipose capsule**, and are enclosed in a tough, fibrous **renal capsule**.

3. Locate the **adrenal glands**, superior to the kidneys and close to the aorta. They are supplied by the suprarenal arteries.

4. Free each kidney by carefully removing the **peritoneum** and the perirenal fat pads.

5. Identify the following structures that pass through the **hilus**, the concave medial surface of the kidney: renal artery, renal vein, and **ureter**.

6. The ureters are cream-colored tubes which descend posteriorly along the dorsal body wall to drain urine into the **urinary bladder**. The bladder is a muscular sac connected to both lateral and ventral walls by suspensory ligaments.

7. The large, expanded region of the bladder is the **fundus**. Posteriorly, the bladder narrows into the **neck** and continues as the **urethra** through which urine passes to the exterior of the body. The bladder must be pulled anteriorly in order to observe the neck and urethra portions.

8. With your scalpel blade, make a frontal (coronal) section through the kidney such that the section passes through the middle of the hilum. Locate the outer, lighter, **cortex** and the inner, darker, **medulla.**

9. Observe the funnel-shaped expansion of the ureter, the **renal pelvis**, which occupies the hollow interior of the kidney, the **renal sinus**, internal to the opening of the hilum. Follow the renal pelvis to a funnel-shaped major calyx and a narrower minor calyx. Locate a single **renal papilla**, the rounded projection of the **renal pyramid** of the medulla in the renal pelvis.

URINARY SYSTEM CHECKLIST

This is a list of bold terms presented in Exercise 45. Use this list for review purposes after you have completed the laboratory activities.

GROSS ANATOMY

• Figures 45.1, 45.2 and 45.6

renal capsule

adipose capsule

renal fascia

hilus

renal sinus

cortex

medulla

renal pyramids

renal columns

minor calyx

major calyces

renal papilla

renal pelvis

ureters

trigone

urinary bladder

rugae

detrusor muscle

internal urethral sphincter

external urethral sphincter

urethra

THE NEPHRON

• Figures 45.3 and 45.4

cortical nephrons

juxtamedullary nephrons

renal corpuscle

glomerulus

capillary endothelium

basement membrane

Bowman's capsule

parietal layer

visceral layer

podocytes

pedicels

filtration slits (slit pores)

capillary endothelium

basement membrane (lamina densa)

capsular space

renal tubule

proximal convoluted tubule

loop of Henle

thin descending limb

thick ascending limb

distal convoluted tubule

collecting duct

papillary ducts

juxtaglomerular apparatus

macula densa

BLOOD SUPPLY TO KIDNEY

• Figure 45.5

renal arteries

segmental arteries

interlobar arteries

arcuate arteries

interlobular arteries

afferent arterioles

glomerulus

efferent arterioles

peritubular capillaries

vasa recta

interlobular veins

arcuate veins

interlobar veins

renal vein

URINARY SYSTEM OF THE CAT

• Figure 45.7

kidneys

adipose capsule

renal capsule

adrenal glands

peritoneum

hilus

ureter

urinary bladder

fundus

neck

urethra

cortex

medulla

renal pelvis

renal sinus

renal papilla

renal pyramid

Name _____   Date _____

Section _____   Number _____

ANATOMY OF THE URINARY SYSTEM
Laboratory Report Exercise 45

A. Matching

   Match the anatomy term on the left with the correct description on the right.

   1. _____ renal papilla   A. drains into collecting duct
   2. _____ cortex   B. located at base of pyramid
   3. _____ Bowman's capsule   C. covers surface of kidney
   4. _____ loop of Henle   D. surrounds glomerulus
   5. _____ renal pelvis   E. surrounds renal pelvis
   6. _____ distal convoluted tubule   F. entrance for blood vessels
   7. _____ efferent arteriole   G. extends into minor calyx
   8. _____ renal sinus   H. tissue between pyramids
   9. _____ renal pyramid   I. U-shaped tubule
   10. _____ arcuate artery   J. transports urine to bladder
   11. _____ hilus   K. functional unit of kidney
   12. _____ renal columns   L. drains glomerulus
   13. _____ renal capsule   M. outer layer of kidney
   14. _____ nephron   N. comprises the medulla
   15. _____ ureter   O. drains major calyces

B. Short-Answer Questions

   1. Describe the components of the renal corpuscle.

   2. List the layers in the renal corpuscle that filtrate must pass through to enter the capsular space.

   3. Why are both essential and waste products removed from the blood during filtration?

4. Describe the tubular components of the nephron.

5. Trace a drop of blood from the abdominal aorta, through a kidney, and into the inferior vena cava.

6. Trace a drop of filtrate from the glomerulus to the renal papillae.

7. Trace a drop of urine from a minor calyx to the urinary bladder.

Physiology of the Urinary System

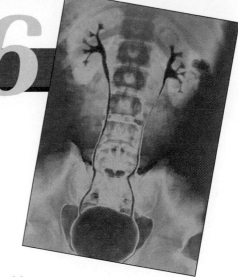

OBJECTIVES

On completion of this exercise, you should be able to:

- Define glomerular filtration, tubular reabsorption, and tubular secretion.
- Describe the normal physical characteristics of urine.
- Recognize normal and abnormal urine constituents.
- Conduct a urinalysis test with special dip sticks.
- Define several urinary conditions.

WORD POWER

bilirubinuria (bili—bile; rubin—red; uria—urine)
hematuria (hemato—blood)
diabetes (diabetes—siphon)

mellitus (mellitus—sweetened with honey)
pyuria (py—pus)

MATERIALS

500-ml urine cups
urinometer
dip sticks (Chemstix or Multistix)
BioHazard disposal container

INTRODUCTION

The kidneys maintain the chemical balance of body fluids by removing metabolic wastes, excess water, and electrolytes from the blood plasma. Three physiological processes occur in the nephrons to produce urine: filtration, reabsorption, and secretion. Each process is highlighted in Figure 46.1. **Filtration** occurs in the renal corpuscle as blood pressure in the glomerulus forces water, small solutes, and ions out of the blood plasma and into the capsular space. The resulting fluid in the capsular space is called **filtrate**. Solutes small enough to pass through the filtration slits surrounding the glomerulus are present in the filtrate. This results in the removal of both wastes and essential solutes from the blood. During **reabsorption**, tubular cells of the nephron reclaim essential ions, nutrients, and water for return to the blood plasma. **Secretion** adds excess ions from the blood and peritubular fluid to the filtrate. The tubular cells actively transport ions, often using countertransport mechanisms that result in the reabsorption of necessary ions and the secretion of harmful ions. As the filtrate passes through the tubules of the nephron, reabsorption and secretion occur. The modified filtrate is processed into **urine** which drips out of the renal papillae into the minor calyxes.

As the filtrate moves through the proximal convoluted tubule (PCT), 60–70% of the water and ions are reabsorbed into the blood. All the organic nutrients, such as glucose and amino acids, are reabsorbed by the PCT. The simple cuboidal epithelium in this part of the nephron has microvilli to increase the surface area for reabsorption. The loop of Henle conserves water and salt while concentrating the filtrate for modification by the distal convoluted tubule (DCT). Reabsorption in the DCT is controlled by two hormones, aldosterone and antidiuretic hormone (ADH). Most of the secretion occurs by the tubular cells of the DCT.

The kidneys filter 25% of the body's blood each minute, producing on average 125 ml/min of filtrate. About 180 liters of filtrate is formed by the glomerulus per day,

Step 1. Glomerular filtration produces a filtrate resembling blood plasma but containing a few plasma proteins. This filtrate has the same solute concentration as plasma or interstitial fluid.

Step 2. In the proximal convoluted tubule, 60–70% of the water and almost all of the dissolved nutrients are reabsorbed. The osmolarity of the tubular fluid remains unchanged.

Step 3. In the PCT and descending loop of Henle, water moves into the surrounding interstitial fluid, leaving a small fluid volume (roughly 20% of the original filtrate) of highly concentrated tubular fluid.

Step 4. The ascending limb is impermeable to water and solutes. The tubular cells actively pump sodium and chloride ions out of the tubular fluid. Because only sodium and chloride ions are removed, urea now accounts for a higher proportion of the solutes in the tubular fluid.

Step 5. The final composition and concentration of the tubular fluid are determined by the events under way in the DCT and the collecting ducts. These segments are impermeable to solutes, but ions can be actively transported into or out of the filtrate under the control of hormones such as aldosterone.

Step 6. The concentration of urine is controlled by variations in the water permeabilities of the DCT and the collecting ducts. These segments are impermeable to water unless exposed to antidiuretic hormone (ADH). In the absence of ADH, no water reabsorption occurs, and the individual produces a large volume of dilute urine. At high concentrations of ADH, the collecting ducts become freely permeable to water, and the individual produces a small volume of highly concentrated urine.

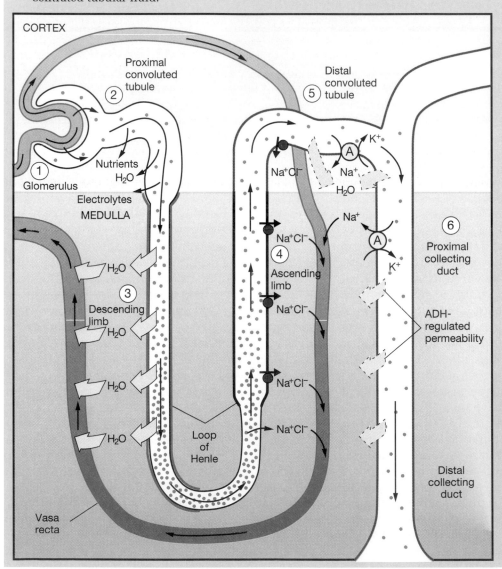

Figure 46.1
Major Steps in Urine Production.

which eventually results in the production of an average daily output of 1.8 liters of urine. The composition of urine can change on a daily basis depending on one's metabolic rate and urinary output. Water accounts for about 95% of the volume of urine. The other 5% contains excess water-soluble vitamins, drugs, electrolytes, and nitrogenous wastes.

A. Composition of Urine

Normal constituents of urine include water, urea, creatinine, uric acid, many electrolytes, and possibly small amounts of hormones, pigments, carbohydrates, fatty acids, mucin, and enzymes.

Urea is produced during **deamination (dē-am-i-NĀ-shun)** reactions that remove ammonia (NH_3) from amino acids. The ammonia combines with CO_2 and forms urea (CH_4ON_2). About 4,600 mg of urea is produced daily, approximately 80% of the nitrogen waste in urine. **Creatinine** is formed from the breakdown of creatine phosphate, an energy molecule in muscle tissue. Uric acid is produced from breakdown of the nucleic acids DNA and RNA from foods or cellular destruction.

Several **inorganic ions** and molecules are also found in urine. Their presence is a reflection of diet and general health. Calcium, potassium, and magnesium ions are cations that form salts with chloride, sulfate, and phosphate ions. Na^+ and Cl^- are ions from sodium chloride, the principal salt of the body. Excretion of salt varies with dietary intake. **Ammonium** (NH_4^+) is a product of protein catabolism and must be removed from the blood before it reaches toxic concentrations. Many types of ions bind with sodium and form a buffer in the blood and urine to stabilize fluid pH. Other substances such as hormones, enzymes, carbohydrates, fatty acids, pigments, and mucin occur in small quantities in the urine.

Certain materials in the urine suggests a disease process or an injury to the kidneys. Excessive consumption of a substance may cause it to saturate the filtrate and overload the transport mechanisms of reabsorption. Because the tubular cells cannot reclaim all of the substance, it appears in the urine.

Ketones in the urine (**ketosis**) may be the result of starvation, diabetes mellitus, or a very low-carbohydrate diet. When carbohydrate concentration in the blood is low, cells begin to catabolize fats. The products of fat catabolism are glycerol and fatty acids. Liver cells convert these fatty acids to ketone bodies which diffuse out of the liver cells and into the blood where they are filtered by the kidneys. In diabetes mellitus, commonly called sugar diabetes, sufficient amounts of glucose cannot enter cells. As a result, cells use fatty acids to produce ATP. This increase in fatty acid catabolism results in ketone bodies appearing in the urine.

Glucose in the urine (**glucosuria**) is usually an indication of diabetes mellitus, commonly called sugar diabetes. Since diabetic individuals do not have the normal cellular intake of glucose, the blood plasma concentration of sugar is abnormally high. This results in more glucose being filtered out of the blood than the tubular cells of the nephrons can reabsorb. Glucose reabsorption is incomplete, and it appears in the urine.

Albumin is a large protein that normally cannot pass through the filtering membranes of the glomerulus. A trace amount of albumin in the urine is considered normal. Excessive albumin in the urine, a condition called **albuminuria**, suggests an increase in the permeability of the glomerular membrane. Reasons for increased permeability can be the result of an injury, high blood pressure, disease, or the effect of bacterial toxins on the cells of the kidneys.

Erythrocytes in the urine (**hematuria**) are usually an indication of an inflammation or infection of the urinary tract with leakage of blood into the urinary tract. Blood in the urine can also indicate a pathological condition such as kidney disease, trauma, or kidney stones (**renal calculi**). **Leukocytes** in the urine **pyuria (pī-Ū-rē-uh)** indicate a urinary tract infection.

CLINICAL APPLICATION

Crystals can form in the urinary tract and are usually voided with the urine. This sediment consists of **casts,** which usually are small clots of blood, tissue, or crystals of mineral salts (see Figure 46.3). Complete blockage of the urinary tract can occur due to the formation of kidney stones, or **renal calculi,** solid pebbles of urinary salts containing calcium, magnesium, or uric acid. Calculi can occur in the kidneys, ureters, bladder, or urethra and cause severe pain. If a stone completely blocks the urinary tract and will not pass out of the body, it must be removed surgically. Lithotripsy is a nonsurgical procedure that uses sound waves in an attempt to shatter the stone so it can pass. The patient is immersed in water, and the wave energy is directed to the area overlying the stone to destroy the calculi.

Bilirubin in large amounts in the urine, or **bilirubinuria**, is a result of the breakdown of hemoglobin from old red blood cells being removed from the circulatory system by phagocytic cells in the liver.

Urobilinogen in the urine is called **urobilinogenuria**. Small amounts of urobilinogen in the urine are normal. It is a product of the breakdown of bilirubin by the intestines and is responsible for the normal brown color of feces. Greater than trace levels in the urine may be due to infectious hepatitis, cirrhosis, congestive heart failure, or a variety of other diseases.

Microbes in large numbers in the urine usually indicate an infection. Generally, urine is sterile, but microbe content is possible in a urine sample for several reasons. Microbes may contaminate a urine sample when presence at the urethral opening and in the urethra.

The average **pH** of urine is 6.0 and normally ranges from 4.5 to 8.0, and is greatly affected by diet. High vegetable fiber diets result in an alkaline pH, while high protein diets contribute to an acidic pH.

The **specific gravity** of a fluid is a comparison of the density of that fluid to the density of water. The density of a fluid is the ratio of the weight of the solutes compared to the volume of the solvent. The specific gravity of water is 1.000. The average specific gravity for a normal urine sample is between 1.003 and 1.030, a density greater than water. Normal constituents of urine include water, urea, creatinine, sodium, chloride, potassium, sulfates, phosphates, and ammonium salts. These solutes determine the specific gravity of urine. A high fluid intake results in frequent urination of a dilute urine with a lower specific gravity, low fluid intake results in a more concentrated urine with a higher specific gravity. Excessively concentrated urine results in the crystallization of solutes into renal calculi or kidney stones.

B. Urinalysis

Abnormal substances in urine can usually be detected by urinalysis, an analysis of the chemical and physical properties of urine. Several brands of test strips provide a fast, inexpensive method of detecting the chemical composition of urine. Multistix and Chemstix are two popular brands used for this test. These sticks may contain from one to nine testing pads. Single tests strips usually are used to determine whether glucose or ketones are in the urine. The multiple test strips are used for a detailed analysis of the urine's chemical content.

During this urinalysis you will examine only your own urine sample. Physical characteristics such as volume, color, cloudiness, and odor will be observed. The test strips will determine the chemical characteristics of your sample. Both physical and chemical properties of urine vary according to fluid intake and diet.

Alternatively, your laboratory instructor may elect to provide your class with a mock urine sample for analysis. This artificial sample will probably include several abnormal urine constituents for instructional purposes.

LABORATORY ACTIVITY   URINALYSIS

MATERIALS

500-ml sample cup
Chemstix or Multistix test strips
BioHazard disposal container

SAFETY NOTICE

Handle and test only your own urine. Dispose of all urine-contaminated materials in the BioHazard disposal container as outlined by your laboratory instructor. If a urine spill occurs, wear gloves and clean your work space with a mild bleach solution.

PROCEDURE I - PHYSICAL CHARACTERISTICS

1. Obtain a sterile 500-ml cup for the collection of your urine sample. A midstream sample should be collected. By avoiding collection of the first few milliliters of urine, you can prevent contamination of the sample by materials such as bacteria and pus from the urethra.
2. Observe the physical characteristics of the sample and record your observations in Table 46.1.

TABLE 46.1   **Physical Observations of Urine Sample**

Characteristic	*Observation*
Volume	
Color	
Turbidity	
Odor	
Specific gravity	

- **Volumes** less than 500 ml or greater than 2 liters per day are considered abnormal. The volume of a single urine sample is variable depending on an individual's hydration and fluid intake.
- The **color** of urine varies from clear to amber. **Urochrome** is a by-product of the breakdown of hemoglobin that gives urine its yellowish color. Color also varies because of ingested food. Vitamin supplements, certain drugs, and the amount of solutes also influence the color of urine. A dark-red or brown color indicates blood in the urine.
- **Turbidity** or cloudiness is related to the amount of solids in the urine. Contributing factors include bacteria, mucus, cell casts, crystals, and epithelial cells. Observe and describe the turbidity of the urine sample. Use descriptive words such as "clear," "clouded," "hazy," and so on.
- Freshly voided urine has no **odor**. As chemical breakdown of substances occurs, the sample may acquire an ammonia odor. Odor serves as a diagnostic tool for freshly voided urine. Starvation causes the body to break down fats and produce ketones that give urine a fruity or acetone-like smell. Individuals with diabetes

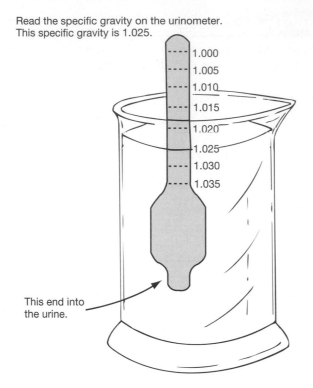

Read the specific gravity on the urinometer.
This specific gravity is 1.025.

1.000
1.005
1.010
1.015
1.020
1.025
1.030
1.035

This end into
the urine.

Figure 46.2
Reading a Urinometer.

mellitus often produce sweet-smelling urine. To smell the sample, place it approximately 12 in. from your face and "wave" your hand over the sample toward your nose.

- **Specific gravity** is the ratio of the weight of a volume of a substance to the weight of an equal volume of distilled water. The specific gravity of water is 1.000. Urine contains a given amount of solutes and solids which directly affect its specific gravity. The amount of fluids ingested affects the volume of urine excreted and therefore the amount of solutes and solids per given volume. A high fluid intake results in more frequent urination of a more dilute urine containing fewer solutes and solids per given volume and therefore leads to a lower specific gravity. A low fluid intake results in less frequent urination of a more concentrated urine with a higher specific gravity. Normal constituents of urine include water, urea, creatinine, sodium, chloride, potassium, sulfates, phosphates, and ammonium salts. Excessively concentrated urine results in the crystallization of solutes, usually salts, into insoluble stones called renal calculi or kidney stones. The average specific gravity for a random sample should fall between 1.001 and 1.030.

  Use a urinometer, as in Figure 46.2, to determine the specific gravity of your urine sample. A urinometer consists of a small glass cylinder and a urine hydrometer. The glass cylinder is used to hold the urine sample being tested. The hydrometer is a calibrated float that has been calibrated against water. Inside the stem of the hydrometer is a scale which is used to interpret the specific gravity of the sample.

  a. Swirl the sample in the collection cup to suspend any materials that may have settled after collection.

  b. Fill the glass cylinder at least two-thirds full with the urine sample.

  c. Carefully lower the urine hydrometer into the cylinder of urine. If the hydrometer does not float, you need to add more urine to the cylinder until it does float.

d. Observe the scale in the stem of the urine hydrometer. The urine will adhere to the walls of the cylinder and form a trough called a meniscus. The scale is read from the bottom of the meniscus where it intersects the lines of the scale. Record the specific gravity of the sample in Table 46.1.

PROCEDURE II - CHEMICAL CHARACTERISTICS - DIP STICK METHOD

Dip sticks are a fast, inexpensive method of determining the chemical composition of urine. Multistix and Chemstix are two popular brands used for this test. These sticks may contain from one to nine testing pads. Single tests strips are usually used to determine whether glucose or ketones are in the urine. Multiple test strips are used to give a more informative evaluation of the urine's chemical content.

1. Swirl your sample of urine before doing this test.
2. Holding a test strip by the end that does not have any test pads, immerse all the pads of the dip stick into the urine and then withdraw the dip stick. Lay the strip on a clean, dry paper towel to absorb any excess urine that may form droplets on the dip stick.
3. Reading of the tests varies from immediately to 2 min. Use the color chart on the side of the bottle to interpret the tests.
4. Record your data from the test strip in Table 46.2.

TABLE 46.2	**Chemical Evaluation of Urine Sample**
Substance	*Remark*
pH	
Specific gravity	
Glucose	
Ketone	
Protein	
Erythrocytes	
Bilirubin	
Other	

- The **pH** of normal urine ranges from 4.5 to 8.0. Diet has a large effect on the pH of urine. High vegetable fiber diets result in a urine pH that tends to be alkaline, whereas high protein diets result in an acidic pH. Be careful not to let urine from other pads drain onto the pad, or false results may occur.
- The presence of **glucose** in the urine, glucosuria, may indicate diabetes mellitus. This may also be the result of a very high-carbohydrate meal which produces a condition of glucose in the urine called temporary glucosuria. Production of epinephrine in response to stress results in the conversion of glycogen to glucose and its release from the liver. The elevated levels of glucose may then be secreted into the urine.
- Usually the urine contains trace amounts of **proteins** that may be too small for normal testing procedures to determine. Proteins are normally too large to pass through the endothelial-capsular membrane. Proteins in the urine, albuminuria,

may occur as the result of increased blood pressure, increased permeability of the endothelial-capsular membrane, or irritation of the tubules from bacterial toxins. Positive results are due to the presence of albumin, a plasma protein.

- The presence of **ketones** in the urine (ketoacidosis) may be the result of diabetes mellitus or a starvation diet, a very low carbohydrate diet that causes the body to break down fats.

- **Nitrites** in the urine indicate a possible urinary tract infection.

- **Bilirubin** is a pigment in bile. When red blood cells are removed from the blood by the liver, the globin portion of the hemoglobin molecule is split off the molecule and the heme portion is converted to biliverdin. The biliverdin is then converted to bilirubin which is a major pigment in bile. A high level of bilirubin in the urine is called bilirubinuria.

- **Urobilinogen** is a by-product of the breakdown of bilirubin. Greater than trace levels in the urine may be due to infectious hepatitis, cirrhosis, congestive heart failure, or a variety of other diseases.

- **Red Blood Cells and Hemoglobin**—Erythrocytes in the urine (hematuria) is usually an indication of an inflammation or infection of the urinary tract with leakage of blood into the urinary tract. Causes for red blood cells and hemoglobin in urine can be a disease condition, irritation of the renal tubules from the formation of kidney stones, trauma such as a hard blow to the kidney, blood from menstrual flow, and possible tumor formation. Care should be used in making sure the urine sample was not contaminated with menstrual blood from the vaginal area. When hemolysis of erythrocytes occurs, the hemoglobin molecules break down into their a and b chains. These two chains are filtered by the kidneys and are excreted in the urine. If a large number of erythrocytes are being broken down in the circulation, the urine develops a dark-brown to reddish color. The presence of hemoglobin in the urine is called hemoglobinuria.

- **Leukocytes**—White blood cells in the urine (pyuria) usually indicate an infection of the urinary tract. Problems can include cystitis, urethritis, or kidney infection.

C. Microscopic Examination of Urine

Examination of the sediment of a centrifuged urine specimen reveals the solid components of the sample. This can be a valuable test to determine or confirm the presence of abnormal contents in the urine. A wide variety of solids occur in urine including cells, crystals, and mucus (see Figure 46.3).

MATERIALS

test tube rack	conical centrifuge tubes
wax pencil	glass slides
coverslips	iodine or sediment stain
Pasteur pipette with bulb	compound microscope

PROCEDURES

1. Mark the test tube two-thirds up from the bottom. Swirl the urine sample to suspend any solids that have settled. Fill the test tube to the wax pencil mark.

2. Place your tube in the centrifuge opposite another tube so the centrifuge remains in a balanced state. This step is very important for protection of the centrifuge.

3. Centrifuge the sample for 8–10 min.

4. Pour off the supernatant and with a Pasteur pipette remove some of the sediment. Place one drop of sediment on the glass slide, add one drop of iodine or sediment

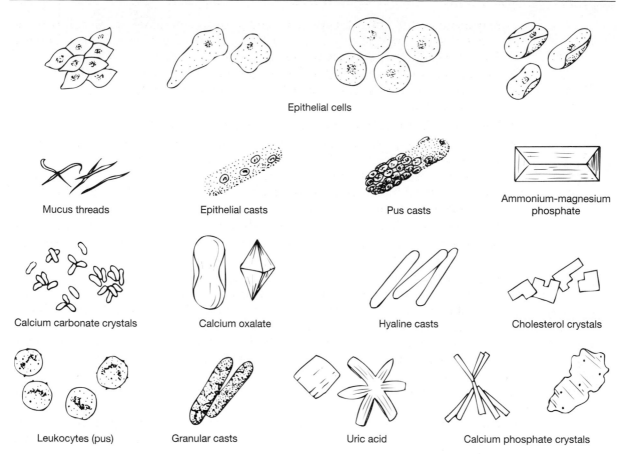

Epithelial cells

Mucus threads   Epithelial casts   Pus casts   Ammonium-magnesium phosphate

Calcium carbonate crystals   Calcium oxalate   Hyaline casts   Cholesterol crystals

Leukocytes (pus)   Granular casts   Uric acid   Calcium phosphate crystals

Figure 46.3
Sediments Found in the Urine.

stain (to make viewing cells, crystals, etc., easier), and place a cover glass on top of the specimen.

5. Begin viewing the stained sediment under 10X. Look for epithelial cells, red blood cells, white blood cells, crystals, and microbes. A high number of bacteria usually indicates a urinary tract infection. Mucin threads and casts may also be seen in the sediment. Mucin is a complex glycoprotein secreted by unicellular exocrine glands such as goblet cells. In water, mucin becomes mucus, a slimy coating that lubricates and protects the lining of the urinary tract. Casts are usually cylindrical and composed of proteins and cells. Crystals of the salts are also found in urine. Renal calculi or kidney stones consist of calcium oxalate, uric acid, and calcium phosphate crystals. They may form and lodge anywhere in the urinary tract and can cause pain. Compare the contents of your urine sample with Figure 46.3.

6. Dispose of all used paper towels and pipettes in the autoclave bag provided. Rinse your test tubes in bleach water and then wash and dry your test tubes and urinometer apparatus. Place glass slides and cover glasses in a beaker of 10% chlorine.

URINARY SYSTEM CHECKLIST

This is a list of bold terms presented in Exercise 46. Use this list for review purposes after you have completed the laboratory activities.

INTRODUCTION

• Figure 46.1
filtration
filtrate
reabsorption
secretion
urine

COMPOSITION OF URINE

• Figure 46.3
urea
  deamination
    creatinine
inorganic ions
  ammonium
ketones
  ketosis
glucose
  glucosuria

albumen
  albuminuria
erythrocytes
  hematuria
    renal calculi
    leukocytes
    pyuria
biliruben
  bilirubinuria
urobilinogen
  urobilinogenuria
pH
microbes
specific gravity
volume
color
  urochrome
turbidity

PHYSIOLOGY OF THE URINARY SYSTEM
Laboratory Report Exercise 46

A. Matching

Match each term listed on the left with the correct description on the right.

1. _____ pyuria		**A.** product of fat metabolism
2. _____ hematuria		**B.** yellow pigment in urine
3. _____ glucosuria		**C.** product of bilirubin breakdown
4. _____ secretion		**D.** leukocytes in the urine
5. _____ albuminuria		**E.** removal of materials from the blood
6. _____ bilirubin		**F.** fluid in minor calyx of kidney
7. _____ urochrome		**G.** removal of ammonia from amino acid
8. _____ renal calculi		**H.** glucose in the urine
9. _____ deamination		**I.** fluid passing through nephron tubules
10. _____ reabsorption		**J.** molecule from breakdown of hemoglobin
11. _____ ketone		**K.** kidney stones
12. _____ filtration		**L.** excess albumin in urine
13. _____ filtrate		**M.** addition of materials to the filtrate
14. _____ urine		**N.** returning materials from the filtrate to the blood
15. _____ urobilinogen		**O.** blood in the urine

B. Short-Answer Questions

1. What is the normal pH range of urine?

2. What is the specific gravity reading of a normal sample of urine?

3. List five abnormal components of urine.

4. What substances in the urine might be an indicator that a person has diabetes?

5. Explain why ketones may appear in the urine.

6. What might affect the odor, color, and pH of a sample of urine?

7. Describe the three physiological processes of urine production.

Reproductive System

OBJECTIVES

On completion of this exercise, you should be able to:

- Describe the location of the male gonad.
- Identify the ducts and accessory glands in males.
- Describe the composition of semen.
- Identify the three regions of the male urethra.
- Identify the structures of the penis.
- Trace the formation and transport of semen in the male reproductive tract.
- Identify the structures of the female reproductive tract.
- Identify the ovaries, ligaments, uterine tubes, and uterus.
- Describe and recognize the three main layers of the uterine wall.
- Identify the vagina and the features of the vulva.
- Identify the structures of the mammary glands.

WORD POWER

gametes (gamete—husband, wife)
sperm (sperm—seed)
epididymis (epi—upon; didymos—twin)
corpus (corpus—body)
glans penis (glans—acorn)
corona (corona—crown)
cremaster (suspender)
corpus luteum (luteum—yellow)
fimbriae (fimbrae—fringe)
vulva (vulva—wrapper or covering)
mons pubis (mons—mountain)

MATERIALS

male urogenital organs model and chart
male pelvis model
microscope
prepared slide: testis
female reproductive organs model and chart
female pelvis model
breast model
prepared slides:
  ovary
  uterus

INTRODUCTION

Whereas all the other systems of the body function to support the continued life of the organism, the reproductive system functions to ensure continued perpetuation of the species. The primary sex organs or **gonads (GŌ-nads)** of the male and female are the testes and ovaries, respectively. They function in the production of **gametes (GAM-ēts)**, the sperm and eggs (ova). Gonads secrete hormones that support maintenance of the male and female sex characteristics. Since these glands secrete their gametes into ducts and their hormones into the blood, they have both exocrine and endocrine functions. The ducts function in receiving and storing gametes. There are several accessory glands that secrete products to protect and support the gametes.

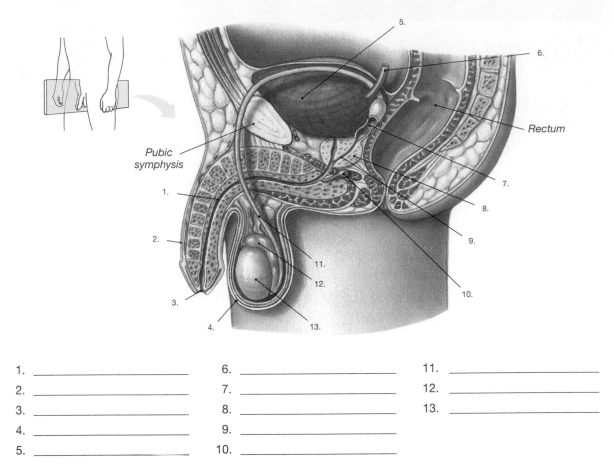

1. _____   6. _____   11. _____
2. _____   7. _____   12. _____
3. _____   8. _____   13. _____
4. _____   9. _____
5. _____   10. _____

Figure 47.1
Components of the Male Reproductive System.
The male reproductive organs as seen in sagittal section.

I. THE MALE REPRODUCTIVE SYSTEM

The male reproductive system, shown in Figure 47.1, consists of a pair of testes, ducts that convey sperm and fluids secreted by accessory glands, and the penis. The **scrotum (SKRŌ-tum)** is a cutaneous pouch that houses the testes. It is divided into two compartments, each containing a testicle. The **dartos (DAR-tōs)** muscle forms part of the septum separating the testes and is responsible for the wrinkling of the skin of the scrotum. The **cremaster (krē-MAS-ter) muscle** encases the testes. It raises or lowers the testes to maintain a cooler temperature for optimum sperm production. Locate the dartos and cremaster muscles in Figure 47.2.

A. The Testes

The **testes (TES-tēz)** or testicles are a pair of oval organs about 5 cm (2 in.) in length and 2.5 cm (1 in.) in diameter. They are the primary sex organ of the male and function in the production of sperm and the hormone testosterone. Compartments in the testicle called **lobules** contain highly coiled **seminiferous (se-mi-NIF-e-rus) tubules**, shown in Figure 47.2. Millions of spermatoza are produced each day by the seminiferous tubules in a process called **spermatogenesis (sper-ma-tō-JEN-e-sis)**. This process is presented in detail in Exercise 48. Follicle-stimulating hormone (FSH) from the anterior pituitary gland regulates spermatogenesis. Between the seminiferous tubules are small clusters of cells called **interstitial cells** which secrete **testosterone**, the male sex hormone. Testosterone is responsible for the development and maintenance of the male secondary sex characteristics and the male sex drive.

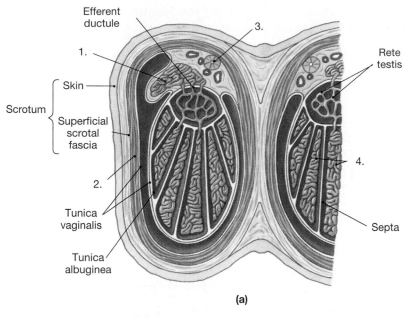

Efferent ductule

1.

Scrotum {

Skin

Superficial scrotal fascia

2.

Tunica vaginalis

Tunica albuginea

3.

Rete testis

4.

Septa

(a)

Figure 47.2
Structure of the Testes.

(a) Diagrammatic sketch and anatomical relationships of the testes. **(b)** A section through a coiled seminiferous tubule. (LM ×786) **(c)** Cellular organization of a seminiferous tubule.

1. _____

2. _____

3. _____

4. _____

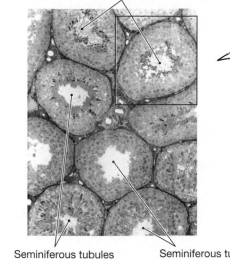

Seminiferous tubules containing nearly mature spermatozoa about to be released into the lumen

Seminiferous tubules containing late spermatids

Seminiferous tubules containing early spermatids

(b)

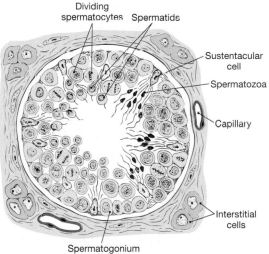

Dividing spermatocytes   Spermatids

Sustentacular cell

Spermatozoa

Capillary

Interstitial cells

Spermatogonium

(c)

LABORATORY ACTIVITY   THE TESTES

MATERIALS

male urogential model and chart
compound microscope
prepared slide: testis

PROCEDURES

1. Locate the scrotum, testes, and associated anatomy on the laboratory models.
2. Examine a microscope slide of the testis in transverse section. Use the micrograph in Figure 47.2 for reference. Scan the slide at low magnification and

observe the many seminiferous tubules. Increase the magnification and locate the interstitial cells between the tubules. In the space below, draw and label a section of the testis.

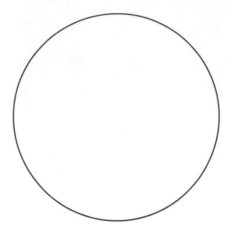

B. **The Epididymis and the Ductus Deferens**

After spermatozoa are produced in the seminiferous tubules they move into the **epididymis (ep-i-DID-i-mus)**, a highly coiled tubule located on the posterior of the testis, visible in Figure 47.1. Spermatozoa mature in the epididymis and are stored until ejaculation out of the male reproductive system. Peristalsis of the smooth muscle of the epididymis propels the sperm into the ductus deferens, the duct that empties into the urethra.

The **ductus deferens (DUK-tus DEF-e-renz)** (see Figure 47.1), or vas deferens ascends as part of the **spermatic cord** into the pelvic cavity. Other structures in the spermatic cord include blood and lymphatic vessels, nerves, and the cremaster muscle. The duct is 46–50 cm (approximately 18–20 in.) long and is lined with pseudostratified columnar epithelium. Peristaltic waves propel spermatozoa toward the urethra. The ductus deferens passes through the **inguinal (ing-gwi-nal) canal** in the lower abdominal wall to enter the body cavity. This canal is a weak area and is frequently injured. An **inguinal hernia** occurs when portions of intestine protrude through the canal and slide into the scrotum. The ductus deferens continues around the posterior of the urinary bladder and widens into the **ampulla (am-PŪL-uh)** before joining the seminal vesicle at the ejaculatory duct.

LABORATORY ACTIVITY   THE EPIDIDYMIS AND DUCTUS DEFERENS

MATERIALS

male urogential model and chart

PROCEDURES

1. Locate the epididymis and ductus deferens on the laboratory models.
2. Locate the spermatic cord, the cremaster muscle, and the inguinal canal.

C. **The Accessory Glands**

Three accessory glands, the seminal vesicles, prostate, and bulbourethral glands, produce fluids that nourish, protect, and support the spermatozoa. Locate these glands in Figures 47.1 and 47.3. The sperm and fluids from the seminal vesicles, prostate gland,

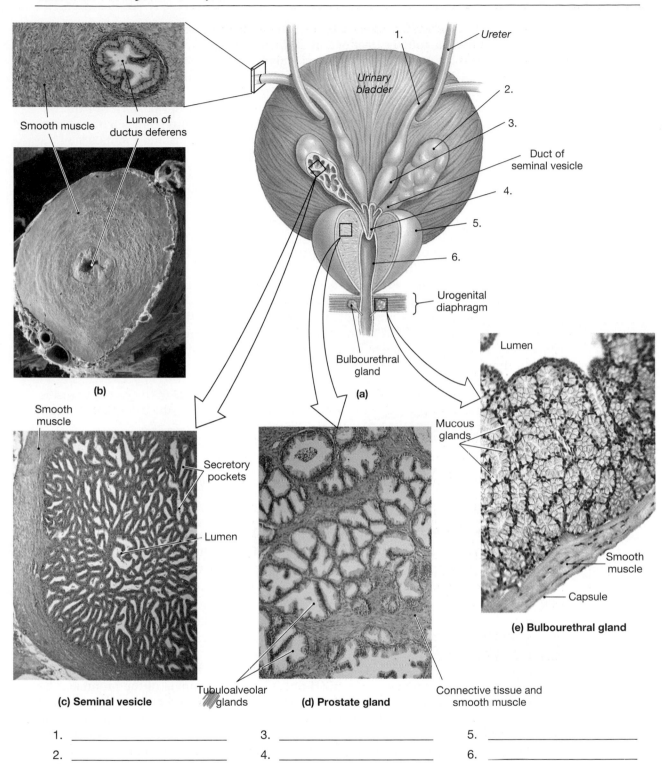

Smooth muscle

Lumen of
ductus deferens

(b)

Smooth
muscle

Secretory
pockets

Lumen

(c) Seminal vesicle

Tubuloalveolar
glands

(d) Prostate gland

1.

Ureter

*Urinary
bladder*

2.

3.

Duct of
seminal vesicle

4.

5.

6.

Urogenital
diaphragm

Bulbourethral
gland

(a)

Lumen

Mucous
glands

Smooth
muscle

Capsule

(e) Bulbourethral gland

Connective tissue and
smooth muscle

1. _____   3. _____   5. _____

2. _____   4. _____   6. _____

Figure 47.3
The Ductus Deferens and Accessory Glands.

(a) A posterior view of the prostate gland, showing subdivisions of the ductus deferens in relation to surrounding structures. **(b)** The ductus deferens, showing the smooth muscle around the lumen. (LM [top] × 34,
SEM × 42) (© R. G. Kessel and R. H. Kardon, *Tissues and Organs: A Text-Atlas of Scanning Electron
Microscopy,* W. H. Freeman & Co., 1979. All rights reserved.) Sections of **(c)** the seminal vesicle (LM × 44),
(d) the prostate gland (LM × 49), and **(e)** a bulbourethral gland. (LM × 177)

and bulbourethral glands form a mixture called **semen (SĒ-men)**. The semen is propelled by peristalsis and contractions of skeletal muscles through the urethra of the penis and expelled into the vagina as the **ejaculate**. The average volume of an ejaculate is between 2 and 5 ml. The average number of sperm per milliliter of semen is between 50 and 150 million.

The **seminal (SEM-i-nal) vesicles** are a pair of glands on the posterior of the urinary bladder. Each gland is approximately 15 cm (6 in.) long and merges with the ductus deferens into an **ejaculatory duct**. The seminal vesicles contribute about 60% of the total volume of semen. They secrete a viscous, alkaline fluid containing the sugar fructose. The alkaline nature of this fluid neutralizes the acidity of the male urethra and the female vagina. The fructose provides the sperm an energy source for beating of the flagellum.

The **prostate (PROS-tāt) gland** is a single gland about 1.5 in. in diameter at the base of the urinary bladder. The ejaculatory duct passes into the prostate gland and empties into the first segment of the urethra, the **prostatic urethra**. The prostate gland secretes a milky white, acidic fluid that contains clotting enzymes to coagulate the semen. These secretions contribute about 30% of the semen volume.

The prostatic urethra exits the prostate gland and passes through the floor of the pelvis as the **membranous urethra**. A pair of **bulbourethral (bul-bō-ū-RĒ-thral)** (Cowper's) **glands** on each side of the membranous urethra add an alkaline mucus to the semen. Before ejaculation, the bulbourethral secretions neutralize the acidity of the urethra and lubricate the end of the penis for sexual intercourse. These glands contribute about 5% of the volume of semen.

LABORATORY ACTIVITY   THE ACCESSORY GLANDS

MATERIALS

male urogential model and chart

PROCEDURES

1. Locate each accessory gland on the laboratory models.
2. Locate the urinary bladder and the prostatic and membranous urethra.

D. The Penis

The **penis (PĒ-nis)**, detailed in Figure 47.4, is the male copulatory organ and serves to deliver semen into the vagina. It is cylindrical and has an enlarged, acorn-shaped head called the **glans**. Around its base is a margin, the **corona** (crown). On an uncircumcized penis, the glans penis is covered with a loose-fitting skin called the **prepuce (PRĒ-pūs)** or **foreskin**. **Circumcision (ser-kum-SIZH-un)** is surgical removal of the prepuce. The **penile urethra** transports semen or urine through the penis and ends at the **external urethral meatus** in the tip of the glans. The **root** of the penis anchors the penis to the pelvis. The **body** consists of three cylinders of erectile tissue, the dorsal **corpora cavernosa (KOR-po-ruh ka-ver-NŌ-suh)**, and the ventral **corpus spongiosum (spon-jē-Ō-sum)**. During sexual arousal, the three erectile tissues become engorged with blood and cause the penis to stiffen into an **erection**.

LABORATORY ACTIVITY   THE PENIS

MATERIALS

male urogential model and chart

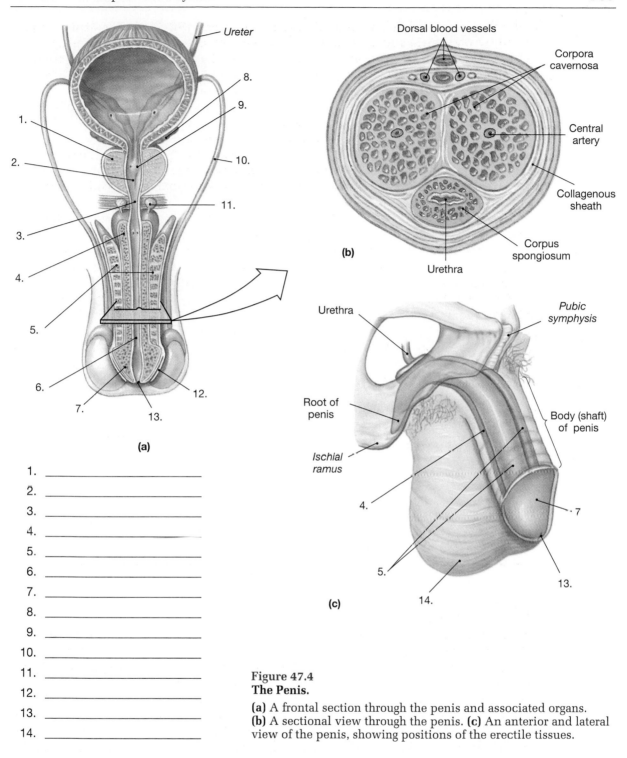

(a)

(b)

(c)

Figure 47.4
The Penis.

(a) A frontal section through the penis and associated organs.
(b) A sectional view through the penis. **(c)** An anterior and lateral
view of the penis, showing positions of the erectile tissues.

1. _____
2. _____
3. _____
4. _____
5. _____
6. _____
7. _____
8. _____
9. _____
10. _____
11. _____
12. _____
13. _____
14. _____

PROCEDURES

1. Identify the glans, body, and root of the penis on the laboratory models.

2. Examine a model of the penis in transverse section and identify the corpora cavernosa and the corpus spongiosum.

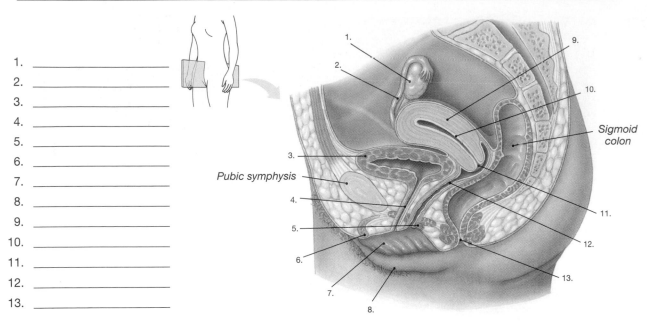

1. _____
2. _____
3. _____
4. _____
5. _____
6. _____
7. _____
8. _____
9. _____
10. _____
11. _____
12. _____
13. _____

Figure 47.5
The Female Reproductive System.

II. THE FEMALE REPRODUCTIVE SYSTEM

The female reproductive system, highlighted in Figure 47.5, includes the ovaries, uterine tubes, uterus, vagina, external genitalia, and mammary glands. **Gynecology** is the branch of medicine that deals with the care and treatment of the female reproductive system.

A. The Ovaries

The **ovaries** are the primary sex organs of the female. They are suspended in the pelvic cavity on each side of the uterus by several ligaments. A double-layered fold of peritoneum called the **mesovarium (mes-ō-VAR-ē-um)** holds the ovaries to the **broad ligament** of the uterus (Figure 47.7). The **suspensory ligaments** hold the ovaries to the wall of the pelvis, and the **ovarian ligament** holds the ovaries to the uterus.

Each almond-sized ovary contains from 100,000 to 200,000 undeveloped eggs clustered in **egg nests**. Within the nests are **primordial follicles** consisting of **oocytes (Ō-ō-sits)** or eggs surrounded by follicular cells. Figure 47.6 details the monthly **ovarian cycle**, during which hormones stimulate follicular cells to proliferate and produce several **primary follicles**. These follicles increase in size, and a few become **secondary follicles**. Eventually, one secondary follicle develops into a **tertiary follicle**, also called a **Graafian (GRAF-ē-an) follicle**. This follicle fills with fluid and ruptures, casting out the ovum in a process called **ovulation**. The Graafian follicle secretes **estrogen** which stimulates rebuilding of the spongy lining of the uterus. After ovulation the tertiary follicle becomes the **corpus luteum (LOO-tē-um)** and primarily secretes the hormone **progesterone (prō-JES-ter-ōn)** which prepares the uterus for pregnancy. If fertilization of an ovum does not occur, the corpus luteum degenerates into the **corpus albicans (AL-bi-kanz)** and most of the rebuilt lining of the uterus is shed as the menstrual flow. Observe the prepared slide of the ovary. Locate primary and secondary follicles and a tertiary follicle.

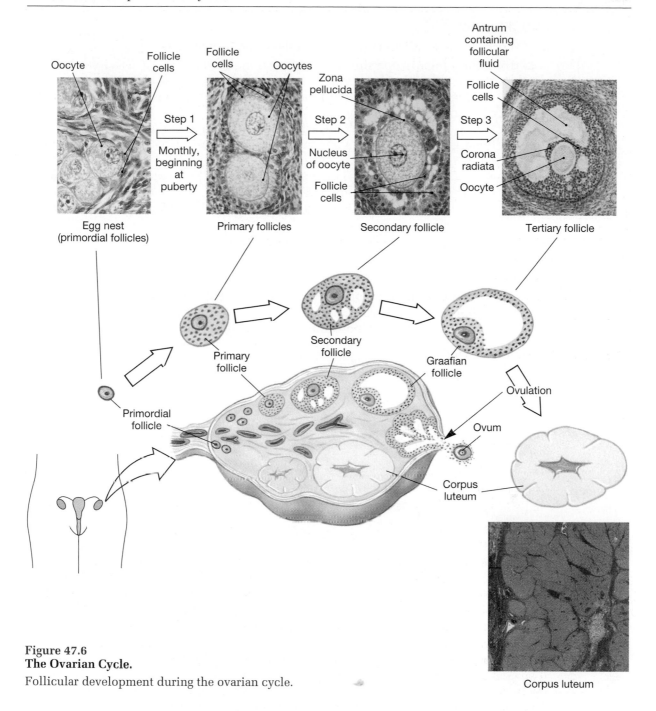

Figure 47.6
The Ovarian Cycle.
Follicular development during the ovarian cycle.

Corpus luteum

LABORATORY ACTIVITY   THE OVARIES

MATERIALS

female reproductive system model and chart
compound microscope
prepared slide: ovary

PROCEDURES

1. Locate the ovaries and ligaments on the laboratory models.
2. Examine a microscope slide an ovary section. Use Figure 47.6 as a reference and locate the following structures.
   - Scan the slide at low magnification and locate an egg nest along the periphery of the ovary.
   - Identify the primary follicles which are larger than the primordial follicles in the egg nests. The ovum has increased in size and is surrounded by follicular cells.
   - Secondary follicles are larger than primary follicles and have a separation between the outer and inner follicular cells.
   - Tertiary follicles are easy to distinguish by their large, fluid-filled space called the **antrum**.

B. **The Uterine Tubes and Uterus**

The ends of the **uterine** or fallopian **tubes** have fingerlike projections called **fimbriae (FIM-brē-ē)** (Figure 47.7). These fimbriae sweep over the surface of the ovary to capture an ovum released during ovulation and draw it into the expanded **infundibulum**. Once inside the uterine tube, ciliated epithelium transports the ovum toward the uterus. The tube widens midway along its length in the **ampulla** and then narrows at the **isthmus (IS-mus)** to enter the uterus. Fertilization of the egg usually occurs in the upper third of the uterine tube.

The **uterus (Ū-ter-us)** is a pear-shaped muscular organ located between the urinary bladder and the rectum. It serves as a site for implantation of a fertilized ovum and sub-

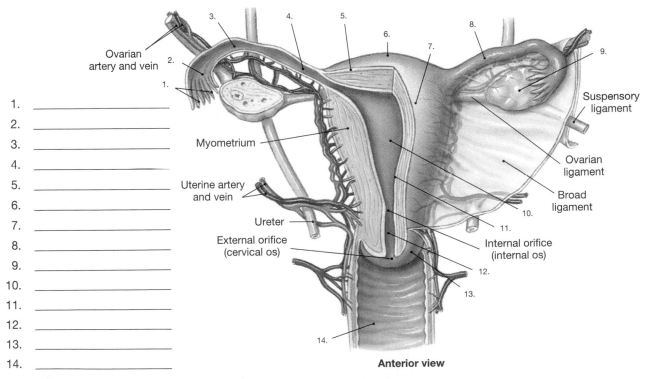

1. _____
2. _____
3. _____
4. _____
5. _____
6. _____
7. _____
8. _____
9. _____
10. _____
11. _____
12. _____
13. _____
14. _____

Anterior view

Figure 47.7
The Ovaries and Uterus.
Anterior view with right portion of uterus, uterine tube, and ovary shown in section.

Figure 47.8
The Uterine Wall.

(a) The basic structure of the endometrium. (LM ×32) **(a)** A sectional view of the uterine wall, showing the endometrial regions and the circulatory supply to the endometrium.

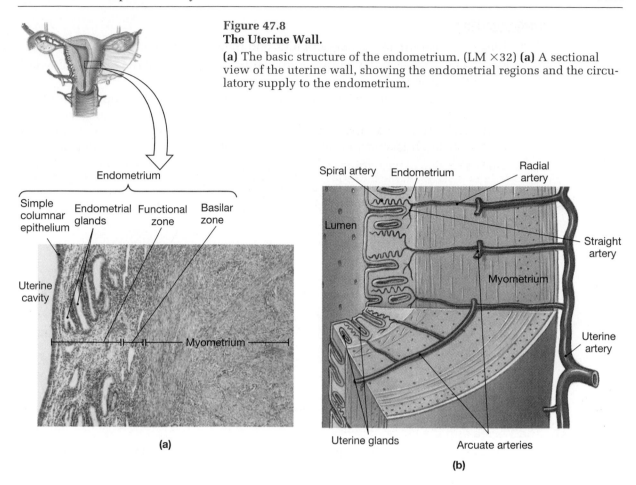

(a)

(b)

sequent development of the fetus during pregnancy. The uterus consists of three major regions, the fundus, the body, and the cervix. The superior dome-shaped portion of the uterus is the **fundus**. Most of the uterus is called the **body**; the inferior narrow portion is called the **cervix (SER-viks)** and opens into the vagina. Within the uterus is a space called the **uterine cavity** which narrows to the **cervical canal** in the cervix. Locate these regions of the uterus in Figure 47.7.

The uterine wall consists of three main layers: the perimetrium, the myometrium, and the endometrium (see Figure 47.8). The **perimetrium** is the outer covering of the uterus. It is an extension of the visceral peritoneum and is therefore also called the serosa. The thick middle layer, the **myometrium (mī-ō-MĒ-trē-um)**, is composed of three layers of smooth muscle and is responsible for the powerful contractions during labor. The **endometrium (en-dō-MĒ-trē-um)** consists of two layers, the **basilar zone** and the **functional zone**. The basilar zone covers the myometrium and produces a new functional zone each month. The functional zone is very glandular and is highly vascularized to support an implanted embryo. This is also the layer that is shed each cycle during menstruation. Observe the prepared slide of the uterus and identify the three major uterine layers.

LABORATORY ACTIVITY   THE UTERINE TUBE AND THE UTERUS

MATERIALS

female reproductive system model and chart
compound microscope
prepared slide: uterus

PROCEDURES

1. Identify the uterine tubes, the ampulla, and the isthmus on the laboratory models.
2. Distinguish between the fundus, body, and cervix of the uterus on available laboratory models.
3. Using Figure 47.8 for reference, examine a prepared microscope slide of the uterus in transverse section. Scan the slide at low and medium magnifications and locate the thick myometrium composed of smooth muscle tissue. The endometrium lines the uterine cavity and is rich in blood vessels and glands.
4. Draw a section of the uterine wall from your microscopic studies in the space below.

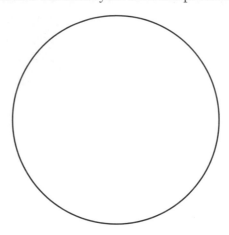

C. **The Vagina and the Vulva**

The **vagina (va-JĪ-nuh)** is a muscular tube approximately 10 cm (4 in.) long. Locate the vagina in the various views in Figures 47.5, 47.7, and 47.9. It is lined with stratified squamous epithelium and serves as the female's copulatory organ, the pathway for menstrual flow, and the lower birth canal. The **vaginal orifice** is the external opening of the vagina. This opening may be partially or totally occluded by a thin fold of vascularized mucus membrane called the **hymen (HĪ-men)**. On either side of the vaginal orifice are openings of the **greater vestibular glands** which produce a mucous secretion that lubricates the vaginal entrance for sexual intercourse. These glands are similar to the bulbourethral glands of the male. A vaginal pocket called the **fornix (FOR-niks)** is formed where the cervix protrudes into the vagina.

The **vulva (VUL-vuh)** consists of the external **genitalia (jen-i-TĀ-lē-uh)** of the female (see Figure 47.9). It includes the following structures:

*   The **mons pubis** is a fatty pad over the pubic symphysis. It is covered with skin and pubic hair and serves as a cushion for the pubic symphysis during sexual intercourse.
*   The **labia (LĀ-bē-uh) majora** are two fatty folds of skin extending from the mons pubis and continuing posteriorly. They are homologous to the scrotum of the male. They usually have pubic hair and contain many sudoriferous (sweat) and sebaceous (oil) glands.
*   The **labia minora (mi-NOR-uh)** are two smaller parallel folds of skin containing many sebaceous glands. This pair of labia lacks hair.
*   The **clitoris (KLI-to-ris)** is a small cylindrical mass of erectile tissue similar to the penis. It also contains a small fold of skin that covers it called the **prepuce**. The exposed portion of the clitoris is called the glans.
*   The **vestibule** is the area between the labia minora that contains the vaginal orifice, hymen, and external urethral orifice.

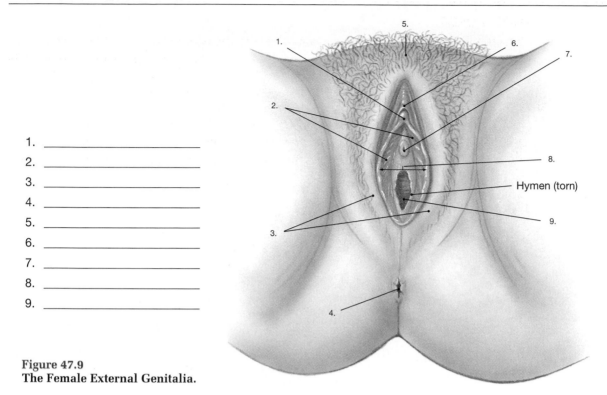

1. _____
2. _____
3. _____
4. _____
5. _____
6. _____
7. _____
8. _____
9. _____

Figure 47.9
The Female External Genitalia.

LABORATORY ACTIVITY   THE VAGINA AND THE VULVA

MATERIALS

female reproductive system model and chart

PROCEDURES

1. Locate the vagina and vaginal orifice on the laboratory models. Examine the fornix where the cervix and vagina connect.

2. Locate each component of the vulva. How is the urethra positioned compared with the vagina and clitoris?

D. The Mammary Glands

The **mammary glands** (see Figure 47.10) are modified sweat glands that produce milk, a process called **lactation,** to nourish the newborn infant. At puberty, the release of estrogens stimulates an increase in the size of these glands. Fat deposition is the major contributor to the size of the breast. Development of the ducts of the mammary glands also occurs. The mammary glands consist of 15 to 20 **lobes** separated by fat and connective tissue. Each lobe contains smaller compartments called **lobules** which contain milk-secreting cells called **alveoli. Lactiferous (lak-TIF-e-rus) ducts** drain milk from the lobules toward the **lactiferous sinuses**. These sinuses empty milk at the raised portion of the breast called the **nipple**. A circular pigmented area called the **areola (a-RĚ-ō-luh)** surrounds the nipple.

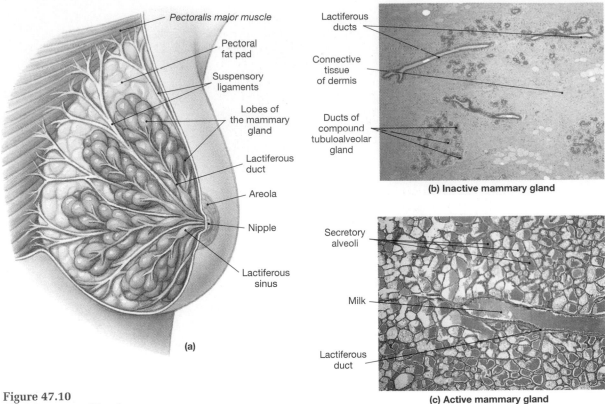

Figure 47.10
The Mammary Glands.

(a) The mammary gland of the left breast. (b) An inactive mammary gland of a nonpregnant woman. (LM × 53) (c) An active mammary gland of a nursing woman. (LM × 119)

LABORATORY ACTIVITY   THE MAMMARY GLANDS

MATERIALS

breast model

PROCEDURES

1. Examine the models and charts of the mammary glands. Trace the pathway of milk from the lobules to the surface of the nipple.

III. DISSECTION OF THE REPRODUCTIVE SYSTEM OF THE CAT

This exercise complements the study of the human reproductive system. The female cat reproductive system is different than that found in humans. Some differences will be noted during the description of the system.

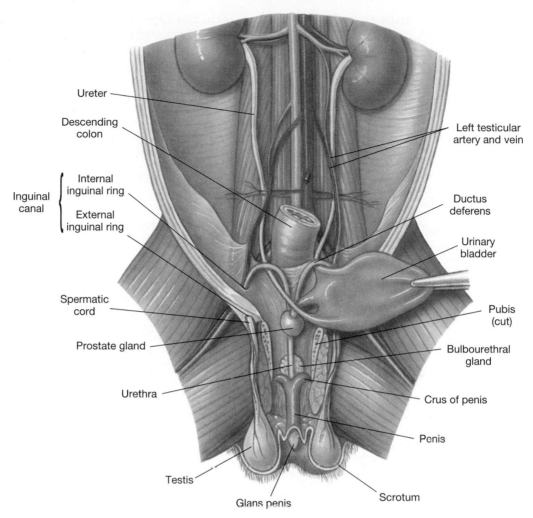

Figure 47.11
Reproductive System of the Male Cat.

LABORATORY ACTIVITY	DISSECTION OF THE REPRODUCTIVE SYSTEM OF THE CAT

MATERIALS

cat
dissecting tools
gloves, safety glasses

PROCEDURES

A. Male cat

Refer to Figure 47.11, showing the male cat, during your dissection.

1. Identify the skin covering the testes, a double-chambered sac located ventral to the anus, termed the **scrotum**. Identify the **penis**, ventral to the scrotum. Carefully make an incision through the skin and expose the **testes**. The testes produce the male gamete, sperm, and are covered by a peritoneal capsule, the **tunica vaginalis**.

2. On the lateral surface of each testis, extending from the cranial to the caudal end of each testis, lies the **epididymis**, a comma-shaped convoluted tubule which stores sperm.

3. Locate and identify the connective tissue–covered **spermatic cord**, which consists of the **spermatic artery**, **spermatic vein**, **spermatic nerve**, and **ductus deferens (vas deferens).** The ductus deferens carries sperm from the epididymis to the urethra for transport out of the body.

4. Trace the spermatic cord through the **inguinal canal** and **ring** into the abdominal cavity.

5. When the spermatic cord components are loosened from their connective tissue wrapping, the ductus deferens is observed to loop over the ureter and pass posterior to the bladder.

    Note: To observe the remaining structures, you must cut through the pelvic muscles and the midline of the pelvic bone (pubic symphysis) with bone cutters. Cut carefully because the urethra is immediately dorsal to the bone. After cutting, split the public bone apart by spreading the thighs apart. This action exposes the structures within the pelvic cavity. Tease and remove any excess connective tissue.

6. Locate the **prostate gland**, a large, hard mass of tissue surrounding the urethra. Trace the ductus deferens from the prostate to its merging with the spermatic cord structures. Compared to human males, cats lack seminal vesicles.

7. Trace the urethra to the proximal end of the penis. The urethra consists of three parts: the **prostatic urethra** (passing through the prostate), the **membranous urethra** (between the prostate gland and the penis), and the **spongy** or **penile urethra** (passing through the penis).

8. The **bulbourethral glands** are located on either side of the membranous urethra.

B. Female cat

Refer to Figure 47.12, showing the female cat, during your dissection.

1. Reflect the abdominal viscera to one side and locate the paired, oval **ovaries**, lying on the dorsal body wall lateral to the kidneys.

2. On the surface of the ovaries locate the small, coiled **uterine tubes**. Their funnel-like opening, the **infundibulum**, curves around the ovary to partially cover it. This ensures the capture of ova released from the ovary.

3. The uterine tubes lead into a single large tube, the **uterine horns**. The horns of the uterus are the location for fertilized egg development. Unlike the pear-shaped uterus of the human, the uterus of the cat is Y-shaped (bicornate) and consists of two uterine horns diverging from a single **uterine body**.

4. Identify the **broad ligament** which aids in anchoring the uterine horns to the body wall.

    Note: To observe the remaining structures, you must cut through the pelvic muscles and the midline of the pelvic bone (pubic symphysis) with bone cutters. Cut carefully because the urethra and vagina are immediately dorsal to the bone. After cutting, split the public bone apart by spreading the thighs apart. This action exposes the structures within the pelvic cavity. Tease and remove any excess connective tissue.

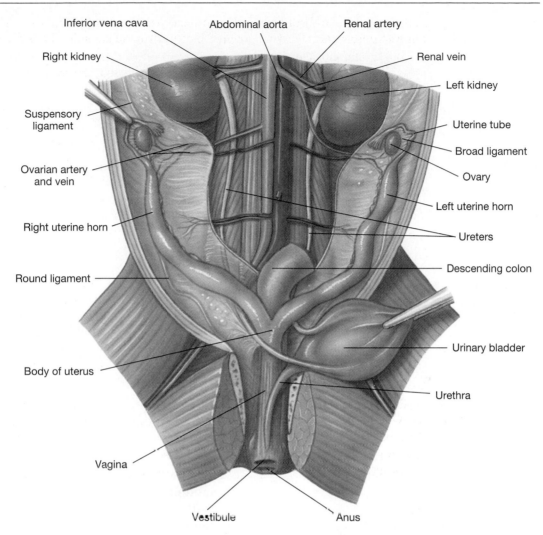

Inferior vena cava   Abdominal aorta   Renal artery

Right kidney

Renal vein

Left kidney

Suspensory
ligament

Uterine tube

Broad ligament

Ovarian artery
and vein

Ovary

Left uterine horn

Right uterine horn

Ureters

Round ligament

Descending colon

Urinary bladder

Body of uterus

Urethra

Vagina

Vestibule   Anus

Figure 47.12
Reproductive System of the Female Cat.

5. Return to the uterine body and follow it caudally into the pelvic cavity where it is continuous with the **vagina**.

6. Locate the urethra which has emerged from the bladder. The vagina is dorsal to the urethra. At its posterior end, the vagina and urethra unite at the **urethral orifice** to form the **urogenital sinus (vestibule)**, the common passage for the urinary and reproductive systems.

7. The urogenital sinus opens to the outside at the urogenital aperture. It is bordered by skin folds, the **labia majora**. Together the urogenital aperture and labia constitute the **vulva** (external genitalia).

REPRODUCTIVE SYSTEM CHECKLIST

This is a list of bold terms presented in Exercise 47. Use this list for review purposes after you have completed the laboratory activities.

MALE

• Figures 47.1–47.4
gonads, gametes
scrotum
dartos and cremaster muscles
testes
lobules
interstitial cells, testosterone
seminiferous tubules
spermatogenesis
testosterone
epididymis
ductus (vas) deferens
spermatic cord
inguinal canal, inguinal hernia
ampulla
semen, ejaculate
seminal vesicle
ejaculatory duct
prostate gland
prostatic urethra
membranous urethra
bulbourethral gland
penis, glans, corona
prepuce (foreskin)
penile urethra
external urethral meatus
root, body
circumcision
corpora cavernosa, corpus spongiosum
erection

FEMALE

• Figures 47.5–47.10
gynecology, ovaries
mesovarian, broad ligament
suspensory ligament
ovarian ligament
ovarian cycle, egg nests
primordial follicle, primary follicle, secondary follicle, tertiary (Graafian) follicle
estrogen, antrum, ovulation
corpus luteum, progesterone
corpus albicans, uterine tubes

fimbriae, infundibulum
ampulla, isthmus
uterus, uterine cavity
fundus, body, cervix
cervical canal
perimetrium, myometrium
endometrium
basilar zone, functional zone
vagina, vaginal orifice, hymen
greater vestibular glands
fornix, vulva, mons pubis
labia majora, labia minora
clitoris, prepuce, vestibule
mammary glands
lobes, lobules, alveoli
lactiferous ducts and sinuses
nipple, areola

MALE CAT

• Figure 47.11
scrotum
penis
testes
tunica vaginalis
epididymis
spermatic cord, spermatic artery, spermatic vein, spermatic nerve
ductus (vas) deferens
inguinal canal and ring
prostate gland, prostatic urethra
membranous urethra
bulbourethral glands

FEMALE CAT

• Figure 47.12
ovaries
uterine tubes
infundibulum
uterine horns, uterine body
broad ligament
vagina
urethral orifice
urogenital sinus
labia majora
vulva

HUMAN REPRODUCTIVE SYSTEMS

Laboratory Report Exercise 47

A. Matching
Match each structure listed on the left with the correct description on the right.

1. _____ epididymis   A. site of sperm storage

2. _____ ductus deferens   B. site of sperm production

3. _____ bulbourethral glands   C. enlarged tip of penis

4. _____ glans   D. paired erectile cylinder

5. _____ corpora cavernosa   E. small glands in pelvic floor

6. _____ prepuce   F. first segment of urethra

7. _____ prostatic urethra   G. also called the foreskin

8. _____ scrotum   H. transports sperm to urethra

9. _____ membranous urethra   I. pouch containing testes

10. _____ seminiferous tubules   J. urethra in pelvic floor

B. Fill in the Blanks

1. The primary sex organs of the male are the _____.

2. The _____ respond to changes in temperature by raising or lowering the testes.

3. The _____ produce the male sex hormone _____ .

4. The three regions of the male urethra are:

   a. _____

   b. _____

   c. _____

5. The _____ is formed by union of the ampulla of the ductus deferens and the duct from the seminal vesicle.

6. List the three accessory glands of the male reproductive system.

   a. _____

   b. _____

   c. _____

C. **Matching**
    Match each structure listed on the left with the correct description on the right.

1. _____ labia minora   **A.** space between labia minora

2. _____ myometrium   **B.** flared end of uterine tube

3. _____ mons pubis   **C.** domed portion of uterus

4. _____ fundus   **D.** female external genitalia

5. _____ infundibulum   **E.** uterine protrusion into vagina

6. _____ isthmus   **F.** narrow portion of uterine tube

7. _____ vestibule   **G.** small fold lacking pubic hair

8. _____ labia majora   **H.** fatty cushion

9. _____ vulva   **I.** muscular layer of uterine wall

10. _____ cervix   **J.** large fold with pubic hair

D. **Fill in the Blanks**

1. After ovulation the tertiary follicle becomes the _____ .

2. List the three layers of the uterus from superficial to deep.

    a. _____

    b. _____

    c. _____

3. The pigmented portion of the breast is called the _____.

4. List the components of the vulva.

    _____

    _____

    _____

Development

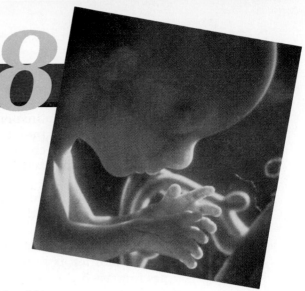

OBJECTIVES

On completion of this exercise, you should be able to:

- Discuss the mechanics of meiosis, and compare gamete formation in males and females.
- Describe the process of fertilization and early cleavage to the blastocyst stage.
- Describe the process of implantation and placenta formation.
- List the three germ layers and the embryonic fate of each.
- List the four extraembryonic membranes and the function of each.
- Describe the general developmental events of the first, second, and third trimesters.
- List the three stages of labor.

WORD POWER

gamete (gamet—spouse)
zygote (zygo—a yoke)
meiosis (meio—less)
diploid (diplo—double)
haploid (haplo—single)
spermatogenesis (...genesis—origin)
blastomere (blast—a sprout)
morula (morul—small mulberry)

trophoblast (tropho—food)
syncytial (syn—without, cytial—cell)
ectoderm (ecto—outer, derm—skin)
endoderm (endo—inner)
mesoderm (meso—middle)
chorion (chorio—a membrane)
placenta (placent—round, flat cake)
decidua (decid—falling off)
parturition (partur—to give birth)

MATERIALS

embryology models
embryology charts
placenta model
parturition model
compound microscope
prepared slides
    seminiferous tubules

INTRODUCTION

An intriguing topic in biology is human development. Two cells called **gametes** from the parents fuse and create a new life, a single cell called the **zygote**, which has its own genetic identity. A phenomenal number of cell divisions and migration of cells occur during the early development of the embryo. Tissue layers and organ systems form in an elaborate sequence designed to progress in complexity until a functional human is born. This exercise will survey the major stages of early development. Some individuals debate when life begins; however, biologically, there is no question: life begins at conception.

The figures in this exercise provide a visual guide to the developmental process. If available, supplement your lab studies with observations of development models to gain a three-dimensional representation of each developmental process.

A. Meiosis: The Formation of Gametes

Sexual reproduction involves the union of gametes from different individuals. Human somatic cells (nonreproductive cells) contain 23 pairs of chromosomes for a total of 46 individual chromosomes. If the number of chromosomes in gametes is not reduced by one-half, each new generation of humans would have twice the number of chromosomes as their parents. Cell division, called **meiosis (mī-Ō-sis)**, produces gametes that are **haploid (HAP-loyd)**; that is, they have a chromosome from each pair of chromosomes. Cells that have both chromosomes from all pairs are **diploid (DIP-loyd)** cells. A human gamete has 23 individual chromosomes, 1 chromosome from each pair. When fertilization occurs, the gametes fuse their genetic material, and the diploid condition is restored in the zygote. The zygote does not represent half the genes from each parent; the zygote has a unique genetic constitution. In many ways, the process of meiosis is remarkably similar to mitosis, cell division that produces identical diploid daughter cells. If necessary, review mitosis in Exercise 5 before continuing in this exercise

Males produce spermatozoa in a process called **spermatogenesis (sper-ma-tō-JEN-e-sis)**. Four haploid sperm cells are produced from each cell that started in meiosis. Formation of the female gamete, the ovum, is called **oogenesis (ō-ō-JEN-e-sis)** and results in only one viable haploid cell. Figure 48.1 is a comparative overview of spermatogenesis and oogenesis. For simplicity, the figure uses a cell with 3 diploid chromosomes instead of 23 diploid as in humans. As in mitosis, cells prepare for meiosis by duplicating their genetic material. After replication, each chromosome consists of two **chromatids**. Meiosis begins as the nuclear membrane dissolves and the chromosomes become visible. The replicated chromosomes match into pairs in a process called **synapsis**, and the four chromatids of the pair are collectively called a **tetrad**. Notice in Figure 48.1 that in the first cell, a primary spermatocyte, the chromosome pairs have joined to form tetrads. Since each chromosome has replicated, a total of four potential chromosomes, each called a **chromatid**, align in a tetrad. Because each chromatid in a tetrad belongs to the same chromosome pair, genetic information may be exchanged between chromatids. This **crossing over** or mixing of the genes increases genetic diversity of the population.

When a male reaches puberty, or sexual maturity, hormones stimulate the testes to begin the process of spermatogenesis (see Figures 48.1 and 48.2). **Spermatogonia (sper-ma-tō-GO-nē-uh)** cells in the walls of the seminiferous tubules divide by mitosis and produce diploid **primary spermatocytes (sper-MA-tō-sits)**. During **meiosis I**, chromosome pairs synapse and form tetrads along the equatorial plane of the primary spermatocytes. Chromosome pairs separate, and the cell divides into two haploid **secondary spermatocytes**. This step is called the **reduction division** of meiosis because haploid cells are produced. Each secondary spermatocyte still possesses chromosomes consisting of 2 chromatids and must go through meiosis II, which results in 4 **spermatids**, each with 23 single-stranded chromosomes. The spermatids undergo a maturation process and develop into **spermatozoa**. The entire process from spermatogonia to mature spermatozoa takes approximately ten weeks.

The process of oogenesis begins in the female's ovaries before she is born. **Oogonia (ō-ō-GŌ-nē-uh)** divide and produce **primary oocytes (Ō-ō-sits)**, which remain suspended in this stage until puberty, when, each month, a primary oocyte divides into two **secondary oocytes**. One of the secondary oocytes is much smaller than its sister cell and is a nonfunctional cell called the **first polar body** (see Figure 48.1). The remaining secondary oocyte will continue to divide if fertilization occurs. Then, the secondary oocyte goes through meiosis II for the separation of double-stranded chromosomes. When this cell divides, another polar body is formed, the **second polar body**. The remaining cell is the fertilized ovum, the zygote. Females produce only a single functional cell by oogenesis, whereas in males, spermatogenesis results in four spermatozoa.

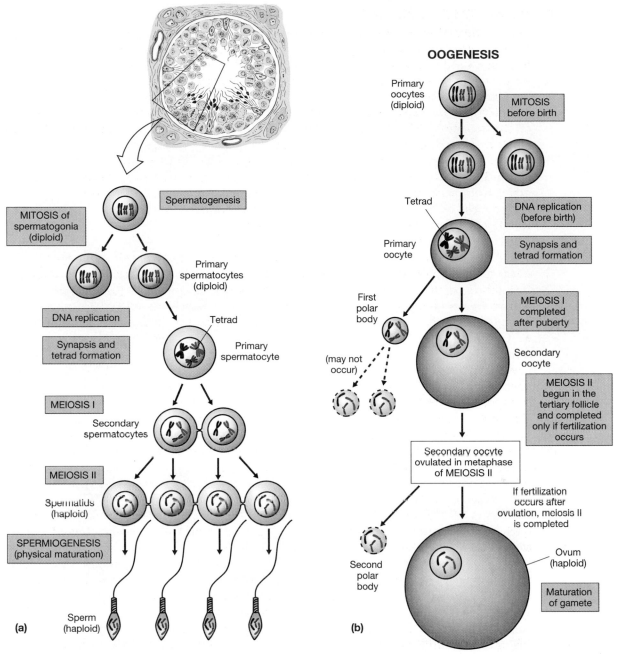

Figure 48.1
Comparison of Spermatogenesis and Oogenesis.

(a) Each diploid primary spermatocyte that undergoes meiosis produces four haploid spermatids. Each spermatid then differentiates into a spermatozoon. **(b)** Diagrammatic view of meiosis and the production of an ovum.

LABORATORY ACTIVITY   MEIOSIS

MATERIALS

      meiosis models
      compound microscope
      prepared slides of seminiferous tubules

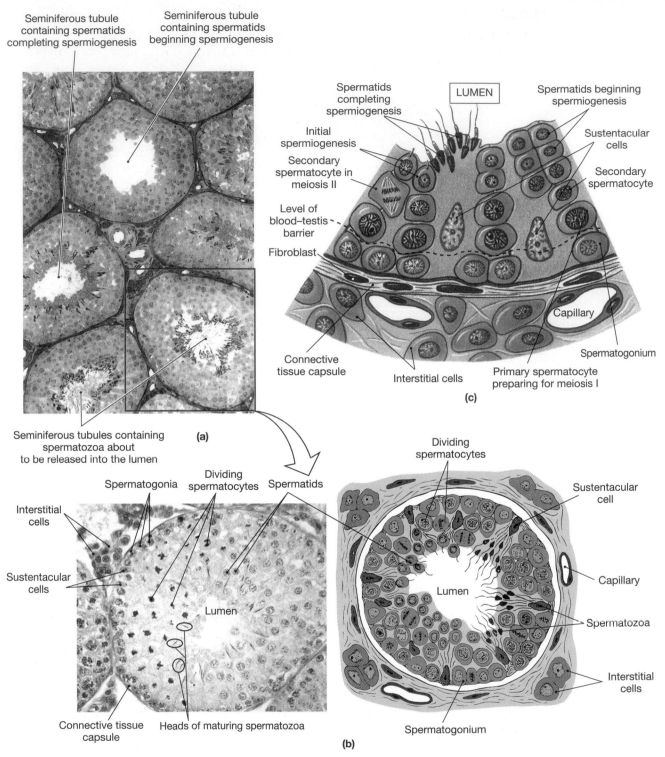

Figure 48.2
The Seminiferous Tubules.

(a) A transverse section through a coiled seminiferous tubule. (b) A cross section through a single tubule. (LM × 786) (c) The lining of a tubule. Sustentacular cells surround the stem cells of the tubule and support the developing spermatocytes and spermatids.

PROCEDURES

1. Label and review the process of meiosis in Figures 48.1 and 48.2.

2. Examine the models of meiosis. When does the reduction division occur to produce haploid cells? Why must meiosis II occur if the cells from meiosis I are already haploid?

_____

_____

3. Obtain a slide of mammalian seminiferous tubules prepared from the testis. Start at low magnification, and locate an oval seminiferous tubule that has distinct cells on the interior. Increase magnification, and using Figure 48.2 as a guide, identify the spermatogonia, spermatocytes in various stages of division, and spermatids. Do you see any spermatozoa in the middle of the tubule?

_____

_____

4. In the space provided, draw a section of a seminiferous tubule, and label the spermatogonia, primary spermatocytes, secondary spermatocytes, spermatids, and spermatozoa.

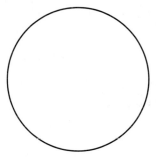

B. Survey of Human Development: The First Trimester

In humans, the prenatal period of development occurs over a nine-month **gestation (jes-TĀ-shun)**. Human gestation is divided into three trimesters, each three months in length. During the first trimester, the **embryo** develops cell layers that are precursors to organ systems. Special membranes, such as the amnion, support the growing embryo and fetus until birth. By the end of the second month, most organ systems have started to form, and the embryo is then called a **fetus**. The second trimester is characterized by growth in length, weight gain, and the appearance of functional organ systems. In the third trimester, increases in length and weight occur, and all organ systems become functional, or are prepared to become operational at birth. After 38 weeks of gestation, the uterus begins to rhythmically contract to deliver the fetus into the world. Although maternal changes occur during the gestation period, this exercise will focus on the development of the fetus.

Development is the process of morphogenesis, growth, and development of offspring to achieve maturity. Morphogenesis involves the division of cells to achieve specialized cells and the migration of cells to produce anatomical form and function. Ultimately, organ systems appear and become functional as the offspring gets closer to birth.

Fertilization is the act of the sperm and egg joining their haploid nuclei to produce a diploid **zygote (ZĪ-gōt)**, a genetically unique cell that develops into an individual. The male ejaculates approximately 300 million spermatozoa into the female's reproductive tract during intercourse. Once exposed to the female's reproductive tract, the sperm complete a process called **capacitation (ka-pas-i-TĀ-shun)**, during which the spermatozoa increase their motility and become capable of fertilizing an egg. Most sperm do not survive the journey through the vagina and uterus, and only an estimated 100

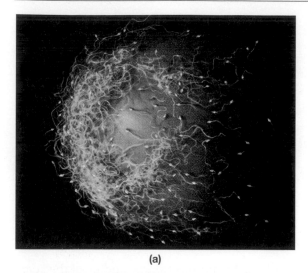

(a)

Figure 48.3
Fertilization.
(a) An oocyte at the time of fertilization: note the difference in size between the gametes.
(b) Fertilization and the preparations for cleavage.

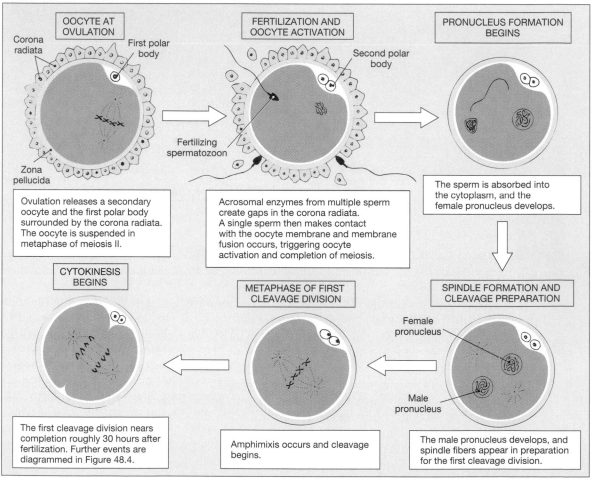

OOCYTE AT OVULATION

Corona radiata

First polar body

Zona pellucida

Ovulation releases a secondary oocyte and the first polar body surrounded by the corona radiata. The oocyte is suspended in metaphase of meiosis II.

FERTILIZATION AND OOCYTE ACTIVATION

Second polar body

Fertilizing spermatozoon

Acrosomal enzymes from multiple sperm create gaps in the corona radiata. A single sperm then makes contact with the oocyte membrane and membrane fusion occurs, triggering oocyte activation and completion of meiosis.

PRONUCLEUS FORMATION BEGINS

The sperm is absorbed into the cytoplasm, and the female pronucleus develops.

SPINDLE FORMATION AND CLEAVAGE PREPARATION

Female pronucleus

Male pronucleus

The male pronucleus develops, and spindle fibers appear in preparation for the first cleavage division.

METAPHASE OF FIRST CLEAVAGE DIVISION

Amphimixis occurs and cleavage begins.

CYTOKINESIS BEGINS

The first cleavage division nears completion roughly 30 hours after fertilization. Further events are diagrammed in Figure 48.4.

(b)

sperm reach the upper uterine tube where fertilization occurs. Normally, only a single egg is released from a single ovary during one ovulation cycle.

Figure 48.3 illustrates the fertilization process. Ovulation releases a secondary oocyte from the ovary. The oocyte is surrounded by a layer of cells called the **corona radiata (kō-RŌ-nuh rā-dē-A-tuh)** that the sperm must pass through to reach the cell membrane of the oocyte. Spermatozoa swarming around the oocyte release an enzyme

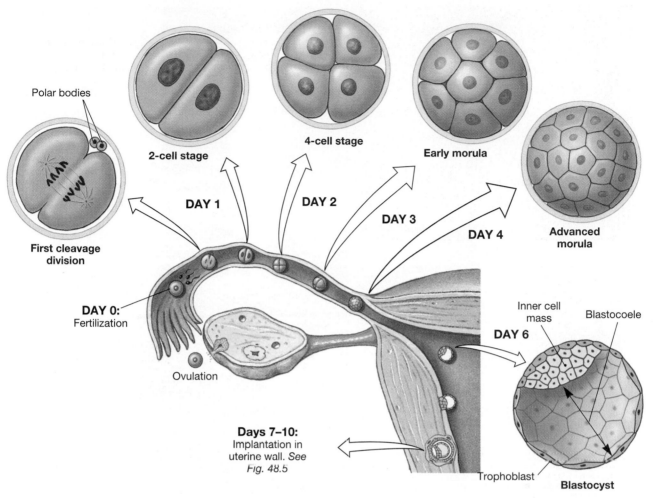

Figure 48.4
Cleavage and Blastocyst Formation.

called hyaluronidase from their acrosomal caps. The combined action of enzymes contributed by all the spermatozoa eventually creates a gap between some coronal cells, and a single spermatozoan slips into the oocyte. The membrane of the oocyte instantly undergoes chemical and electrical changes that prevent additional sperm from entering the cell. The oocyte, suspended in meiosis II since ovulation, now completes meiosis while the sperm prepares the paternal chromosomes for the union with the maternal chromosomes. Each set of nuclear material is called a **pronucleus**. Within 30 hours of fertilization, the male and female pronuclei come together in **amphimixis (am-fi-MAK-sis)** and undergo the first **cleavage (KLĒ-vij)**, a mitotic division resulting in two cells, each called a **blastomere (BLAS-tō-mēr)**. During cleavage, the existing cell mass of the egg is subdivided by each mitosis.

As the zygote slowly descends in the uterine tube toward the uterus, cleavages occur approximately every 12 hours. By the third day, the blastomeres are organized into a solid ball of nearly equal cells called a **morula (MOR-ū-la)**, shown in Figure 48.4. Around day six, the morula has entered the uterus and changed into a **blastocyst (BLAS-tō-sist)**, a hollow ball of cells with an internal cavity called the **blastocoele (BLAS-tō-sel)**. The process of **differentiation** or specialization begins. The blastomeres are now different sizes and have migrated into two regions. Cells on the outside compose the **trophoblast (TRŌ-fō-blast)**, which will burrow into the uterine lining and eventually form part of the placenta. Cells clustered inside the blastocoele form the **inner cell mass**, which will develop into the embryo.

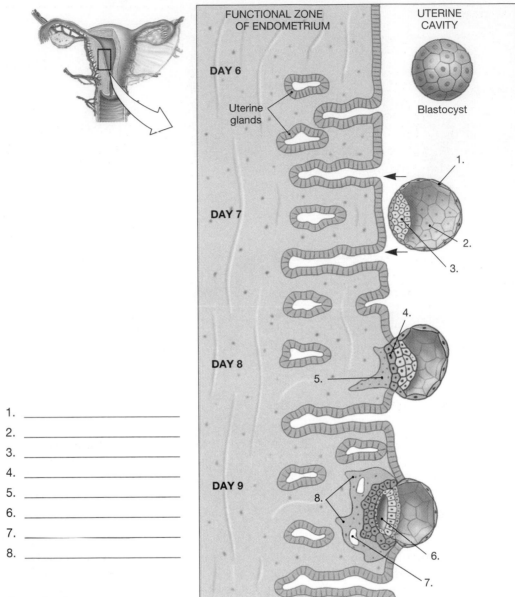

1. _____

2. _____

3. _____

4. _____

5. _____

6. _____

7. _____

8. _____

Figure 48.5
Stages in the Implantation Process.

Implantation begins on day seven or eight when the blastocyst touches the spongy uterine lining. The trophoblast near the inner cell mass faces the uterus and burrows into the functional zone of the endometrium. The plasma membranes of these trophoblast cells dissolve, and the cells mass together as a cytoplasmic layer of multiple nuclei called the **syncytial (sin-SISH-al) trophoblast**. The cells secrete hyaluronidase to erode a path for implantation of the blastocyst (see Figure 48.5). Implantation continues until the embryo is completely covered by the stratum functionalis of the endometrium, which occurs by the 14th day. To establish a diffusional link with maternal circulation, the syncytial trophoblast grows extensions called **villi** into spaces in the endometrium called **lacunae**. Maternal blood from the endometrium seeps into the lacunae and bathes the villi with nutrients and oxygen. These materials diffuse into the blastocyst to support the inner cell mass. Beneath the syncytial layer is the **cellular trophoblast**, which will soon help form the placenta.

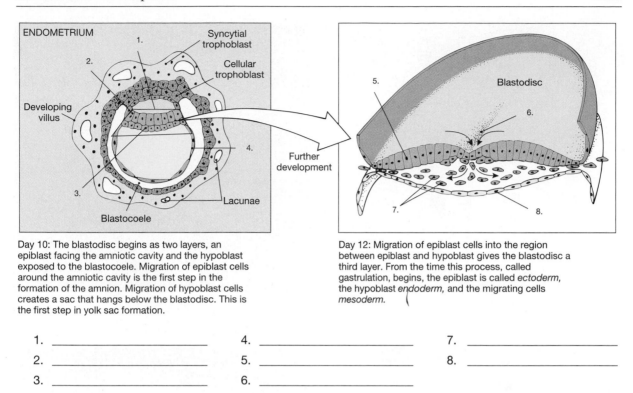

Day 10: The blastodisc begins as two layers, an epiblast facing the amniotic cavity and the hypoblast exposed to the blastocoele. Migration of epiblast cells around the amniotic cavity is the first step in the formation of the amnion. Migration of hypoblast cells creates a sac that hangs below the blastodisc. This is the first step in yolk sac formation.

Day 12: Migration of epiblast cells into the region between epiblast and hypoblast gives the blastodisc a third layer. From the time this process, called gastrulation, begins, the epiblast is called *ectoderm,* the hypoblast *endoderm,* and the migrating cells *mesoderm.*

1. _____   4. _____   7. _____

2. _____   5. _____   8. _____

3. _____   6. _____

Figure 48.6
Blastodisc Organization and Gastrulation.

On the ninth or tenth day, the middle layers of the inner cell mass drop away from the upper layer next to the cellular trophoblast. This movement of cells forms the **amniotic (am-nē-OT-ik) cavity.** The inner cell mass organizes into a **blastodisc (BLAS-tō-disk),** which contains two cell layers, the **epiblast (EP-i-blast)** on top and the **hypoblast (HĪ-pō-blast)** facing the blastocoele. The blastodisc is the early embryo and is illustrated in Figure 48.6.

Within the next few days a process of cell migration called **gastrulation (gas-troo-LĀ-shun)** begins (see Figure 48.6). Cells of the epiblast move toward the medial plane of the blastodisc to a region known as the **primitive streak.** As cells arrive at the primitive streak, infolding, or **invagination,** occurs and cells are liberated between the epiblast and the hypoblast, producing three cell layers in the embryo. The epiblast becomes the **ectoderm,** the hypoblast is now the **endoderm,** and the cells proliferating between the two layers form the **mesoderm.** These three layers, called **germ layers,** each produce specialized tissues that contribute to the formation of the organ systems. For example, the ectoderm forms the nervous system, the skin, hair, and nails. The mesoderm contributes to the development of the skeletal and muscular systems, and the endoderm forms part of the lining of the respiratory and digestive systems.

By the end of the fourth week of development, the embryo is distinct and has a **tail fold** and a **head fold.** The dorsal and ventral surfaces and the right and left sides are well defined. The process of **organogenesis** begins as organ systems develop from the germ layers. Observe Figure 48.7a, a fiber-optic view of the embryo at four weeks. The heart is clearly visible and has beat since the 21st day in utero. **Somites (so-MĪ-tis),** embryonic precursors of skeletal muscles, appear. Elements of the nervous system are also developing. Buds for the arms and legs and small discs for the eyes and ears are also present. By week eight, individual fingers and toes are present (see Figure 48.7b). At the end of the second month, the embryo is called a **fetus.** At the end of the third month, the first trimester is completed, and each organ system has appeared in the fetus.

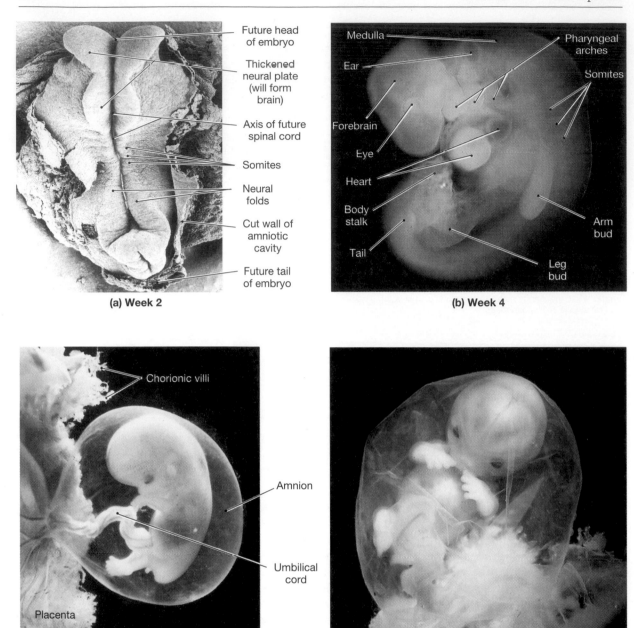

(a) Week 2

Future head
of embryo

Thickened
neural plate
(will form
brain)

Axis of future
spinal cord

Somites

Neural
folds

Cut wall of
amniotic
cavity

Future tail
of embryo

(b) Week 4

Medulla

Ear

Forebrain

Eye

Heart

Body
stalk

Tail

Pharyngeal
arches

Somites

Arm
bud

Leg
bud

(c) Week 8

Chorionic villi

Amnion

Umbilical
cord

Placenta

(d) Week 12

Figure 48.7
The First Trimester.

(a) An SEM of the superior surface of a monkey embryo after 2 weeks of development. A human embryo at this stage would look essentially the same. **(b–d)** Fiber-optic views of human embryos at 4, 8, and 12 weeks.

LABORATORY ACTIVITY   THE FIRST TRIMESTER

MATERIALS

embryology models
embryology charts

PROCEDURES

1. Label and review the developmental processes illustrated in Figures 48.3–48.7.
2. Examine the models of fertilization and early embryologic development. Describe the fertilization process and the early cleavages that form the morula. Draw a sequence of cells from the zygote to the morula.

3. Identify the structures of the blastocyst at the time of implantation. How does the amniotic cavity form? What are the two cell layers called below the amniotic cavity? Do the models show the distinction between the cellular and syncytial trophoblast? In the space provided, draw a blastocyst with each trophoblast layer.

4. If available, examine a model or chart of a 12-day-old embryo. Identify each germ layer if possible. Briefly list the fate of each germ layer.

C. Extraembryonic Membranes and the Placenta

Four extraembryonic membranes develop from the germ layers: the yolk sac, amnion, chorion, and allantois. These membranes lie outside the embryonic disc and provide protection and nourishment for the embryo and fetus. The **yolk sac** is the first membrane to appear around the tenth day (see Figures 48.6 and 48.8). Initially, cells from the hypoblast form a pouch under the blastodisc. Mesoderm reinforces the sac, and blood vessels appear. As the syncytial trophoblast develops more villi, the yolk sac's importance in providing nourishment for the embryo diminishes.

While the yolk sac is forming, cells in the epiblast migrate to line the inner surface of the amniotic cavity with a membrane called the **amnion (AM-nē-on)**, the "water bag." As with the yolk sac, the amnion is soon reinforced with mesoderm tissue. Early embryonic growth continues, and the amnion mushrooms and envelops the entire embryo in a protective environment of **amniotic fluid**, as illustrated in Figure 48.8e.

The **allantois (a-LAN-tō-is)** develops from the endoderm and mesoderm near the base of the yolk sac (see Figure 48.8b). The allantois forms part of the urinary bladder and contributes to the **body stalk**, the tissue between the embryo and the developing chorion. Blood vessels pass through the body stalk and into the villi protruding into the lacunae of the endometrium.

The outer extraembryonic membrane is the **chorion (KOR-ē-on)**, formed by the cellular trophoblast and mesoderm. The chorion completely encases the embryo and the blastocoele. In the third week of growth, the chorion extends **chorionic villi** and blood vessels into the endometrial lacunae to establish the structural framework for the development of the **placenta (pla-SENT-uh)** (see Figure 48.8).

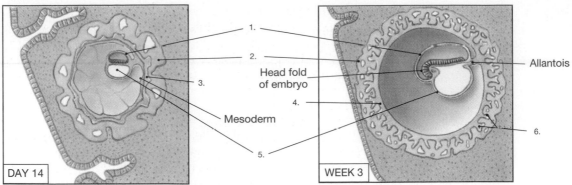

(a) Migration of mesoderm around the inner surface of the trophoblast creates the chorion. Mesodermal migration around the outside of the amniotic cavity, between the ectodermal cells and the trophoblast, creates the amnion. Mesodermal migration around the endodermal pouch below the blastodisc creates the definitive yolk sac.

(b) The embryonic disc bulges into the amniotic cavity at the head fold. The allantois, an endodermal extension surrounded by mesoderm, extends toward the trophoblast.

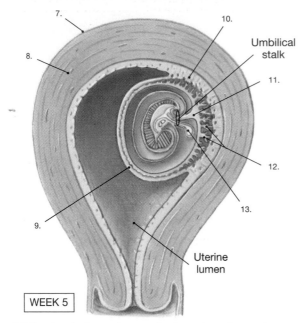

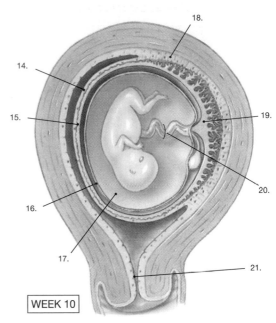

(c) The developing embryo and extraembryonic membranes bulge into the uterine cavity. The trophoblast pushing out into the uterine lumen remains covered by endometrium, but no longer participates in nutrient absorption and embryo support. The embryo moves away from the placenta, and the body stalk and yolk stalk fuse to form an umbilical stalk.

(d) The amnion has expanded greatly, filling the uterine cavity. The fetus is connected to the placenta by an elongate umbilical cord that contains a portion of the allantois, blood vessels, and the remnants of the yolk stalk.

1. _____   8. _____   15. _____
2. _____   9. _____   16. _____
3. _____   10. _____   17. _____
4. _____   11. _____   18. _____
5. _____   12. _____   19. _____
6. _____   13. _____   20. _____
7. _____   14. _____   21. _____

Figure 48.8
Embryonic Membranes and Placenta Formation.

The placenta, detailed in Figure 48.8e, is a temporary organ through which nutrients, blood gases, and wastes are exchanged between the mother and the embryo. The embryo is connected to the placenta by the body stalk constructed of the allantois and blood vessels. The **yolk stalk** where the yolk sac attaches and the body stalk together form the **umbilical cord**. Inside the umbilical cord are two **umbilical arteries**, which transport deoxygenated blood into the placenta, and a single **umbilical vein**, which returns oxygenated blood to the embryo.

The placenta does not completely surround the embryo. By the fourth week of development, the chorionic villi have enlarged only where they face the uterine wall, and villi that face the uterine cavity become insignificant. The placenta is in contact with the area of the endometrium called the **decidua basalis (dē-SID-ū-a ba-SA-lis)**, where the chorionic villi occur. The rest of the endometrium, where villi are absent, isolates the embryo from the uterine cavity and is called the **decidua capsularis (kap-sū-LA-ris)**.

LABORATORY ACTIVITY   EXTRAEMBRYONIC MEMBRANES AND THE PLACENTA

MATERIALS

embryology models
embryology chart
placenta model or biomount

PROCEDURES

1. Label and review the extraembryonic membranes in Figure 48.8.
2. Examine a model or chart that details the extraembryonic membranes. Locate the yolk sac and the amnion. How do each of these membranes form, and what is the function of each?

_____

_____

3. Locate the allantois and the chorion. How do each of these membranes form? Describe the chorionic villi and their significance to the embryo.

_____

_____

4. Examine the model or biomount of a placenta. How do the surfaces of the placenta appear? Is the amniotic membrane attached to the placenta?

_____

_____

5. Examine the umbilical cord attached to the placenta. Describe the vascular anatomy in the cord.

_____

_____

_____

D. The Second and Third Trimesters and Birth

Growth during the second trimester is fast, and the fetus doubles in size and substantially increases its weight. As the fetus grows, the uterus expands and displaces the maternal abdominal organs. Figure 48.9 details the uterine size compared with the gestational month. Quarters for the fetus become cramped toward the end of pregnancy.

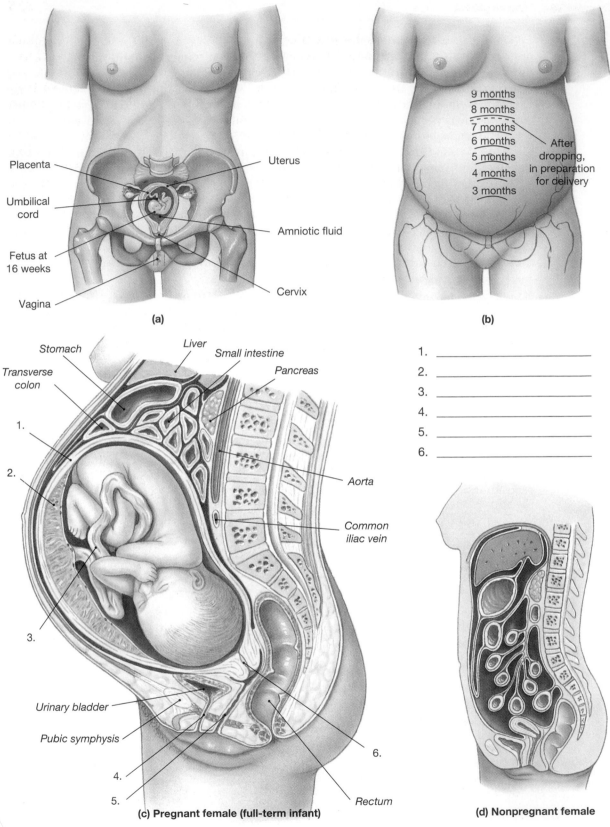

Placenta

Umbilical cord

Fetus at 16 weeks

Vagina

Uterus

Amniotic fluid

Cervix

(a)

9 months
8 months
7 months
6 months
5 months
4 months
3 months

After dropping, in preparation for delivery

(b)

Stomach

Transverse colon

1.

2.

3.

Urinary bladder

Pubic symphysis

4.

5.

Liver

Small intestine

Pancreas

Aorta

Common iliac vein

6.

Rectum

(c) Pregnant female (full-term infant)

1. _____
2. _____
3. _____
4. _____
5. _____
6. _____

(d) Nonpregnant female

Figure 48.9
Growth of the Uterus and Fetus.

Pregnancy at 16 weeks, showing the positions of the uterus, fetus, and placenta. **(b)** Pregnancy at 3 months months (full term), showing the position of the uterus within the abdomen. **(c)** Pregnancy at full term. Note osition of the uterus and full-term fetus within the abdomen and the displacement of abdominal organs, as red with **(d)**, a sectional view through the abdominopelvic cavity of a woman who is not pregnant.

The fetus begins to move as the muscular system becomes functional, and articulations begin to form in the skeleton. The nervous system organizes the neural tissue that developed in the first trimester, and many sensory organs complete their formation. During the third trimester, weight increases, and all organ systems complete their development and become partially operational. The fetus responds to sensory stimuli such as a hand rubbing across the mother's swollen abdomen.

Birth, or **parturition (par-tū-RISH-un)**, involves muscular contractions of the uterine wall to expel the fetus into the outside world. Delivering the fetus is much like pulling on a turtle neck sweater. Muscle contractions must stretch the cervix of the uterus over the fetus's head, pulling the uterine wall thinner as the fetus passes into the lower birth canal, the vagina. Once true labor contractions begin, positive feedback mechanisms operate to increase the frequency and force of uterine contractions. Stretching of the uterine wall causes secretion of **oxytocin (oks-i-TŌ-sin)** from the posterior pituitary gland. Oxytocin stimulates contraction of the myometrium of the uterus. The hormone **relaxin** causes the symphysis pubis to become less rigid.

Labor is divided into three stages: the dilation, expulsion, and placental stages, each illustrated in Figure 48.10. The **dilation stage** begins at the onset of true labor contractions. The cervix of the uterus dilates, and the fetus begins to travel down the cervical canal. To be maximally effective at dilation, the contractions must be less than ten minutes apart. Each contraction lasts approximately one minute and spreads from the upper cervix downward to efface or thin the cervix for delivery. Contractions usually rupture the amnion, and amniotic fluid flows out of the uterus and the vagina.

The **expulsion stage** occurs when the cervix is dilated completely, usually to 10 cm, and the fetus passes through the cervix and the vagina. The expulsion stage usually lasts less than two hours and results in the **delivery** or birth of the newborn. Once the baby is breathing on its own, the umbilical cord is cut, and the newborn must now rely on its own organ systems to survive.

During the **placental stage**, the uterus continues to contract to break the placenta free of the endometrium and deliver it out of the body as the **afterbirth**, illustrated in Figure 48.10c. This stage is usually short, and many women deliver the afterbirth within five to ten minutes after the birth of the fetus.

LABORATORY ACTIVITY   THE SECOND AND THIRD TRIMESTERS AND BIRTH

MATERIALS

> second and third trimester models
> parturition models

PROCEDURES

1. Label Figures 48.9 and 48.10.
2. Examine the models of a second and third trimester fetus. Describe how the fetus is positioned in the uterus. If shown, describe the location of the amnion and the placenta.

   _____

   _____

3. Examine the models of parturition. Describe the contractions that force the fetus out of the uterus.

   _____

   _____

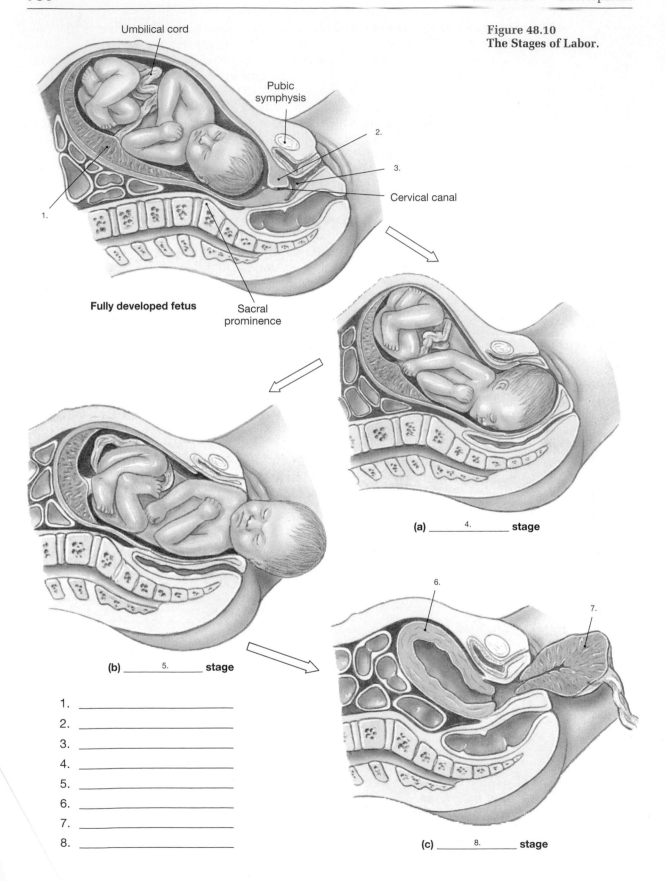

**Figure 48.10
The Stages of Labor.**

Umbilical cord

Pubic symphysis

2.

3.

Cervical canal

1.

Fully developed fetus

Sacral prominence

(a) _____ 4. _____ **stage**

(b) _____ 5. _____ **stage**

6.

7.

(c) _____ 8. _____ **stage**

1. _____
2. _____
3. _____
4. _____
5. _____
6. _____
7. _____
8. _____

DEVELOPMENT CHECKLIST

This is a list of bold terms presented in Exercise 48. Use this list for review purposes after you have completed the laboratory activities.

MEIOSIS

- Figures 48.1 and 48.2

gamete
haploid
diploid
meiosis
spermatogenesis
oogenesis
chromatid
synapsis
tetrad
crossing over
spermatogonia
primary spermatocytes
secondary spermatocytes
spermatids
spermatozoa
meiosis I
reduction division
meiosis II
oogonia
primary oocytes
first polar body
secondary oocyte
second polar body
ovum

FIRST TRIMESTER

- Figures 48.3–48.8

gestation
embryo
fetus
corona radiata
pronucleus
cleavage
blastomere
morula
blastocyst
blastocoele
trophoblast
inner cell mass
implantation
syncytial trophoblast
villi
lacunae
cellular trophoblast
amniotic cavity
blastodisc
epiblast
hypoblast
gastrulation
primitive streak
invagination
ectoderm
endoderm
mesoderm
germ layer
tail fold
head fold
organogenesis
somites
differentation

EXTRAEMBRYONIC MEMBRANES & PLACENTA

• Figures 48.6 and 48.8

yolk sac
amnion
amnionic fluid
allantois
body stalk
chorion
chorionic villi
placenta
yolk stalk
umbilical cord
umbilical arteries
umbilical vein
decidua capsularis
decidua basalis

SECOND & THIRD TRIMESTERS AND BIRTH

• Figures 48.9 and 48.10

parturition
oxytocin
relaxin
dilation
expulsion
placental stages
afterbirth

HUMAN DEVELOPMENT

Laboratory Report Exercise 48

A. Matching

Match each structure on the left with the correct description on the right.

1. _____ blastomere		A.	has villi
2. _____ amnion		B.	differentiates between epiblast and hypoblast
3. _____ allantois		C.	cavity of blastocyst
4. _____ morula		D.	forms part of urinary bladder
5. _____ blastocoele		E.	becomes the ectoderm
6. _____ haploid		F.	migration of cells
7. _____ chorion		G.	water bag
8. _____ mesoderm		H.	birth
9. _____ endoderm		I.	cell produced by early cleavage
10. _____ epiblast		J.	produces lining of respiratory tract
11. _____ parturition		K.	solid ball of cells
12. _____ gastrulation		L.	cell with half the number of chromosomes

B. Short-Answer Questions

1. Describe the process of fertilization. Where does it occur? Why are so many spermatozoa required for the process?

2. Compare spermatogenesis and oogenesis.

3. How does the amniotic cavity form?

4. List the three stages of labor.

5. Discuss the formation of the three germ layers.

6. Describe the structure of the blastocyst.

7. How does a fetus obtain nutrients and gases from the maternal blood?

8. List the four extraembryonic membranes and their specific functions.

Weights and Measures

TABLE A.1	The U.S. System of Measurement		
Physical Property	**Unit**	**Relationship to Other U.S. Units**	**Relationship to Household Units**
Length	inch. (in.)	1 in. = 0.083 ft	
	foot (ft)	1 ft = 12 in. = 0.33 yd	
	yard (yd)	1 yd = 36 in. = 3 ft	
	mile (mi)	1 mi = 5,280 ft = 1,760 yd	
Volume	fluidram (fl dr)	1 fl dr = 0.125 fl oz	
	fluid ounce (fl oz)	1 fl oz = 8 fl dr = 0.0625 pt	= 6 teaspoons (tsp) = 2 tablespoons (tbsp)
	pint (pt)	1 pt = 128 fl dr = 16 fl oz = 0.5 qt	= 32 tbsp = 2 cups (c)
	quart (qt)	1 qt = 256 fl dr = 32 fl oz = 2 pt = 0.25 gal	= 4 c
	gallon (gal)	1 gal = 128 fl oz = 8 pt = 4 qt	
Mass	grain (gr)	1 gr = 0.002 oz	
	dram (dr)	1 dr = 27.3 gr = 0.063 oz	
	ounce (oz)	1 oz = 437.5 gr = 16 dr	
	pound (lb)	1 lb = 7000 gr = 256 dr = 16 oz	
	ton (t)	1 t = 2000 lb	

TABLE A.2   The Metric System of Measurement

Physical Property	Unit	Relationship to Standard Metric Units	Conversion to U.S. Units	
Length	nanometer (nm)	1 nm = 0.000000001 m (10^{-9})	= 3.94×10^{-8} in.	25,400,000 nm = 1 in.
	micrometer (µm)	1 µm = 0.000001 m (10^{-6})	= 3.94×10^{-5} in.	25,400 mm = 1 in.
	millimeter (mm)	1 mm = 0.001 m (10^{-3})	= 0.0394 in.	25.4 mm = 1 in.
	centimeter (cm)	1 cm = 0.01 m (10^{-2})	= 0.394 in.	2.54 cm = 1 in.
	decimeter (dm)	1 dm = 0.1 m (10^{-1})	= 3.94 in.	0.25 dm = 1 in.
	meter (m)	standard unit of length	= 39.4 in.	0.0254 m = 1 in.
			= 3.28 ft	0.3048 m = 1 ft
			= 1.093 yd	0.914 m = 1 yd
	kilometer (km)	1 km = 1000 m	= 3280 ft	
			= 1093 yd	
			= 0.62 mi	1.609 km = 1 mi
Volume	microliter (µl)	1 µl = 0.000001 l (10^{-6}) = 1 cubic millimeter (mm^3)		
	milliliter (ml)	1 ml = 0.001 l (10^{-3}) = 1 cubic centimeter (cm^3 or cc)	= 0.0338 fl oz	5 ml = 1 tsp
				15 ml = 1 tbsp
				30 ml = 1 fl oz
	centiliter (cl)	1 cl = 0.01 l (10^{-2})	= 0.338 fl oz	2.95 cl = 1 fl oz
	deciliter (dl)	1 dl = 0.1 l (10^{-1})	= 3.38 fl oz	0.295 dl = 1 fl oz
	liter (l)	standard unit of volume	= 33.8 fl oz	0.0295 l = 1 fl oz
			= 2.11 pt	0.473 l = 1 pt
			= 1.06 qt	0.946 l = 1 qt
Mass	picogram (pg)	1 pg = 0.000000000001 g (10^{-12})		
	nanogram (ng)	1 ng = 0.000000001 g (10^{-9})	= 0.000000015 gr	66,666,666 mg = 1 gr
	microgram (µg)	1 µg = 0.000001 g (10^{-6})	= 0.000015 gr	66,666 mg = 1 gr
	milligram (mg)	1 mg = 0.001 g (10^{-3})	= 0.015 gr	66.7 mg = 1 gr
	centigram (cg)	1 cg = 0.01 g (10^{-2})	= 0.15 gr	6.67 cg = 1 gr
	decigram (dg)	1 dg = 0.1 g (10^{-1})	= 1.5 gr	0.667 dg = 1 gr
	gram (g)	standard unit of mass	= 0.035 oz	28.4 g = 1 oz
			= 0.0022 lb	454 g = 1 lb
	dekagram (dag)	1 dag = 10 g		
	hectogram (hg)	1 hg = 100 g		
	kilogram (kg)	1 kg = 1000 g	= 2.2 lb	0.454 kg = 1 lb
	metric ton (kt)	1 mt = 1000 kg	= 1.1 t	
			= 2205 lb	0.907 kt = 1 t

Temperature	Centigrade	Fahrenheit
Freezing point of pure water	0°	32°
Normal body temperature	36.8°	98.6°
Boiling point of pure water	100°	212°
Conversion	°C → °F: °F = (1.8 × °C) + 32	°F → °C: °C = (°F − 32) × 0.56

Photo and Illustration Credits

Exercise 1 **CO** Ward's Natural Science Establishment, Inc.

Exercise 2 **2.6a,b,c** Custom Medical Stock Photo, Inc.

Exercise 3 **CO** Ralph T. Hutchings

Exercise 4 **4.1 and 4.2** Science Vu / Visuals Unlimited

Exercise 5 **CO** CNR Science Source / Photo Researchers, Inc. **5.5a** Ed Reschke / Peter Arnold, Inc.

Exercise 6 **CO** Lennart Nilsson / Albert Bonniers Forlay AB **6.5a,b,c** David M. Phillips / Visuals Unlimited **6.6** M Sahliwa / Visuals Unlimited

Exercise 7 **CO** Frederic H. Martini **7.2** Frederic H. Martini **7.3** Ward's Natural Science Establishment, Inc. **7.4** Fred Hosslier / Visuals Unlimited **7.5** Michael J. Timmons

Exercise 8 **CO** Custom Medical Stock Photo, Inc. **8.3a** Ward's Natural Science Establishment, Inc. **8.3c** Frederic H. Martini **8.4a,b,c** Michael G. Wood **8.6a,b,c** Michael G. Wood

Exercise 9 **9.2** Ward's Natural Science Establishment, Inc. **9.3a** Science Source / Photo Researchers, Inc. **9.3b** Frederic H. Martini **9.3c** John D. Cunningham / Visuals Unlimited **9.4** Ward's Natural Science Establishment, Inc. **9.5a,b,c,d** Michael G. Wood **9.7** Frederic H. Martini

Exercise 10 **10.1a,b** G.W. Willis, MD / Biological Photo Service **10.2, 10.3 and 10.4** Michael G. Wood

Exercise 11 **11.2** Michael G. Wood

Exercise 12 **12.2** John D. Cunningham / Visuals Unlimited **12.3** Frederic H. Martini **12.4a,b** Frederic H. Martini

Exercise 13 **CO** Ralph T. Hutchings **13.2** Phototake NYC **13.3** Ralph T. Hutchings

Exercise 14 **14.2** Ralph T. Hutchings

Exercise 15 **CO** Ralph T. Hutchings **15.2** Ralph T. Hutchings **15.3** Ralph T. Hutchings **15.4** Ralph T. Hutchings **15.6a,b** Ralph T. Hutchings **15.7a,b** Ralph T. Hutchings **15.8a,b** Ralph T. Hutchings **15.8c** Michael J. Timmons **15.9a,b** Ralph T. Hutchings **15.10a,b** Ralph T. Hutchings **15.11** Ralph T. Hutchings **15.12a** Ralph T. Hutchings **15.13** Michael Siegfried, MD **15.17a,b,c** Ralph T. Hutchings **15.18a,b,c** Ralph T. Hutchings **15.19a,b,c** Ralph T. Hutchings **15.20a,b,c** Ralph T. Hutchings **15.21a.** Ralph T. Hutchings **15.22a,b** Ralph T. Hutchings

Exercise 16 **CO** Ralph T. Hutchings **16.1a,b,c** Ralph T. Hutchings **16.2b** Bates / Custom Medical Stock Press, Inc. **16.2c** Ralph T. Hutchings **16.3a,b** Ralph T. Hutchings **16.4** Ralph T. Hutchings **16.6a,b** Ralph T. Hutchings **16.8a,b** Ralph T. Hutchings **16.9a,b** Ralph T. Hutchings **16.10a,b** Ralph T. Hutchings **16.11a,b** Ralph T. Hutchings

Exercise 18 **18.3a,b** J. J. Head / Carolina Biological Supply Company/Phototake **18.4** Fred Hossler / Visuals Unlimited

Exercise 19 **19.3** Ralph T. Hutchings

Exercise 20 **20.3a** Mentor Networks Inc. **20.3b** Ralph T. Hutchings

Exercise 21 **21.4a,b,c** Mentor Networks, Inc.

Exercise 22 **22.4a,b** Mentor Networks, Inc.

Exercise 24   CO Fred Hossler / Visuals Unlimited   **24.1L,R** J. J. Head/ Carolina Biological Supply Company/Phototake

Exercise 24A   All photos Courtesy of BIOPAC Systems, Inc.

Exercise 25   CO David Scott / Phototake NYC   **25.3 and 25.6** Michael G. Wood   **25.7** Fred Hossler / Visuals Unlimited

Exercise 26   CO Ralph T. Hutchings   **26.3** Ralph T. Hutchings   **26.4** Michael J. Timmons   **26.6** Ralph T. Hutchings   **26.7b** Ralph T. Hutchings   **26.8, 26.9, and 26.10** Michael G. Wood

Exercise 27   CO Ralph T. Hutchings   **27.1a,b** Ralph T. Hutchings   **27.2** Ralph T. Hutchings **27.5a** Michael J. Timmons   **27.5b** Pat Lynch / Photo researchers, Inc.   **27.6** Michael G. Wood **27.7** Ralph T. Hutchings

Exercise 28   CO Michael J. Timmons   **28.2** Michael J. Timmons   **28.3** Ralph T. Hutchings

Exercise 30   CO Frederic H. Martini   **30.2b** Manfred Kage / Peter Arnold, Inc.   **30.4b,d** Frederic H. Martini   **30.5b,c** Frederic H. Martini **30.6c,d** Frederic H. Martini   **30.7** Ward's Natural Science Establishment, Inc.   **30.8b** Ward's Natural Science Establishment, Inc.   **30.9** Don W. Fawcett, M.D. Harvard Medical School   **30.10** Frederic H. Martini

Exercise 31   31.2c G.W. Willis/Terraphotographics / Biological Photo Service

Exercise 32   CO Ralph T. Hutchings   **32.1a** Ralph T. Hutchings   **32.4a** Ed Reschke / Peter Arnold., Inc.   **32.4c** Custom Medical Stock Photo, Inc.

Exercise 34   CO Lennart Nilsson / Lennart Nilsson / Bonnier Alba alba AB, BEHOLD MAN, pp. 206-207   **34.4** Ward's Natural Science Establishment, Inc.

Exercise 36   CO David Scharf / Peter Arnold, Inc.   **36.1** Martin M. Rotker   **36.2ab** David Scharf / Peter Arnold, Inc.   **36.3a,b,c,d,e** Alfred Owczarzak / Biological Photo Service   **36.6 and 36.7** Michael G. Wood

Exercise 37   CO © Lennart Nilsson, BEHOLD MAN, Little, Brown and Company   **37.2** Peter Arnold, Inc.   **37.3b1,c** Ralph T. Hutchings **37.4b,d** Ralph T. Hutchings   **37.10a,b** Michael G. Wood   **37.11a,b** Michael G. Wood

Exercise 38   CO Ralph T. Hutchings   **38.2** Biophoto Associates / Photo Researchers, Inc.   **38.4** Ralph T. Hutchings   **38.9b** Ralph T. Hutchings

Exercise 39   CO Photodisc, Inc.   **39.3b** Photodisc, Inc.

Exercise 39A   All photos Courtesy of BIOPAC Systems, Inc.

Exercise 39B   All photos Courtesy of BIOPAC Systems, Inc.

Exercise 40   CO Frederic H. Martini   **40.3b and 40.6** Frederic H. Martini

Exercise 41   CO Ralph T. Hutchings   **41.04** John D. Cunningham / Visuals Unlimited   **41.5 and 41.6** Ralph T. Hutchings   **41.7c** Ward's Natural Science Establishment, Inc.   **41.7b** Micrograph by P. Gehl, from Bloom & Fawcett, "Textbook of Histology," W. B. Saunders Co.

Exercise 42   CO Ralph T. Hutchings /PhotoEdit **42.3** Michael G. Wood

Exercise 43   43.3 Phototake NYC   **43.5b** Frederic H. Martini   **43.7** Ralph T. Hutchings   **43.9a** P. Motta/Dept. Of Anatomy/University "La Sapienza," Rome/Science Photo Library/Photo Researchers, Inc.   **43.9b** John D. Cunningham / Visuals Unlimited   **43.10** Ralph T. Hutchings **43.11a** Ward's Natural Science Establishment, Inc.   **43.11b and 43.14** Michael J. Timmons **43.16** Frederic H. Martini

Exercise 45   CO David M. Phillips / Visuals Unlimited

Exercise 46   CO Photo Researchers, Inc.

Exercise 47   CO David M. Phillips / Visuals Unlimited   **47.3a** Ward's Natural Science Establishment, Inc.   **47.3c,d,e** Frederic H. Martini **47.6a,b,c,d** Frederic H. Martini   **47.6e** G.W. Willis, MD / Biological Photo Service   **47.8** Ward's Natural Science Establishment, Inc. **47.10b** Fred E. Hossler / Visuals Unlimited **47.10c** Frederic H. Martini

Exercise 48   CO Petit Format/ Nestle / Photo Researchers, Inc.   **48.2b** Ward's Natural Science Establishment, Inc.   **48.3** Francis Leroy Biocosmos / Custom Medical Stock Photo, Inc.   **48.7a** Dr. Arnold Tamarin   **48.7b,c,d** Photo Lennart Nilsson / Albert Bonniers Forlag